Mathematical Models of Non-Linear Excitations, Transfer, Dynamics, and Control in Condensed Systems and Other Media

Edited by

Ludmila A. Uvarova
Moscow State University of Technology "STANKIN"
Moscow, Russia

Arkadii E. Arinstein
N. N. Semenov Institute of Chemical Physics
Russian Academy of Sciences
Moscow, Russia

and

Anatolii V. Latyshev
Moscow State Pedagogical University
Moscow, Russia

Kluwer Academic / Plenum Publishers
New York, Boston, Dordrecht, London, Moscow

Proceedings of a symposium entitled Mathematical Models of Non-Linear Excitations, Transfer, Dynamics, and Control in Condensed Systems and Other Media, held June 29–July 3, 1998, in Tver, Russia

ISBN 0-306-46133-1

©1999 Kluwer Academic / Plenum Publishers, New York
233 Spring Street, New York, N.Y. 10013

10 9 8 7 6 5 4 3 2 1

A C.I.P. record for this book is available from the Library of Congress.

All rights reserved

No part of this book may be reproduced, stored in a retrieval system, or transmitted in any form or by any means, electronic, mechanical, photocopying, microfilming, recording, or otherwise, without written permission from the Publisher

Printed in the United States of America

Mathematical Models
of Non-Linear Excitations,
Transfer, Dynamics, and
Control in Condensed Systems
and Other Media

PREFACE

Everything that is interesting is nonlinear.
Everything that isn't nonlinear is noninteresting.

The following articles were taken from the Third International Scientific Conference titled *Mathematical Models of Nonlinear Excitations, Transport, Dynamics, and Control in Condensed System and other Media*"*. This conference was held at the State Technical University in Tver (an ancient Russian town situated 160 kilometers from Moscow on the Volga River) from June 29 to July 3, 1998. Topics from the conference included: nonlinear excitations in condensed matter systems, evolution of complex systems, dynamics and structure of molecular and biomolecular systems, mathematical models of transfer processes in nonlinear systems and numerical modeling and algorithms.

At present mathematical modeling may be considered one component of research methodology, determined by application of mathematical problems and by wide possibility of computed experiments. The book contains articles in which mathematical modeling methods are used for the investigation of many complicated problems, including transport theory, physical chemical, and biophysical problems. These problems are united by the nonlinearity of the research processes and phenomena. In addition, some works dedicated to the discussion of concrete mathematical modeling methods. Throughout this book, the key words are "nonlinearity" and "modeling", which characterize the present scientific paradigm. The prism of nonlinearity and modeling permit us to open new facets of the surrounding world.

The articles which form this book are united in the following four sections in accordance with their problems:

I.   Mathematical models for nonlinear phenomena and properties (kinetic and dynamic processes, mass and heat transfer);
II. Numerical methods and computer simulations;
III. Nonlinear phenomena in physics;
IV. Nonlinear models in chemical physics and physical chemistry.

In the first section, problems such as evolution of the dynamic dissipative systems, solitary waves in absorbing media, nonlinear dynamics of DNA, self-organization of nonlinear systems, nonlinear heat and mass transfer induced by the electromagnetic radiation, and physical kinetics problems, among others are discussed.

---

* The conference was supported by Russian Foundation of Fundamental Research (Grant N98–01–10049).

The second section is devoted to the use of numerical modeling methods for the description of complicated phenomena such as methane burning in surroundings, nonstationary flow of viscous liquid under highly general conditions, and others.

The third and fourth sections contain articles in which the models of nonlinear physical and chemical phenomena and their effects are discussed. Models are proposed for dust particles interaction, echo–phenomenon in ferroelectric crystals, collection drops evaporation; and multiple disintegration of solids under intensive stress action such as compression and shear. In addition, quantum–chemical models for the active centers of transition metals biologically active complexes are discussed, among others.

The topics introduced in this book, in our opinion, are of fundamental importance, And are of interest for many researchers specializing in the range of mathematical modeling of condensed systems, transport theory, numerical methods, physics, physical chemistry, and biophysics.

We decided to include some topics containing disputable points of view in connection with discussions that took place at the conference. Reports of such kind can lead to new discussions, in our opinion, each author has the right to voice his or her personal point of view, and bears the responsibility for his or her own ideas.

On the authors' behalf we would like to thank Kluwer Academic/Plenum Publishers, which afforded us the opportunity to publish this book. We would also like to thank the secretary of the conference organizing committee, Dr T.V. Naumovich, whose role both in the conference organization and the preparation of this book cannot be overestimated.

We hope that the readers of this book will find here new and interesting information for their research.

Ludmila A. Uvarova
Arkadii E. Arinshtein
Anatolii V. Latyshev

CONTENTS

I. Mathematical models for nonlinear phenomena and properties (kinetic and dynamic processes, mass and heat transfer)

A.V. Latyshev, A.A. Yushkanov. Temperature jump and weak evaporation in a polyatomic gas 3

S.F. Timashev, Ye.Yu. Budnikov, V.L. Klochikhin, I.G. Kostuchenko, S.G. Lakeev, A.V. Maximychev. Evolution of the dynamic dissipative systems as a temporal "colour" fractal 17

J. Chen, W. Greenberg, R.L. Bowden. Existence of solutions and dynamical models of Chandrasekhar H-equations 51

P.B. Dubovskii. Fluid dynamic limit of the Boltzmann kinetic equation arising in the coagulation–fragmentation dynamics 71

V.P. Shutyaev. Necessary and sufficient conditions for solvability of the initial-boundary value transport problem 77

Y.V. Kistenev, A.V. Shapovalov. Solitary waves in two-component resonantly absorbing media 85

L.V. Yakushevich. Nonlinear vector model of the internal DNA dynamics 93

D.S. Chernavskii, O.D. Chernavskaya, A.V. Scherbakov, B.A. Suslakov, N.I. Starkov. The dynamics of the economic society structure 103

L.A. Uvarova. Mathematical model for the heat mass transfer in the systems with the nonlinear properties induced by the electromagnetic radiation 121

M.A. Smirnova. Mathematical model nonlinear heat transfer in an inhomogeneous dispersible system 129

II. Numerical methods and computer simulations

B.N. Chetverushkin, E.V. Shilnikov. Unsteady viscous flow simulation based on QGD system 137

B.N. Chetverushkin, M.V. Iakobovski, M.A. Kornilina, K.Yu. Malikov, N.Yu. Romanukha. Ecological after-effects numerical modelling under methane combustion 147

L.V. Pletnev, N.I. Gamayunov, V.M. Zamyatin. Computer simulation of evaporation process into the vacuum 153

Ju.V. Elyseeva. On oscillation and nonoscillation domains for difference Riccati equation 157

III. Nonlinear phenomena in physics

A.P. Nefedov, V.D. Lakhno. Dusty particle interaction in plasma placed in magnetic field 171

D.S. Cernavskii, N.M. Chernavskaya, L.A. Uvarova. Tunnel transport of electrons at anharmonic accepting mode 181

Sh. Takeno, H. Matsueda. Atomic operator formalism of elementary gates for quantum computation and impurity-induced exciton quantum gates 195

Kh.Kh. Muminov, V.K. Fedyanin. Nonlinear spin waves and magnet-acoustic resonance in the model of Heisenberg magnet 205

V.A. Fedirko, S.V. Polyakov. Modeling of $2d$ electron field emission from silicon microcathode 221

S.S. Lapushkin, A.R. Kessel. Echo-phenomenon in ferroelectric solid and liquid crystals 229

A.L. Bondareva, G.I. Zmievskaya. Investigations of blistering in solids using stochastic model 241

N.I. Gamayunov. Electric potentials distribution for particles located in solution 251

Yu.G. Ionov, A.Yu. Ionov. Statement and solution of a boundary value problem in a model of a plasma generator as controlled system 257

IV. Nonlinear models in chemical physics and physical chemistry

L.I. Manevich. Complex representation of dynamics of coupled nonlinear oscillators ... 269

E.S. Shikhovtseva. Conducting channels structure and dielectric-metal switching stability in thin amorphous films 301

A.E. Arinstein. Phenomenological description for process of multiple disintegration of solids under intensive stress action such as compression & shear 311

A.B. Nadykto, E.R. Shchukin. Vaporization and growth of aerosol particles, given internal heat release and radiant heat exchange 325

E.R. Shchukin, A.B. Nadykto. Vaporization and growth of large and moderately large particles at considerable differences of gaseous component concentrations 329

E.R. Shchukin, A.B. Nadykto. Diffusive vaporization and growth of assembly of N-large particles 339

V.A. Kaminsky, M.V. Egorov. Analysis of the phenomenon of autoacceleration in free radical polymerization 369

L.Yu. Vasil'eva. Quntum-chemical models for the active centers of transition metals biologically active complexes. Interconnection of the active centers structure and the functions 385

Yu.G. Papulov, M.G. Vinogradova. Relations between the properties of substances and their molecular structure: phenomenological study of substituted methanes and their analogs 399

Index 409

Mathematical Models for Non-Linear Phenomena and Properties (Kinetic and Dynamic Processes, Mass and Heat Transfer)

TEMPERATURE JUMP AND WEAK EVAPORATION IN A POLYATOMIC GAS

Anatolii V. Latyshev[1] and Alexander A. Yushkanov[2]

[1]Mathematical Analysis Department
[2]Theoretical Physics Department
Moscow Pedagogical University
107005 Moscow, Russia

INTRODUCTION

Much attention has been paid in the past to so called Smoluchowski problem (SP) on the temperature jump[1,2]. This is due to the fact that the problem at issue is not only of great theoretical importance but it has numerous practical applications (see, e.g., refs[3,4]). For a one-atomic (simple) gas SP was investigated both with the help of analytical methods[5], and approximate or numerical methods applying to the original Boltzmann equation[6,8]. Along with the SP there has been a great deal of interest in study of boundary conditions under a weak evaporation from a surface[9–11].

Previous treatments mostly relied on a case of one-atomic gas however the majority of real gases refers to the polyatomic (or molecular) ones Therefore study of the processes under consideration in the case of molecular gases is of undeniable interest. It is evident, that kinetic processes in molecular gases differ by greater complexity in comparison with the simple gases[12]. In molecular gases, for instance, many elementary processes, such as elastic and inelastic collisions of molecules are scantily known to analyze quantitatively the exact Boltzmann equation. In this connection the role of model integrals of collisions increases.

It is worth noting that for many molecular gases there is a wide temperature range (from tens up to thousands degrees Kelvin), where, on the one hand, oscillatory degrees of freedom are "frozen", (i.e. they don't contribute in thermodynamic and kinetic properties of the gas) and, on the other hand, rotary degrees of freedom can be considered as quasiclassical[13]. This very case we shall also consider here.

For the latter case model integrals of collisions were investigated in a number of papers[14,17]. Model integral of collisions for two-atomic gas was considered in work[18] and the problem on temperature jump was solved there. In the present paper a generalization of the approach proposed in ref.[18] is applied to describe the case of polyatomic gases, i.e. gases, which molecules containing more than two atoms.

Mathematical Models of Non-Linear Excitations, Transfer, Dynamics, and Control in Condensed Systems and Other Media, edited by Uvarova *et al.*, Kluwer Academic / Plenum Publishers, New York 1999.

1. FORMULATION OF THE PROBLEM AND BASIC EQUATION

It is well known that the Bhatnagar, Gross and Krook (BGK) model kinetic equation for the steady state has the following form[1,2]

$$v\nabla f = \nu(f_{eq} - f), \qquad (1)$$

here $f(v)$ is a velocity of gas molecules, $f(r,v)$ is the distribution function and ν is a frequency of collisions between molecules.

In the given work we treat the case, when the frequency of collisions does not depend on the molecules velocity[2].

The function f_{eq} is the equilibrium distribution function depending on the average values,

$$f_{eq} = n_{eq}\left(\frac{m}{2\pi k T_{eq}}\right)^{3/2} \exp\left[-\frac{m}{2kT_{eq}}(v - u_{eq})^2\right],$$

where m is mass of a molecule, k is the Boltzmann constant.

The quantities n_{eq}, T_{eq}, u_{eq} are determined by the relevant moments of the distribution function f, i.e.

$$n_{eq} = \int f d^3v, \quad u_{eq} = \frac{1}{n_{eq}}\int v f d^3v, \quad T_{eq} = \frac{2}{3k n_{eq}} \int \frac{m}{2}(v - u_{eq})^2 f d^3v.$$

The integration in these equations is performing over the whole velocity space.

It follows directly from the definition of n_{eq}, u_{eq}, T_{eq}, that they are functions on spatial variables $f(r)$.

In the case, when the internal degrees of freedom may be treated as quasiclassical, the model kinetic equation is of the form (1) with the following equilibrium distribution function

$$f^P_{eq} = n_* \frac{(m^3 J_1 J_2 J_3)^{1/2}}{(2\pi k T_*)^3} \exp\left[-\frac{m}{2kT_*}(v - u_*)^2 - \frac{J_1\omega_1^2 + J_2\omega_2^2 + J_3\omega_3^2}{2kT_*}\right],$$

where J_1, J_2, J_3 are the main moments of inertia of a molecule, $\omega = (\omega_1, \omega_2, \omega_3)$ is the relevant vector of angular velocity of its rotation. The quantities n_*, u_*, T_* are determined by the corresponding moments of the distribution function $f(r,v,\omega)$:

$$n_* = \int f d^3v d^3\omega, \qquad u_* = \frac{1}{n_*}\int v f d^3v d^3\omega,$$

$$T_* = \frac{1}{3kn_*}\int\left[\frac{m}{2}(v - u_*)^2 + \frac{1}{2}(J_1\omega_1^2 + J_2\omega_2^2 + J_3\omega_3^2)\right] d^3v d^3\omega.$$

Here the integration is performing over the whole space of translational and angular velocities of a molecule.

Hereafter we shall be interested in the case, when the thermodynamic quantities gradients are small. We shall believe that speed of gas in system of readout connected to a surface of the unit. For the problems at issue, these conditions may be written as follows:

$$\lambda \nabla T \ll 1, \sqrt{\frac{m}{2kT_0}} u_* \ll 1,$$

where λ is the mean free path of the gas molecules, T_0 is the temperature of gas at a points of the considered region (for example, in the origin of coordinates).

Under the above conditions the distribution function may be represented in the form

$$f = f_0(1+\varphi) \text{ as } |\varphi| \ll 1,$$

here

$$f_0 = n_s \frac{(m^3 J_1 J_2 J_3)^{1/2}}{(2\pi k T_s)^3} \exp[-\frac{mv^2}{2kT_s} - \frac{J_1\omega_1^2 + J_2\omega_2^2 + J_3\omega_3^2}{2kT_s}],$$

T_s is the temperature of a surface, n_s is the concentration of the saturated vapour of a surface material at the temperature of the surface.

The linearization of the model integral of collisions in (1) leads to the following kinetic equation

$$(v\nabla\varphi) + v\varphi = v[\frac{\delta n}{n_s} + (\frac{mv^2}{2kT_s} + \frac{J_1\omega_1^2 + J_2\omega_2^2 + J_3\omega_3^2}{2kT_s})\frac{\delta T}{T_s} + \frac{m}{kT_s}u_*v], \qquad (2)$$

where

$$\delta n = n_* - n_s = \int f_0 \varphi d^3v d^3\omega, \quad u_* = \frac{1}{n}\int f_0 \varphi v d^3v d^3\omega, \quad \delta T = T_* - T_s,$$

$$\frac{\delta T}{T_s} = -\frac{\delta n}{n_s} + \frac{2}{3knT_s}\int f_0 \varphi[\frac{mv^2}{2} + \frac{1}{2}(J_1\omega_1^2 + J_2\omega_2^2 + J_3\omega_3^2)]d^3v d^3\omega.$$

Hereafter we shall be interested in the SP and a weak evaporation from the flat surface. Let us introduce the Cartesian coordinates with the origin on the surface of the condensed phase such that axis x, is directed along the normal to the surface. Assume also that there is a temperature gradient in the gas at infinity, i.e. $T = T_e + Ax$ as $x \to \infty$ here $A = \left(\frac{dT}{dx}\right)_{x=\infty}$.

The process of evaporation or condensation occurs on the surface, and therefore there is the gas velocity outwards or towards the surface. The quantity $\Delta T = T_e - T_s$ is the temperature jump. Denote the gas velocity far from the surface as U. In case of linear approximation one has

$$\varepsilon_t = \frac{T_e - T_s}{T_s} = N_1 U + M_1 A, \varepsilon_n = N_2 U + M_2 A.$$

We need to find quantities $N_i, M_i (i=1,2)$ from the solution of the kinetic equation.

It follows from the statement of the problem, that the distribution function depends on one spatial coordinate x, molecular velocity v and the absolute value of the angular velocity ω. It is convenient to introduce the following dimensionless variables

$$x' = v\sqrt{\frac{m}{2kT_s}}x, \xi = \sqrt{\frac{m}{2kT_s}}v, U' = \sqrt{\frac{m}{2kT_s}}U, \omega' = \sqrt{\frac{J_1\omega_1^2 + J_2\omega_2^2 + J_3\omega_3^2}{2kT_s}}\omega.$$

The prime on these new variables will be omitted below to simplify the notation. Then the kinetic equation (2) may be rewritten in terms of dimensionless variables

$$\xi_x \frac{\partial \varphi}{\partial x} + \varphi(x,\xi,\omega) = \frac{4}{\pi^2}\int e^{-\xi'^2 - \omega'^2} k(\xi,\omega;\xi',\omega')\varphi(x,\xi',\omega')\omega'^2 d^3\xi' d^3\omega', \qquad (3)$$

where

$$k(\xi,\omega;\xi',\omega') = 1 + 2\xi\xi' + \frac{1}{3}(\xi^2 + v^2 - 3)(\xi'^2 + v'^2 - 3)$$

is the kernel of the kinetic equation. Consider the case when the molecules are diffusively reflected from the wall, i.e. $f(0,\xi,\omega) = f_0$, if $\xi_x > 0$. Hence we have for the function φ the following condition:

$$\varphi(0,\xi,\omega) = 0, if \xi_x > 0. \qquad (4)$$

Far from the wall the distribution function should approach the Chepmen-Enskog function, i.e.

$$\varphi(x,\xi,\omega) = \varphi_{as}(x,\xi,\omega) + o(1), as\, x \to \infty, \xi_x < 0, \qquad (5)$$

where

$$\varphi_{as} = \varepsilon_n + 2U\xi_x + \varepsilon_t(\xi^2 + \omega^2 - 3) + K(x - \xi_x)(\xi^2 + \omega^2 - 4),\ K = A/T_s.$$

Thus, the Smoluchowski problem in the case of weak evaporation is completely posed as the solution of the equation (3) under the boundary conditions (4) and (5). Note that in the above relation the quantities of temperature and concentration jumps ε_t and ε_n, respectively, are unknown. Let us expand the function φ along the two orthogonal directions $e_1 = 1$ and $e_2 = \xi^2 + \omega^2 - 3$, i.e.

$$\varphi = h_1(x,\xi_x) + \gamma(\xi^2 + \omega^2 - 3)h_2(x,\xi_x), \gamma^2 = \frac{1}{3}. \qquad (6)$$

The orthogonality here is in sense of the following scalar product:

$$(f,g) = \frac{4}{\pi^2}\int \exp(-\xi^2 - \omega^2) f(\xi,\omega) g(\xi,\omega)\omega^2 d\omega d^3\xi. \qquad (7)$$

Provided one substitutes expansion (6) into the equation (3), one gets the following set of the kinetic equations:

$$\xi_x \frac{\partial h_1}{\partial x} + h_1(x,\xi_x) = (1,h_1) + \gamma(\xi_x^2 - \frac{1}{2}, h_1) + 2\xi_x(\xi_x,h_1) + 2\gamma(\xi_x^3 - \frac{1}{2}\xi_x, h_2),$$

$$\xi_x \frac{\partial h_2}{\partial x} + h_2(x,\xi_x) = \gamma(\xi_x^2 - \frac{1}{2}, h_1) + \gamma^2(\xi_x^4 - \xi_x^2 + \frac{11}{4}, h_2).$$

The scalar product (7) in these equations becomes simpler and has the form

$$(f,g) = \frac{1}{\sqrt{\pi}} \int_0^\infty \exp(-\xi_x^2) f(\xi_x) g(\xi_x) d\xi_x.$$

Introduce a column vector $h = [h_1, h_2]^t$. Then setting $\mu = \xi_x$ the obtained set of the equations may be represented in a vector form

$$\mu \frac{\partial h}{\partial x} + h(x,\mu) = \frac{1}{\sqrt{\pi}} \int_{-\infty}^\infty \exp(\mu'^2) K_0(\mu,\mu') h(x,\mu') d\mu', \qquad (8)$$

where matrix function $K(\mu,\mu') = K_0(\mu') + 2\mu\mu' L(\mu')$ is the kernel of the equation (8),

$$K(\mu) = \begin{bmatrix} 1 & l(\mu) \\ l(\mu) & l^2(\mu) + 5/6 \end{bmatrix}, L(\mu) = \begin{bmatrix} 1 & l(\mu) \\ 0 & 0 \end{bmatrix}, l(\mu) = \gamma(\mu^2 - 1/2).$$

The boundary conditions (4) and (5) with the help of the expansion (6) will be transformed to the following:

$$h(0,\mu) = \begin{bmatrix} 0 \\ 0 \end{bmatrix}, \quad \mu > 0, \qquad (9)$$

$$h(x,\mu) = h_{as}(x,\mu) + o(1), x \to +\infty, \mu < 0, \qquad (10)$$

where

$$h_{as}(x,\mu) = \begin{bmatrix} \varepsilon_n + 2U\mu - K(x-\mu) \\ [\varepsilon_t + K(x-\mu)/\gamma] \end{bmatrix}$$

Note here, that the function $h_{as}(x, \mu)$ is a linear combination of the particular solutions of the equation (8):

$$h^{(1)}(x,\mu) = \begin{bmatrix} 1 \\ 0 \end{bmatrix}, h^{(2)}(x,\mu) = \begin{bmatrix} 0 \\ 1 \end{bmatrix}, h^{(3)}(x,\mu) = \mu \begin{bmatrix} 1 \\ 0 \end{bmatrix}, \qquad (11)$$

$$h^{(4)}(x,\mu) = (x - \mu) \begin{bmatrix} -1 \\ 1/\gamma \end{bmatrix}.$$

2. EXPANSION OF THE SOLUTION BY EIGENVECTORS

Separating variables in the equation (8), we obtain continuum family of its particular solutions

$$h_\eta(x,\mu) = \exp(-x/\eta)\Phi(\eta,\mu),$$

where η is the spectral parameter, or parameter of separation which is, generally speaking, a complex number, and the function $\Phi(\eta,\mu)$ satisfies the vector characteristic equation

$$(\eta - \mu)\Phi(\eta,\mu) = \frac{1}{\sqrt{\pi}}\left[n_0(\eta) + 2\mu\begin{bmatrix}1 & 0 \\ 0 & 0\end{bmatrix}n_1(\eta)\right], \qquad (12)$$

here

$$n_0(\eta) = \int_{-\infty}^{\infty}\exp(-\mu^2)K_0(\mu)\Phi(\eta,\mu)d\mu, \qquad (13)$$

$$n_1(\eta) = \int_{-\infty}^{\infty}\exp(-\mu^2)\mu K_0(\mu)\Phi(\eta,\mu)d\mu.$$

Multiplying the characteristic equation (12) from the left by a matrix $\exp(-\mu^2)K_0(\mu)$ we obtain the equation, which gives

$$n_1(\eta) = \begin{bmatrix}0 \\ 0\end{bmatrix}.$$

Hence, the characteristic equation reduces to the form

$$(\eta - \mu)\Phi(\eta,\mu) = \frac{1}{\sqrt{\pi}}\eta n_0(\eta). \qquad (14)$$

From the equations (14) and (13) for $-\infty < \eta < \infty$ in the generalized functions space we find[19] eigen vectors of a continuous spectrum $\Phi(\eta,\mu) = F(\eta,\mu)n_0(\eta)$. Here

$$F(\eta,\mu) = \eta P\frac{1}{\eta - \mu}I + \exp(\eta^2)K^{-1}(\eta)\Lambda(\eta)\delta(\eta - \mu)$$

is the eigen matrix corresponding to the "unit" normalization, symbol Px^{-1} means principal value of integral for x^{-1}, $\delta(x)$ is Dirac's delta function,

$$\Lambda(z) = I + \frac{1}{\sqrt{\pi}}\int_{-\infty}^{\infty}\exp(-\mu^2)K_0(\mu)\frac{d\mu}{\mu - z}$$

is the dispersion matrix, and

$$\int_{-\infty}^{\infty} \exp(-\mu^2) K_0(\mu) F(\eta,\mu) d\mu = I.$$

Calculating the dispersion matrix in an explicit form we derive:

$$\Lambda(z) = \lambda_c(z) K_0(z) + \frac{1}{2} Q(z), \qquad Q(z) = \begin{bmatrix} 0 & 1 \\ 1 & \gamma(z^2 + 1/2) \end{bmatrix},$$

here

$$\lambda_c(z) = 1 + \frac{1}{\sqrt{\pi}} z \int_{-\infty}^{\infty} \exp(-t^2) \frac{dt}{t-z}$$

is the Cercignani's dispersion function. The determinant of the dispersion matrix is called the dispersion function, which is the square trinomial with respect to $\lambda_c(z)$:

$$\lambda(z) = \det \Lambda(z) = \frac{1}{2} \gamma^2 [5\lambda_c^2(z) - (z^2 - 3/2)\lambda_c(z) - 1/2].$$

The dispersion function we shall represent as the product

$$\lambda_c(z) = \frac{5}{2} \gamma^2 \Omega_1(z) \Omega_2(z),$$

where

$$\Omega_\alpha(z) = \lambda_c(z) - \frac{1}{10}[z^2 - 3/2 + (-1)^\alpha r(z)], \quad \alpha = 1, 2, \tag{15}$$

$$r(z) = \sqrt{q(z)}, \quad q(z) = z^4 - 3z^2 + 49/4.$$

Let $\theta_\alpha(\mu) = \arg \Omega_\alpha^+(\mu)$ is a single-valued branch of the argument of function $\Omega_\alpha^+(\mu)$, fixed by the condition $\theta_\alpha(0) = 0, \alpha = 1,2$. Expansion

$$\lambda(z) = \frac{1}{3z^4} + o(z^{-4})(|z| \to \infty)$$

means, that the dispersion function at infinite point $z = \infty$ has the zero of the fourth order. This zero is corresponds to the four particular solutions (11) of the initial equation (8). It is so-called hydrodynamical modes.

Theorem. The boundary value problem (8)–(10) has the unique solution. This solution may be presented as expansion

$$h(x,\mu) = h_{as}(x,\mu) + \int_0^\infty \exp(-x/\eta) F(\eta,\mu) a(\eta) d\eta. \tag{16}$$

The expansion (16) is understood in the classical sense:

$$h(x,\mu) = h_{as}(x,\mu) + \frac{1}{\sqrt{\pi}} \int_0^\infty \exp(-x/\eta) \frac{\eta a(\eta) d\eta}{\eta - \mu} + \exp(\mu^2 - x/\mu) P(\mu) a(\mu) \vartheta(\mu), \quad (17)$$

here $P(\mu) = K_0^{-1}(\mu) \Lambda(\mu) \vartheta_+(\mu)$ is characteristic function of the positive semiaxis.

Proof. With the help of the boundary conditions (9) and (10) we proceed from expansion (17) to the vector characteristic singular integral equation with the Cauchy kernel [20]:

$$\frac{1}{\sqrt{\pi}} \int_0^\infty \frac{\eta a(\eta) d\eta}{\eta - \mu} + \exp(\mu^2) P(\mu) a(\mu) + h_{as}(0,\mu) = 0, \mu > 0. \quad (18)$$

Let us introduce an auxiliary function

$$N(z) = \frac{1}{\sqrt{\pi}} \int_0^\infty \frac{\eta a(\eta) d\eta}{\eta - \mu}. \quad (19)$$

Multiplying both parts of the equation (18) from the left by the matrix

$$\Lambda^+(\mu) - \Lambda^-(\mu) = 2\sqrt{\pi} i \mu K_0(\mu) \exp(-\mu^2),$$

we have:

$$\Lambda^+(\mu)[N^+(\mu) + h_{as}(0,\mu)] = \Lambda^-(\mu)[N^-(\mu) + h_{as}(0,\mu)], \mu > 0. \quad (20)$$

Then multiplying both parts of the equation (20) from the left by the matrix $K_0^{-1}(\mu)$ we obtain the vector Riemann–Hilbert boundary value problem:

$$P^+(\mu)[N^+(\mu) + h_{as}(0,\mu)] = P^+(\mu)[N^+(\mu) + h_{as}(0,\mu)], \mu > 0. \quad (21)$$

Note, that the matrix $P(z) = \lambda_c(z) I + M(z)$, where

$$M(z) = \frac{1}{5} \begin{bmatrix} 1/2 - z^2 & \gamma(3 - z^2) \\ 1/\gamma & 1 \end{bmatrix}$$

has a jump which is proportional to a unit matrix with transition of a point z through the real axis. This very fact is crusial for the analytical solution of the vector boundary value problem (21) by its reduction to a diagonal form and subsequent factorization of its matrix coefficient $G(\mu) = [P^+(\mu)]^{-1} P^-(\mu)$.

The factorization problem of matrix coefficient we shall consider as

$$G(\mu) = X^+(\mu)[X^-(\mu)]^{-1}, \quad (22)$$

where $X(z)$ is unknown matrix function, analytical everywhere in the complex plane, except a cut along the real semiaxis and with possible singular points for the zeros of polynomial $q(z) = z^4 - 3z^2 + 49/4$.

In order to reduce matrix $P(z)$ to the diagonal form it is enough to diagonalize a matrix $M(z)$. Let us solve now the eigen values problem for $M(z)$. After some algebra we find that the matrix $S(z)$, which reduces $P(z)$ to the diagonal form, has the following form

$$S(z) = \begin{bmatrix} 1/\gamma & 1/2[r(z) + z^2 - 1/2] \\ -1/\gamma & 1/2[r(z) - z^2 + 1/2] \end{bmatrix}.$$

By definition

$$S^{-1}(z)P(z)S(z) = diag\{\Omega_1(z), \Omega_2(z)\}$$

(the functions $\Omega_\alpha(z)$ are introduced by the equality (15)).

Let us connect by segments Γ_1 and Γ_2 polynomial zeroes $q(z)$, laying in the up and down semiplanes accordingly; and denote $\Gamma = \Gamma_1 \cup \Gamma_2$. We shall look for the matrix $X(z)$ in the form

$$X(z) = S(z)U(z)S^{-1}(z), \qquad (23)$$

where $U(z) = diag\{U_1(z), U_2(z)\}$ is unknown diagonal matrix. Substituting (23) into (22), we shall get the matrix boundary value problem

$$\Omega^+(\mu)U^+(\mu) = \Omega^-(\mu)U^-(\mu), \mu > 0,$$

which is equivalent to the following two scalar problems:

$$\Omega_\alpha^+(\mu)U_\alpha^+(\mu) = \Omega_\alpha^-(\mu)U_\alpha^-(\mu), \mu > 0. \qquad (24)$$

For the one-valuedness of the matrix $X(z)$ we should require, that the following condition holds along the cut Γ:

$$X^+(\tau) = X^-(\tau), \tau \in \Gamma,$$

i.e.

$$U^+(\tau)T(\tau) = T(\tau)U^-(\tau), \tau \in \Gamma, \qquad (25)$$

where

$$T(\tau) = [S^+(\tau)]^{-1}S^-(\tau), \tau \in \Gamma.$$

The methods of the solution of problems of the form (24), (25) were proposed in the previous papers[5,11,18]. Therefore their solutions we present here without a derivation:

$$U_\alpha(z) = exp\{-A(z) + (-1)^\alpha r(z)[B(z) + R(z)]\}, \quad \alpha = 1,2 \qquad (26)$$

Here

$$A(z) = \frac{1}{2\pi} \int_0^\infty [\theta_1(\tau) + \theta_2(\tau) - 2\pi] \frac{d\tau}{\tau - z},$$

$$B(z) = \frac{1}{2\pi} \int_0^\infty \frac{\theta_1(\tau) - \theta_2(\tau)}{r(\tau)(\tau - z)} d\tau, \qquad R(z) = \int_0^{\mu_0} \frac{d\tau}{r(\tau)(\tau - z)}.$$

The point μ_0 is determined by the equation

$$\frac{1}{2\pi} \int_0^\infty \frac{\theta_1(\tau) - \theta_2(\tau)}{r(\tau)} d\tau + \int_0^{\mu_0} \frac{d\tau}{r(\tau)} = 0,$$

which is a special case of Jacoby problem for the elliptic integrals.

Thus, the matrix $X(z)$ according to the formulae (23) and (26) is constructed. Now we return to the solution of the problem (21), which with the help of the problem (22) we represent as a problem on the determination of a vector function on its zero jump on the semiaxis:

$$[X^+(\mu)]^{-1}[N^+(\mu) + h_{as}(0,\mu)] - [X^-(\mu)]^{-1}[N^-(\mu) + h_{as}(0,\mu)] = 0, \mu > 0.$$

The general solution of this problem has the form

$$N(z) = -h_{as}(0,z) + X(z)\Phi(z), \qquad (27)$$

$$\Phi(z) = \begin{bmatrix} \Phi_1(z) \\ \Phi_2(z) \end{bmatrix} = \begin{bmatrix} \alpha_1 z + \alpha_0 + \alpha_{-1}/(z - \mu_0) \\ \beta_1 + \beta_0 + \beta_{-1}/(z - \mu_0) \end{bmatrix}.$$

Here α_i and $\beta_i (i = -1, 0, 1)$ are arbitrary constants. Coefficient of the continuous spectrum $a(\eta)$ with the aid of Sokhotski formula for the vector function $N(z)$, is determined by equality (19), where we should substitute the general solution (27). Thus we have:

$$2\sqrt{\pi} i \eta a(\eta) = [N^+(\eta) - N^-(\eta)]\Phi(\eta).$$

Point out, that the solution (27) has poles of the second order at points $z = 0$ and $z = \mu_0$ and also the simple pole at the infinite point $z = \infty$. Eliminating these singularities and equating to zero the limit of the function $N(z)$ at the infinite point, due to the choice of free parameters in the solution, we can obtain the set of eight linear equations as regards to unknown values $\alpha_i, \beta_i (i = -1, 0, 1), \varepsilon_t, \varepsilon_n$. As it follows from below, this set of equations has the unique solution. The uniqueness of expansion (16) (or (17)) is proved by contradiction using the impossibility of nontrivial decomposition the zero vector into eigen vectors of the characteristic equation. The theorem is completely proved.

The elements of the factor–matrix according to (23) are determined by the following expressions:

$$X_{\alpha\alpha}(z) = \frac{1}{2}\{U_1(z) + U_2(z) + (-1)^\alpha \frac{z^2 - 1/2}{r(z)}[U_1(z) - U_2(z)]\}, \alpha = 1,2,$$

$$X_{12}(z) = -\gamma \frac{z^2 - 1}{r(z)}[U_1(z) - U_2(z)], X_{21}(z) = \frac{U_1(z) - U_2(z)}{r(z)}.$$

Let

$$U_1(z) = p_0 + p_{-1}\frac{1}{z} + \ldots, U_2(z) = q_0 + q_{-1}\frac{1}{z} + \ldots$$

be the asymptotic expansion in a vicinity of the infinite point, where

$$p_0 = \exp(-B_{-2} - R_{-2}), q_0 = 1/p_0,$$

$$p_{-1} = p_0[-A_{-1} - B_{-3} - R_{-3}], q_{-1} = q_0[-A_{-1} + B_{-3} + R_{-3}].$$

Here

$$A_{-1} = -\frac{1}{2\pi}\int_0^\infty [\theta_1(\tau) + \theta_2(\tau) - 2\pi]d\tau, R_{-2} = -\int_0^{\mu_0} \frac{\tau d\tau}{r(\tau)},$$

$$B_{-2} = -\frac{1}{2\pi}\int_0^\infty \tau[\theta_1(\tau) - \theta_2(\tau)]\frac{d\tau}{r(\tau)}, R_{-3} = -\int_0^{\mu_0}\frac{\tau^2 d\tau}{r(\tau)},$$

$$B_{-3} = -\frac{1}{2\pi}\int_0^\infty [\theta_1(\tau) - \theta_2(\tau)]\frac{\tau^2 d\tau}{r(\tau)}.$$

Expanding the solution (27) into the asymptotic series in a vicinity of the infinite point and equating to zero all factors with $z^k (k=0,1)$, we obtain the set of four equations, which in there turn lead to the relations

$$\varepsilon_t = \gamma(\beta_0 p_0 + \beta_{-1} p_{-1}) = K\frac{p_{-1}}{p_0} + \gamma\beta_0 p_0,$$

$$\varepsilon_n = 2U\frac{q_{-1}}{q_0} + K\frac{p_{-1}}{p_0} + \alpha_0 q_0,$$

$$\beta_1 = -\frac{1}{\gamma}Kq_0, \alpha_1 = 2Up_0 + Kq_0.$$

The pole of the second order in the origin of coordinates for the solution (27) is eliminated by the condition:

$$S'^{-1}_{21}(z)\Phi_1(z) + S^{-1}_{22}(z)\Phi_2(z) = O(z^2)(z \to 0),$$

where $S^{-1}_{ij}(z)$ are elements of the matrix $S^{-1}(z)$. Using this condition one gets two equations

$$\alpha_{-1} - \alpha_0 p_0 + 2\gamma(\beta_0 \mu_0 - \beta_{-1}) = 0, \qquad \alpha_{-1} - \alpha_1 \mu_0^2 + 2\gamma(\beta_1 \mu_0 - \beta_{-1}) = 0.$$

Now in the solution (27) we eliminate poles of the second order at the point μ_0. It is possible to perform if:

$$S^{-1}_{11}(z)\Phi_1(z) + S^{-1}_{12}(z)\Phi_2(z) = O(z - \mu_0)(z \to \mu_0).$$

The latter condition gives two equations, which lead the relations

$$\alpha_{-1} = \alpha\beta_{-1}, \qquad \alpha_1\mu_0 + \alpha_0 - \alpha(\beta_1\mu_0 + \beta_0) - \alpha'\beta_{-1} = 0.$$

Here

$$\alpha = \alpha(\mu_0), \qquad \alpha' = \alpha'(\mu_0), \qquad \alpha(\mu) = -\frac{1}{2}\gamma[r(\mu) + \mu_0^2 - 1/2].$$

The desired formulas for the temperature and concentration jumps may be rewritten now in the following form

$$\varepsilon_t = \delta_t K + \gamma_t 2U, \qquad \varepsilon_n = \delta_n K + \gamma_n 2U. \tag{28}$$

Here

$$\delta_t = -\frac{p_{-1}}{p_0} + \mu_0(1 + 3D), \qquad \gamma_t = \mu_0^2 p_0^2 D.$$

and

$$\delta_n = \frac{p_{-1}}{p_0} + \mu_0 q_0^2 3(2 - 3D) - \mu_0(1 + 3D), \qquad \gamma_n = \frac{q_{-1}}{q_0} + \mu_0(1 + 3D) - \mu_0 p_0^2 D,$$

$$D = 4\frac{3\mu_0^2/2 - 49/4 - 7r(\mu_0)/2}{r(\mu_0)[r(\mu_0) + \mu_0^2 + 7/2]}.$$

Thus the theorem is proved. Let us rewrite the formulas (28) into the dimensional form. We take the mean free path as $l = \eta\sqrt{\pi m/2kT_0}/\rho$, where η and ρ gas viscosity and density respectively. This relationship may be rewritten in the equivalent form $l = \vartheta \Pr\sqrt{\pi m/2kT_0}$, where ϑ is the thermal diffusivity coefficient for the gas under consideration and $\Pr$ is Prandtl number. Using the these relations, we obtain

$$\varepsilon_t = \delta_T i K + \gamma_T 2U, \qquad \varepsilon_n = \delta_N K + \gamma_N 2U,$$

where

$$\delta_T = \frac{2}{\sqrt{\pi}\,\mathrm{Pr}}\delta_t,\ \delta_N = \frac{2}{\sqrt{\pi}\,\mathrm{Pr}}\delta_n$$

are coefficients of temperature and concentration jumps, and U is the dimensional velocity of evaporation.

Results of the numerical calculations which have been carried out with the help of the results obtained in the present work and using the results of previous papers[18,5], we represent in the following table

Table

Problem	Coefficient	1-atomic gas	2-atomic gas	Polyatomic gas
Smoluchowski problem	δ_T	2.20262	2.05647	1.87224
Smoluchowski problem	δ_N	-1.23035	-1.07247	-1.22067
Problem on weak evaporation	δ_T	-0.22436	-0.15478	-0.13423
Problem on weak evaporation	δ_N	-0.84350	-0.85118	-0.96176

The numerical calculations show that the temperature jump decreases monotonously when the number of atoms in the molecule increases (provided Prandtl number is a constant).

CONCLUSIONS

Thus, the model kinetic equation describing behaviour of molecular (polyatomic) gases, for which the oscillatory degrees of freedom are "frozen", and rotary are considered as quasiclassical, is constructed. The analytical solution (in explicit form) of the problem on the temperature jump and weak evaporation of a flat surface in semispace is derived. The distribution function for both problems is represented in terms of expansion into the singular eigen generalized functions of the characteristic equation corresponding to the initial kinetic equation. The numerical calculations were carried out and the comparison with the previous results is presented.

It is important to note that the proposed method of the analytical solution may be applied to a wide class of linear boundary problems for polyatomic gases.

REFERENCES

1. M.N. Kogan, *Dynamics of Rarefied Gas. Kinetic Theory*. Nauka, Moscow (1967).
2. C. Cercignani, *Theory and Application of the Boltzmann equation*. Scottish Acad. Press. Edinburgh and London (1978).
3. O.A. Kolenchitz, *Thermal Accomodation of Systems Gas–Solid Body*. Nauka i technika. Minsk (1977).
4. B.V. Derjaguin and Yu.I. Yalamov, The theory of thermophoresis and diffusiophoresis of aerosol particles and their experimental testing. in *Topics Current Aerosol Research*. Pergamon. 3: 1 (1972).
5. A.V. Latyshev, Application of Case' method to the solution the linear kinetic B.G.K. equation in the problem about temperature jump problem. *Appl. math. and mech.* 54: 581 (1990).

6. E.P. Gross, E.A. Jackson and S. Ziering, Kinetic models and the linearized Boltzmann equation. *Phys. Fluids.* 2: 432 (1959).
7. S.K. Loyalka, Slip and jump coefficients for rarefied gas flows: variational results for Lennard–Jones and $n(r)$-potentials. *Physica A*. 163: 813 (1990).
8. E.G. Mayasov, A.A. Yushkanov and Yu.I. Yalamov, About thermophoresis non-volatile spherical particle in rarefied gas with small Knudsen numbers. *Letters to Journal of Tech. Phys.* 4: 498 (1988).
9. Y. Sone and Y. Onishi, Kinetic theory of evaporation and condensation. *J.Phys. Soc. Japan.* 35: 1773 (1973).
10. Y. Onishi and Y. Sone, Kinetic theory of evaporation and condensation: Hydrodynamics equation and slip boundary condition. *J. Phys. Soc. Japan.* 44: 1981 (1978).
11. E.B. Dolgosheina, A.V. Latyshev and A.A. Yushkanov, Exact solutions of model Boltzmann B.G.K. equation in prblems about jumps of temperature and weak evaporation. *Izvestiya AN USSR. Ser. M. Zh. G.* 1: 163 (1992).
12. V.M. Zhdanov and M.Ya. Alievskii, *Transfer and Relaxation Processes in Molecular Gases.* Nauka, Moscow (1989).
13. L.D. Landau and E.M. Lifshitz, *Statistical Physics.* Nauka, Moscow (1964).
14. V.A. Rykov, Model kinetic equation for gas with rotary degrees of freedom. *Isvestiya AN USSR. Ser. M. Zh. G.* 6: 105 (1975).
15. I.N. Larina and V.A. Rykov. Research of a flow of sphere by two-nuclear rarefied gas. in *Numerical methods in dynamics of rarefied gases*, VZ AN USSR, Moscow 4: 52 (1978).
16. K. Barwinkel and U. Thelker, Definition and evaluation of a BGK model with internal degrees of freedom. in *20th Int. Symp. On Rarefied Gas Dynamics*, Book of Abstracts. 3: (1996).
17. I.N. Larina and V.A. Rykov, About boundary conditions of gases on a surface of a body. *Izvestiya AN USSR. Ser. M. Zh.G.* 5: 141 (1986).
18. A.V. Latyshev and A.A. Yushkanov, Analytical solution of a problem of temperature jump in gas with rotary degrees of freedom. *Theor. and mathem. physics.* 95: 530 (1993).
19. V.S. Vladimirov, *Generalized Functions in Mathematical Physics.* Nauka, Moscow (1976).
20. F.D. Gahkov, *Boundary Value Problems.* Nauka. Moscow (1977).

EVOLUTION OF THE DYNAMIC DISSIPATIVE SYSTEMS AS A TEMPORAL "COLOUR" FRACTAL

Serge F. Timashev, Yegor Yu. Budnikov,
Vladimir L. Klochikhin, Irina G. Kostuchenko,
Sergey G. Lakeev and Alexander V. Maximychev.

Karpov Institute of Physical Chemistry,
103064 Moscow, Russia

INTRODUCTION

New possibilities in elaborating the evolution of dynamical dissipative system evolution are associated with the ideas of nonlinear system dynamics and deterministic chaos[1-4]. In this paper a new phenomenological method for analysis of non-linear dissipative system dynamics, which may be called flicker-noise spectroscopy[5-8], is presented. An algorithm is developed which allows us to obtain as many phenomenological parameters – "passport data" – as necessary for the description and characterization of the dynamic system's state and the changes of its state during evolution. We demonstrate the application of this approach for analysis of the time series by describing various phenomena. Spatial space structures may be examined in a similar manner.

METHODOLOGY OF THE ANALYSIS OF DYNAMIC DISSIPATIVE SYSTEMS' STATES [5,6]

Complexity and non-regularity in the changes in the measured dynamic variables of complex systems and structures of different origin, as well as self-similarity in these non-regularities, are the main features of the evolution of non-linear dynamic dissipative systems. In the case of time series elaboration, this statement is manifested in the observation of very sharp jumps and other irregularities in the observed dynamic variable $V(t)$ which display self-similarity in large temporal intervals (Fig. 1). This phenomenon allows us to propose the hypothesis that not all points on the time axis but a sequence of the discrete points carries the main information about the state and the evolutionary features of the system under consideration. This hypothesis suggests that there is a set of short-time δ_i-intervals, for every i-th spatial-temporal level of the evolution, in which most of the information about the evolution $V(t)$ is contained. The time intervals between the short-time δ_i-intervals also contain the information about the evolution of the system. The latter time intervals are not "empty": they host information that is contained in the smaller δ_{i+1}-intervals, etc. There are definite correlation links between the individual δ_i-intervals in the introduced sequences (see below). These links are characterized by a set of parameters, which carry information about the dynamical state of the system under consideration. Below the parameters will be introduced.

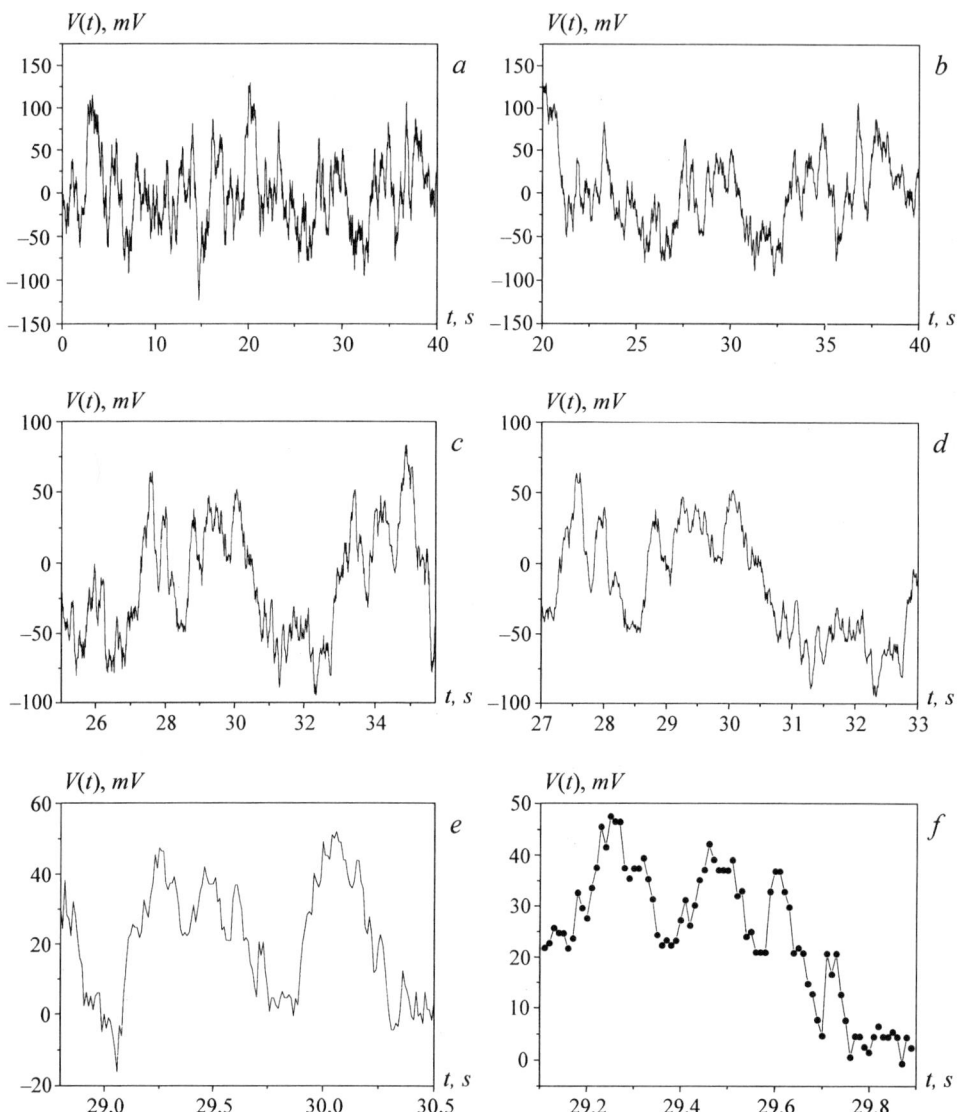

Figure 1. The time series of electric voltage fluctuations in the electrochemical system with cation exchange membrane at "over-limiting current density" condition when the relation of the current density to the corresponding limiting current density value is equal j/j_{lim} =7 (10 mM NaCl solution, membrane diameter is 0.1 cm)[22]: *a*) – voltage fluctuation realization for the time 40 s (4000 points); *b*)–*f*) – fragments of the realization. The appeared discretization in the panel *f*) is a result of the equipment resolving power.

The transition of the system from one δ_i-interval to the adjacent one must be accompanied by a change of the structure-energetic characteristics of the system's state. The observed non-regularities of the dynamic variable $V(t)$ manifest this regularity in evolution. According to the "Triest theory" of von Weizsäcker[9], an event happens when irreversibility is realized. This means[10] that irreversibility must have emerged in the δ_i-intervals, while the intervals between the adjacent δ_i-intervals, which are "empty" on the i-level, may be called "*Now*". This image combines "the past" which has happened and "the future" which potentially exists.

Let us idealize this image of evolution and contract the δ_i-intervals to their corresponding points. Let every "new δ_i-interval" with zero duration carries the information about the structure-energy system state, which was contained in the "old δ_i-intervals" with finite duration. Zero duration of these "new δ_i-intervals" means that the meaning of the dynamic variable $V(t)$ within each point must contain either "real" or "potential" singularities. "Real" singularity means that $V(t)$ is a generalized function of compact support in $\{0\}$ (supp $V(t) = \{0\}$)[11]. "Potential" singularity at some point means that the meaning of the dynamic variable $V(t)$ suffers a jump and the discontinuity of the derivatives of some order in the neighborhood of this point. Obviously, the "potential" singularities transform to "real" ones after the differentiation of $V(t)$.

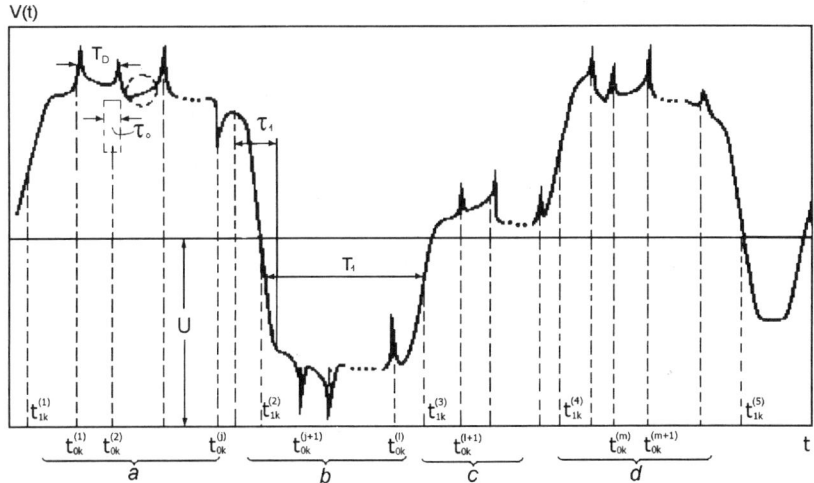

Figure 2. Intermittent evolution of a dynamic variable $V(t)$.

The introduced idealized $V(t)$-image (Fig. 2) is of the type of the "model-caricature" according to Frenkel's definition[12]. Frenkel wrote that it was just a simple "model-caricature" which had to be introduced in order to obtain simple equations to describe complicated real processes. Therefore, we postulate a generalized intermittency dynamics[6,7] for the i-th spatial-temporal level of $V(t)$ signal. This dynamics is characterized by relatively weak changes in the values extending in long-time intervals ("laminar" phases) with characteristic time T_0^i, which are interrupted by "bursts" of short duration (τ_0^i; $\tau_0^i \ll T_0^i$) against the indicated background variations. We note t_{0k}^i as the time of the appearance of the burst of number k ($-\infty < t_{0k}^i < +\infty$). Note (see Fig. 2) that jumps of the $V(t)$ function are present at the same t_{0k}^i points. We denote these jumps as "first type". In addition to this type of evolution[6,7] we assume that the $V(t)$ value may drastically (in a short time τ_1^i, like the Heaviside function) change its "laminar background" values (the mean values of the "laminar" regions). We denote T_1^i as a characteristic time interval between these jumps ("second type") as well as t_{1k}^i as the time of the appearance of the jump of number k

($-\infty < t_{1k}{}^i < +\infty$) and assume $\tau_1{}^i \ll T_1{}^i$. We denote the time moments $t_{0k}{}^i$ and $t_{1k}{}^i$ as "$0k$" and "$1k$" points respectively. We also assume that the irregularities display self-similarity, i.e. the image of the evolution $V(t)$ presented in Fig. 2, is reproduced on a smaller scale in the dotted circle drawn into in Fig. 2. This allows us to present the evolution of a non-linear dissipative system as a temporal fractal and obtain the simple phenomenological equations which may be used as a basis for studying the experimental time series and for retrieving the straight phenomenological information about the state of the system under consideration.

We suppose that the process considered is a stationary stochastic one, whose statistical characteristics do not change in the course of time. This means[6,7] that the autocorrelation function $\psi(\tau) = <V(t)V(t+\tau)>$, where the brackets denote time averaging, depends only on τ and $\psi(\tau) = \psi(-\tau)$. In this case we receive the following equation for the Fourier transformation $S(f)$ of the autocorrelation function or the power spectrum (in accordance with the Wiener-Khinchin theorem):

$$S(f) = 2\int_0^\infty \cos(2\pi f \tau)\psi(\tau)d\tau, \tag{1}$$

$$\psi(\tau) = 2\int_0^\infty \cos(2\pi f \tau)S(\tau)d\tau.$$

We introduce also the average difference moments $\Phi_{(p)}(\tau)$ (so-called "structural function" in the theory of developed turbulence) of different order p:

$$\Phi_{(p)}(\tau) = \left\langle |V(t) - V(t+\tau)|^p \right\rangle, \tag{2}$$

where τ is a delay time value. It is known that the spectral dependence $S(f)$ as well as the dependence $\Phi_{(p)}(\tau)$ on the parameter τ indicate the presence of correlating links between the previous and following events. In the case of "white niose", such correlating links are absent, and $S(f)$ as well as $\Phi_{(p)}(\tau)$ are constant at $f > 0$ and $\tau > 0$ respectively. However, what is the meaning of "the events" in every time series $V(t)$ considered? We shall demonstrate that, in the approach presented here, these "events" are marked by the introduced non-regularities of the $V(t)$.

At the first stage of the power spectrum $S(f)$ calculation, we decompose the signal $V(t)$ into two terms: the singular $V_S(t)$, which is formed by the bursts or spikes of the evolution (see Fig. 2), and so-called "regular" $V_R(t)$ term, which is formed by $V(t)$ without bursts:

$$V(t) = V_R(t) + V_S(t); \quad V_S(t) \equiv V(t) - V_R(t). \tag{3}$$

The second term in Eq. (3) is non-zero in the regions of the bursts only. In accordance with our hypothesis, we consider $V_S(t)$ as a generalized function of compact support in $\{0\}$, and use the well-known theorem of the theory of the generalized function[11]:

$$V_S(t) = \sum_{k,p} c_k^p \delta^{(p)}(t - t_{0k}), \tag{4}$$

where $\delta(t)$ and $\delta^{(p)}(t)$ are the Dirac delta-function and its time derivatives of p-order, which form the total set of the generalized functions of compact support in $\{0\}$; $c^p{}_k$ are the corresponding coefficients of the expansion of $V_S(t)$ in the complete system of the functions. Note that we have omitted in Eq. (4) the i-index of the considered space-temporal level (scale). This is due to the fact that we shall find the conditions when the calculated value of $S(f)$ does not depend on the spatial-temporal scale, which means that self-similarity is realized.

For calculating $S(f)$ it is necessary to insert Eq. (3) in Eq. (1). Then $S(f)$ is expressed as the sum of the following four terms: $S_{RR}(f) = \langle V_R(t)V_R(t+\tau)\rangle$,

$S_{SS}(f) = \langle V_S(t)V_S(t+\tau)\rangle$,

$S_{RS}(f) = \langle V_R(t)V_S(t+\tau)\rangle$,

$S_{SR}(f) = \langle V_S(t)V_R(t+\tau)\rangle$.

The simple transformation results in:

$$S(f) = S_{RR}(f) + 2\sum_{n=-\infty}^{n=+\infty} C_n \cdot Cos(2\pi ft_n) ,\qquad(5)$$

$$C_n = A_n + 2U_R B_n;\quad A_n \equiv \sum_{k=-\infty}^{k=+\infty} c_k^0 c_{k+n}^0;\quad B_n \equiv \sum_{k=1}^{N_1} c_{k+n}^0;\qquad(6)$$

$$U_R \equiv \frac{1}{N}(V_{R1} + V_{R2} + \ldots + V_{Rm});\quad N = N_1 + N_2 + \ldots + N_m.$$

Here t_n is the time of the n-th burst appearance corresponding with the i-th – spatial-temporal level of $V(t)$ signal; T_0 is a characteristic interval between the adjacent bursts; N_k is the number of bursts at the k-th ($k = 1, 2, \ldots, m$) interval with characteristic duration, T_1, from the sequence $a, b, c, d, \ldots$ for the signal $V(t)$ (see Fig. 2); m is the total amount of these intervals during the time of observation; U_R is the mean value of the model signal $V_R(t)$ after averaging over all m intervals; V_{Rk} is a meaning of the $V_R(t)$ function at the k-th interval.

Let us suppose for the sake of simplicity that $U = 0$ and therefore, $C_n = A_n$ (it is possible to consider a more general case). In the case $2\pi f T_0 \ll 1$, in the second term of the right part of Eq. (5), it is possible to replace the summation by integration over the dimensionless variable $\xi = 2\pi f T_0 n$ ($n = 1, 2, 3, \ldots$), so that $d\xi = 2\pi f T_0 \Delta n \ll 1$ if $\Delta n = 1$ for the indexes of adjacent bursts. In this case[6]:

$$S(f) = S_R(f) + \frac{1}{T_0^i f\pi}\int_0^\infty \Phi_0^i\left(\frac{\xi}{2T_0^i f\pi}\right) Cos\xi\, d\xi =$$

$$= S_R(f) + 2\int_0^\infty \Phi_0^i(\eta)\cdot Cos(2T_0^i f\eta\pi)d\eta,\qquad(7)$$

where the function Φ_0^i characterizes the effective density of the burst sequence and takes into account the bursts' amplitudes.

In accordance with the dimension [sec] of the function $\Phi_0^i(z)$ we may present this in the form:

$$\Phi_0^i(\eta) = \frac{g}{4\pi} T_0^i \chi_0(b_0^i \eta), \qquad (8)$$

where $\chi_0(b_0^i \eta)$ is a dimensionless function, b_0^i is a dimensionless scale parameter of the i-th spatial-temporal level and g is a constant. Then we have:

$$S(f) = S_{RR}(f) + I_S(f);$$

$$I_S(f) \equiv \frac{g}{K_0 \pi} \int_0^\infty \chi_0(x) \cos Zx \, dx; \qquad K_0 \equiv \frac{b_0^i}{T_0^i}, \qquad Z \equiv \frac{2 f\pi}{K_0}. \qquad (9)$$

Here, K_0 is the scale-invariant parameter, which may be considered as a universal parameter of the system.

It is convenient to use various approximations of the function $\chi_0(b_0^i \eta)$. For instance, possible types of relaxation, including the retarded non-exponential ones of non-linear systems with rearrangement of structure elements during relaxation under above-threshold excitation, are analyzed using these approximations:

(i) $\chi(z) = z^{-\mu}$, $(0 < \mu < 1)$ – the "flicker-noise" approximation; (10)

(ii) $\chi(z) = \exp[-(\lambda_0 z)^s]$, $(0 < s < 2)$ – the Lèvy approximation, (10a)

where μ, λ_0 and s are parameters. The approximations (10) are responsible for the self-similar equations for the functions $I(f)$, i.e., $I(f)$ may be considered as a temporal fractal. For the Lèvy approximation, the dependencies $I_S(f)$ may be obtained exactly if $s = 0.5$ and $s = 1$ (see Ref.[6]). Specifically, in the case of $s = 1$ we obtain:

$$I_S = \frac{4g(\lambda_0 K_0)^{-1} q^2}{\left(q^2 + \frac{1}{16}\right)}, \qquad (11)$$

where $q \equiv \lambda_0 K_0 /(8\pi f)$.

From (9) – (11), it follows that $I_S(f) \to 0$ if there are no correlations in the burst sequence (the Poisson distribution), i.e., for $\mu \to 0$, $s \to 0$, $\lambda_0 \to 0$. Obviously, as the frequency f decreases, the inequality $f \ll 1/(2\pi T_0^i)$ is realized for increasing spatial-temporal scales. The case (i) corresponds to flicker-noise, or, in other words, to the realization of the "infinite memory" of the system. The Lèvy approximation (ii) corresponds to the "shot noise" case when the power spectrum does not depend on the frequency in the limit of $f \ll f_0 \equiv \lambda_0 K_0 /(2\pi)$ (the case of the "losing memory" at the time exceeding the value $t_M \sim f_0^{-1}$), and falls with the frequency increasing as f^{-n}, where $n = s + 1$ and $1 < n < 3$ in the frequency interval: $f_0 \ll f \ll K_0 /(2\pi)$. The latter means that the parameter $\lambda_0 K_0$, which characterizes a "non-linear relaxation" in the correlating sequence of the bursts generated in the non-linear dissipative system by external energy sources, may be considered as a component of the Kolmogorov K-entropy vector[6]. Note that the dimensionless parameters $(1 - \mu)$ and $n \equiv (s + 1)$ characterize the content of the correlating links ("memory") in the correlating sequence of the bursts at the relatively small $(\Delta t \ll t_M \sim f_0^{-1})$ time interval.

In the general case, the interpolation relation which characterizes the correlation links in the correlating sequence of the δ-like bursts of the "model-caricatures" $V(t)$, may be presented as:

$$S(f) \approx g_0 \frac{\varepsilon^{a_1}(\lambda_0 K_0)^{a_2} U^{n-1}}{(\lambda_0 K_0/2\pi)^n + f^n} + S_{n0}. \tag{12}$$

Here the phenomenological parameters are introduced: ε is the specific energy dissipation rate; U is the mean velocity of the energy flow with which the wave number $k = f/U$ may be connected; n is above-mentioned parameter; g_0 is a dimensionless function which connects the physical parameters of the system under consideration with the introduced phenomenological parameters ε and $\lambda_0 K_0$; S_{n0} is a constant; a_1 and a_2 are indices of a power which are determined on the basis of the dimension in question.

Eq. (12) may be used for describing different power spectra of flicker-noise or shot-noise type, which are calculated on the basis of various experimental time series data. Here, it should be emphasized in particularly that these dependencies are formed by the singular terms of the $V(t)$ evolution. This suggests that the nonspecific character of power laws in natural sciences as well as in the humanities[6,7] are a result of irregularities in the corresponding evolution changes. It means also that the $S_{RR}(f)$ dependencies are changed more slowly at the low frequency limit and may manifest more specific features of the evolution in the region of higher frequencies. Of course, this is a hypothetical conclusion. It is necessary to present some additional arguments to support it. Among the possible experiments is research into the dependence of the observed power spectra on the "whole volume" υ (or its analogue) of the systems under consideration: $S(f) \sim \upsilon^{-m}$ ($m > 0$). Indeed, if the considered system has quite a large volume, the correlating sequence of the bursts will be generated in various parts of this system. And there will not be any correlating links between the different sequences.

Exactly the same effect takes place in semiconductor systems or thin metal films[13]. This is the so-called Hooge phenomenological law: $S(f) \sim \upsilon^{-m} f^{-n}$ ($n \sim 1$, $m \sim 1$). It is very likely that analogous dependencies may manifest themselves in other kinds of systems where evolution changes of dynamic variable $V(t)$ are characterized by irregular behavior and where correlation links between the dynamic fluctuations exist.

Note that the numerical values of the introduced parameters $\lambda_0 K_0$ and n depend on the internal structure of the system under consideration and the ability of this structure to local structure rearrangements as a result of internal links under the influence of internal and external energy fluxes. The classic flicker-noise in electroconductive systems which is expressed by the law $S(f) \sim 1/f$ corresponds to a partial case $n \to 1$ and $\lambda_0 K_0 \to 0$. This case means that the system has a memory at "all times" and may be described (12), if $a_2 = n - 1 \to 0$ and ε has a dimension $\left[\{\dim(S(f))\cdot Hz\}^{1/a_1}\right]$. In the case of fully developed turbulence (Kolmogorov theory), when the parameter ε [cm^2/s^3] has a sense of the mean energy dissipation rate, U is the mean velocity of the hydrodynamic flow, and the dimension of $S(f)$ equals [cm^2/s], we have[5]: $a_1 = (3-n)/2$ and $a_2 = (3n-5)/2$. This means that the classic flicker-noise which is characterized by "infinite memory" in hydrodynamic flow may be described by Eq. (12) if $\lambda_0 K_0 \to 0$ and $n \to 5/3$ (the Kolmogorov-Obukhov law).

It was shown[6] that the dependencies $\Phi_{(p)}(\tau)$ are largely formed by jumps of the dynamic "model-caricature" $V(t)$, which are described by the Heaviside Θ-function. This is easy to understand because only the changes in the dynamic variable in the adjacent "laminar" regions, as a result of the jumps, contribute to $\Phi_{(p)}(\tau)$ at every point of "$0k$" or "$1k$" type. The values of the burst amplitudes do not contribute to $\Phi_{(p)}(\tau)$ values. In order to find the dependencies for $\Phi_{(p)}(\tau)$, we introduce a more general definition of the "difference moment of order p" and rewrite Eq. (2) in the form:

$$\Phi^{(1)}_{(p)}(\tau) = \left\langle \left| \int_t^{t+\tau} \frac{dV(z)}{dz} dz \right|^p \right\rangle, \qquad (2a)$$

where the upper index of the difference moments indicates that the integrand is the first derivative of the $V(t)$ function. Obviously the definition (2a) coincides with (2) if $dV(t)/dt$ is a regular differentiable function in the whole range of the argument variation. Let us pick out the "regular" and "singular" parts of the derivative $dV(t)/dt$. The corresponding regular part which is denoted as $[dV(z)/dz]_R$ includes the jumps of $dV(t)/dt$ which are determined by the above-mentioned discontinuities of the derivatives of the $V_R(t)$ function in the "0k" points. The singular part of $(dV/dz)_S$ is defined by the sum of the Dirac delta-functions and their time derivatives which are the result of the differentiation of the $V(t)$ function in the regions of its "1k" and "0k" points respectively.

The corresponding equations for the $\Phi_{(p)}(\tau)$ function were obtained in[6]. Here, we present the interpolation equations for the difference moment of order p for the case that the contribution by the jumps at the "1k" points is negligibly small:

$$\Phi^{(1)}_{(p)}(\tau) \approx g_1(p) U^p \left[1 - \Gamma^{-1}(\nu_1) \Gamma(\nu_1, \lambda_1 K_1 \tau) \right]^p . \qquad (13)$$

Here, K_1 is the second scale-invariant parameter of the process; ν_1 and λ_1 are parameters of non-linear relaxation which determine the rate of jump-like changes in the dynamic variable after perturbation at small and large time intervals respectively; $\Gamma(x)$ is the gamma-function and $\Gamma(s, x)$ is the incomplete gamma-function ($x \geq 0$ and $s > 0$). It is easy to obtain particular cases:

$$\Phi^{(1)}_{(p)}(\tau) \approx g_1 \Gamma^{-p}(\nu_1 + 1) U^p (\lambda_1 K_1 \tau)^{\nu_1 p}, \qquad \lambda_1 K_1 \tau \ll 1;$$
$$\qquad (13a)$$
$$\Phi^{(1)}_{(p)}(\tau) \approx g_1 U^p \left[1 - \Gamma^{-1}(\nu_1)(\lambda_1 K_1 \tau)^{\nu_1 - 1} exp(-\lambda_1 K_1 \tau) \right]^p, \qquad \lambda_1 K_1 \tau > 1$$

Eqs. (13) and the more general[6] may be used for obtaining information about the phenomenological parameters ν_1 and $\lambda_1 K_1$ of the system under consideration by comparing these equations with the corresponding expression obtained on the basis of time series data. Note that the jumps at the "1k" points must contribute to $\Phi_{(p)}(\tau)$ dependence at large time intervals when $t \gg T_1$ (see Fig.2). In this case, the additional parameters ν_2 and $\lambda_2 K_2$ have to be introduce[6].

It is necessary to note we distinguish the parameters, which are obtained from the power spectra and from the difference moments of the 2nd order. In the frame of the traditional approach these parameters carry out the same information, and $2\nu_1 \approx n - 1$ (if $n > 1$), and $\lambda_1 K_1 \approx 2\pi f_0$. The latter conclusion may be understand by considering the obvious link between $\Phi_{(2)}^{(1)}(\tau)$ and $S^{(0)}(f)$:

$$\Phi^{(1)}_{(2)}(\tau) = 4 \int_0^\infty \left[1 - Cos(2\pi f \tau) \right] S^{(0)}(f) df \qquad (14)$$

and inserting Eq. (12) in Eq. (14). In the particular case $n = 2$, the indicated links are obtained as a rigorous result, in other case the links are approximate. However the values of the mentioned parameters which are found under time series processing do not obey the pointed out relations very often[8]. This means that the interpolative Eq. (12) which is used for finding the main parameters of power spectra does not contain some features of the $S^{(0)}(f)$ dependence which form the $\Phi_{(2)}^{(1)}(\tau)$ dependence [Eq. (13)] in accordance with Eq. (14). This circumstance is an additive argument that may be used for justification of the proposal approach. According to this statement, the power spectra and the difference moments of the different orders are formed by the non-regularities of the different types, which are the sequences of the dynamical burst and the dynamical jumps respectively. In the general case the above-mention relations between n and ν_1 as well as between f_0 and $\lambda_1 K_1$ may be realized in a particular case[7] if the $V(t)$ values of the jumps and the bursts at every point of irregularities are proportional to each other.

Additional information about the evolution process is contained in the time series generated by derivatives $d^m V/dt^m$ ($m \geq 1$). The latter time series may be obtained from the original $V(t)$ time series by differentiation. Each new time series may produce new dependencies, namely, the Fourier spectra $S^{(m)}(f)$ of the corresponding auto-correlation function and the difference moments $\Phi_{(p)}^{(m)}(\tau)$. We do not intend to discuss here the mathematical difficulties which arise under the realization of the differentiation procedures. We would like to call attention to the possibility of finding as many parameters of the process under consideration as required. All parameters introduced are marked by non-regularities of different types at the "$0k$" and "$1k$" points. As a result of this fact, analysis of the time series formed by the first derivative dV/dt provides new information concerning the correlation of the jumps (by taking into account the $S^{(1)}(f)$ dependencies) as well as some new features of the evolution near the bursts (by taking into account the $\Phi_{(p)}^{(1)}(\tau)$ dependencies). The procedure of differentiation has to converge as the order of the derivatives increases because the increase of the derivative order leads to identity ($0 \equiv 0$) at the certain step of the differentiation.

Applying of wavelet transformation or localized spectral analysis[13] open new possibilities in studies of time series and allow to obtain principally new information about complex system dynamics. Wavelet analysis is a decomposition of studied signal with the basis of wavelets that are soliton like functions well localized both in time and in frequency space. To construct full basis wavelet functions of different time scale and shifted along time axis are used. Wavelet transformation of a signal $V(t)$ with chosen basic function $\varphi(t)$ is given as

$$W_V(a,b) = |a|^{-1/2} \int_{-\infty}^{\infty} V(t) \varphi\left(\frac{t-b}{a}\right) dt, \qquad (15)$$

where $W_V(a, b)$ is wavelet coefficient, a is a scale coefficient and b is a shift parameter. The result of transformation is the two-argument function $W_V(a, b)$, where a and b parameters have the sense of time scale or period and time accordingly. It is convenient to express a result of wavelet transformation like wavelet map, where time or b coefficient is presented at the horizontal axis, and time scale or a coefficient is at vertical one. Values of wavelet coefficients $W_V(a, b)$ are represented by color scale map (14 halftones of grayscale in this work).

A wide range of functions may be used as basic wavelets, it make possible to emphasize one or another signal feature. So Daubechies and MHAT (Mexican hat) wavelets to a greater extent emphasize evolution singularities (bursts, jumps, derivative breaks and fractures etc.), Morlet wavelets displays resonance's excited. As a basic function we used MHAT wavelet (Mexican hat) that is the Gaussian 2nd derivative:

$$\varphi(z) = (1-z^2)\exp(-z^2/2),\qquad(16)$$

and complex Morlet wavelet:

$$\varphi_{k_0}(z) = \exp(ik_0 z)\exp(-z^2/2),\qquad(17)$$

where k_0 is a specific parameter, which defines the number of this wavelet oscillations (we use $k_0 = 6$).

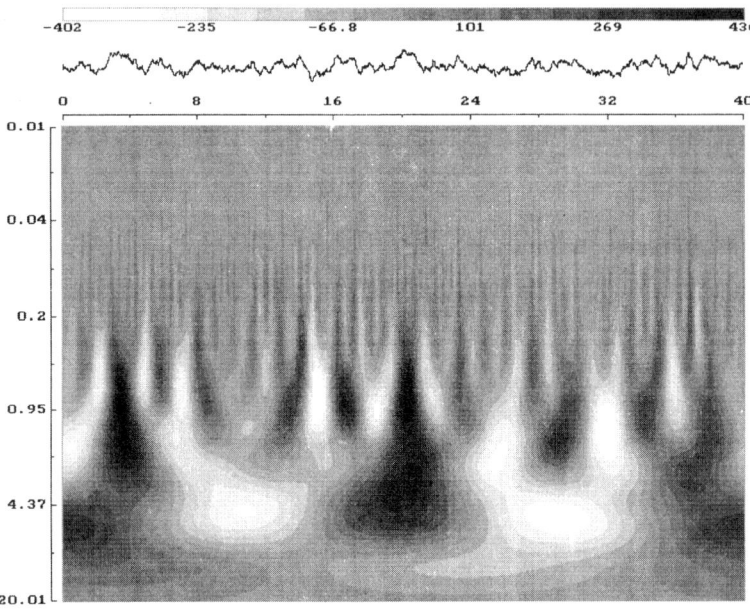

Figure 3. MHAT-wavelet map corresponding to the time series presented in Fig. 1a. The horizontal coordinate is the time (in seconds). The vertical coordinate corresponds to the logarithmic values of the wavelet scales (in seconds). Grayscale (legend on the top) reflects the values of the wavelet coefficients. Time series is shown over the map.

The map of found coefficients of MHAT wavelet transformation for time series presented in Fig. 1a is shown in Fig. 3. Time in seconds is represented at the abscissa axis. Wavelet time scale is at the logarithmic ordinate axis. Differences in values of wavelet transformation coefficients are reproduced by grayscale, its legend is shown over the map. Darker fields correspond to larger coefficient values. Coefficients squares correspond to energy at the given time scale and at the given time moment. So displayed wavelet map characterizes energy redistribution dynamics in the system at the evolution expressed by "full" time series (Fig. 1a). There is manifested hierarchy structure in such maps that is reproduced on different time scales. Branching points in the wavelet map correspond to scale splitting up. Presented by wavelet maps structure that characterized the energy redistribution in the system and reproduced on different scales point to certain correlation of studied membrane potential fluctuations and can be characterized by fractal dimension determined from[13] dependence of local extremes number on scale magnitude. This value is appeared 1.17 in the considered case.

Wavelet analysis is truly called a mathematical microscope. The wavelet maps reveal local non-stationary regions in different temporal intervals where remarkable energy redistribution takes place. It is obvious that for each of these intervals n and v_1 parameters introduced above are different. Thus it is possible to describe quantitatively the local non-stationarity degree of the process, which can be considered at large time intervals as the stationary one. In a case when the process is principally non-stationarity at large time intervals it is reason to believe that local non-stationarity character should be dependent on such non-stationarity and can be considered as early precursor of coming changes in the system. Certainly the last conclusion should be built on preliminary obtained information which may be considered as *a priori* knowledge about system evolution. In this way when time series length increases, the "time behavior" of introduced parameters can become *the methodological base for prognostication of complex system evolution* with determination its possible direction under external influences.

INFORMATION CONTENT OF THE INTRODUCED PARAMETERS[5]

The basis of the presented methodology is a postulate about the information content of the non-regularities in the measured dynamic variables (temporal or spatial). It corresponds with the abstract theory of information (ATI)[14]. According to this theory *"knowledge is based on the possibility of distinction. This remark reflects the fact that distinguishability – the possibility of making distinctions – presumably represents a precondition to the cognitive abilities of any rational living being. Without the possibility of distinction, we would not be able to perceive, to form concepts, or to speak. ... Empirical knowledge presupposes temporality. The difference between the past and the future is a precondition of experience. ... Distinguishability and temporality are always interwoven. Any temporal transition can be looked upon as a change of distinguishabilities "*[14].

In the framework of our approach, all requirements and the basic principles of the ATI are fulfilled. The introduced non-regularities are characterized by *distinguishability* (in accordance with the types of non-regularities) and the *temporal fractal regulation*. The latter definition replaces and makes more specific the vague term *temporality* introduced in the ATI. In this case, the idea of the fractalness acquires a pragmatic meaning and illustrates the metaphysical image of the "realism of ideas"[15]. In fact, the entire information corresponding to the concrete changes in evolution of every introduced "color – non-regularity" may be extracted by using the concrete calculating procedures – by comparing Eqs. (9)–(13) with the power spectra and the difference moments of different orders obtained from the time series data. In other words, we realize the fundamental cognition (with obtaining the corresponding numerical characteristics) of the distinguishable non-regularities of the observed dynamic variable.

In the framework of the proposed approach, two sequences of the introduced dynamic parameters may be formulated – dimensionless (n, v_1, v_2, ...) and dimensional ($\lambda_0 K_0$, $\lambda_1 K_1$, $\lambda_2 K_2$, ...). These parameters may be considered as "passport data" of the dynamic system under consideration. The second set of the dimensional parameters has a meaning of the components of the Kolmogorov K-entropy. Usually, the Kolmogorov entropy[1,2] is considered as a scalar. In our approach, the introduced parameters characterize the rates of losing information which depends on the type of the "color – non-regularities". These parameters may differ for different types of non-regularities. We call the developed approach to time series analysis flicker-noise spectroscopy (FNS)[6–8]. Obviously, the FNS provides a possibility to obtain more complete information compared to the information contained in the traditional Lyapunov exponents as well as in the dynamic entropy.

It is necessary to remark that the "polychromism" of the dynamic variable changes during the evolution may be correlated with the image of "topo-chronology" introduced by

D. Bohm[16]. "The choice of this word is intended to signify that not only must we emphasize spatial topology, i.e., the study of the order of placing one thing in relationship to another, but the study of how one event or moment acts physically in another"[16]. In other words, the image of "topo-chronology" means, in accordance with our approach, that the correlation links between the structure-energetic states of the system may be realized not only at adjacent moments in time but within different temporal intervals.

Obviously, the presented phenomenological approach does not reveal the concrete physical essence of the introduced parameters, but this approach may be used for the analysis of various systems dynamics because of its universality and nonspecificity. As an illustration, time series corresponding to the fluctuations of hydrodynamic turbulent flows as well as to the electric voltage fluctuations in the electromembrane process (see Fig. 1) and in very thin polycrystal threads (whiskers), have been analyzed [8] and the corresponding parameters ("passport data") which characterize these system states were determined. This time series analysis may be used for studying various processes in geophysics and astrophysics, in ecology and medicine, in economics, etc. The approach under consideration may be used for discovering the main features of the spatial structures or image (imprint) profiles of different origin (surface roughness, content of the imprint color and darkness, etc.) for the purpose of describing them by introducing a set of "passport parameters". In these cases, we divide the space interval along the space axis (a Cartesian co-ordinate) of the structure or images (imprints) into small equal intervals. Thus, we have a set of points (the number of points must be quite large, more than several hundred) and the corresponding values of the measured dynamic variables under consideration (the heights of the roughness, the color nuance parameters, etc.). These "time series" which reflect the complexity of the non-linear correlated processes of the structure formation ("evolution") may be examined in order to obtain various parameters of evolution.

This methodology may be used for obtaining the parameters of various natural structures in geophysics as well as the "passport data" of the medical images (tomography, mammography) and DNA sequences. This analysis may be fruitful for studying the correlation structures in music, novels, painting and other art objects. Therefore, it provides a possibility "to promote a cross-fertilization between the humanities and the natural sciences, to interlace the world of the arts with the world of natural laws and theories". FNS provides new possibilities in the analysis of literary and musical texts, in the finding of correlation links in fragments of painted canvases as well as of temporal sequences of historical events and of the sound signals of animals[6,7,17,18]. Note that the introduction of "color" in FNS analysis makes it much more informative and provides the possibility to reach the high degree of considering subject personification.

Obviously, the introduced "model-caricature" $V(t)$ is an *a priori* non-observable, purely ideal (the compact support!) metaphysical image[5]. It is necessary to point out that it is precisely the highest abstract images which form the basis for gnoseology[9]. Indeed, "hypothetical suppositions about non-observable essences and about hidden mechanisms of natural phenomena"[19] form the basis for the general physical theories (quantum mechanics, the inflation model of the formation of the Universe, string theory, etc.). It is easy to understand that the secrecy about the "first mechanisms" as well as the difference between the essence of the processes and the process manifestation during the measuring procedure[20], depend on the inertia of real objects as well as the inevitability of experimental errors. In effect, we have this situation described by Plato, where a person in a cave has to draw a conclusion about the essence of an object by watching the shadow of this object on the cave wall only. Plato's image is a universal one. We are in a similar situation whenever we research whatever phenomenon. This is why any scientific construction has to start from metaphysical principles[21], and the metaphysical arguments must accepted to be used as a legitimate tool for elaborating new ideas in physics and mathematics[16].

APPLICATION OF THE FLICKER-NOISE SPECTROSCOPY TO THE COMPLEX DYNAMIC SYSTEM ANALYSIS.

Potential difference fluctuations in the electrochemical system with cation exchange membrane[22,23]

Time series of the potential difference fluctuations in the electrochemical system with cation exchange membrane and its fragments are presented in Fig. 1. The time series is obtained under following conditions: current density over critical diffusion current density (CCD) is $j/j_{lim}=7$, solution is 10 mM *NaCl*, membrane diameter is 0.1 cm, cation exchange polyamide homogeneous membrane with statically ion exchange capacity of 1.3 meq/g was used[22].

Power spectra and 2nd order difference moments of membrane potential time series obtained under the same conditions but different current densities are shown on Fig. 4. Membrane potential fluctuations increase drastically at current density close to the CCD. The form of noise spectrum appears to be strongly dependent on the density of the current passed through a membrane. Fluctuations of the membrane potential at currents near the CCD are characterised by line spectra (Fig. 4a left). An increase of the current results in broadening of high frequency discrete peaks and simultaneous increase of low-frequency noise spectral density. Thus the linear picture transforms into a smooth flicker noise spectrum (Fig. 4d left) Difference moment changing has the similar character: from function consisted of periodic components to more smooth one (Fig. 4 a-d right). Parameters of power spectra and 2nd order difference moments are shown in table 1.

Table 1. Parameter values for power spectra and 2nd order difference moments shown in Fig. 4.

j/j_{lim}	0,5	1,2	2,7	7
Power spectra parameters				
n	1,36	1,56	2,57	2,43
$\lambda_0 K_0$, Hz	2,0	0,94	2,0	3,0
Difference moment parameters				
$2\nu_1$	–	–	1,3	1,2
$\lambda_1 K_1$, Hz	–	~1	~1	~1

MHAT wavelet coefficient map for fluctuation time series at high current characterized by smooth flicker-noise spectrum (Fig. 4d) is shown in Fig. 3. The values of wavelet coefficients increase sequentially from small to large time scales. This corresponds to the flicker-noise dependence of the power spectrum. It's impossible to point strictly scale hierarchy, there is no particular time scales. Typical branching self-similar structure of wavelet distribution appears on any time scales. The absence of particular scales and existing of self-similar structures allow us to interpret observed membrane potential fluctuations as a time fractal.

Wavelet pictures in the case of low current density time series characterized by resonant power spectrum (Fig. 4a) are shown in Fig. 5. Comparison of the maps obtained for the same time series with using Morlet (Fig. 5a) and MHAT (Fig. 5b) wavelets demonstrate known advantage of Morlet wavelet in showing up periodic processes. Clear horizontal strips in Fig. 5a correspond to time scales stable appeared in time series. Some short periodic regions are shown by "islands". Studying of periodic processes with Morlet wavelet has some advantages over Fourier transformation. First mentioned method will not give sub-harmonics and will show both time and frequency behavior. Using of MHAT wavelet (Fig. 5b) allow emphasizing in a larger degree the "scale invariant" time dynamics of signal and its local features. Self-similar structures appear in time series already at low currents but periodic processes are dominating.

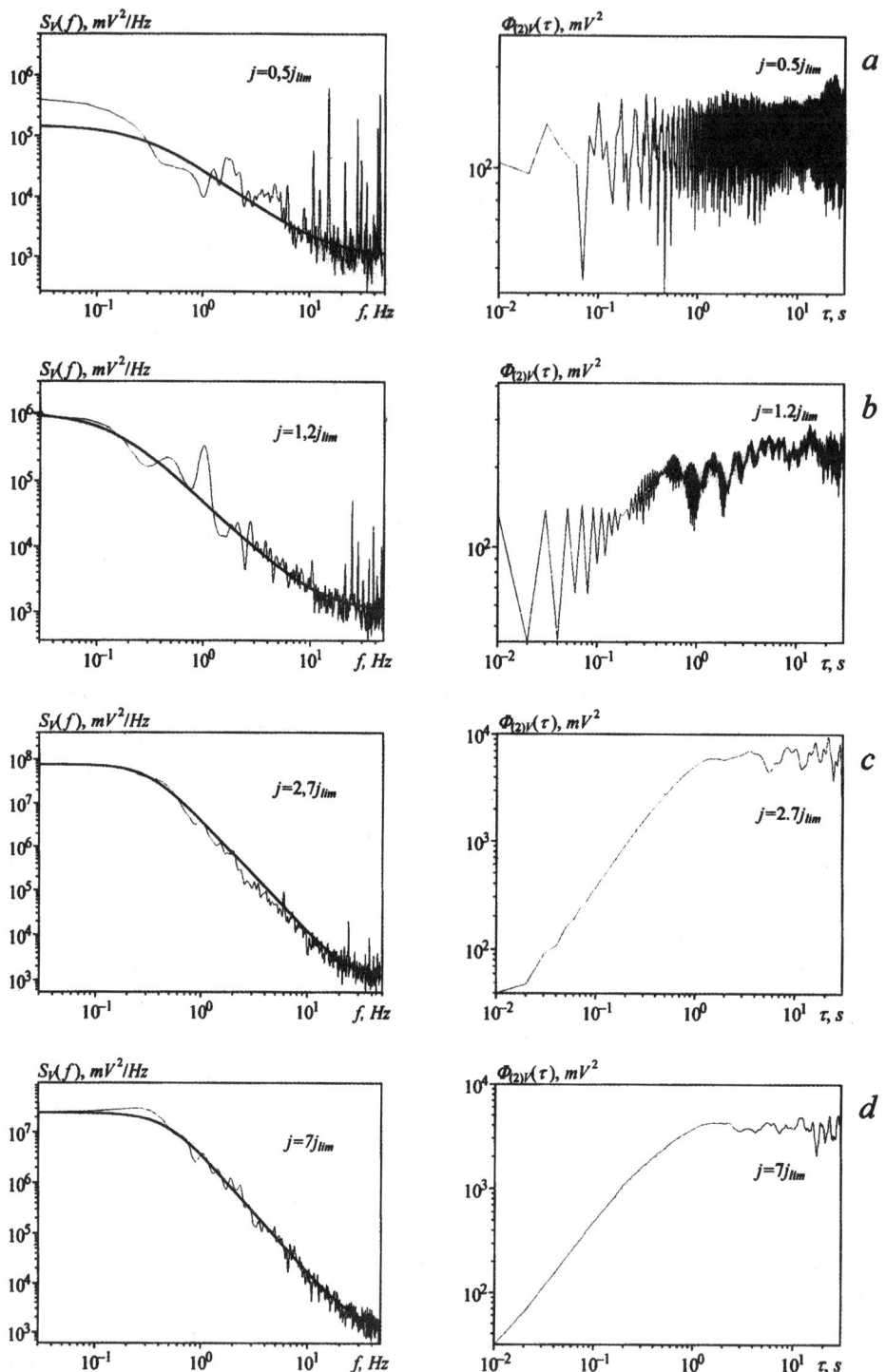

Figure 4. Power spectra (left figures) and 2nd order difference moments (right figures) for time series of potential difference fluctuations in the electrochemical system with cation exchange membrane at different current density over limiting current density ratios j/j_{lim}: a – 0.5; b – 1.2; c – 2.7; d – 7. Thick lines (left figures) show spectra approximations with formula (12). Parameter values are in the Table 1.

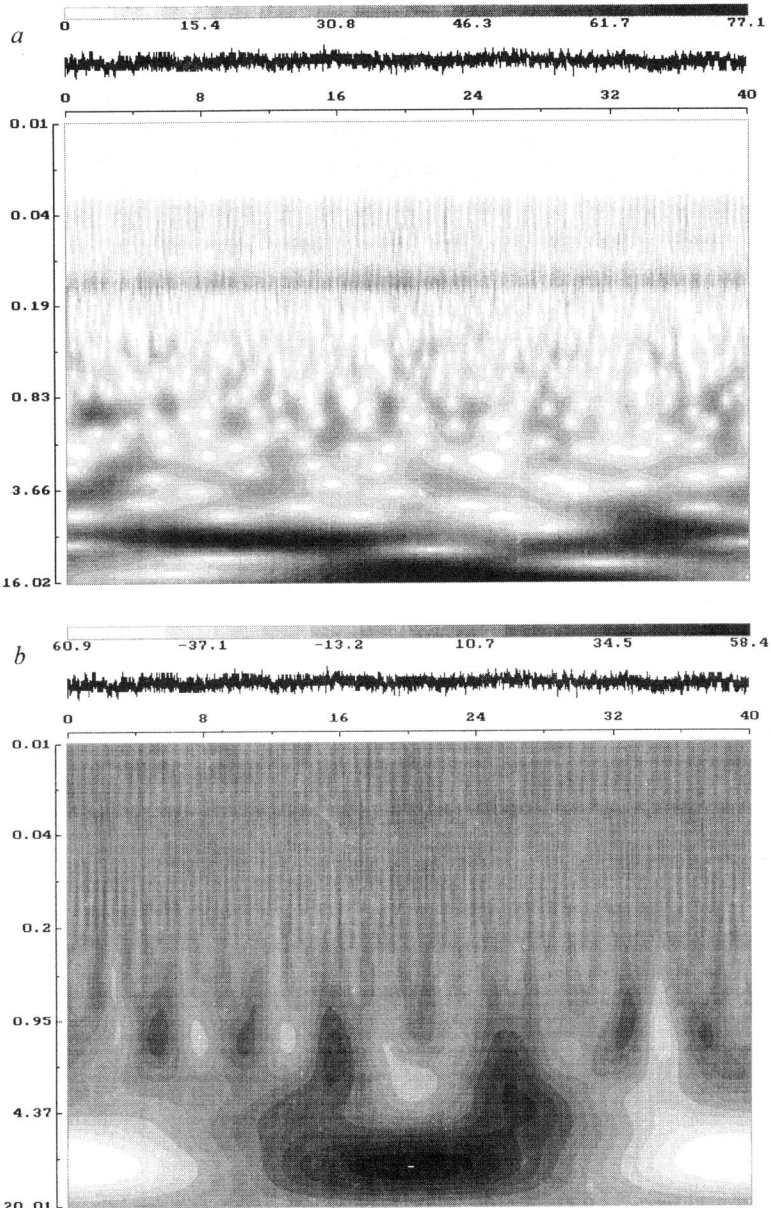

Figure 5. Morlet (*a*) and MHAT (*b*) wavelet maps for time series of membrane potential fluctuations at j/j_{lim}=0.5. The horizontal coordinate is the time (in seconds). The vertical coordinate corresponds to the logarithmic values of the wavelet scales (in seconds). Grayscale (legend on the top) reflects absolute values of wavelet coefficients for Morlet wavelet map (*a*) and coefficients values for MHAT one (*b*). Time series is shown over the map.

Fluctuations of the membrane potential drop at constant current density in the electrochemical system with cation exchange membrane were studied to clarify a mechanism of ion transport in such a system at current densities below diffusion critical current density[22,23]. Fluctuations occurring over CCD were examined as a function of current density, electrolyte concentration, viscosity of the solution, the membrane aperture, chemical structure and homogeneity of the membrane. As a result, a convective model of ion transport in the electromembrane system at different current densities has been proposed[23].

Electric current fluctuations under cathode hydrogen releasing[24]

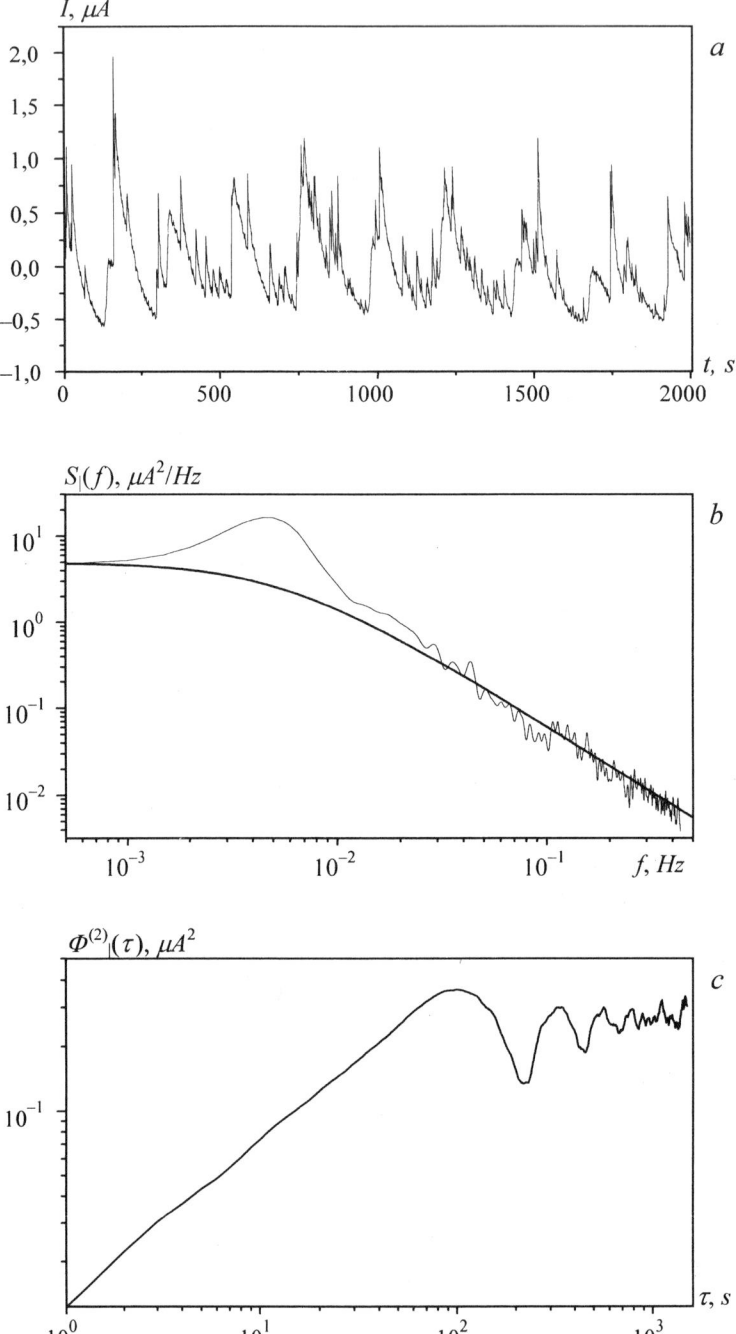

Figure 6. Time series (*a*), its power spectrum (*b*), and its 2nd order difference moment (*c*) for current fluctuations during electrolytic cathode hydrogen releasing. Spectrum approximation with formula (12) is shown by thick line in the panel (*b*). Parameter values are in the text.

The process of electrolytic decomposition of water in the cell with Pt wire vertical oriented cathode and Pt cylindrical net anode was studied[24]. Cathode area was 0.2 mm² at butt-end downward oriented surface and 0.8 mm² at the cylindrical side. Anode area was considerably larger that cathode one. Cell volume was 25 ml. All the experiments were carried out at 293 K temperature. There was used 0.5 M sodium sulfate with 0.01 M sulfur acid solution. Current fluctuations were measured under constant potential conditions.

The time series of current fluctuations at potential $\varphi = -0,8$ V (vs. Ag/AgCl reference electrode) and mean current density $j_0 = 0.7$ mA/cm² is shown on Fig. 6a. Discretization frequency is $f_d = 1$ Hz. The time series shows quasi-cyclic dependence with period $T = 220$ s and different scale fluctuations. Power spectrum of the current fluctuations presented in Fig. 6b. The extremum at $f_1 = 0.0045$ Hz corresponds to the quasi-cyclic process. The flicker-noise part of the spectrum was described by using parameters $n = 1.5$ and $\lambda_0 K_0 = 0.034$ Hz. The difference moment of the 2nd order dependence is shown in Fig.6c. The corresponding parameters are $\nu_1 = 0.75$ and $\lambda_1 K_1 = 0.01$ Hz.

The current fluctuations in this electrochemical system are due to hydrogen bulb realizing at the cathode surface. There are some "active" regions of the cathode surface where bulbs appear and grow. Because of the bulbs screen the cathode surface, an effective electric resistance increases and current is falls down. The corresponding decrease is described by the dependence $I \sim I_o(1 - \gamma t^{-1/3})$. When bulbs become quite large, they "recognize each other" and interact. The result of interactions consists in joining bulbs into largest one, which leaves the electrode surface and cleans it from small bulbs by an avalanche-like process. When the surface is cleaned the current jumps up at this moment. We believe that bulbs interaction in different scale avalanches is the reason of the quasi-periodic variations and fluctuations of electric current on the downward oriented cathode surface. Estimation of some process parameters such as number of bulbs, their sizes and diffusion layer thickness shows applicability of this simple model.

Solar activity fluctuation dynamics[25]

Describing of the Observational Data. It is well known that solar activity is not stable in time but undergoes significant variations on various time-scales. The most prominent are complex quasi-periodic variations with average period almost 11 years (solar cycle). In spite of understanding in principle of the mechanism of solar magnetic cycles generation in the frames of theories of the solar dynamo, the problem of genesis of significant variability of its amplitude and period as well as the nature of appearance of short term nonperiodic variations is still open. The progress in developing of methods of nonlinear time series analysis has resulted in the increase in understanding of chaotic deterministic behavior in process of sunspot activity[26,27]. Here we present the results of the application of described approach for analysis of solar activity phenomena in wide range of its manifestations.

We analyzed following time series data for solar activity.

Data describing solar magnetic variability at the photosphere layer: daily and monthly values of sunspots (Wolf numbers) (1900–1990 yy) and daily values of mean solar magnetic field provided by the Standford Solar Observatory (1975–1990 yy).

Data describing solar irradiance: total energy emitted by the Sun over all wavelengths – the "solar constant" (Nimbus-7 satellite observations, 1978–1989 yy), the hydrogen Layman-alpha emission at 121.5 nm from chromosphere (1982–1989 yy), the coronal index (emission at 530.3 nm) from solar corona (1964–1983 yy), the solar radioflux in different wavelengths generated at different layers of the chromosphere and the corona.

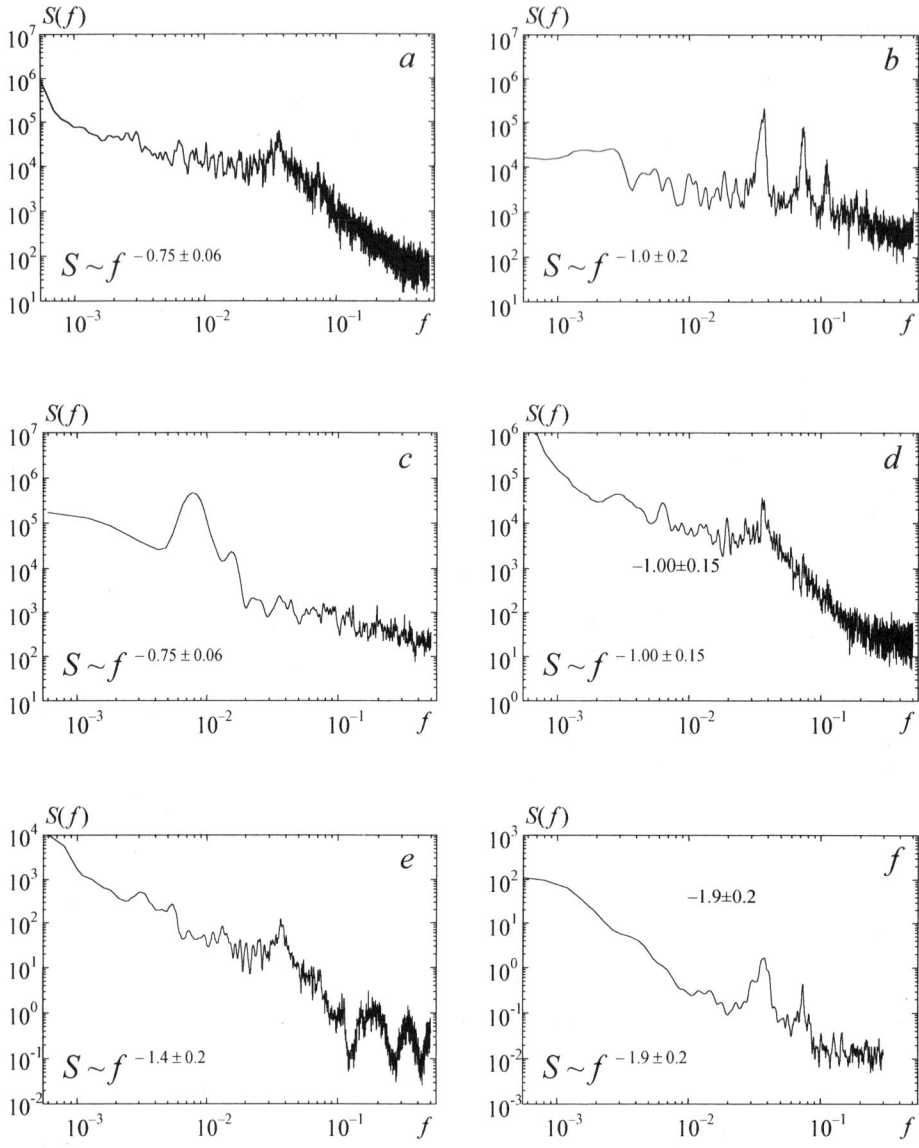

Figure 7. Power spectra $S(f)$ of: *a*) daily sunspot numbers; *b*) daily values of solar mean magnetic field; *c*) monthly sunspot numbers; *d*) daily values of 2800 MHz solar radioflux; *e*) daily values of emission of solar corona at $\lambda = 530.3$ nm; *f*) daily values of hydrogen $L\alpha$ emission at $\lambda = 121.5$ nm. All plots are in double logarithmic scale. Frequency (f) are in inverted intervals of measurements.

The analyzed time series were taken from the NOAA's World Data Center. They cover wide range of time scales and include only short interruptions, which were filled using the linear interpolation. These data describe local manifestations of the solar activity in the outer solar layers: photosphere, chromosphere and corona, summed over all visible solar disk.

We also used the unique time series characterizing activity in the solar core measured by the Homestake solar neutrino observatory[28,29]. These data describe mean over 2–3 months values of solar neutrino capture rate and cover a period 1970–1994 yy.

Results and discussion. Power spectra of analyzed time series are shown in Fig. 7, Fig. 8. One can see, that all presented spectra demonstrate common behavior: the increase of spectral density as $1/f^n$ with different values of spectral index n. Hence, the values of observational variables in all studied processes are not random but correlated. It means that every examined time series describes a global dynamic process with specific for the corresponding solar layer "memory time". Peaks of the spectral density which correspond to the 11 years solar cycle period and to approximately 27 days solar rotation period are visible.

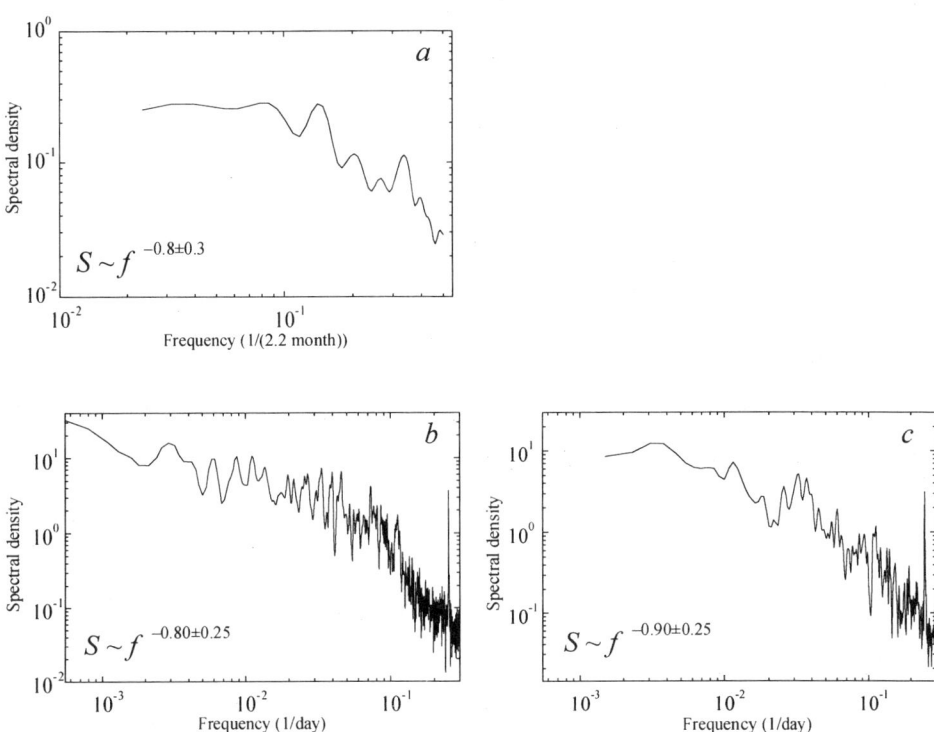

Figure 8. Power spectra of solar neutrino capture rate series (*a*); the solar total irradiance time series before (*b*) and after (*c*) exclusion of the sunspot area influence. All plots are in a double logarithmic scale.

We should emphasize one detail in the spectra of daily time series – the change of the spectrum slope in the region of the solar rotation frequency. To explain this peculiarity we should take into account that active regions are not distributed homogeneously on the solar sphere and at every moment one can see only half of this rotating sphere. That is why the "artificial" dynamic caused by the solar rotation is added to the "natural" dynamic of solar activity. This leads to an increase of the effective rate of relaxation of fluctuations within the frequency range of more than 1/(27 days). As was shown earlier (Eq. 9–11), in this case the spectrum should become steeper. The comparison of spectra for daily and monthly values of the same variable confirms this suggestion. For example, the spectrum of daily time series of Wolf numbers in the frequency range $f \gg 1/(27\ \text{days})$ has the slope ($n = 0.75\pm0.6$) identical to the slope of the monthly time series. On the other hand for $f < 1/(27\ \text{days})$ the spectrum is much steeper (n is about 3). Hence for obtaining information about the solar process itself, one should use the frequency range $f > 1/(27\ \text{days})$ in corresponding spectra of daily time series.

The values of the n-parameter were determined for all spectra for the frequency range between the solar cycle frequency and the solar rotation period. It was established that for analyzed time series the uncertainty in value of n-parameter is determined by the following factors: experimental errors, procedure of spectrum smoothing and length of time series. To estimate the impact of the first factor we analyzed 3 nonoverlapping parts of the sunspot time series with length of every part 2^{14} points. We find that low smoothing (over 2–3 points) gives more stable result. We got $n = 0.75\pm0.06$ for these spectra in agreement with[27]. For the spectra of all other time series the error in value of n-parameter caused by short length of time series was estimated by comparison of n calculated for long sunspot time series and for its parts of corresponding lengths. For time series of the length of about solar cycle the contribution of trend in the value of n was estimated in every case and included into the error. The determined values of n turned out to be almost the same for processes of the sunspot activity evolution ($n = 0.75\pm0.06$) and for the formation of mean solar magnetic field ($n = 1.0\pm0.2$). These n values are typical for slow nonexponential relaxation and thus for a long memory time in these processes. The stability of the spectrum slope till frequency 1/(50 months) for sunspot activity monthly time series indicates, that memory time of this process is not less then 50 months. The fact that relaxation time turned out to be similar for both processes argues, that the solar magnetic variability over all spatial scales at the photosphere level reflects a single process of complex turbulent convection in the convective zone.

Unfortunately, the time series of the solar neutrino flux is not sufficiently long (only 109 points), measurements are not completely regular, there are the gap in data and the experimental errors are 30–100%. These data were regularized using linear interpolation to obtain 128 equidistant in time points. Note that this procedure makes data not only uniform but also smoothed. That is why we did not use points in the frequency range more than 1/(6 months) in the corresponding spectrum for estimation of the n-parameter. The spectrum of neutrino flux time series is presented in Fig. 8a. One can see that it has a significant slope. To estimate the uncertainty in value of n-parameter due to the experimental errors in this case we constructed twenty modeling time series using the following procedure: every value of initial time series was displaced randomly according Gauss distribution and the correspondent experimental error. The calculated standard deviation of n-parameter is equal ±0.2. To check the influence of nonuniformity of the data and the gap we used the time series taken from[28] that contains values of the sunspot numbers averaged for the same intervals of measurement as neutrino flux. We obtained the value of n for this case $n = 0.85$ that differ from that defined for the original long sunspot time series on 0.1. We estimated the total error in n to be ± 0.3 including the uncertainty in determination of the mean date of the measurements. Thus determined $n = 0.8\pm0.3$. Hence we can conclude that the observed variations (with typical time of order a few months) of the solar neutrino flux are not random but reflect some real nonlinear process. The possible sources for the cyclic variations of the neutrino capture rate were listed in Ref.[30] in the discussion of the discovered anticorrelation of the solar neutrino capture rate with the solar surface magnetic flux measured near the solar equator. The same reasons could lead to the observed flicker-noise in the power spectrum of the short scale solar neutrino capture rate variations. The first reason could be the variations of the neutrino production rate. In this case the nuclear fusion reaction zone is not stable and uniform but nonlinear processes with long relaxation time occur there. Paper[31] discuss the possible sources for such processes. The second reason could be a modulation of the solar neutrino flux by the solar magnetic field[32]. It is interesting to note that the spectra of the solar neutrino flux and the solar mean magnetic field have the similar slopes in the low frequency range. This could be an indication that variations in both time series are caused by the same process.

The solar total irradiance time series and its spectrum are of a special interest. This variable describes the energy emitted by the Sun over all wavelengths and its measurements cover the period of almost the whole solar activity cycle (10.5 years). The solar total irradiance undergoes small (approximately 2%) but significant long-scale variations from maximum to minimum of the solar activity, which correlate with long-scale variations of other considered variables. On the other hand short-scale variations of the solar total irradiance reveal strong anticorrelation with the total sunspots square because of blocking of irradiance from the sunspot area. The mechanism of such influence is clear so it is possible to exclude this influence using regression. The influence of the sunspots should result in a break in the power spectrum due to the solar rotation as was discussed above. After removal of this influence we expect no break in the spectrum because it is natural to assume that the Sun emits the energy homogeneously from the whole sphere. The spectra of the observed total irradiance time series before and after excluding of the sunspot area influence are presented in Fig. 8b-c. Indeed, the spectrum of original time series has a break of the slope in the region corresponding to the solar rotation period, while the spectrum of time series where impact of sunspots was excluded has no such pronounced break. This demonstrates that the influence of the sunspots is removed effectively. Since the obtained spectrum still has a considerable slope ($n = 0.90\pm0.25$) we can conclude that short-scale variations of total solar irradiance are not determined only by the sunspot activity but probably reflect the variations of energy production in the solar core or short-term irregularities in the process of the energy transport from the core to the photosphere level through the convective zone.

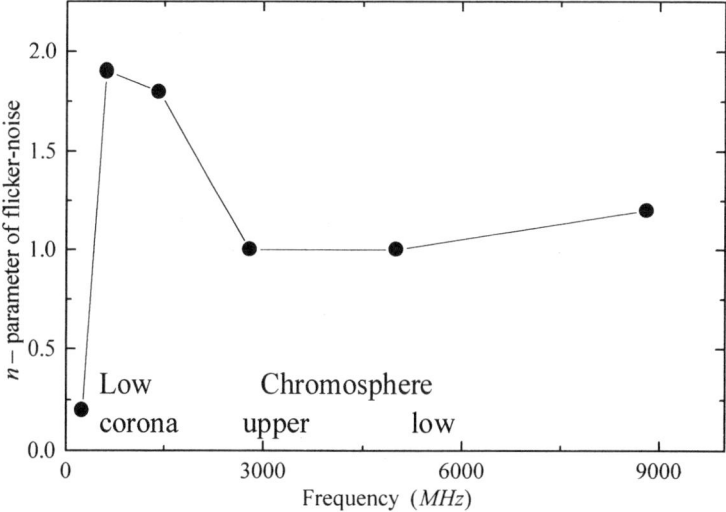

Figure 9. The dependence of the flicker-noise n-parameter for the processes of solar radioflux emission on the radioflux frequency.

For variables characterizing the processes in the solar atmosphere (in the chromosphere and the corona) the spectra turned out to be different. The solar atmosphere on the one hand is pierced by solar magnetic field that generally forms its structure and on the other hand has low density and positive temperature gradient. This determines the specifics of processes here. All examined data characterizing this solar region describe various emissions from diverse atmospheric layers. Their values are functions of temperature and density of the solar matter in a corresponding layer[33] and, hence, the variations of its intensity are caused by variations of the solar atmosphere parameters. The spectrum of time series of hydrogen Layman-alpha emission (Fig. 7*f*) which is originated in the upper chromosphere has $n = 1.9\pm0.2$. This value of *n* corresponds to a quicker relaxation in comparison with the variables reflecting the processes in the convective zone and deeper. The spectrum of time series of the coronal emission at 530.3 nm (Fig. 7*e*) which is originated in a more external solar layer has $n = 1.4\pm0.2$. The solar radioflux comes from the different layers of the solar atmosphere: the higher frequency of radioflux the lower layers where it comes from. The variations of radioflux intensity are determined by the processes in the solar atmosphere. It is remarkable that *n* remains of ~1 till the level of the upper chromosphere and increases for more remote layers of the atmosphere as can be seen in Fig. 9 (with the exception of the outermost point where the value of *n* is close to zero which corresponds to the case of "shot" noise, that is the relaxation time in this layer is already comparable with the measurement interval). Hence, we can suggest that the processes in the solar atmosphere till the upper chromosphere are mostly governed by underlying regions but at the upper layers local processes (random or with quick relaxation time) become more and more prominent.

Summarily, we revealed the flicker-noise in a wide variety of the solar processes in the different solar layers. This indicates their global nature. Using the developed approach we determined the dynamical characteristics of all studied processes. The variables describing these processes have absolutely distinct origin, however, it turned out that the short-term (with typical time less than ~ 5 years) variations of their intensity have the similar time scale. Their dynamical characteristics are almost the same for the processes that are observed at the level of the photosphere and almost gradually modify for more remote layers. This can occur when the variations of conditions responsible for variations of observational variables in all solar layers are governed by a single global source. It is naturally to suggest that this source is the turbulent convection in the convective zone, which is established to cause the short-term magnetic variability[27]. From this point of view all Sun layers is a single dynamical system. The revealed flicker-noise with $n = 0.8\pm0.3$ in the spectrum of the solar neutrino capture rate demonstrates that the short-term fluctuations in this rate are not random but caused by some dynamical mechanism. Existence of the similar flicker-noise in the total solar irradiance indicates that this mechanism can be connected with possible variations in the energy production in the solar core. However, it is not possible to exclude other reasons for these variations like neutrino flux modulation by the solar magnetic field and the irregularities in the process of energy transport through the convective zone.

Monte Carlo modeling of nonlinear dissipate processes (flicker-noise in chemical kinetics)[34]

This part is devoted to analysis of the chemical reaction kinetics in active medium on the basis of developed methodology. The macroscopic kinetics of condense phase chemical processes and, particularly, the homogenous or heterogeneous catalytic reactions, depends in a great degree on the media microscopic dynamics in relatively extended (as compared with the dimensions of typical reactive transition state) spatial regions (from 1 to 100 nm). This can be observed, particularly, in the dynamics of dissociative absorption of molecular oxygen on platinum[35]. The high resolution electron microscopy also provide data on the temporal evolution of nanocrystals (4–10 nm) of lead[36], that confirm the possibility of dynamic rearrangement of solid particles.

Chemical processes in such dynamic systems, through which the relatively great flows of energy pass, can be simultaneously statistically "steady" (stationary) and kinetically "non-steady". In the first case we say about the system macrokinetics, averaged over the relatively long period of time, and in the latter case we say about the analysis of its fluctuations, determined by non-equilibrium excitations with corresponding distribution functions. The both types of kinetics bear an information about microscopic parameters of the system and the elementary stages of its transformations, but despite of it, the object of investigation in chemical kinetics, as a rule, is only the first one – the macroscopic averaged kinetics.

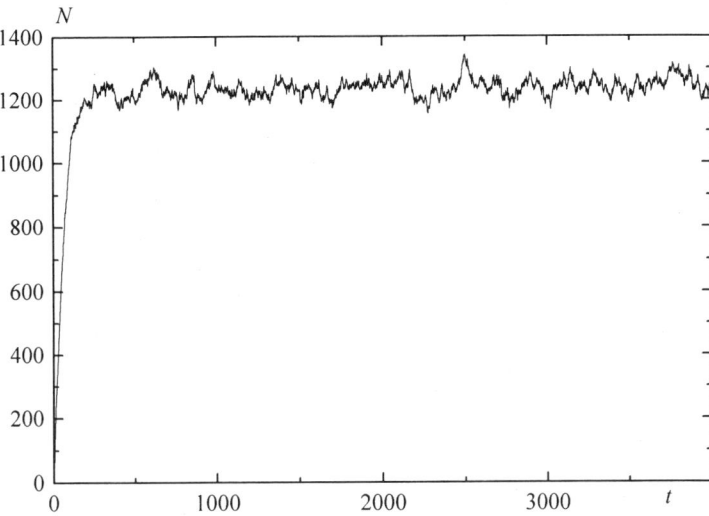

Figure 10. Typical time series of the number of free radicals in the 3-D lattice in the stationary process of their generation/recombination. The time t is measured in units of $\Delta\tau$ – the mean lifetime of free radicals random walks to the adjacent nodes of 3D $101\times101\times101$ l^3 lattice.

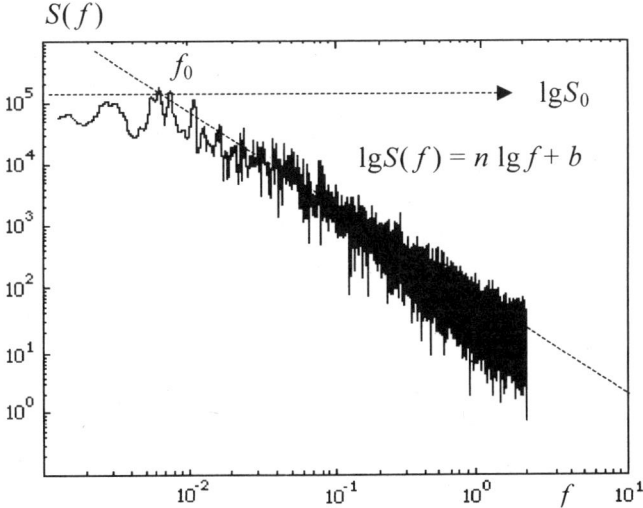

Figure 11. Power spectrum for the typical time series of 8000 points (periodogramm after smoothing with frequency filter). The characteristic parameters (in dimensionless units) are the frequency $f_0 = \tau_M^{-1}$ corresponding to the time of the system memory loss (τ_M) and the spectrum slope n in the region of the flicker-noise.

Below we demonstrate that the analysis of fluctuational component of chemical kinetics can provide very valuable information about the system microscopic parameters, which determine the experimentally observed phenomena. It is especially important in the case of stationary chemical processes, the average parameters of which in principle do not depend on time. We will carry out Monte Carlo simulation of one of the classic problems – the kinetics of generating and recombination of reactive particles in solids under the impact of external source of energy. Monte Carlo method of computer simulation allows one to produce time series of any variable with arbitrary length and time scale, *a priori* having the full knowledge about model system. Recently this approach was used for computer modeling and spectral analysis of fluctuations in chemical kinetics[37,38].

The fluctuation kinetics of complicated systems, which temporal evolution is measured as temporal series of some variable $V(t)$ in the simplest case has a character of the "Brownian" motion around the mean value V_m. Notice, that even in this simplest case the high-frequency branch of the power spectrum does not have the "white" noise character. This means that there exists the correlation between the variable magnitudes at short intervals of time (system has a "memory") and the immediate relaxation of fluctuations is impossible.

Monte Carlo model. Let us consider the model of the process of simultaneous generating and diffusion-controlled recombination of free radicals under the impact of ionizing radiation, developed in the papers[39,40]. Reactive particles (free radicals) make random walks on 3-dimension cubic lattice from site to adjacent site with the mean time of jump $\Delta\tau$. The lattice has dimensions $101\times101\times101\ l^3$, where l is the elementary cell dimension. The reaction A+A→C is modeled as an annihilation of particles when they meet in neighboring sites. Such sites closest to the site (0, 0, 0) are not only (± 1, 0, 0), (0, ± 1, 0), (0, 0, ± 1), but also sites (0, ± 1, ± 1), (± 1, 0, ±1), (± 1, ± 1, 0), (± 1, ± 1, ± 1) – all together $3^3 - 1 = 26$ sites. In this proposition, depending on the geometry of mutual co-ordination, the distance between neighboring sites equals l, $\sqrt{2}\,l$, or $\sqrt{3}\,l$, i.e. the reaction happens when particles approach each other at the distance less than $R_p < 2l$ ("black sphere" radius)[39,40]. The diffusion constant, evidently, equals $D = l^2/6\Delta\tau$, and the bimolecular rate constant is $k = 8\pi DR_p$ (doubled as compared with the standard expression $k = 4\pi DR_p$ numerical coefficient corresponds to the fact that the both reacting particles are mobile, and their complementary diffusion coefficient equals $2D$). We use the cyclic boundary condition: after a particle walks out the cube it appears on its opposite side. The space distribution of generated particles can be homogenous or as clusters (spurs). In the latter case we throw first the coordinates of spur center and than throw the coordinates of required number of particles inside the sphere of the radius R_s. Further we take l as a length unit and $\Delta\tau$ as a time unit. This allows introducing the dimensionless units for the model parameters. The total number of particles was considered as the system dynamic variable.

The typical temporal series of the total number of particles $N(t)$ in the volume $101\times101\times101\ l^3$ is presented in Fig. 10. The number of radicals is growing from zero to some stationary level around which the fluctuations at different time scales take place.

Results and discussion. Below we present the results of the detailed spectral analysis of this signal within the range of stationary kinetics. Fourier spectrum of the autocorrelation function $<N(t)N(t+\tau)>$ has been calculated for the time series of $N(t)$ – the variations of the total number of the particles at stationary kinetic conditions. We use smoothing with the frequency filter for obtaining precise parameters of the Fourier spectrum. Table 2 shows the values of the parameters n, f_0 and S_0 for the next generation-recombination regimes:

a) homogeneous particle generation and diffusion-controlled recombination inside the cube 101 on edge;

b) homogeneous particle generation and recombination inside the cube from 29 to 101 on edge (for the revealing of the border effects);

c) homogeneous generation and recombination inside the cube 101 on edge in case of mobility anisotropy with parameter γ which is equal ratio of the particle mobility along one of the coordinate axes Z to mobility along all other directions ($\gamma \equiv D_Z/D_X = D_Z/D_Y \neq 1$, where $\gamma = 0.1-10$).

d) Inhomogeneous generation of free radicals in the spurs of radius 3, 4, 5 grid constants and the numbers of the throw particles (4 or 5).

We can estimate specific frequency $f_0 = \tau_M^{-1}$ of the memory loss from the crossing point of the spectral function low-frequency limit S_0 values (Table 2) and the linear approximation of the spectral dependence decreasing region (Fig. 11)

Table 2. Monte Carlo simulation parameters (the first ten columns) and corresponding parameters of the power spectra for different temporal series of free radicals generating/recombination process (the last four columns).

Experiment number	N_s	R_s	L	G	N_{av}	$C \cdot 10^3$	$C_s \cdot 10^3$	γ	lg S_0	n	b	$-\lg f_0$
1.	1		101	5	764	0.74		1.0	4.58	1.92	0.67	2.04
2.	1		101	5	999	0.97		0.1	4.73	1.92	0.67	2.12
3.	1		101	5	891	0.86		0.33	4.88	1.94	0.66	2.17
4.	1		101	5	495	0.48		10	4.31	1.92	0.65	1.91
5.	1		101	5	630	0.61		3	4.62	1.93	0.66	2.04
6.	1		29	1	46	1.89			2.97	1.83	0.03	1.61
7.	1		35	2	86	2.01			3.24	1.84	0.28	1.61
8.	1		41	3	136	1.97			3.59	1.84	0.46	1.70
9.	1		51	6	259	1.95			3.60	1.83	0.76	1.55
10.	1		61	2	220	0.97			4.10	1.92	0.27	2.00
11.	1		71	17	738	2.06			4.23	1.85	1.20	1.63
12.	1		81	5	538	1.01			4.48	1.93	0.65	1.97
13.	1		91	36	1538	2.04			4.43	1.82	1.54	1.58
14.	1		101	50	2121	2.06			4.83	1.80	1.70	1.74
15.	1		101	50	2117	2.05			4.54	1.83	1.68	1.56
16.	1		101	40	1941	1.88			4.75	1.82	1.59	1.73
17.	5	3	101	10	1368	1.33	44.21		4.89	1.75	1.52	1.93
18.	5	3	101	1	439	0.43	44.21		4.74	1.82	0.49	2.34
19.	5	4	101	10	1570	1.52	18.65		4.91	1.78	1.60	1.85
20.	5	4	101	1	513	0.50	18.65		4.64	1.85	0.60	2.19
21.	5	5	101	10	1746	1.69	9.55		4.82	1.77	1.66	1.78
22.	5	5	101	1	563	0.55	9.55		4.67	1.86	0.64	2.16
23.	4	3	101	10	1306	1.27	35.37		4.89	1.76	1.48	1.94
24.	4	3	101	1	418	0.41	35.37		4.67	1.83	0.47	2.29
25.	**4**	**4**	**101**	**10**	**1505**	**1.46**	**14.92**		**4.62**	**1.76**	**1.57**	**1.73**
26.	4	4	101	1	485	0.47	14.92		4.74	1.84	0.54	2.29
27.	4	5	101	10	1643	1.59	7.64		4.69	1.84	1.55	1.70
28.	4	5	101	1	539	0.52	7.64		4.77	1.87	0.56	2.26

In the table: N_s is the number of radicals to be generated in a spur, R_s is spur radius, in lattice elementary periods, L is system extension, in lattice elementary periods, G is intensity of generating, spurs/s, N_{av} is mean number of radicals in the system, C is mean concentration of particles ($C = N_{av}/L^3$), C_s is initial local concentration of radicals in a spur ($C_s = 3N_s/4\pi R_s^3$), γ is the anisotropy parameter($D_Z = \gamma D_Y = \gamma D_X$), lg S_0 is the low-frequency limit of the power spectrum in logarithmic scale n is the power spectrum slope in the region of flicker-noise, b is a parameter of the dependence lg $S(f) = n \lg f + b$, lg f_0 is the logarithm of the system "memory loss" frequency ($f_0 = \tau_M^{-1}$)

Analysis of the data obtaining under different simulation conditions shows that in isotropic case ($\gamma = 1$) and the same recombination constant the memory lost time correlate with the only parameter – average particle concentration C independent from all other system characteristics (cf. Fig. 12). Correlation dependence lg τ_M from lg C in the experiments with anisotropic mobility ($\gamma = 0.1–10$) and therefore with different recombination constants k_γ, has opposite sign (open circles at Fig. 12). One can explain these facts in the next way.

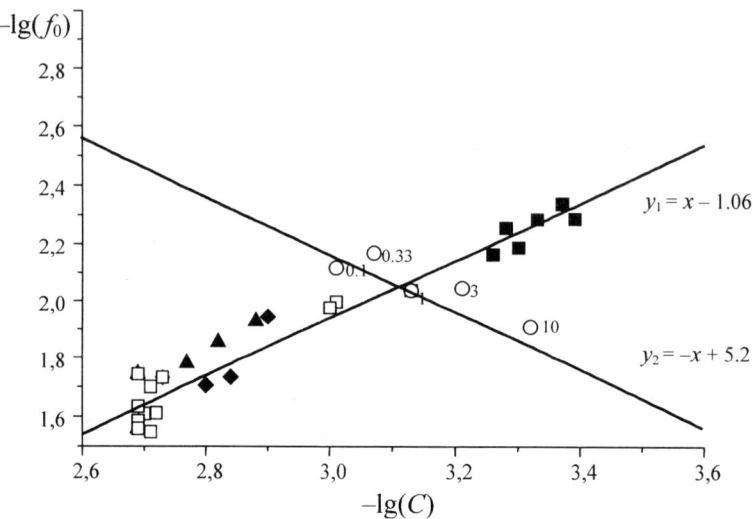

Figure 12. The correlation between the frequency f_0 of the system "memory loss" and the mean concentration of particles C in the volume. Open circles – simulation of the process with the anisotropy diffusion with correlation dependence $y_2(x)$ (numbers near circles are the multiplication factors γ of the diffusion along Z-axis). Other points are the realizations of the isotropic process with the same diffusion and bimolecular rate constant for different power and inhomogeneity generation with correlation dependence $y_1(x)$.

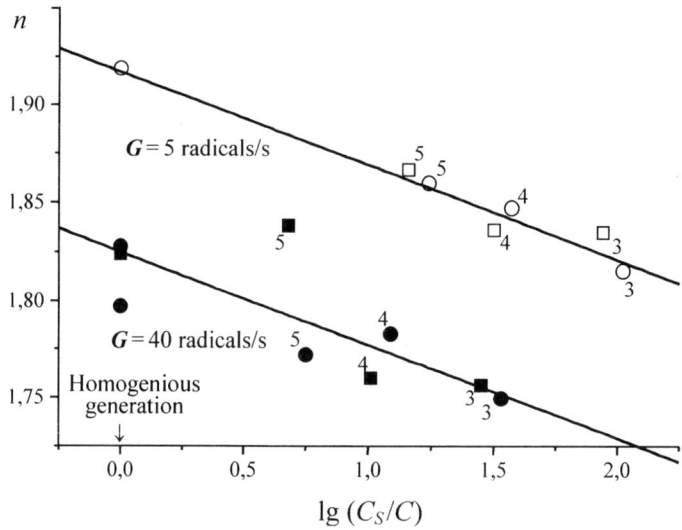

Figure 13. Correlation between the power spectrum slope n and the degree of spatial inhomogeneity of free radicals generation $\lg(C_S/C)$ at different intensities of generation G. Here C is the mean spatial concentration, C_S is the local concentration of free radicals in spurs. Numbers near points corresponds to the spur radius, open points – the lower intensity of generation ($G = 5$ radicals/s), filled points – the higher one ($G = 40$ radicals/s).

a) Isotropic case.

The relaxation of small ($\delta N \ll N$) positive fluctuation of the spatial density of particles, as the probability of survival of individual particle, are determined by the pseudo-first law $\delta N \sim \exp(-kCt)$, where k – is the bimolecular rate constant of recombination. From this it follows that characteristic relaxation time, or the memory loss time, should be estimated as $\tau_M \approx (kC)^{-1}$, from where we get lg $\tau_M^{theor} \approx -\lg C - \lg 8\pi D R_p \approx -\lg C - 0.92$ (here D – is the individual diffusion constant, R_p – reaction radius, see the model description above). Comparing the estimations made with this formula with the correlation dependence in the Fig. 12, one can see that the proposed explanation is in a good agreement with the result of computer experiment (lg $\tau_M^{exper} = -\lg C - 1.06$).

b) Anisotropic case.

In this case the mobility of particles and bimolecular rate constant depend on the anisotropy parameter γ. Let us limit ourselves by consideration of the rate constant estimate following from the condition of stationarity: $(G/L^3) = k C^2$. For this experimental series the generating intensity is constant and equals $G = 5$ radical/s. Calculating k from above equation and substituting it in the estimations of relaxation time $\tau_M \approx (kC)^{-1}$, we find lg $\tau_M^{theor} \approx \lg C - \lg(G/L^3) \approx \lg C + 5.3$. This prediction is also in good agreement with the experimental correlation in Fig. 12 (lg $\tau_M^{exper} = \lg C + 5.2$).

Let us consider another important parameter of fluctuation kinetics – the frequency slope of the power spectrum n. For homogenous low-intensity generating the magnitude of this parameter approaches to value 2 (Fig. 13), that corresponds to the exponential law ($\sim\exp\{-kCt^{n-1}\}$) of fluctuations decreasing. In another case of great generating intensity and/or its spatial non-homogeneity n is substantially less than 2 (reach the value 1.75), demonstrating the decreasing relaxation rate at longer times.

Possible reasons of non-exponential relaxation kinetics can be the temporal dependence of Smoluchowsky-Weit effective bimolecular rate constant at the initial non-stationary stage and the time-dependent rate of spur reactions[39,40]. At the long times scale one observes the stationary reaction kinetics. The peculiarities concerned with the spatial non-homogeneity of just now generated particles are valuable in the short times scale – the reaction rate seems to be more rapid. This dependence on the time scale provides the change of the slope n of the power spectrum.

No substantial dependence of parameter n on system dimensions and anisotropy parameter γ was found. Consequently, the boundary effects and diffusion anisotropy does not influence on the temporal dependence of relaxation rate.

Another way for the analysis of these time series is a wavelet transform. The logarithm of module of continuous wavelet transform coefficients for the time series No 25 are shown in the Fig. 14. We use Mexican hat as a mother wavelet function for the calculation of this wavelet map. Each next fragment of this map is the one forth in area of the previous picture. Comparison of the wavelet map fragments at different scales (Fig. 14) shows fractal, self-similar structure of the signal at time scales differs more than one hundred times. This conclusion is confirmed on the base of analysis of the dependence of the number of extremums (cf. Ref.[13]) in the picture of the wavelet transform coefficients module vs. scales (Fig. 15). The tangent of the slope of this function is about 1 in the time scale ranges from 2 to 300 in dimensionless units of measurement, and we can say that this is time fractal.

Fig. 16 shows structure functions – second order difference moment $\Phi_{(2)}(\tau)$ for the time series No 1 (average value $N_{av} = 764$) and No 25 (average value $N_{av} = 1505$) from Table 2. The tangents of the slope $(2\nu_1)$ of the left parts of these functions are equal 0.96 and 0.85 sequentially. These parameters are related to values n (1.92 and 1.88) for spectral density functions: $2\nu_1 = n - 1$, what means that both the spectral density and the second order difference moment contain in this case the same information. The longer time series are used for the calculation of the second order difference moment, the smaller fluctuations are observed as a consequence of the average (cf. Fig. 16, curves 2 (500 points) and 4 (8000 points)).

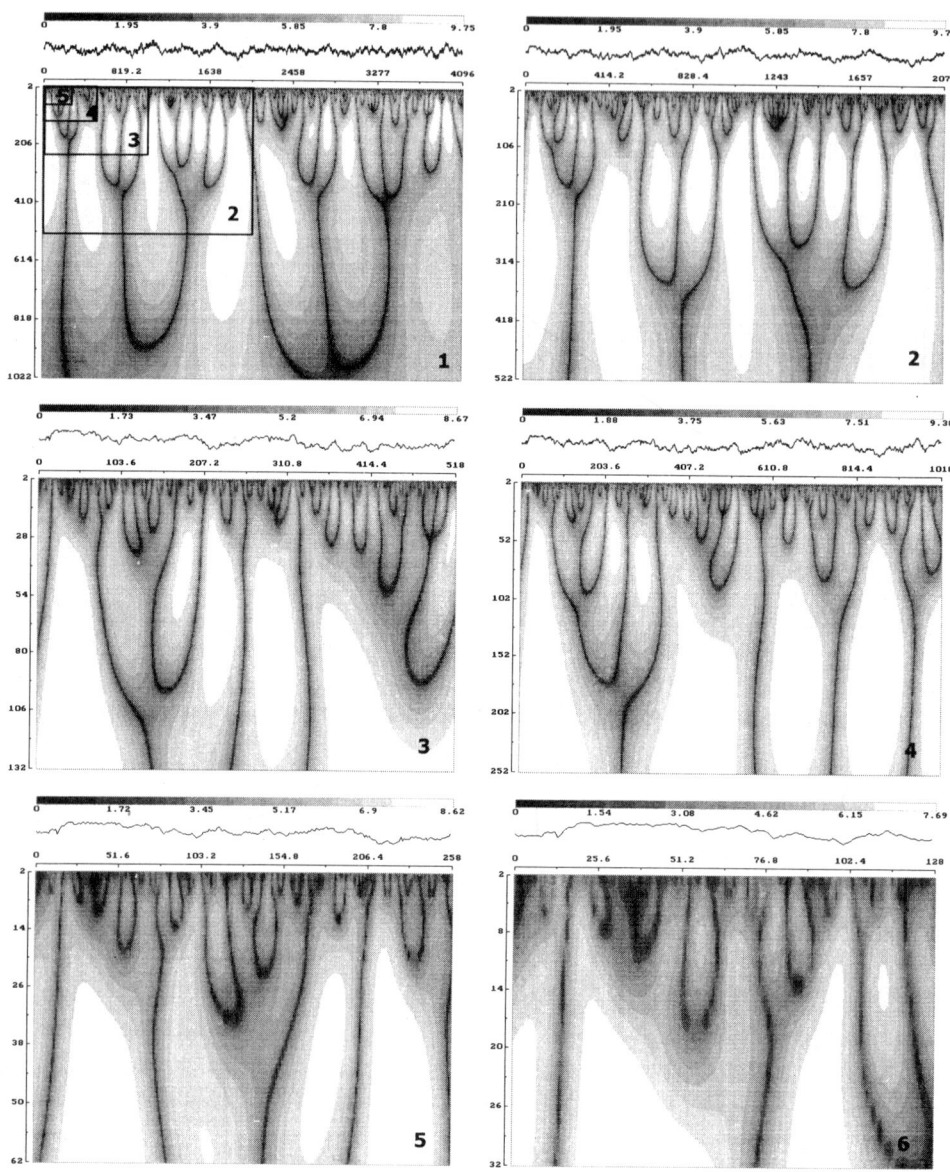

Figure 14. The logarithms of module of continuous wavelet transform coefficients (with Mexican hat as a mother wavelet) for the time series No 25. Each next fragment of this map is the one forth in area of the previous picture. The horizontal coordinate is the time (in dimensionless units). The vertical coordinate corresponds to the values of the wavelet scales (in dimensionless units). Grayscale (legend on the top) reflects logarithm of absolute values of wavelet coefficients. Time series is shown over the maps.

So, this approach to chemical process fluctuation analysis can help us to obtain direct information about its basic elementary processes. The other stationary chemical process mechanisms can also be simulated by Monte Carlo method, and the same methods for obtaining fundamental parameters, which determine fluctuation kinetics, can be used.

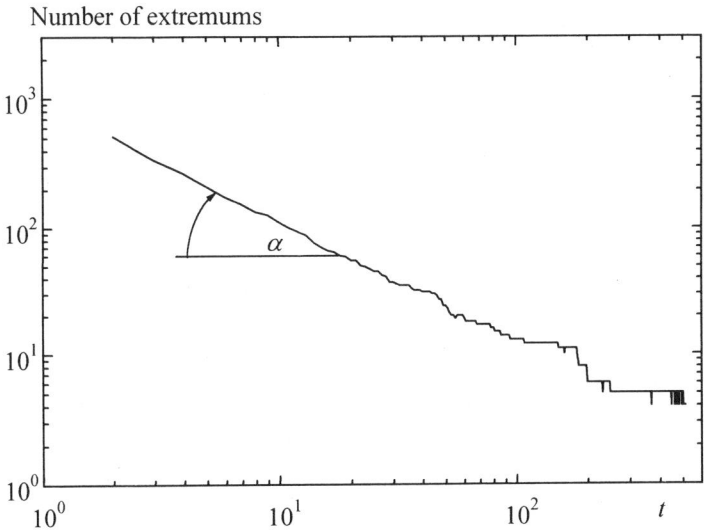

Figure 15. Dependence of the extremes number in the picture of the wavelet transform coefficients module vs. scales (time series No 25 from Table 2). The tangents of the slope of these function $tg(\alpha) \sim 1$.

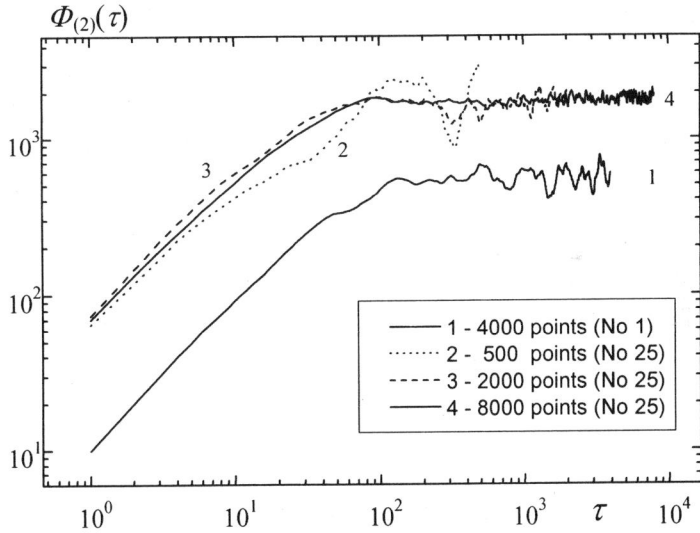

Figure 16. Structure functions – second order difference moment $\Phi_{(2)}(\tau)$ for the time series No 1 (average value $N_{av} = 764$) and No 25 (average value $N_{av} = 1505$) from Table 2 (see the legend inside the picture).

Monte Carlo method allows also involving a different arbitrary radical generating regimes. Particularly, applying to simulation the bimodal one with two generation intensities (their ratio $g = G_{min}/G_{max}$) and random switching between them (with probability w), we can produce time series very similar to the real natural processes. For example, Fig. 17 shows comparison between real (left side) and simulated (right side) time series. One of the simulation runs looks like electric current fluctuations under cathode hydrogen releasing (Fig. 17(1a-1b); for the sake of convenience we reproduce here a part of the Fig. 6a), others resemble voltage fluctuation in viskers[8] (Fig. 17(2a-2b)) and burst-kinetics of single ion channels electric current in biological membranes[34] (Fig. 17(3a-3b)). However, in this work we do not consider correspondence between the parameters of calculated with time series and the real process characteristics. Here we only want to show the principle possibility of modeling substantially different natural physical, chemical, biological and other processes with Monte Carlo method.

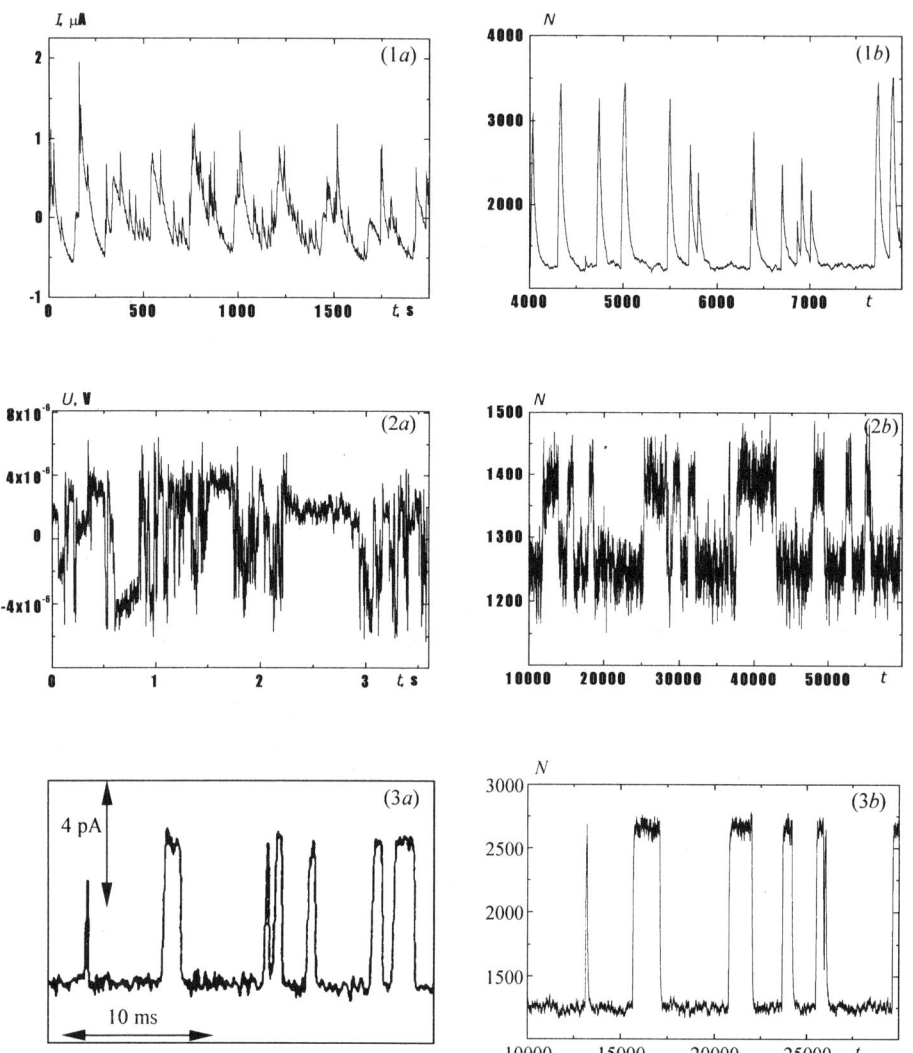

Figure 17. Comparison between real time series (left column) and simulated ones by Monte Carlo method (right column) with different parameters (g and w) of bimodal generation regime. 1. Electric current fluctuations under cathode hydrogen releasing (1a) and simulation (1b) with $g = 0.1$, $w = 0.001$. 2. Voltage fluctuation in viskers[8] (2a) and simulation (2b) with $g = 0.8$, $w = 0.0001$. 3. Burst-kinetics of single ion channels electric current in biological membranes[34] (3a) and simulation with $g = 0.2$, $w = 0.0001$ (3b). Here g is ratio min to max magnitudes of generating intensity in bimodal regime, w is the probability of switching between regimes for the next generating event.

CONCLUDING REMARKS

Besides the above-mentioned problems the proposed methodology may help to solve some of urgent problems in the frame of the sustainable development paradigm. Below we indicate several of them.

1. How much would the anthropogenic contribution to the total ozone depletion for the last 20 years be a result of the chlorofluocarbon use or the other anthropogenic (the aerosol concentration rise) and non-anthropogenic (aerosol from the volcano eruption, the endogenous fluxes of the reducing fluids[42,43]) factors which affect the atmosphere dynamics? May the total ozone depletion phenomenon be connected to the natural processes in the thermodynamically open Earth-Atmosphere system[44]? How much would the contribution of the carbon dioxide emitted to the atmosphere because of the humanity economic activity be a real factor of global warming?
2. What is a possible role of the outer for the Earth factors in initiating the global natural phenomena like El Nino, large earthquakes and volcano eruptions? Among these outer factors are the complex dynamics of the planet movement relative to the Solar system center of gravity[42,43], the variability of the solar activity as well as Galactic Cosmic Ray fluxes[45]. Is it possible to predict the above-mentioned dangerous phenomena?
3. How much would the various natural system states be estimated adequately? How much would the natural resources be used effectively in the production of material values? May the developed methodology be used for solving the concrete problems of agriculture like introduction of the "passport data" for various plough-land, the correction of the market prices, etc.?
4. How much would the undergoing environment as well as the external endogenous and exogenous factors determine the historical development of ethnoses, state systems? How much would the genetic and bioenergetic factors as well as the number of population[46] be considered as adequate parameters of the mankind evolution? In other way: may the theoretical history be[46,47]?
5. How much would the proposed methodology be used for creating new methods of the medical diagnostics? In other ways: how much would the long-range correlations in measured medical data (dynamics of cardiac arrhythmia, electroencephalograph as well as the cardiac beat-to-beat intervals data, age-structured populations with genetic mutations, etc.) be considered as an important index of the personal functional activity?
6. In which manner may the genom state of the humanity as well as other living organisms be estimated adequately? What is the role of various anthropogenic factors in the permanent genetic variations? The civilization future fortune will depend on this knowledge.

Acknowledgments

The authors thank Dr. Susie Vrobel for interesting discussions and large help in preparing this paper.

The work was supported by the Russian Foundation for Basic Researches (Grants 96-03-33998 and 96-15-97608).

REFERENCES

1. H.G. Schuster, *Deterministic Chaos. An Introduction.* Physik-Verlag, Weinheim (1984).
2. P. Berge, Y. Pomeau and C. Vidal, *L'Ordre dans le Chaos*, Hermann, Paris (1988).
3. P. Bak, *How Nature works. The Science of Self-Organized Criticality*, Oxford University Press, Oxford (1997).

4. S.P. Kurdyumov, G.G. Malinetskii and A.B. Potapov, Nonstationary structures, dynamic chaos, cellular automata, in: *The Novel in Synergetic. Enigmas of the Non-Equilibrium Structure World*, Nauka, Moscow (1996) (in Russian).
5. S.F. Timashev, Complexity and the evolution law for natural systems. *Proceedings of the Int. Workshop Tempos in Science and Nature (Siena, 23–26 Sep. 1998) Annals of the New York Academy of Sciences.* (1998) (In press)
6. S.F. Timashev, Flicker-Noise as an Indicator of the "Time Arrow". Methodology of the time series analysis on the base of the deterministic chaos theory. *Mendeleev Chemistry Journal*, 41, No 3:17 (1997).
7. S.F. Timashev, The principles of the nonlinear system evolution. *Mendeleev Chemistry Journal*, 42,. No 3:18 (1998).
8. S.F. Timashev, *et al*. 1998. The methodology of time series analysis on the base of the deterministic chaos theory. in: *Atlas of Temporal Variations of Natural, Anthropogenic and Social Processes. Vol. 2. Cyclical Dynamics in the Nature and Society*, Scientific World, Moscow (1998).
9. C.F. Weizsäcker, von, Time-empirical mathematics – quantum theory. in: *Time, Temporality, Now. Experiencing Time and Concept of Time in an Interdisciplinary Perspective*, Harald Atmanspacher, Eva Ruhnau, eds., Springer-Verlag, Berlin, Heidelberg, New York (1997).
10. E. Ruhnau, The deconstruction of time and the emergence of temporality, in: *Time, Temporality, Now. Experiencing Time and Concept of Time in an Interdisciplinary Perspective*, Harald Atmanspacher, Eva Ruhnau, eds., Springer-Verlag, Berlin, Heidelberg, New York (1997).
11. V.S. Vladimirov, *Generalized Functions in Mathematical Physics*, Mir, Moscow (1979) (in Russian).
12. Ya.I. Frenkel', *The Dawn of the Novel Physics*, Nauka, Leningrad (1970) (in Russian).
13. N.M. Astafieva. Wavelet analysis: base of the theory and the examples of applications, *Uspekhi Fizicheckih Nauk (Russian)* 166: 1145 (1996).
14. H. Lyre, Time and information, in: *Time, Temporality, Now. Experiencing Time and Concept of Time in an Interdisciplinary Perspective*, Harald Atmanspacher, Eva Ruhnau, eds., Springer-Verlag, Berlin, Heidelberg, New York (1997).
15. S.I. Gessen, Mystics and Metaphysics. in: *Logos. Book 1*, Musaget, Moscow (1910).
16. B.J. Hiley and M. Fernandes, Process and time, in: *Time, Temporality, Now. Experiencing Time and Concept of Time in an Interdisciplinary Perspective*, Harald Atmanspacher, Eva Ruhnau, eds., Springer-Verlag, Berlin, Heidelberg, New York (1997).
17. J.D. Barrow. *The Artful Universe*. Penguin Books, London (1995).
18. B.A. Trubnikov. The law of the competitor distribution. *Nature (Russian)*. No 11:3 (1993).
19. E.A. Mamchur and Yu. Cao N'jan. The Story of the XX Century's Field Conceptions. *Problems of Philosophy (Russian)*, 4:150 (1998).
20. J. Klose, Whitehead's theory of perception, in: *Time, Temporality, Now. Experiencing Time and Concept of Time in an Interdisciplinary Perspective*, Harald Atmanspacher, Eva Ruhnau, eds., Springer-Verlag, Berlin, Heidelberg, New York (1997).
21. G.J. Dalenoort, Cognitive aspects of the representation of time, in: *Time, Temporality, Now. Experiencing Time and Concept of Time in an Interdisciplinary Perspective*, Harald Atmanspacher, Eva Ruhnau, eds., Springer-Verlag, Berlin, Heidelberg, New York (1997).

22. A.V. Kolyubin, A.V. Maximychev, S.F. Timashev, The use of flicker-noise spectroscopy for studying the mechanism of the overlimiting current in a system with cation exchange membrane, *Russian Electrochemistry,* 32:227 (1996).
23. Ye. Yu. Budnikov, A.V. Maximychev, A.V. Kolyubin, V.G. Merkin, S.F. Timashev Wavelet analysis in the appliance to investigation of overlimiting current nature in the electrochemical system with cation exchange membrane, *Russ. J. Chem. Phys.*, 73:No 2 (1999).
24. Ye. Yu. Budnikov, S.V. Kozlov, A.V. Kolyubin, S.F. Timashev, Fluctuation phenomena analysis during the process of the electrochemical isolation of hydrogen on a platinum, *Russ. J. Chem. Phys.*, 73:No 3 (1999).
25. I.G. Kostuchenko, S.F. Timashev, Flicker-noise in processes of solar activity, *Int. J. Bifurcation and Chaos*, 8:No 4-5 (1998)
26. M.D. Mundt, W.B. Maguire and R.R. Chase, Chaos in sunspot cycle: analysis and prediction, *J.Geophys.Res.*, 96:1705 (1991).
27. J.K. Lawrence, A.C. Cadavid and A.A. Ruzmaikin, Turbulent and chaotic dynamics underlying solar magnetic variability, *Astrophys. J.* 455:366 (1995).
28. K. Lande, B. Cleveland, T. Daily *et al.* Results from the Homestake solar neutrino observatory. *Proc. 1990 year Int. Conf. on High Energy Phys.* Singapore (1992).
29. K. Lande, The Homestake solar electron neutrino detector program: clorine and iodine, *"Neutrino-96" Proceedings of the 17th Int. Conf. on Neutrino Phys. and Astrophys.*, Finland, (1997).
30. D.S. Oakley, H.B. Snodgrass, R.K. Ulrich and T.L. Van De Kop, On the correlation of solar surface magnetic flux with solar neutrino capture rate, *Astrophys. J.* 437:L63 (1994).
31. S. Turck-Chieze, W. Dappen, E. Fossat *et al.*, The solar interior, *Phys. Reports* 230:57 (1993).
32. M.B. Voloshin, M.I. Vysotskii, and L.B. Okun, Neutrino electrodynamics and possible effects for solar neutrinos, *Soviet Phys. JETP*, 64:446 (1986).
33. E.G. Gibson, *The quiet Sun*, NASA, Washington (1973).
34. V.L. Klochikhin, S.G. Lakeev and S.F. Timashev . Flicker-noise in chemical kinetics (microscopic kinetic and fluctuations in a stationary chemical processes). *Russian J. Phys. Chem.* 73:No2 (1999)
35. T. Zambelli, J.V. Barth, J. Wintterlin and G. Ertl, Complex pathways in dissociative adsorption of oxygen on platinum, *Nature*, 390:495 (1997).
36. T. Ben-David, Y. Lereah, G. Deutscher *et.al.*, Correlated orientations in nanocrystal fluctuations, *Phys. Rev. Lett.*, 78:2585 (1997).
37. E. Clément, P. Leroux-Hugon and L.M. Sander, Exact results for a chemical reaction model, *Phys. Rev. Lett.*, 67:1661(1991).
38. K. Fichthorn, E. Gulari and R. Ziff, Noise-induced bistability in a Monte Carlo surface-reaction model, *Phys. Rev. Lett.*, 63:1527 (1989).
39. V.L. Klochikhin and S.Ya. Pshezhetskii, Monte Carlo modeling of solid phase kinetics of free radicals recombination, *High Energy Chemistry (Russian)* 19: 44 (1985).
40. V.L. Klochikhin and S.Ya. Pshezhetskii, Diffusion controlled kinetics of system contaning a finite number of reacting particles, *Chem. Phys.*, 122: 279(1988).
41. *Single-Channel Recording*, B. Sakmann and E. Neher, eds., Plenum Press, New York and London (1983).
42. S.F. Timashev. The role of chemical factors in the evolution of natural systems (chemistry and ecology). *Russian Chemical Reviews.* 60: 1183 (1991).

43. S.F. Timashev. Physicochemical principles of global ecology. *Mendeleev Chemistry Journal.* 40:155 (1996).
44. V.I. Naidenov, V.I. Shveikina. Problems of the nonlinear hydrology. *Mendeleev Chemistry Journal.* 42:102 (1998).
45. S.F. Timashev, S.P. Perov and E.E. Gutman. Problems of the physical chemistry of the ozone layer. *Russian J. Phys. Chem.* 68:1231 (1994).
46. S.P. Kapitza, S.P. Kurdjumov and G.G. Malinetsky. *Synergetics and the Forecast of the Future.* Nauka, Moscow (1997). (in Russian)
47. S.F. Timashev. May the theoretical history be? in: *Gumilev's Studies: the experience of giving a meaning. The second Gumilev's reading.* Russian Academy of Science, Moscow (1998). (in Russian)

EXISTENCE OF SOLUTIONS AND DYNAMICAL MODELS OF CHANDRASEKHAR H-EQUATIONS

J. Chen, W. Greenberg and R.L. Bowden

Center for Transport Theory & Mathematical Physics
Virginia Tech, Blacksburg, VA 24061, U.S.A.

INTRODUCTION

A variety of linear transport equations may be solved in terms of a scattering function, which itself is expressible as a product of so-called H-functions. These functions, first introduced by S. Chandrasekhar [1], satisfy nonlinear integral equations of the form

$$H(\mu) = 1 + \mu H(\mu) \int_0^1 \frac{\psi(\mu')}{\mu + \mu'} H(\mu') d\mu', \qquad (1)$$

where $\psi(\mu)$ is a characteristic function. It is well known that this equation does not have a unique solution. However, the *physical* solution of (1) is subject to constraints at the zeroes of the dispersion function. Chandrasekhar [2] provided a solution formula with constants depending upon the roots of a characteristic equation, although such roots are difficult to calculate.

Traditionally, one obtained values for the H-equation by iteration, for example,

$$H_{n+1}(\mu) = 1 + \mu H_n(\mu) \int_0^1 \frac{\psi(\mu')}{\mu + \mu'} H_n(\mu') d\mu'. \qquad (2)$$

Bittoni et al [3] showed that a solution of the H-equation can be obtained by iteration for nonnegative ψ. Subsequently, Bowden and Zweifel [4] showed that the solution so obtained satisfied the constraints of the *physical* solution. These results correspond to the characteristic functions $\psi(\mu) \geq 0$. We consider both nonnegative and nonpositive ψ.

1. EXISTENCE OF THE SOLUTION FOR THE H-EQUATION

We wish to consider the H-equation (1) in $C[0,1]$ with $\psi \in L_1[0,1]$ satisfying $\int_0^1 \psi(t)dt \leq \frac{1}{2}$. From the equation (1) we can see its solution $H(x)$ should satisfy

$H(0) = 1$ and $H(x) \neq 0$. This implies the continuous solution of the H-equation must be positive.

Lemma 1 *(i) If $H(x) \in C[0,1]$ is a solution of (1), then*

$$\int_0^1 \psi(t)H(t)dt = 1 - [1 - 2\int_0^1 \psi(t)dt]^{\frac{1}{2}} \tag{3}$$

or

$$\int_0^1 \psi(t)H(t)dt = 1 + [1 - 2\int_0^1 \psi(t)dt]^{\frac{1}{2}}; \tag{4}$$

(ii) If $H \in C[0,1]$ is a solution of (1), then $\int_0^1 \psi(t)dt \leq \frac{1}{2}$ and $H(x) \geq 1$, for $x \in [0,1]$;
(iii) The necessary and sufficient conditions of the positive function $H(x) \in C[0,1]$ satisfying

$$H(x)^{-1} = [1 - 2\int_0^1 \psi(t)dt]^{\frac{1}{2}} + \int_0^1 \frac{t}{x+t}\psi(t)H(t)dt \tag{5}$$

are that $H(x)$ satisfies (1) and (3), where $\psi(x) \geq 0$ or $\psi(x) \leq 0$;
(iv) For $\psi(t) \leq 0$, there is no solution of (1) and (4) in $C[0,1]$.

Proof: (i) and (ii) are easy to prove. Let's prove (iii) as follows. For the necessary condition, let $H(x)$ be a positive solution of (5). Then

$$1 = \sqrt{1 - 2\int_0^1 \psi(t)dt}\, H(x) + \int_0^1 \frac{t}{x+t}\psi(t)H(t)H(x)dt.$$

Multiplying by $\psi(x)$, integrating both sides and simplifying, we will have

$$\int_0^1 \psi(t)H(t)dt = \pm 1 - \sqrt{1 - 2\int_0^1 \psi(x)dx}. \tag{6}$$

Since $\psi(t) \geq 0$ or ≤ 0, $H(t)$ satisfies (3). So we should take the positive sign in (6). In fact, for $\psi(t) \geq 0$ it is easy to see, and for $\psi(t) \leq 0$, substituting (6) to (5) and simplifying we have

$$H(x)^{-1} = \pm 1 - H(x)\int_0^1 \frac{x}{x+t}\psi(t)H(t)dt.$$

That $H(x)$ is positive and $H(0) = \pm 1$ implies that we should take the positive sign. Substituting (3) in (5) we find that $H(x)$ satisfies (1) and the necessary condition is proved.

To prove the sufficient condition, using the fact

$$\int_0^1 \frac{x}{x+t}\psi(t)H(t)dt = \int_0^1 \psi(t)H(t)dt - \int_0^1 \frac{t}{x+t}\psi(t)H(t)dt$$

we can obtain (5) from (1) and (3).

To prove (iv), note that $\psi(t) \leq 0$ and $H(t) \geq 0$ imply $\int_0^1 \psi(t)H(t)dt \leq 0$. But this is a contradiction to (4). ♯

Definition 1 *T is an increasing operator on $C[0,1]$ if for $h_1, h_2 \in C[0,1]$ with $h_1(x) \leq h_2(x), x \in [0,1]$, then $Th_1(x) \leq Th_2(x)$; T is a decreasing operator on $C[0,1]$ if for $h_1, h_2 \in C[0,1]$ with $h_1(x) \leq h_2(x), x \in [0,1]$, then $Th_1(x) \geq Th_2(x)$.*

Let $\rho = \sqrt{1 - 2\int_0^1 \psi(t)dt}$ and $A_\rho = \{h \in C[0,1] : h(x) \geq \rho, x \in [0,1]\}$. For $\rho > 0$, we define operators, $S : A_\rho \longmapsto C[0,1]$ and $T : A_\rho \longmapsto C[0,1]$ by

$$Sh(x) = 1 + h(x) \int_0^1 \frac{x}{x+t} \psi(t) h(t) dt, h \in A_\rho$$

and

$$Th(x) = \rho + \int_0^1 \frac{t}{x+t} \psi(t) h(t)^{-1} dt, h \in A_\rho.$$

It is easy to see that for $\psi \geq 0$ S and T are continuous and S is increasing, T is decreasing. $h(x)$ is a solution of (1) iff $Sh = h$. From Lemma 1 $Th = h$ iff h^{-1} is a solution of (1) and (3). For $\psi(t) \leq 0$, T is increasing.

Let $C_d[0,1] = \{f \in C[0,1], |f| > d\}$ with a positive constant d.

Lemma 2 Let $\psi \in L_1[0,1]$. (i) For given $d > 0$, $TC_d[0,1]$ is equicontinuous; (ii) For $\psi \geq 0$, or $\psi \leq 0$, if there is a solution of (1), let $H(\mu)$ be the solution. Then the series $\{S^{(n)}1\}_{n=1}^\infty$ is equicontinuous where 1 represents the constant function.

Proof: (i) Let $h \in C_d[0,1]$. Since $\psi \in L[1,0]$, for any $\epsilon > 0$ there is γ such that

$$\int_0^\gamma |\frac{\psi(t)}{h(t)}| dt < \frac{\epsilon}{4}$$

where γ small enough. For $x, y \in [0,1]$ and $|x - y| < \delta$ with

$$\delta = \frac{d\gamma^2 \epsilon}{2\int_\gamma^2 |\psi(t)|dt},$$

we have $|Th(x) - Th(y)| \leq 2\frac{\epsilon}{4} + \delta \int_\gamma^1 \frac{1}{\gamma^2} |\psi(t)| \frac{1}{d} dt \leq \frac{\epsilon}{2} + \frac{\epsilon}{2} = \epsilon$. (ii) For $\psi \geq 0$, $x, y \in [0,1]$ let $E = \{h \in C[0,1] : 1 \leq h \leq H\}$, and $R : E \longmapsto C[0,1]$ $Rh(x) = \int_0^1 \frac{x}{x+t} \psi(t) h(t) dt, x \in [0,1], h \in E$, $|Rh(x) - Rh(y)| = |\int_0^1 [\frac{x}{x+t} - \frac{y}{y+t}] \psi(t) h(t) dt|$, and for any $\epsilon > 0$, there is δ, $0 < \delta < 1$ such that $\int_0^\delta |\psi(t) H(t)| dt < \frac{\epsilon}{4}, \int_\delta^1 |\psi(t) H(t)| dt > 0$. So for $x, y \in [0,1]$ $|\int_0^\delta [\frac{x}{x+t} - \frac{y}{y+t}] \psi(t) H(t) dt| \leq 2\int_0^\delta |\psi(t)| dt \leq \frac{\epsilon}{2}$. Then for $|x - y| \leq \delta_1$ with $\delta_1 = \frac{\epsilon}{2}\delta^2(\int_\delta^1 |\psi(t) H(t)| dt)^{-1}$ we have $|Rh(x) - Rh(y)| \leq \frac{\epsilon}{2} + \int_\delta^1 \frac{|x-y|}{\delta^2} |\psi(t) H(t)| dt = \epsilon$.

For $h(x) \in E$, $x \in [0,1]$, from $Rh(x) \leq \int_0^1 \psi(t) H(t) dt = 1 - \sqrt{1 - 2\int_0^1 \psi(t) dt} \leq 1$, we see that there is a β, $0 < \beta < 1$, such that for any $h(x) \in E$ and $x \in [0,1]$, $Rh(x) < \beta$. For $\epsilon_0 > 0$, from equicontinuity of $R(E)$, there is $\delta_0 > 0$ such that for $|x - y| < \delta_0$, and for $g \in R(E)$ $|g(x) - g(y)| < \|H\|^{-1}(1 - \beta)\epsilon_0$. Since $S(1)$ is continuous, there is δ_1, $0 < \delta_1 < \delta_0$, such that for $|x - y| < \delta_1$, $|S^k(1)(x) - S^k(1)(y)| < \epsilon_0$ for $k = 1, 2, ..., n$. For such δ_1 and $|x - y| < \delta_1$, we have $|S^{n+1}1(x) - S^{n+1}1(y)| < \epsilon_0$ In fact, by the definition of S we know $S(1), S^2(1)..., ...S^n(1) \in E$. Let $h(x) = S^n(1)(x)$, then $h(x) \in E$.

Letting $|x - y| < \delta_1$, we have $|Sh(x) - Sh(y)| < \|H\|\|H\|^{-1}(1 - \beta)\epsilon_0 + \beta\epsilon_0 = \epsilon_0$. So $|x - y| < \delta_1$ implies $|S^{n+1}(1)(x) - S^{n+1}(1)(y)| < \epsilon_0$. Hence for $\epsilon_0 > 0$, there is $\delta_1 = \delta_1(\epsilon_0)$ such that for $|x - y| < \delta_1$ and for $n = 1, 2, ...$, we have $|S^n(1)(x) - S^n(1)(y)| < \epsilon_0$ i.e., $\{S^n(1)(x)\}_{n=1}^\infty$ is equicontinuous.

In the same manner, we can prove the same result, for $\psi \leq 0$ and $E = \{h \in C[0,1] : H \leq h \leq 1\}$

♯

Theorem 1 *For $\psi(t) \geq 0 \in L_1[0,1]$ there is a solution $H(x)$ of (1) and (3) iff*

$$\int_0^1 \psi(t)dt \leq \frac{1}{2} \qquad (7)$$

and $\{S^n(1)\}_{n=0}^{\infty}$ converges to $H(x)$. If there is a strict inequality in (7), then $\{T^n(\rho)\}_{n=0}^{\infty}$ converges to $H(x)^{-1}$ with

$$|H(x)^{-1} - T^n \rho(x)| \leq |T^n \rho(x) - T^{n+1} \rho(x)|, \ x \in [0,1]. \qquad (8)$$

Proof: It is easy to prove the necessary condition from Lemma 1. To prove the sufficient condition, suppose $\int_0^1 \psi(t)dt < \frac{1}{2}$. Then $\rho > 0$. By $\psi(t) \geq 0$, we have $\rho \leq T\rho$ and $\rho \leq T^2 \rho$. Since T is decreasing, we have $\rho \leq T^2 \rho \leq T\rho$, $\rho < T^2 \rho \leq T^3 \rho \leq T\rho$, ..., ... $\rho \leq T^2 \rho \leq T^4 \rho \leq T^6 \rho \leq ..., ... \leq T^5 \rho \leq T^3 \rho \leq T\rho$. Using Lemma 2, we know $\{Th : \rho \leq h \leq T\rho\}$ is equicontinuous. Using the Ascoli-Arzera Theorem, the function series

$$\{T^{2n}(\rho)\}_{n=1}^{\infty} = \{T(T^{2n-1}\rho)\}_{n=1}^{\infty} \ and \ \{T^{2n+1}(\rho)\}_{n=0}^{\infty} = \{T(T^{2n}\rho)\}_{n=0}^{\infty}$$

have subsequences which converge to u and v. Since T^{2n} is increasing, $T^{2n+1}\rho$ is decreasing and T is continuous, we have $\lim_{n \to \infty} T^{2n}\rho = u$, $\lim_{n \to \infty} T^{2n+1}\rho = v$, with $\rho \leq u \leq v$ and $Tu = v$, $Tv = u$. By $min\{T^{2n}\rho\}_{n=0}^{\infty} \geq \rho > 0$ there is a constant $a = max\{k : kv \leq u\}$ with $0 < a \leq 1$. For $a = 1$ we have $u = v$. Now, letting $a < 1$ and $T_1 h = Th - \rho$, $h \in D(T)$, we have

$$u = \rho + T_1 v > \rho + T_1(a^{-1}u) = (1-a)\rho + av \geq bv + av = (a+b)v,$$

where b is a positive constant. This is in contradiction with the definition of a. So $Tu = u = v$, and the function $H = u^{-1}$ is a solution of (1) and (3) with $H^{-1} = \lim_{n \to \infty} T^n \rho$. From $T^{2n}\rho \leq H^{-1} \leq T^{2n+1}\rho$, $n = 1, 2, ..., ...$ it is easy to get (8).

Let us consider the case $\int_0^1 \psi dt = \frac{1}{2}$. Let $\{k_n\}_{n=1}^{\infty}$ be a series with $0 < k_i < K_{i+1}$, $(i = 1, 2, ..., ...)$ and $k_i \to 1$ for $i \to \infty$. Since $\int_0^1 k_n \psi dt = \frac{1}{2} k_n < \frac{1}{2}$, we have that there is a positive solution $H_n \in C[0,1]$ of

$$H(x) = 1 + H(x) \int_0^1 \frac{x}{x+t} k_n \psi dt \ n = 1, 2, ...$$

Then $H_n(x) \geq 1$ for $x \in [0,1]$, and

$$(H_n(x))^{-1} = \sqrt{1 - 2\int_0^1 k_n \psi(t)dt} + \int_0^1 \frac{t}{x+t} k_n \psi(t) H_n(t) dt \geq k_1 \int_0^1 \frac{t}{x+t} \psi(t)dt.$$

Hence there is $a > 0$ such that $(H_n)^{-1} \geq a$, for $x \in [0,1], n = 1, 2, ...$

Let $B = \{h \in C[0,1] : a \leq h(x) \leq 1, x \in [0,1]\}$. It implies $H_n^{-1} \in B$, $n = 1, 2, ...$ Let $T : B \longmapsto C[0,1]$ $Th(x) = \int_0^1 \frac{t}{x+t} \psi(t)(h(t))^{-1} dt$, $h \in B$. It is easy see $T(B)$ is bounded and equicontinuous. Letting $h_n = H_n^{-1}$, $n = 1, 2, ...$, we have $h_n(x) = \sqrt{1 - 2\int_0^1 k_n \psi(t)dt} + k_n (Th_n)(x)$. Since $Th_n \in T(B)$, there are subsequences $\{Th_{n_j}\}_{j=1}^{\infty}$ of $\{Th_n\}_{n=1}^{\infty}$ such that $\lim_{n \to \infty} Th_{n_j} = h_0 \in C[0,1]$. From

$$h_{n_j}(x) = \sqrt{1 - 2\int_0^1 k_{n_j}\psi(t)dt} + k_{n_j}(Th_{n_j})(x), \qquad (9)$$

we have $\lim_{j\to\infty} h_{n_j} = h_0$, with $h_0(x) = \sqrt{1 - 2\int_0^1 \psi(t)dt} + \int_0^1 \frac{t}{x+t}\psi(t)(h_n(t))^{-1}dt$. Using Lemma 1 we obtain that the function h_0^{-1} satisfies (1) and (3) (Since $\int_0^1 \psi(t)dt = \frac{1}{2}$, (3) is the same as (4)). So we have proved that if ψ satisfies (7) there is a positive continuous function satisfying (1) and (3).

Now we want to prove if (7) is satisfied, the solution of (1) and (3) is unique. Letting the function H be a solution of (1) and (3). we have $1 \leq H$, and $1 \leq S(1)$. From the increasing operator S we have $1 \leq S(1) \leq S^2(1) \leq S^3(1) \leq ... \leq H$. So $\{S^n(1)\}_{n=1}^{\infty}$ is uniform bounded. From Lemma (2), there is a subsequence $\{S^{n_k}(1)\}_{k=1}^{\infty}$ such that $\lim_{k\to\infty} S^{n_k}(1) = h \leq H$. Since $S^n(1) \geq S^{n+1}(1)$, we have $\lim_{n\to\infty} S^n(1) = h$. Then $S(h) = h$, i.e., h satisfies (3) or (4). Since $0 < h < H$, h satisfies (3), and

$$(h(x))^{-1} \leq \sqrt{1 - 2\int_0^1 \psi(t)dt} + \int_0^1 \frac{t}{x+t}\psi(t)H(t)dt = (H(x))^{-1},$$

i.e., $h^{-1} \leq H^{-1}$. But from $h \leq H$ we have $h = H$. So H is a unique solution of (1) and (3) and $\lim_{n\to\infty} S^n(1) = H$. ♯

Lemma 3 For $\psi \in L_1[0,1]$, letting $h_{n+1} = Th_n$ we have

$$(\frac{1}{h_{n+1}} - \frac{1}{h_n})h_{n+1} = -\frac{1}{h_n}\int_0^1 \frac{t}{x+t}\psi(t)(\frac{1}{h_n} - \frac{1}{h_{n-1}})dt \quad n = 0, 1, 2, ...$$

If $\psi \leq 0$, and $h_1 \geq h_0$ we have $\rho \geq h_{n+1} \geq h_n \geq h_0 \geq 0$.

Proof: By the definition of T we have

$$1 = \frac{\rho}{h_n} + \frac{1}{h_n}\int_0^1 \frac{t}{x+t}\psi(t)\frac{1}{h_{n-1}(t)}dt, \quad 1 = \frac{\rho}{h_{n+1}} + \frac{1}{h_{n+1}}\int_0^1 \frac{t}{x+t}\psi(t)\frac{1}{h_n(t)}dt.$$

Subtraction of the first equation from second above yields

$$0 = h_{n+1}(\frac{1}{h_{n+1}} - \frac{1}{h_n}) + \frac{1}{h_n}\int_0^1 \frac{t}{x+t}\psi(t)(\frac{1}{h_n(t)} - \frac{1}{h_{n-1}(t)})dt.$$

Then

$$h_{n+1}(\frac{1}{h_{n+1}} - \frac{1}{h_n}) = -\frac{1}{h_n}\int_0^1 \frac{t}{x+t}\psi(t)(\frac{1}{h_n(t)} - \frac{1}{h_{n-1}(t)})dt.$$

If $h_1(t) \geq h_0 > 0$ and we suppose for $n = N$ the relation $\rho \geq h_N \geq h_{N-1} \geq 0$ is true, then by the definition of h_{N+1} and $\psi \leq 0$ we obtain

$$h_N = \rho + \int_0^1 \frac{t}{x+t}\psi(t)\frac{1}{h_{N-1}(t)}dt \leq \rho + \int_0^1 \frac{t}{x+t}\psi(t)\frac{1}{h_N(t)}dt = h_{N+1} < \rho.\sharp$$

From Lemma 3 if we can find h_0 such that $h_1 \geq h_0$, then it yields an increasing series $\{h_n\}_0^{\infty}$ for index n. From Lemma 2 we know this series converges to a continuous function $h(x)$ such that $h(x) = Th(x)$. Then from Lemma 1 $h(x)$ is a solution of (1) and (3). The following Theorem gives a sufficient condition of the existence of a solution.

Theorem 2 For $\psi \leq 0 \in L_1[0,1]$ and $-\frac{1}{2} < \int_0^1 \psi(t)dt$ there is an increasing series $\{h_n\}_0^{\infty}$ converging to a continuous solution of (1) and (3).

Proof: From the definition of T and ρ we have $h_0 - h_1 = h_0 - Th_0 = \frac{1}{h_0}[h_0^2 - \rho h_0 - \int_0^1 \frac{t}{x+t}\psi(t)dt] \leq \frac{1}{h_0}[h_0^2 - \rho h_0 - \int_0^1 \psi(t)dt] = \frac{1}{h_0}[h_0 - \frac{1}{2}\rho - \frac{1}{2}\sqrt{1+2\int_0^1 \psi(t)dt}][h_0 - \frac{1}{2}\rho + \frac{1}{2}\sqrt{1+2\int_0^1 \psi(t)dt}]$. Taking positive h_0 such that $\frac{1}{2}\rho - \frac{1}{2}\sqrt{1+2\int_0^1 \psi(t)dt} < h_0 < \frac{1}{2}\rho + \frac{1}{2}\sqrt{1+2\int_0^1 \psi(t)dt}$ we obtain $h_1 > h_0$. From Lemma 3 this implies that there is a series $\{h_n\}_0^\infty$ increasing and converging to a continuous solution of (1) and (3). ♯

Theorem 3 For $\psi(x) \leq 0 \in L_1[0,1]$, there is a $b_0 > 0$ such that for $\gamma \geq b_0$,

$$T^2\gamma \leq \gamma, \tag{10}$$

$$\frac{1}{\rho} < T\gamma \leq \gamma, \quad \rho = \sqrt{1 - 2\int_0^1 \psi(t)dt} \tag{11}$$

Proof: From the definition of T, $Th(x) = \rho + \int_0^1 \frac{t}{x+t}\psi(t)\frac{1}{h(t)}dt$, T is increasing for h. For $\rho > 0$, we have

$$\gamma - T\gamma \geq \gamma - \rho - \frac{1}{\gamma}\int_0^1 \frac{t}{1+t}\psi(t)dt = \frac{1}{\gamma}[\gamma(\gamma - \rho) - \int_0^1 \frac{t}{1+t}\psi(t)dt]$$

Let $b_0 = \frac{\rho}{2} + \frac{1}{2}\sqrt{\rho^2 + 4\int_0^1 \frac{t}{1+t}\psi(t)dt}$ if the square root is real. Otherwise let b_0 be any constant greater than $\frac{\rho}{2}$. In both cases $b_0 \geq \frac{\rho}{2}$.

For $\gamma \geq b_0$ we have $\gamma(\gamma - \rho) - \int_0^1 \frac{t}{1+t}\psi(t)dt \geq 0$, and then $\gamma - T\gamma \geq \frac{1}{\gamma}[\gamma(\gamma - \rho) - \int_0^1 \frac{t}{1+t}\psi(t)dt] > 0$ i.e., $\gamma > T\gamma$. By the definition of T we also have $T\gamma = \frac{1}{\rho}[\rho^2 + \frac{\rho}{\gamma}\int_0^1 \frac{t}{x+t}\psi(t)dt] \geq \frac{1}{\rho}[\rho^2 + \frac{\rho}{\gamma}\int_0^1 \psi(t)dt] > \frac{1}{\rho}$. (Note: $\gamma > b_0 \geq \frac{\rho}{2}$)

On the other hand, the estimation $\rho - \gamma + \int_0^1 \frac{t}{t+x}\psi(t)\frac{1}{T\gamma}dt < \rho - \gamma + \frac{1}{\gamma}\int_0^1 \frac{t}{t+x}\psi(t)dt = T\gamma - \gamma < 0$ (using $\gamma > T\gamma$) implies $T^2\gamma = \rho + \int_0^1 \frac{t}{t+x}\psi(t)\frac{1}{T\gamma}dt = \gamma + [(\rho - \gamma) + \int_0^1 \frac{t}{t+x}\psi(t)\frac{1}{T\gamma}dt]$. Hence $T^2\gamma < \gamma$. ♯

Theorem 4 For $\frac{1}{2} \leq \psi(x) \leq 0 \in L_1[0,1]$, let $b_1 > 0$ such that $\rho - b_1 \geq b_0$ and let

$$\frac{b_1 - \sqrt{b_1^2 + 4}}{2} \leq \rho \leq \frac{b_1 + \sqrt{b_1^2 + 4}}{2}, \tag{12}$$

If

$$\rho - b_1 \leq \gamma \leq \rho + b_1, \tag{13}$$

then

$$\rho - b_1 \leq \frac{1}{\rho} \leq T\gamma \leq \rho + b_1. \tag{14}$$

Proof: Taking $b_0 \leq \gamma$ in Theorem 3 we have $\frac{1}{\rho} \leq T\gamma < \rho$. The condition (12) implies $\rho - b_1 \leq \frac{1}{\rho}$, so $\rho - b_1 \leq T\gamma < \gamma \leq \rho + b_1$. ♯

Theorem 5 For $\psi(x) \leq 0 \in L_1[0,1]$ and $\frac{1}{2} < \int_0^1 \psi(t)dt$ there is a unique solution of (1) and (3), $H(x) \in C[0,1]$ with $0 < H(x) \leq 1$, $x \in [0,1]$.

Proof: Taking b_0 and b_1 satisfying the conditions (12) and (13), from (4) we have $\rho - b_0 \leq \frac{1}{\rho} \leq T\gamma \leq \gamma < \rho + b_0$. Using the fact that T is increasing, we know $\frac{1}{\rho} < T\frac{1}{\rho} < T^n\gamma < T^{n-1}\gamma..., ... < \gamma$. So the series $\{T^n\gamma\}_{n=0}^{\infty}$ is uniform bounded. In the same way as the proof of Lemma 2 we have $T^n\gamma_{n=0}^{\infty}$ is equicontinuous. Hence, from the Ascoli-Arzera Theorem and the fact that T is increasing, there is $h_0 \geq 0 \in C[0,1]$ such that $h_0(x) = \lim_{n\to\infty} T^n\gamma$. Since T is continuous we have $h_0(x) = \lim_{n\to\infty} T^n\gamma = \lim_{n\to\infty}(TT^{n-1}\gamma) = Th_0(x)$. i.e., $h_0(x)$ is a positive solution of (5). Using (iii) of Lemma 1, $h_0(x)^{-1}$ is a positive solution of (1) and (3).

If there is another $h(x)$ which is a positive continuous solution of $h(x) = Th(x)$, then $0 \leq h(x) < \rho$. From the definition of T, we have $T\rho < \rho$, and using that T is increasing, we know $Th(x) = h(x) \leq T\rho < \rho$. Hence $h(x) < T^n\rho < T^{n-1}\rho..., ... < T\rho < \rho$. Then there is $h_1(x) \in C[0,1]$ such that $h_1(x) = \lim_{n\to\infty} T^n\rho$ in $C[0,1]$. Therefore $Th_1(x) = h_1$ and $h(x) \leq h_1$. Since $h(x) = Th(x)$ and $h_0(x) = Th_0(x)$ we know $H(x) = \frac{1}{h(x)}$ and $H_0(x) = \frac{1}{h_0(x)}$ are solutions of (1).

Since $h(x) \leq h_0(x)$ we have $H(x) \geq H_0(x)$. But S is decreasing, so $SH(x) \leq SH_0(x)$. The solutions of (1), $H(x)$ and H_0, and their relation above imply $H(x) = SH(x) \leq SH_0(x) = H_0(x)$, thus $H(x) = H_0(x)$. So the solution of (1) and (3) is unique. Continuity of H(x) with $H(0) = 1$ implies that $H(x)$ is positive. By the equation (1) with $\psi(x) \leq 0$, $H(x)$ must be less than one. ♯

It is possible there are other solutions of (1) satisfying (4). Only when $\int_0^1 \psi(t)dt = \frac{1}{2}$, the solution satisfying (1) and (3) is a unique solution of (1). When $\int_0^1 \psi(t)dt \leq \frac{1}{2}$ we have following theorem.

Theorem 6 *For $\psi(x) \geq 0 \in L_1[0,1]$ and $\int_0^1 \psi(t)dt \leq \frac{1}{2}$, let $H(x)$ be a solution of (1) and (3). Then there is a solution of (1) and (4) iff*

$$\int_0^1 \frac{\psi(t)}{1-t} H(t) > 1. \tag{15}$$

When (15) is satisfied, the function

$$H_1(x) = \frac{1+kx}{1-kx} H(t) \tag{16}$$

is a unique solution of (1) and (4), where k is a unique value satisfying

$$\int_0^1 \frac{\psi(t)}{1-kt} H(t) = 1, \ 0 \leq k \leq 1. \tag{17}$$

Proof: Since $(1-kt)^{-1}$ is increasing for $k \in (0,1)$ we have

$$\lim_{k\to 1} \int_0^1 \frac{\psi(t)}{1-kt} H(t)dt = \int_0^1 \frac{\psi(t)}{1-t} H(t)dt \tag{18}$$

Letting k and $\psi(x)$ satisfy (15) and $f(k) = \int_0^1 \frac{\psi(t)}{1-kt} H(t)dt$, using (3), we have $f(0) = 1 - \sqrt{1 - 2\int_0^1 \psi(t)dt} < 1$. (15) and $f(k)$ increasing strictly for k imply that there is a unique $k \in (0,1)$ such that (17) holds. Taking the function H_1 from (16), we have

$$\int_0^1 \frac{x}{x+t} \psi(t) H_1 dt = \int_0^1 \frac{x}{x+t} \psi(t) \frac{1+kt}{1-kt} H(t)dt = 1 - \frac{1}{H_1(x)}.$$

So H_1 satisfies (1). Then H_1 must satisfies (3) or (4). From $H_1(x) > H(x)$ for $x \in (0,1]$ we know H_1 satisfies (4).

Let $H_1 \in C[0,1]$ and satisfying (1) and (4). We know H_1 satisfies $\int_0^1 \psi(t)H_1(t)dt > 1$ from (4), and $\int_0^1 \frac{\psi(t)}{1+t}H_1(t)dt = 1 - \frac{1}{H_1(t)} < 1$. There is a unique k, $0 < k < 1$, such that $\int_0^1 \frac{\psi(t)}{1+kt}H_1(t)dt = 1$. Let $H_2 \in C[0,1]$, $H_2(x) = \frac{1-kx}{1+kx}H_1$, $x \in [0,1]$. Then $\int_0^1 \frac{x}{x+t}\psi(t)H_2 dt = \int_0^1 \frac{x}{x+t}\psi(t)\frac{1-kt}{1+kt}H_1(t)dt = 1 - \frac{1+kx}{1-kx}\frac{1}{H_1(x)} = 1 - \frac{1}{H_2(x)}$, i.e., H_2 is a solution of (1). So H_2 satisfies (3) or (4). From $H_2(x) < H_1(x)$, $x \in (0,1]$, we obtain that $H_2(x)$ satisfies (3), so $H_2 = H$, and $\int_0^1 \frac{\psi(t)}{1+kt}H(t)dt = \int_0^1 \frac{\psi(t)}{1-kx}\frac{1-kt}{1+kt}H_t dt = 1$. Hence $\int_0^1 \frac{\psi(t)}{1-t}H(t)dt > 1$. ♯

2. STABILITY OF FIXED POINTS

1. Origin of the Problem

To obtain solutions numerically, typically some sort of quadrature scheme is employed. Relatively little is known about the stability of such schemes as dynamical systems.

An iteration model of the Nth approximation of the H-equation (1) may be given by an N-dimensional nonlinear discrete dynamical system,

$$H_{n+1}(\mu_i) = 1 + \mu_i H_n(\mu_i) \sum_{j=1}^{N} \frac{a_j \psi(\mu_j)}{\mu_i + \mu_j} H_n(\mu_j), \quad i = 1, 2, ..., n, \tag{19}$$

where $a_{\pm j}(j=1,..,n)$, $a_j = -a_{-j}$ and $\mu_{\pm j}(j=1,..,n)$, $\mu_j = -\mu_{-j}$ are the weights and divisions appropriate to a quadrature formula in the interval $(-1,+1)$. For equation (1), S. Chandrasekhar[1] found the solution

$$H(\mu) = \frac{1}{\mu_1 \cdots \mu_n} \frac{\prod_{j=1}^{N}(\mu + \mu_j)}{\prod_{i=1}^{N}(1 - k_i \mu)}, \tag{20}$$

where the k_i's$(i=1,...,N)$ are the roots of the associated characteristic equation

$$1 = 2 \sum_{j=1}^{N} \frac{a_j \psi(\mu_j)}{1 - k^2 \mu_j^2}. \tag{21}$$

We can see that there are 2^N fixed points of system (19). Physicists are interested in the unique fixed point of (19) with k_i's non-negative and $\mu \geq 0$.

Little is known about such multidimensional dynamical systems. Initiating a study of the stability of such systems, we wish to present here a survey of results for the case N=2.

With the change of the variable, $a_j \psi(\mu_j) = \frac{c\omega_i}{2}$ $h_i = \frac{c}{4}H(\mu_i)$ the fixed points of (1) for $N=2$ satisfy

$$h_1 = \frac{c}{4} + \omega_1 h_1^2 + \frac{2\mu_1 \omega_2 h_1 h_2}{\mu_1 + \mu_2} \tag{22}$$

$$h_2 = \frac{c}{4} + \frac{2\mu_2 \omega_1 h_1 h_2}{\mu_1 + \mu_2} + \omega_2 h_2^2 \tag{23}$$

Letting $x = \omega_1 h_1 + \omega_2 h_2$ and $y = \mu_2 \omega_1 h_1 + \mu_1 \omega_2 h_2$ we are led to study the iterated discrete dynamical system

$$x(n+1) = (\omega_1 + \omega_2)\frac{c}{4} + x(n)^2 \qquad (24)$$

$$y(n+1) = (\mu_2\omega_1 + \mu_1\omega_2)\frac{c}{4} + \frac{1}{\mu_1 + \mu_2}(\mu_1\mu_2 x(n)^2 + y(n)^2) \qquad (25)$$

2. Fixed Points

The fixed points of the discrete dynamical system (24) and (25) may be worked out algebraically, if tediously. For $c > \frac{1}{\omega_1+\omega_2}$, there is no real fixed point. For $c \leq \frac{1}{\omega_1+\omega_2}$ we have four fixed points which we label $(x_+, y_{+,+})$, $(x_+, y_{+,+})$, $(x_-, y_{-,+})$, $(x_-, y_{-,-})$, where

$$x_\pm = \frac{1}{2}[1 \pm \sqrt{1 - c(\omega_1 + \omega_2)}]$$

$$y_{\pm,\pm'} = \frac{1}{2}(\mu_1 + \mu_2)\{1' \pm' \frac{1}{\mu_1+\mu_2}\sqrt{\mu_1^2 + \mu_2^2 - c(\mu_2^2\omega_1 + \mu_1^2\omega_2) \mp 2\mu_1\mu_2\sqrt{1 - c(\omega_1+\omega_2)}}\}$$

3. Stability of Fixed Points

For the system (24) and (25), the stability of the fixed points can be determined by the eigenvalues of the Jacobi matrix of the system. It is easily found from the Jacobi matrix that the eigenvalues of the fixed points $(x_i, y_{i,j})$, $i, j = +, -$ are $\lambda_i^{(1)} = 2x_i$, $\lambda_{i,j}^{(2)} = \frac{2}{\mu_1+\mu_2} y_{i,j}$, where $i, j = +, -$, i.e.,

$$\lambda_\pm^{(1)} = 1 \pm \sqrt{1 - c(\omega_1 + \omega_2)},$$

$$\lambda_{\pm,\pm'}^{(2)} = 1' \pm' \sqrt{1 - 2\frac{\omega_1\mu_2 + \omega_2\mu_1}{(\omega_1+\omega_2)(\mu_1+\mu_2)}\lambda_\pm^{(1)} + \frac{\mu_2^2\omega_1 + \mu_1^2\omega_2}{(\omega_1+\omega_2)(\mu_1+\mu_2)^2}\lambda_\pm^{(1)2}}$$

Also $\lambda_\pm^{(1)}$ and $\lambda_{\pm,\pm}^{(2)}$ are real for $c \leq \frac{1}{\omega_1+\omega_2}$.

Let $c_i = \frac{1}{\omega_1+\omega_2}(-\xi_i^2 + 2\xi_i)$, $i = 1, 2$, where

$$\xi_1 = 1 + \frac{\mu_1\mu_2(\omega_1+\omega_2)}{\mu_2^2\omega_1 + \mu_1^2\omega_2}$$
$$+ \frac{\mu_1+\mu_2}{\mu_2^2\omega_1 + \mu_1^2\omega_2}\sqrt{4(\mu_2^2\omega_1^2 + \mu_1^2\omega_2^2) + \omega_1\omega_2[3(\mu_1^2+\mu_2^2) + 2\mu_1\mu_2]}$$

$$\xi_2 = \frac{\mu_1+\mu_2}{\mu_2^2\omega_1 + \mu_1^2\omega_2}[\omega_1\mu_2 + \omega_2\mu_1 - \sqrt{(\omega_1\mu_2+\omega_2\mu_1)^2 + 3(\mu_2^2\omega_1+\mu_1^2\omega_2)(\omega_1+\omega_2)}].$$

This implies $c_2 > c_1$. By a study of the eigenvalues $\lambda_i^{(1)}$, $\lambda_{i,j}^{(2)}$, where $i, j = +, -$, the result of the stability of fixed points is given in the following theorem.

Theorem 7 *(1) For $c \leq \frac{1}{\omega_1+\omega_2}$ the fixed point $(x_+, y_{+,+})$ is a repelling point.*
(2) For $c > -\frac{3}{\omega_1+\omega_2}$ the fixed point $(x_-, y_{-,+})$ is a saddle point; for $c < -\frac{3}{\omega_1+\omega_2}$ it is a repelling point.
(3) For $c_1 < c < \frac{1}{\omega_1+\omega_2}$ the fixed point $(x_+, y_{+,-})$ is a saddle point; for $c < c_1$, it is a repelling point.
(4) If $\frac{1}{\omega_1+\omega_2} > c > -\frac{3}{\omega_1+\omega_2}$ and $\frac{1}{\omega_1+\omega_2} > c > c_2$ then the fixed point $(x_-, y_{-,-})$ is an attracting point; if $\frac{1}{\omega_1+\omega_2} > c > -\frac{3}{\omega_1+\omega_2}$ and $c < c_2$, or $c < -\frac{3}{\omega_1+\omega_2}$ and $\frac{1}{\omega_1+\omega_2} > c > c_2$ then the fixed point $(x_-, y_{-,-})$ is a saddle point; if $c < -\frac{3}{\omega_1+\omega_2}$ and $c < c_2$ then the fixed point $(x_-, y_{-,-})$ is a repelling point.

3. QUANTITATIVE PROPERTIES OF ORBITS

For special values of μ_i, ω_i, namely $\omega_1 = \omega_2 = \frac{1}{2}$, $\mu_1 = \frac{1}{2}(1+\frac{1}{\sqrt{3}})$, and $\mu_2 = \frac{1}{2}(1-\frac{1}{\sqrt{3}})$ we have studied some quantitative properties of the iterated discrete system. In this case the dynamical system becomes

$$x_{n+1} = \frac{c}{4} + x_n^2 \tag{26}$$

$$y_{n+1} = \frac{c}{8} + \frac{1}{6}x_n^2 + y_n^2. \tag{27}$$

To discuss the properties of the orbit of this system, we consider a vector field $(\Delta x_n, \Delta y_n)$ given by the system $\Delta x_n = x_{n+1} - x_n = \frac{c}{4} + x_n^2 - x_n$, $\Delta y_n = y_{n+1} - y_n = \frac{c}{8} + \frac{1}{6}x_n^2 + y_n^2 - y_n$, which gives the direction of the orbit at (x_n, y_n). First we want to find the divergence region of the orbit $\{(x_n, y_n)\}_0^n$, i.e. if the orbit starts in this region, then the orbit will diverge.

Theorem 8 *Let $c \leq 1$. For some n_0 if (1) $y_{n_0} > \frac{1}{2} + \frac{1}{2\sqrt{2}}\sqrt{2-c}$ or (2) $|x_{n_0}| > \frac{1}{2} + \frac{1}{2}\sqrt{1-c}$, or (3) $\frac{1}{6}x_{n_0}^2 + y_{n_0}^2 > \frac{1}{2} - \frac{c}{8} + \frac{1}{2\sqrt{2}}\sqrt{2-c}$, then the orbit $\{(x_n, y_n)\}_0^\infty$ is divergent.*

Proof: If $y_{n_0} > \frac{1}{2} + \frac{1}{2\sqrt{2}}\sqrt{2-c}$, we know $\Delta y_{n_0} > 0$. Thus $y_{n_0+1} > y_{n_0} > \frac{1}{2} + \frac{1}{2\sqrt{2}}\sqrt{2-c}$. Then for $n_0 < n$, we have $y_{n+1} > y_n > \frac{1}{2} + \frac{1}{2\sqrt{2}}\sqrt{2-c}$ and $\Delta y_n > \frac{c}{8} - \frac{1}{4} + (y_{n_0} - \frac{1}{2})^2 > 0$. We have $\{y_n\}_0^\infty$ is divergent, so the orbit $\{x_n, y_n\}_0^\infty$ is divergent.

If $x_{n_0} > \frac{1}{2} + \frac{1}{2}\sqrt{1-c}$, for $n > n_0$ $x_{n+1} > x_n \geq x_{n_0+1} > x_{n_0}$ and $\Delta x_n = [x_n - \frac{1}{2}(1+\sqrt{1-c})][x_n - \frac{1}{2}(1-\sqrt{1-c})] > \Delta x_{n_0} > 0$. Hence $\{x_n\}_{n_0}^\infty$ is increasing and divergent. Similar results can be found easily for $x_{n_0} < -\frac{1}{2}(1+\sqrt{1-c})$.

For $\frac{1}{6}x_{n_0}^2 + y_{n_0}^2 > \frac{1}{2} - \frac{c}{8} + \frac{1}{2\sqrt{2}}\sqrt{2-c}$, the same result comes from $y_{n_0+1} = \frac{c}{8} + \frac{1}{6}x_{n_0}^2 + y_{n_0}^2 > \frac{1}{2} + \frac{1}{2\sqrt{2}}\sqrt{2-c}$ and result of the first case in this proof. ♯

Note that if $-c$ is large enough, the orbit will diverge, even if starting from a neighborhood of the origin.

Theorem 9 *For $c < -8$, if for some n_0 $x_{n_0}^2 < -\frac{c}{4} - \frac{1}{2}(1+\sqrt{1-c})$, then the orbit $\{(x_n, y_n)\}_0^\infty$ is divergent.*

Proof: First we note that the neighborhood of the origin $x_{n_0}^2 < -\frac{c}{4} - \frac{1}{2}(1+\sqrt{1-c})$ is not empty since the right side of the inequality is positive for $c < -8$. From the condition, we have $x_{n_0+1} = \frac{c}{4} + x_{n_0}^2 < \frac{c}{4} - \frac{c}{4} - \frac{1}{2}(1+\sqrt{1-c}) = -\frac{1}{2}(1+\sqrt{1-c})$. Using Theorem 8, we have the orbit $\{(x_n, y_n)\}_0^\infty$ is divergent. ♯

We seek a region of the $x-y$ plane such that all orbits $\{(x_n, y_n)\}_0^\infty$ with initial point in the region are boundary. First, we list properties of the one dimensional mapping (26).

Theorem 10 *(1) For $1 \geq c \geq -8$ if $|x_n| \leq |\frac{1}{2}(1-\sqrt{1-c})|$ then $-\frac{1}{2}(1+\sqrt{1-c}) \leq x_{n+1} \leq \frac{1}{2}(1-\sqrt{1-c})$. (2) For $c \leq 1$, if $\frac{1}{2}(1+\sqrt{1-c}) \geq |x_n| \geq \frac{1}{2}(-1+\sqrt{1-c})$ then $\frac{1}{2}(1-\sqrt{1-c}) \leq x_{n+1} \leq \frac{1}{2}(1+\sqrt{1-c})$ and $|x_{n+1}| \leq |x_n|$. (3) For $1 \geq c \geq -8$, $\{x_n\}_0^\infty$ is bounded if and only if $|x_0| \leq \frac{1}{2}(1+\sqrt{1-c})$ (4) For $0 \geq c$, we have $\frac{1}{2}(1+\sqrt{1-c}) \geq |x_n| > \frac{1}{2}\sqrt{2\sqrt{1-c}-2-c}$ iff $\frac{1}{2}(1+\sqrt{1-c}) \geq x_{n+1} > \frac{1}{2}(-1+\sqrt{1-c})$; and $\frac{1}{2}(-1+\sqrt{1-c}) \leq |x_n| \leq \frac{1}{2}\sqrt{2\sqrt{1-c}-2-c}$ iff $|x_{n+1}| \leq \frac{1}{2}(-1+\sqrt{1-c})$.*

Define the mappings $F(x) = \frac{c}{4} + x^2$, $G(x,y) = \frac{c}{8} + \frac{1}{6}x^2 + y^2$. Let the mapping of (26) and (27) be $R(x,y) = (F(x), G(x,y))$. In order to study the boundary orbit region of the system (26) and (27) we divide the domain of the mapping into the following regions

$$D1 = \left\{ \begin{array}{c} (x,y): \quad -\frac{1}{2}(1+\sqrt{1-c}) \leq x < \frac{1}{2}(1-\sqrt{1-c}), \\ -\frac{1}{2}[1-\sqrt{1-(\frac{c}{2}+\frac{2}{3}x^2)}] < y \leq \frac{1}{2}[1+\sqrt{1-(\frac{c}{2}+\frac{2}{3}x^2)}] \end{array} \right\}$$

$$D2 = \left\{ \begin{array}{c} (x,y): \quad |x| \leq \frac{1}{2}|1-\sqrt{1-c}|, \\ -\frac{1}{2}[1-\sqrt{1-(\frac{c}{2}+\frac{2}{3}x^2)}] < y \leq \frac{1}{2}[1+\sqrt{1-(\frac{c}{2}+\frac{2}{3}x^2)}] \end{array} \right\},$$

$$D3 = \left\{ \begin{array}{c} (x,y): \quad -\frac{1}{2}(1-\sqrt{1-c}) \leq x < \frac{1}{2}(1+\sqrt{1-c}), \\ -\frac{1}{2}[1-\sqrt{1-(\frac{c}{2}+\frac{2}{3}x^2)}] < y \leq \frac{1}{2}[1+\sqrt{1-(\frac{c}{2}+\frac{2}{3}x^2)}] \end{array} \right\},$$

$$D4 = \left\{ \begin{array}{c} (x,y): \quad -\frac{1}{2}(1+\sqrt{1-c}) \leq x < \frac{1}{2}(1-\sqrt{1-c}), \\ |y| \leq \frac{1}{2}|1-\sqrt{1-(\frac{c}{2}+\frac{2}{3}x^2)}|\} \end{array} \right\},$$

$$D5 = \left\{ \begin{array}{c} (x,y): \quad |x| \leq \frac{1}{2}|1-\sqrt{1-c}|, \\ |y| \leq \frac{1}{2}|1-\sqrt{1-(\frac{c}{2}+\frac{2}{3}x^2)}| \end{array} \right\},$$

$$D6 = \left\{ \begin{array}{c} (x,y): \quad -\frac{1}{2}(1-\sqrt{1-c}) < x \leq \frac{1}{2}(1+\sqrt{1-c}), \\ |y| \leq \frac{1}{2}|1-\sqrt{1-(\frac{c}{2}+\frac{2}{3}x^2)}| \end{array} \right\},$$

$$D7 = \left\{ \begin{array}{c} (x,y): \quad -\frac{1}{2}(1+\sqrt{1-c}) \leq x < \frac{1}{2}(1-\sqrt{1-c}), \\ -\frac{1}{2}[1+\sqrt{1+(\frac{c}{2}+\frac{2}{3}x^2)}] \leq y \leq \frac{1}{2}[1-\sqrt{1-(\frac{c}{2}+\frac{2}{3}x^2)}] \end{array} \right\},$$

$$D8 = \left\{ \begin{array}{c} (x,y): \quad |x| \leq \frac{1}{2}|1-\sqrt{1-c}|, \\ -\frac{1}{2}[1+\sqrt{1-(\frac{c}{2}+\frac{2}{3}x^2)}] \leq y \leq \frac{1}{2}[1-\sqrt{1-(\frac{c}{2}+\frac{2}{3}x^2)}] \end{array} \right\},$$

$$D9 = \left\{ \begin{array}{c} (x,y): \quad -\frac{1}{2}(1-\sqrt{1-c}) < x \leq \frac{1}{2}(1+\sqrt{1-c}), \\ -\frac{1}{2}[1+\sqrt{1-(\frac{c}{2}+\frac{2}{3}x^2)}] \leq y \leq \frac{1}{2}[1-\sqrt{1-(\frac{c}{2}+\frac{2}{3}x^2)}] \end{array} \right\}.$$

Also we divide part of the image of the mapping $R(x,y)$ into the following regions

$$R1 = \left\{ \begin{array}{c} (x,y): \quad -\frac{1}{2}(1+\sqrt{1-c}) \leq x \leq \frac{1}{2}(1-\sqrt{1-c}), \\ -\frac{1}{2}[1-\sqrt{1-(\frac{c}{3}+\frac{2}{3}x)}] \leq y \leq \frac{1}{2}[1+\sqrt{1+(\frac{c}{3}+\frac{2}{3}x)}] \end{array} \right\},$$

$$R2 = \left\{ \begin{array}{c} (x,y): \quad \frac{1}{2}(1-\sqrt{1-c}) \leq x \leq \frac{1}{2}(1+\sqrt{1-c}), \\ -\frac{1}{2}[1-\sqrt{1-(\frac{c}{3}+\frac{2}{3}x)}] \leq y \leq \frac{1}{2}[1+\sqrt{1+(\frac{c}{3}+\frac{2}{3}x)}] \end{array} \right\},$$

$$R3 = \left\{ \begin{array}{l} (x,y): \quad -\frac{1}{2}(1+\sqrt{1-c}) \leq x \leq \frac{1}{2}(1-\sqrt{1-c}), \\ \qquad -\frac{1}{2}[1+\sqrt{1-(\frac{c}{3}+\frac{2}{3}x)}] \leq y \leq \frac{1}{2}[1-\sqrt{1+(\frac{c}{3}+\frac{2}{3}x)}] \end{array} \right\},$$

$$R4 = \left\{ \begin{array}{l} (x,y): \quad \frac{1}{2}(1-\sqrt{1-c}) \leq x \leq \frac{1}{2}(1+\sqrt{1-c}), \\ \qquad -\frac{1}{2}[1+\sqrt{1-(\frac{c}{3}+\frac{2}{3}x)}] \leq y \leq \frac{1}{2}[1-\sqrt{1+(\frac{c}{3}+\frac{2}{3}x)}] \end{array} \right\}.$$

Theorem 10 gives us the means to study $\{(x_n, y_n)\}_0^\infty$. The iterated form (27) can be written as $y_{n+1} = \frac{1}{4}(\frac{c}{2} + \frac{2}{3}x_n^2) + y_n^2$. We can treat $\frac{c}{2} + \frac{2}{3}x_n^2$ above as c in Theorem 10. We use this idea to find the following result for the mapping $R(x,y)$,

Theorem 11 *The function R maps Di to R(Di), $i = 1, ..., 9$ such that:*

$$R1 \supset R(D2 \cup D8), \tag{28}$$

$$R2 \supset R(D1 \cup D3 \cup D7 \cup D9), \tag{29}$$

$$R3 \supset R(D5), \tag{30}$$

$$R4 \supset R(D4 \cup D6). \tag{31}$$

By the definition of Di and Rj, $i = 1, ..., 9, j = 1, ...4$, we have $D1 \cup D4 \cup D7 \subset R1 \cup R3$, and $D2 \cup D3 \cup D5 \cup D6 \cup D8 \cup D9 \supset R2 \cup R4$. Let

$$K1 = \left\{ \begin{array}{l} (x,y): \quad -\frac{1}{2}(1+\sqrt{1-c}) \leq x \leq \frac{1}{2}(1-\sqrt{1-c}), \\ \qquad \frac{1}{2}[1+\sqrt{1-(\frac{c}{2}+\frac{2}{3}x^2)}] \leq y \leq \frac{1}{2}[1+\sqrt{1-(\frac{c}{3}+\frac{2}{3}x)}] \end{array} \right\}, \tag{32}$$

$$K2 = \left\{ \begin{array}{l} (x,y): \quad -\frac{1}{2}(1+\sqrt{1-c}) \leq x \leq \frac{1}{2}(1-\sqrt{1-c}), \\ \qquad -\frac{1}{2}[1+\sqrt{1-(\frac{c}{2}+\frac{2}{3}x^2)}] \geq y \geq \frac{1}{2}[1+\sqrt{1-(\frac{c}{3}+\frac{2}{3}x)}] \end{array} \right\}. \tag{33}$$

Then we have $K1 \cup K2 = R1 \cup R3 \setminus (D1 \cup D4 \cup D7)$, Since $K1$ and $K2$ are out of the region $\cup_{i=1}^9 Di$, it is possible that part of the image $R(K1 \cup K3)$ is in the divergence region in (3) of Theorem 8. The following theorem specifies a region in the $x - y$ plane such that all orbits $\{(x_n, y_n)\}_0^\infty$ with initial point in the region are boundary.

Theorem 12 *If the initial point (x_0, y_0) of an orbit $\{(x_n, y_n)\}_0^\infty$ is in the region*

$$BD = \left\{ \begin{array}{l} (x,y): \quad |x| \leq \frac{1}{2}|1-\sqrt{1-c}|, \\ \qquad |y| \leq \sqrt{\frac{1}{2}[1+\sqrt{1-(\frac{c}{2}+\frac{2}{3}(\frac{c}{4}+x^2)^2)}] - (\frac{c}{8}+\frac{1}{6}x^2)}; \\ \qquad \frac{1}{2}|1-\sqrt{1-c}| \leq |x| \leq \frac{1}{2}(1+\sqrt{1-c}), \\ \qquad |y| \leq \frac{1}{2}[1+\sqrt{1-(\frac{c}{2}+\frac{2}{3}x^2)}]. \end{array} \right\} \tag{34}$$

then the orbit will remain in this region.

Proof: First we need to find the inverse of the mapping R on K1, i.e. $R^{-1}(K1)$. From (1) of Theorem 10 we know if $(F(x), G(x,y)) \in K1$ then $|x| \leq \frac{1}{2}|1 - \sqrt{1-c}|$, and y satisfies

$$\frac{1}{2}[1+\sqrt{1-(\frac{c}{2}+\frac{2}{3}(\frac{c}{4}+x^2)^2)}] - (\frac{c}{8}+\frac{1}{6}x^2) \leq y^2 \leq \frac{1}{2}[1+\sqrt{1-(\frac{c}{2}+\frac{2}{3}x^2)}] - (\frac{c}{8}+\frac{1}{6}x^2).$$

This implies

$$R^{-1}(K1) = \left\{ \begin{array}{c} (x,y): \quad |x| \leq \frac{1}{2}|1-\sqrt{1-c}|, \\ \frac{1}{2}[1+\sqrt{1-(\frac{c}{2}+\frac{2}{3}(\frac{c}{4}+x^2)^2)}] - (\frac{c}{8}+\frac{1}{6}x^2) \\ \leq y^2 \leq \frac{1}{2}[1+\sqrt{1-(\frac{c}{2}+\frac{2}{3}x^2)}] - (\frac{c}{8}+\frac{1}{6}x^2) \\ = \frac{1}{4}[1+\sqrt{1-(\frac{c}{2}+\frac{2}{3}x^2)}]^2 \end{array} \right\} \quad (35)$$

In the same way we can get the inverse of $K2$,

$$R^{-1}(K2) = \left\{ \begin{array}{c} (x,y): \quad |x| \leq \frac{1}{2}|1-\sqrt{1-c}|, \\ -\frac{1}{2}[1+\sqrt{1-(\frac{c}{2}+\frac{2}{3}(\frac{c}{4}+x^2)^2)}] - (\frac{c}{8}+\frac{1}{6}x^2) \\ \geq y^2 \geq -\frac{1}{2}[1+\sqrt{1-(\frac{c}{2}+\frac{2}{3}x^2)}] - (\frac{c}{8}+\frac{1}{6}x^2) \end{array} \right\} \quad (36)$$

But from the y-inequality above we can see that $R^{-1}(K2)$ is empty. From (1) of Theorem 11 we can see that for positive y in $R^{-1}(K1)$,

$$D2 \setminus R^{-1}(K1) = \left\{ \begin{array}{c} (x,y): \quad |x| \leq \frac{1}{2}|1-\sqrt{1-c}|, \\ -\frac{1}{2}[1-\sqrt{1-(\frac{c}{2}+\frac{2}{3}x^2)}] \leq y \leq \\ \sqrt{\frac{1}{2}[1+\sqrt{1-(\frac{c}{2}+\frac{2}{3}(\frac{c}{4}+x^2)^2)}] - (\frac{c}{8}+\frac{1}{6}x^2)} \end{array} \right\} \quad (37)$$

and for negative y in $R^{-1}(K1)$

$$D8 \setminus R^{-1}(K1) = \left\{ \begin{array}{c} (x,y): \quad |x| \leq \frac{1}{2}|1-\sqrt{1-c}|, \\ \frac{1}{2}[1-\sqrt{1-(\frac{c}{2}+\frac{2}{3}x^2)}] \geq y \\ \geq -\sqrt{\frac{1}{2}[1+\sqrt{1-(\frac{c}{2}+\frac{2}{3}(\frac{c}{4}+x^2)^2)}] + (\frac{c}{8}+\frac{1}{6}x^2)} \end{array} \right\}. \quad (38)$$

From Theorem 11 and the definition of $R^{-1}(K1)$ we have

$$R((D2 \cup D5 \cup D8) \setminus R^{-1}(K1)) \subset D1 \cup D4 \cup D7 \quad (39)$$

Now we want to prove

$$R(D1 \cup D4 \cup D7) \subset (D2 \cup D5 \cup D8) \setminus R^{-1}(K1) \quad (40)$$

In fact, we can see by definitions $D2, D5, D8$ and $R^{-1}(K1) \equiv R_0$

$$R(D1 \cup D4 \cup D7) \subset \left\{ \begin{array}{c} (x,y): \quad |x| \leq \frac{1}{2}|1-\sqrt{1-c}| \\ |y| \leq \frac{1}{2}[1+\sqrt{1-(\frac{c}{2}+\frac{2}{3}x)}] \end{array} \right\}$$

$$(D2 \cup D5 \cup D8) \setminus R_0 = \left\{ \begin{array}{c} (x,y): \quad |x| \leq \frac{1}{2}|1-\sqrt{1-c}| \\ |y| \leq \sqrt{\frac{1}{2}[1+\sqrt{1-(\frac{c}{2}+\frac{2}{3}(\frac{c}{4}+x^2)^2)}] - (\frac{c}{8}+\frac{1}{6}x^2)} \end{array} \right\}$$

So to prove the relation (40) it is sufficient to prove that $\frac{1}{2}[1+\sqrt{1-(\frac{c}{2}+\frac{2}{3}x)}] \leq \sqrt{\frac{1}{2}[1+\sqrt{1-(\frac{c}{2}+\frac{2}{3}(\frac{c}{4}+x^2)^2)}] - (\frac{c}{8}+\frac{1}{6}x^2)}$, i.e., $\frac{1}{2}[1+\sqrt{1-(\frac{c}{2}+\frac{2}{3}x)}] - \frac{1}{4}(\frac{c}{3}+\frac{2}{3}x) \leq \frac{1}{2}[1+\sqrt{1-(\frac{c}{2}+\frac{2}{3}(\frac{c}{4}+x^2)^2)}] - (\frac{c}{8}+\frac{1}{6}x^2)$. To prove this inequality we note from Theorem 10 that for $|x| \leq -\frac{1}{2}[1-\sqrt{1-c}]$, $-\frac{1}{2}[1+\sqrt{1-c}] \leq \frac{c}{4}+x^2 \leq \frac{1}{2}[1-\sqrt{1-c}] \leq x$.

Then we have $\frac{1}{2}\sqrt{1-[\frac{c}{2}+\frac{2}{3}(\frac{c}{4}+x^2)^2]} - (\frac{c}{8}+\frac{1}{6}x^2) \geq \frac{1}{2}\sqrt{1-[\frac{c}{2}+\frac{2}{3}x^2]} - \frac{c}{12} - \frac{1}{6}x \geq \frac{1}{2}\sqrt{1-[\frac{c}{3}+\frac{2}{3}x]} - (\frac{c}{12}+\frac{1}{6})x$. So (40) is proved.

From (39) and (40) we know $R(D1 \cup D4 \cup D7 \cup [(D2 \cup D5 \cup D8) \setminus R^{-1}(K1)]) \subset D1 \cup D4 \cup D7 \cup [(D2 \cup D5 \cup D8) \setminus R^{-1}(K1)]$. Hence for $(x,y) \in D1 \cup D4 \cup D7 \cup [(D2 \cup D5 \cup D8) \setminus R^{-1}(K1)]$ the image of (x,y) is in the region too, $F(x,y) \in D1 \cup D4 \cup D7 \cup [(D2 \cup D5 \cup D8) \setminus R^{-1}(K1)]$. Since for any (x,y)

$$R(x,y) = R(-x,y) = R(-x,y) = R(-x,-y), \quad (41)$$

by the definition of the region BD we know $BD = D1 \cup D4 \cup D7 \cup [(D2 \cup D5 \cup D8) \setminus R^{-1}(K1)]$. So for any $(x_0, y_0) \in BD$, $R(x_0, y_0) \in BD$. Therefore the orbit $\{(x_n, y_n)\}_0^\infty$ is in BD. ♯

From (41) we see that the inverse of the mapping $R(x,y)$ is four to one. It tell us that from an initial point in the region BD the orbit is contained in BD. The following lemma gives us a property of the mapping $R(x,y)$.

Lemma 4 R maps the elliptic region $\frac{1}{6}x^2 + y^2 \leq r_0^2$ to the triangular region:

$$\Delta = \{(x,y) : \frac{c}{4} + 6r_0^2 \geq x \geq \frac{c}{4}, \frac{c}{8} + r_0^2 \geq y \geq \frac{1}{6}x + \frac{c}{12}\}.$$

Proof: Let $0 \leq r \leq r_0$. R maps the ellipse Γ, $\frac{1}{6}x_n^2 + y_n^2 = r^2$, to a line section, $\frac{c}{4} \leq x_{n+1} \leq \frac{c}{4} + 6r^2$, $y_{n+1} = \frac{c}{8} + r^2$. So R maps the region D_0 on the (x_n, y_n) plane to the region on the (x_{n+1}, y_{n+1}) plane with boundary $x_{n+1} = \frac{c}{4} + 6r^2$, $y_{n+1} = \frac{c}{8} + r^2$ with parameter r in $[0, r_0]$, i.e., $\frac{c}{8} + r_0^2 \geq y_{n+1} \geq \frac{1}{6}x_{n+1} + \frac{c}{12}$, $\frac{c}{4} + 6r_0^2 \geq x_{n+1} \geq \frac{c}{4}$. ♯

As r_0 goes to infinity, the Lemma above tells us the image of R is an unbounded triangular region Δ. Using Theorem 12 and the Lemma we have

Theorem 13 For initial point $(x_0, y_0) \in BD$ the orbit $\{(x_n, y_n)\}_0^\infty$ is attracted by the region $BD \cap \Delta$.

Proof: If $(x_0, y_0) \in BD$ from Theorem 12 we know $\{(x_n, y_n)\}_0^\infty \subset BD$. Using Lemma 4 we have that $\{(x_n, y_n)\}_1^\infty \subset \Delta$. Hence $(x_n, y_n) \in BD \cap \Delta$. ♯

4. INVARIANT MANIFOLDS

Now we want to study the invariant manifolds of the system (26) and (27). Let the invariant manifold passing through (u, v) be $(x(t), y(t))$, where $x(t) = u + \sum_{i=1}^\infty a_i t^i$ and $y(t) = v + \sum_{i=1}^\infty b_i t^i$. Then there are β_1 and β_2 such that $\frac{c}{4} + x^2(t) = x(\beta_1 t)$ and $\frac{c}{8} + \frac{c}{6}x^2(t) + y^2(t) = y(\beta_2 t)$. So $\frac{c}{4} + [u + \sum_{i=1}^\infty a_i t^i]^2 = u + \sum_{i=1}^\infty a_i \beta_1^i t^i$. By computing explicitly the coefficient of t^n we can find that $a_1(2u - \beta_1) = 0$. Let $\beta_1 = 2u$ and $a_1 = 1$. For $i = n > 1$, we have the relation

$$2ua_n + \sum_{k=1}^{n-1} a_k a_{n-k} t^i = a_n \beta_1^n. \quad (42)$$

It implies for $u = 0$ or $a_1 = 0$, then $a_k = 0, k = 1, 2, \ldots$.

Theorem 14 (1) If $\beta_1 = 2u = \pm 1$, then $a_k = 0$ $k = 1, 2...$ and $x(t) = u$.
(2) If $2u \neq \pm 1$, the sequence of $a_n, n = 1, 2, ...$, satisfies

$$a_n = \frac{1}{\beta_1^2 - 2u} \sum_{k=1}^{n-1} a_k a_{n-k}, \; n \geq 2 \;.$$

Proof: (1) If $\beta_1 = 2u = 1$, from (42) for $n > 1, a_n$ satisfies $\sum_{k=1}^{n-1} a_k a_{n-k} = 0$. It implies that the coefficients $a_n = 0$.
If $\beta_1 = 2u = -1$, then using relation (42) we have

$$a_{2n} = -\frac{1}{2} \sum_{k=1}^{2n-1} a_k a_{2n-k}, \text{ and } \sum_{k=1}^{2n} a_k a_{2n+1-k} = 0, \; n = 1, 2, ... \quad (43)$$

For $n = 1$ from (43) we have $a_2 = -\frac{1}{2} a_1^2$ and $\sum_{k=1}^{2} a_k a_{3-k} = 2a_1 a_2 = 0$. It implies that $a_1 = a_2 = 0$. If $a_i = 0, i = 1, ..., 2m$, then

$$a_{2J} = -\frac{1}{2} \sum_{k=1}^{2J-1} a_k a_{2J+1-k} = -\frac{1}{2}[2\sum_{k=1}^{J-1} a_k a_{2J-k} + a_J^2] = -\frac{1}{2} a_J^2 = 0, \; J \leq 2m$$

i.e., $a_n = 0$ for $n = 2k \leq 4m$. For $n = 2m - 1$, we need to prove $a_{2m+1} = 0$. Considering $a_{3(2m+1)}$, i.e. $a_{2(3m+1)+1}$ and using (43) and the result above, we can find $a_{2m+1} = 0$. Hence we have $a_k \equiv 0$ for all $k = 1, 2, ...$
(2) It is easy to prove (2) by (42). ♯

If $2u \neq \pm 1, 0$, we can obtain the coefficients of the series $x(t)$ from (2) in Theorem 14. Then can find the coefficients of the $y(t)$ from recursion relations

$$\begin{array}{l} a_n = \frac{1}{\beta_1^n - 2u} \sum_{i=1}^{n-1} a_i a_{n-i}, \\ b_n = \frac{1}{\beta_2^n - 2v}[\frac{1}{6}(2u)^n a_n + \sum_{i=1}^{n-1} b_i b_{n-i}], \; n \geq 2 \end{array} \quad (44)$$

where a_1 and b_1 satisfy $a_1(2u - \beta_1) = 0$, $\frac{1}{3} u a_1 + (2v - \beta_2) b_1 = 0$.

Theorem 15 For the system (26) and (27), if a fixed point is hyperbolic, there are a stable invariant manifold W^s and a unstable invariant manifold W^u passing through the fixed point.

Proof: In (44) let $a_1 = a_1^{(u)} = 1$, $\beta_1 = \beta_2 = 2u$ and $b_1 = b_1^{(u)} = \frac{u}{6(u-v)}$. Let the manifold $M(u) = (x^{(u)}(t), y^{(u)}(t))$ with parameter t be defined by

$$x^{(u)}(t) = u + \sum_{i=1}^{\infty} a_i^{(u)} t^i, \; y^{(u)}(t) = v + \sum_{i=1}^{\infty} b_i^{(u)} t^i \quad (45)$$

where the coefficients of t^n, $a_n^{(u)}$ and $a_n^{(u)}$ are given by (44). It satisfies

$$\frac{c}{4} + x^2(t) = x(2ut), \; \frac{c}{8} + \frac{c}{6} x^2(t) + y^2(t) = y(2ut). \quad (46)$$

It implies that if the initial point (x_0, y_0) of the orbit of the dynamical system (26) and (27) is on the manifold $M(2u)$, i.e., there is t_0 such that $(x_0, y_0) = (x^{(u)}(t_0), y^{(u)}(t_0))$, then the orbit $\{(x_n, y_n)\}_0^{\infty}$, starting from the point (x_0, y_0) on the manifold $M(u)$ satisfies $(x_n, y_n) = (x^{(u)}((2u)^n t_0), y^{(u)}((2u)^n t_0))$, $n = 0, 1, ...$

We can construct another manifold, $M(v) = (x^{(v)}, y^{(v)})$, which passes through the point (u,v), $x^{(v)}(t) = u + \sum_{i=1}^{\infty} a_i^{(v)} t^i$, $y^{(v)}(t) = v + \sum_{i=1}^{\infty} b_i^{(v)} t^i$, with $a_n^{(v)} \equiv 0$. Then $x^{(v)}(t) \equiv u$. Let $b_1^{(v)} = 1$ and $\beta_2 = 2v$. From (44), $b_n^{(v)}$, $n = 2, 3, ...$ are obtained by $b_n^{(v)} = \frac{1}{(2v)^n - 2v} \sum_{i=1}^{n-1} b_i^{(v)} b_{n-i}^{(v)}$. In the same way as for the manifold $M(u)$, we know that for $(x_0, y_0) = (x^{(v)}(t_0), y^{(v)}(t_0))$, then the orbit $\{(x_n, y_n)\}_0^{\infty}$ in the manifold $M(v)$ satisfies $(x_n, y_n) = (x^{(v)}((2v)^n t_0), y^{(v)}((2v)^n t_0))$, $n = 0, 1, ...$.

Due to the hyperbolicity of the fixed point (u, v), the radii of convergence of the power series $x^{(u)}(t)$, $y^{(u)}(t)$, $x^{(v)}(t)$, and $y^{(v)}(t)$ are positive [5].

For any given fixed point (u, v), from (26) and (27) we know the eigenvalues at this fixed point are $2u$ and $2v$. Since we consider a hyperbolically fixed point, one of the eigenvalues is in $(-1, 1)$ and another is outside of $[-1, 1]$. It implies that one of manifolds, $M(u)$ and $M(v)$, is stable and the other is unstable, depending on which absolute value of the eigenvalue at the fixed point is less than 1 and which absolute value of the eigenvalue at the fixed point is greater than 1. In any case, we have the stable manifold W^s and unstable manifold W^u. ♯

5. ATTRACTORS AND CHAOS

The behavior of the higher dimensional system can be related to a one dimensional mapping, called the governing system. By studying bifurcations of the governing system, as a function of the parameter c, we may obtain information on the behavior of the higher dimensional system. First we outline the results for the one dimensional dynamical system (26).

From (26) we can find fixed points, $x_+^{(1)} = \frac{1}{2}(1 + \sqrt{1-c})$ and $x_-^{(1)} = \frac{1}{2}(1 - \sqrt{1-c})$. For $c \in (-3, 1)$, x_+ is unstable and x_- is stable. $c = -3$ is a bifurcation point. For $c < -3$, the two points, $x_+^{(2)} = \frac{-1+\sqrt{-c-3}}{2}$ and $x_-^{(2)} = \frac{-1-\sqrt{-c-3}}{2}$ form a 2-periodic orbit. For $c \in (-5.0025, -3)$, this 2-periodic orbit is an attracting orbit and the fixed points, $x_+^{(1)}, x_-^{(1)}$, become unstable. When $c \in (-5.4516, -5.0025)$, we can find a new 4-periodic orbit which is attracting. The fixed points remain unstable, and the two periodic orbit, $x_+^{(2)}, x_-^{(2)}$, becomes unstable. If we continued, then we would find a number c such that the dynamical system (26) has an attracting 8-periodic orbit. In fact there is a sequence of numbers $c_1 > c_2 > ...$ such that when $c_n > c > c_{n+1}$ the dynamical system has an attracting 2^n-periodic orbit. We already know that $c_1 = -3, c_2 = -5.0025$. But there is more. There are a series of values, c_n, which are all greater than the number -5.6049. The closer c is to -5.6049, the larger the attracting periodic orbit of (26). We get a series of period doubling bifurcations, each closer to -5.6049 than the one before. All of the branches combine to give an attracting 2^n-periodic orbit.

What happens for the two dimensional system (26) and (27)? The following theorem provides the answer.

Theorem 16 *For the two dimensional system (26) and (27), if $c \in (-3, 1]$, there are no bifurcation points. There are only four fixed points, $(x_-^{(1)}, y_{-,-}^{(1)})$, $(x_-^{(1)}, y_{-,+}^{(1)})$, $(x_+^{(1)}, y_{+,-}^{(1)})$, $(x_+^{(1)}, y_{+,+}^{(1)})$, where*

$$y_{\pm,\pm}^{(1)} = \frac{1}{2}[1' \pm' \sqrt{1 - \frac{c}{2} + \frac{2}{3}(x_\pm^{(1)})^2}], \tag{47}$$

and there are no more n-periodic orbits with $n > 1$. For $c \in (-8, 1]$ there is only one bifurcation point, -7.14589034, at which one of the fixed points, $(x_-^{(1)}, y_{-,-}^{(1)})$, bifurcates to a 2-periodic orbit: $(x_-^{(1)}, y_-^{(2)})$, $(x_-^{(1)}, y_+^{(2)})$ where

$$y_\pm^{(2)} = \frac{-1 \pm \sqrt{-(\frac{c}{2} + \frac{2}{3}(x_-^{(1)})^2) - 3}}{2}. \tag{48}$$

Proof: We know that for $c \in (-3, 1]$ by substituting the two fixed points $x_\pm^{(1)}$ of the system (26) to (27) which is

$$y_{k+1} = \frac{1}{4}(\frac{c}{2} + \frac{2}{3}(x_\pm^{(1)})^2) + y_k^2, \tag{49}$$

we can find four fixed points for (26) and (27): $(x_-^{(1)}, y_{-,-}^{(1)})$, $(x_-^{(1)}, y_{-,+}^{(1)})$, $(x_+^{(1)}, y_{+,-}^{(1)})$, and $(x_+^{(1)}, y_{+,+}^{(1)})$.

Now we consider for $c \in (-3, 1]$ if these fixed points bifurcate n-periodic. In the equation (49) we can replace y_k with x_k and $\frac{c}{2} + \frac{2}{3}(x_\pm^{(1)})^2$ with c in (26). Then using the result of the system (26) we have that c satisfies the following inequality

$$\frac{c}{2} + \frac{2}{3}(x_\pm^{(1)})^2 < -3. \tag{50}$$

But it is easy to check that this is not true. Then we have proved that for $c \in (-3, 1]$ the four fixed points can not bifurcate into periodic orbits.

For $c \in (-8, 1]$, we can solve (50) and have $c \in (-8, -7.14589034)$. For $c > -7.14589034$ (50) is not true. Replacing $\frac{c}{2} + \frac{2}{3}(x_\pm^{(1)})^2$ and y in (49) with c and x in (26), respectively, we can get two periodic orbits of the system (49)

$$y_\pm^{(2)} = -\frac{1}{2} \pm \frac{1}{2}\sqrt{-(\frac{c}{2} + \frac{2}{3}(x_-^{(1)})^2) - 3}$$

It means $(x_-^{(1)}, y_{-,-}^{(1)})$ bifurcates to a 2-periodic orbit $(x_-^{(1)}, y_-^{(2)})$, $(x_-^{(1)}, y_+^{(2)})$ at the bifurcation point $c = -7.14589034$. ♯

In general, for given k and c if there is a k-periodic orbit of the one dimensional system (26), what happens for the two dimensional system? We have the following result.

Theorem 17 *For given c and any k, if there is a k-periodic orbit of the one dimensional system (26) then there are at least two k-periodic orbits of the two dimensional system (26) and (27).*

Proof: Let $\{x_n^{(k)}\}_{n=1}^\infty$ be a k-periodic orbit of (26), i.e., $x_{n+k}^{(k)} = x_n^{(k)}$. Let $G_n(y) = \frac{c}{8} + \frac{1}{6}(x_n^{(k)})^2 + y^2$. To find k-periodic orbits of the two dimensional system we want to prove there are solutions of the equation

$$y = G_k \circ G_{k-1} \circ \ldots \circ G_1(y). \tag{51}$$

Let us consider the value of the function G_n at $y = 0$, $G_n = \frac{c}{8} + \frac{1}{6}(x_n^{(k)})^2$. We know from (3) of Theorem 10 that $|x_n^k| \leq \frac{1}{2}(1 + \sqrt{1-c})$. So for $c < -3$ we have $G_k \leq \frac{c}{8} + \frac{1}{6}[\frac{1}{2}(1 + \sqrt{1-c})]^2 = \frac{1}{12}(1 + \sqrt{1-c})(\sqrt{1-c} - 2) < 0$. It is easy to see the mapping of the system (26) and (27) is a four to one mapping. So we can not find a unique inverse. But we can define one inverse function of $G(y)$:

$$G_n^{-1} = \sqrt{y - (\frac{c}{8} + \frac{1}{6}x_n^2)} \qquad (52)$$

such that $G_n \circ G_n^{-1}(y) = y$. Let $y_0 = G_1^{-1} \circ G_2^{-1} \circ ... \circ G_k^{-1}(0)$. Then we have $G_k \circ G_{k-1} \circ ... \circ G_1(y_0) = G_k \circ G_{k-1} \circ ... \circ G_1 \circ G_1^{-1} \circ G_2^{-1} \circ ... \circ G_k^{-1}(0) = G_k \circ G_{k-1} \circ G_{k-1}^{-1}(0) = G_k(0) < 0$. So the function $G_k \circ G_{k-1} \circ ... \circ G_1(y)$ is negative at y_0. By the definition of the inverse function of G_n we have $y_0 \geq 0$. Hence $G_k \circ G_{k-1} \circ ... \circ G_1(y_0) - y_0 < 0$. Since the function $G_k \circ G_{k-1} \circ ... \circ G_1(y)$ is an even polynomial, we have $\lim_{|y| \to \infty} G_k \circ G_{k-1} \circ ... \circ G_1(y) = \infty$. By continuity, there are at least two solutions of (51). Let these solutions be $y_1^{(k)}$ and $y_2^{(k)}$. Then for the two dimensional system, the two orbits with initial points $(x_1^{(k)}, y_1^{(k)})$ and $(x_1^{(k)}, y_2^{(k)})$, respectively, are k-periodic orbits. ♯

We know that for $c \leq -5.6049$ the system (26) has 2^n-periodic orbits. But for what c does the system have n-periodic orbits with any given n? And how about the two dimensional system? We have the following result.

Theorem 18 *(1) The two dimensional system (26) and (27) has n-periodic orbits for any given n iff $c \leq -7$. (2) For $c \in (-8, -7]$ there is an uncountable set $S \subset BD \cap \Delta$ (containing no periodic points), which satisfies the following conditions: for every $p, q \in S$ with $p \neq q$, we have*

$$\lim_{n \to \infty} sup|R^n(p) - R^n(q)| > 0, \qquad (53)$$

$$\lim_{n \to \infty} inf|F^n(p) - F^n(q)| = 0. \qquad (54)$$

for every $p \in S$ and periodic point $q \subset BD \cap \Delta$, we have

$$\lim_{n \to \infty} sup|R^n(p) - R^n(q)| > 0, \qquad (55)$$

where $|(x, y)| = \sqrt{x^2 + y^2}$.

Proof: (1) From Theorem 17 we have that the two dimensional system (26) and (27) has n-periodic orbits for any given n iff there are periodic orbits for the one dimensional system (26). From Sarkovskit [7] we know that if $f : R \to R$ is continuous and $x_{n+1} = f(x_{n+1})$ has a 3-periodic orbit, then it has periodic points of all other periods. Therefore, we know that the system (26) has any periodic orbits iff it has a 3-periodic orbit. Now we want to find 3-periodic orbits of (26).

For given c, to find 3-periodic orbits we need find a real solution of the equation $F^3(x) = x$. Computing numerically, we have that for $c > -7$ the equation above has no real solution which is one point of a three periodic orbit of the one dimension system and when $c \leq -7$ there is a real solution which is one point of a three periodic orbit. So (1) proved.

(2) Using Tien-Yien Li and J.A Yorke Theorem [6], we have that there is a set $S_x \subset [\frac{1}{2}(1 - \sqrt{1-c}), \frac{1}{2}(1 + \sqrt{1-c})]$, such that for every $v, u \in S_x$ with $v \neq u$,

$$\lim_{n \to \infty} sup|F^n(v) - F^n(u)| > 0, \qquad (56)$$

$$\lim_{n \to \infty} inf|F^n(v) - F^n(u)| = 0, \qquad (57)$$

and for every $p \in S$ and periodic point $q \subset BD \cap \Delta$,

$$\lim_{n \to \infty} \sup |F^n(p) - F^n(q)| > 0. \tag{58}$$

Taking $S = \{(x,y) : x \in S_x \text{ and } (x,y) \in BD \cap \Delta\}$ from (56) and (58) we can get (53) and (55) for S_x the x-component of S. ♯

We know for $c \in (-5.6049, 1]$ there is one 2^n-periodic attractor for (26). Hence from Theorem 17 the two dimensional system has at most one 2^n-periodic attractor.

For $-8 < c \leq -5.6049$ we compute numerically for the two dimensional system and find the attractor changes quickly by changing c. We list the results as follows.

Theorem 19 *c decreasing from -5.6049 there are a series bifurcation points of c for the system (26) and (27) : -5.89879, -5.91895, -5.91904, -5.91911, -5.91914, -5.91923, -5.92438, -5.92844, -5.92845, -5.92849,... ... -7, -7.07403, -7.109, ,-7.1096, -7.1114, ... For c in the following intervals there are periodic orbit attractors:*
$(-5.91895, -5.89879]$ attractor is $6-$periodic; $(-5.91904, -5.91895]$ is $7-$periodic;
$(-5.91911, -5.91904]$ is $8-$periodic; $(-5.91914, -5.91911]$ is $9-$periodic;
$(-5.91923, -5.91914]$ is $10-$periodic; $(-5.92438, -5.91923]$ is $11-$periodic;
$(-5.92844, -5.92438]$ is $12-$periodic; $(-5.92845, -5.92844]$ is $13-$periodic;
$(-5.92849, -5.92845]$ is $14-$periodic; $(-5.92849-\delta, -5.92849]$ for small δ is $15-$periodic;
... ... $(-7.07403, -7.00000]$ is $3-$periodic; $(-7.10900, -7.07403]$ is $6-$periodic;
$(-7.1096, -7.10900]$ is $8-$periodic; $(-7.1114, -7.1096]$ is $11-$periodic;
$(-7.1114 - \delta, -7.1114]$ for small δ is $12-$periodic, ...
These attractors are not in sensitive dependence on the initial values in a neighborhood of the origin. When decreasing c approaches -7 or -8 the attractor becomes an infinite point attractor.

Theorem 20 *There are an infinite number of windows in $[-8, -5.6049]$. Except for these windows the dynamical system exhibits chaos.*

REFERENCES

1. Chandrasekhar, S. and Breen, F.H., Astrophys. J. **106**, 145 (1947).
2. Chandrasekhar, S., *Radiative Transfer*, New York: Dover Publications, 1960.
3. Bittoni, E., Casadei, G. and Lorenzutta, S., Boll. U.M.I. **4**, 535 (1969).
4. Bowden, R.L. and Zweifel, P.F., Astrophys. J. **210**, 178 (1976).
5. Simo, C., J. Stat. Phys. **21**, 465-493 (1979).
6. Li, T.Y. and Yorke, J.A., Amer. Math. Mon. **82**, 985-992 (1975).
7. Schroeder, M., *Fractals, Chaos, Power Laws: Minutes from an Infinite Paradise*, New York: W. H. Freeman and Co., 1991.

FLUID DYNAMIC LIMIT OF THE BOLTZMANN KINETIC EQUATION ARISING IN THE COAGULATION–FRAGMENTATION DYNAMICS

Pavel B. Dubovski

Institute of Numerical Mathematics, Russian Academy of Sciences,
117924 Moscow, Russia

PRELIMINARIES

This paper is devoted to establishing a connection between the kinetic coagulation–fragmentation theory and macroscopic dynamics. We derive the fluid dynamic limit and discuss the corollaries. For celebrated Boltzmann equation from gas dynamics, which describes elastic interactions of particles, the fluid dynamics limit is, at some assumptions, just five Euler equations of hydrodynamics. This number of hydrodynamic equations arises due to five conservation laws for solutions of the Boltzmann equation of gas kinetics. In our case particles may merge and/or split and, thus, the interactions are nonelastic. So, the energy conservation law fails. The conservation law for the total number of particles fails, too. The conservation laws for impulse turns into the only conservation law – mass (or density) conservation law expressed by the first moment of the distribution function. So, the aim of this paper is derivation and justification a new equation of macroscopic dynamics for density. The importance of this research is pointed out, e.g., in book [1], p.175, where an attempt to derive the fluid dynamic limit for coagulation processes is done. However, the resulting equations in [1] (they are not justified mathematically) include integrals of the distribution function and, thus, cannot be considered as a closed macrodynamic system. So, the aim of this paper is deriving and justifying a closed macroscopic equation for evolution of mass density of the substance.

1. PROPERTIES OF THE COAGULATION–FRAGMENTATION EQUATION

The Boltzmann transport equation for coagulation–fragmentation dynamics has the following form [2]:

$$\partial_t c(x,z,t) + v(x)\partial_z c(x,z,t) = S(c)(x,z,t), \qquad (1)$$

where collision operator S is defined by

$$S(c)(x,z,t) = \frac{1}{2}\int_0^x K(x-y,y)c(x-y,z,t)c(y,z,t)dy - c(x,z,t)\int_0^\infty K(x,y)c(y,z,t)dy$$
$$+ \int_x^\infty F(x-y,x)c(y,z,t)dy - \frac{1}{2}c(x,z,t)\int_0^x F(x-y,y)dy, \qquad (2)$$

and initial condition is also defined:

$$c(x,z,) = c_0(x,z) \geq 0. \qquad (3)$$

Here $c(x, z, t)$ is the distribution function of particles of mass $x \in [0,+\infty)$ at the time moment $t \geq 0$ at the spatial point $z \in R^3$. The left-hand side of (1) is just the result of acting the common transport operator at function $c(x, z, t)$, here $\partial_z c$ is the z-derivative of c (or, in the multi-dimensional case, its divergence). Nonnegative symmetric functions $K(x, y)$ and $F(x, y)$, which are called coagulation and fragmentation kernels, respectively, are supposed to be known from the physics of the process concerned. They have the most essential influence on the evolution of coagulating systems and characterize the intensity of merging particles of masses x and y and intensity of splitting particles of mass $x + y$ into two particles with masses x and y, respectively; $v(x) \in R^3$ is a known velocity of spatial transfer of particles. The mathematical theory of this kinetic equation and corresponding references can be found in the book [3].

For convenience of notations we denote the integral of any function f(x) with respect to variable x by $<f>$,

$$<f> = \int_0^\infty f(x)dx.$$

The remarkable fact is that for any function c the first moment of the collision operator is equal to zero if all integrals are bounded:

$$<x\, S(c)> = 0.$$

The first moment of a solution means physically the mass density denoted by ρ, $\rho(z, t) = <xc(x, z, t)>$. Integrating (1) over $x \in [0, \infty)$ with weight x in the spatially homogeneous case ($v = 0$) yields the mass (density) conservation law

$$\frac{d}{dt} <xc> = 0.$$

To demonstrate some difficulties arising in the analysis of equation (1), we integrate (1) with weight x over $[0, \infty)$. Then we obtain

$$\partial_t \rho(z,t) + \partial_z \int_0^\infty xv(x)c(x,z,t) = 0. \qquad (4)$$

If $v(x) \neq$ const then equation (4) is not closed with respect to density ρ. If, for example, $v(x) = x$ then we come to the necessity to estimate the second moment of the solution. Multiplying (1) by x^2 and subsequent integrating give us the third moment, and so on. So, this simplest way cannot yield us an equation in a closed form for the mass density $\rho(z, t)$.

The aim of the article is deriving and justifying a closed equation for density ρ. Our approach is motivated by work [4].

2. SCALING EQUATION

We consider the scaling equation

$$\partial_t c_\varepsilon + v \partial_z c_\varepsilon = \frac{1}{\varepsilon} S(c_\varepsilon) \tag{5}$$

with a small parameter $\varepsilon > 0$. Similar approximated equations were used in many papers devoted to fluid dynamic limit for Boltzmann equation from gas kinetics (see, e.g.,[4-10]). Other results related to fluid dunamic limits in kinetic processes for gas and plasma can be found, e.g., in [11-26]. Changing variables $t = \varepsilon \tau$ in (5) yields

$$\partial_t c_\varepsilon + \varepsilon v \partial_z c_\varepsilon = S(c_\varepsilon). \tag{6}$$

From (6) we see that tending ε to zero means passing to the spatially homogeneous case and large times. Consequently, it is naturally to expect that c_ε tends to the solution of equilibrium equation $S(c) = 0$.

3. F–THEOREM FOR CONSTANT COAGULATION AND FRAGMENTATION KERNELS

Let us assume that the coagulation and fragmentation kernels are positive constants. Then it is well known that collision operator $S(c)$ is equal to zero if

$$c(x) = \frac{F}{K} \exp\left(-x\sqrt{F/(K\rho)}\right) \tag{7}$$

Moreover, the following properties are equivalent (F-theorem,[27]):

$S(c) = 0;$ \hfill (i)

$< S(c) \ln(Kc/F) > = 0;$ \hfill (ii)

$equality (7) holds.$ \hfill (iii)

Note also, that time-dependent solutions of spatially homogeneous initial value problem for (1) converge to this equilibrium as $t \to \infty$ (see [27, 28]). This case corresponds to $v = 0$.

Later on we assume that property (ii) can be replaced by the following expression

$$< xS(c) \ln(Kc/F) > = 0. \tag{ii'}$$

We are in a position now to formulate the following basic theorem.

4 FLUID DYNAMIC LIMIT

Theorem. Let integral kernels K and F be constants, and let c_ε, $\varepsilon > 0$, be nonnegative solutions of scaling equation (5). Let there exist a nonnegative function c such that almost everywhere $c_\varepsilon \to c \geq 0$ as $\varepsilon \to 0$. Let also

$$<xc_\varepsilon> \to <xc>,$$
$$<xv(x)c_\varepsilon> \to <xv(x)c>,$$
$$<xc_\varepsilon \ln c_\varepsilon> \to <xc \ln c>,$$
$$<xv(x)c_\varepsilon \ln c_\varepsilon> \to <xv(x)c \ln c>$$

as $\varepsilon \to 0$, and

$$<xS(c)\ln c> \leq 0, \lim_{\varepsilon \to 0} <xS(c_\varepsilon)\ln c_\varepsilon> \leq <xS(c)\ln c>. \tag{8}$$

Let, finally, hypothesis (ii') holds. Then

$$c(x,z,t) = \frac{F}{K}\exp\left(-x\sqrt{F/(K\rho(z,t))}\right) \tag{9}$$

and density $\rho(z, t)$ satisfies the following dynamic equation (fluid dynamic limit)

$$\partial_t \rho(z,t) + \frac{1}{2}\left[\frac{F}{K\rho(z,t)}\right]^{1/2}\left[\int_0^\infty x^2 v(x)\exp\left(-x\sqrt{F/K\rho(z,t)}\right)dx\right]\partial_z \rho(z,t) = 0. \tag{10}$$

Proof. We multiply scaling equation (5) by $\varepsilon x[1+\ln(Kc_\varepsilon/F)]$ and integrate it over $x \in [0, \infty)$. Then we come to the equality

$$<xS(c)\ln c> \geq 0.$$

Passing $\varepsilon \to 0$, we obtain

$$\varepsilon[\partial_t <xc_\varepsilon \ln(Kc_\varepsilon/F)> + \partial_z <xv(x)c_\varepsilon \ln(Kc_\varepsilon/F)>] = <x\ln(Kc_\varepsilon/F)S(c_\varepsilon)>.$$

From condition (8) we see that $<xS(c)\ln c> = 0$. Hence, applying hypothesis (ii'), we obtain

$$c(x,z,t) = \frac{F}{K}\exp\left(-x\sqrt{F/(K\rho(z,t))}\right). \tag{11}$$

Integrating scaling equation (5) with weight x yields

$$\partial_t <xc_\varepsilon> + \partial_z <xv(x)c_\varepsilon> = 0.$$

Again, letting $\varepsilon \to 0$ and the use of conditions of the theorem give us the correspondence

$$\partial_t \rho(z,t) + \partial_z <xv(x)c> = 0. \tag{12}$$

Substituting (11) to (12) enables us to derive basic equation (10), which is actually the desired hydrodynamic limit. This proves the Theorem.

5 CONCLUSIONS

For different prescribed functions $v(x)$ we come from (10) to slightly different hydrodynamic limit equations. If, e.g., $v(x) = x^\gamma$, $\gamma \geq 0$, then (10) yields

$$\partial_t \rho + \frac{1}{2}\left(\frac{K}{F}\right)^{\gamma/2} \Gamma(\gamma+3) \rho^{\gamma/2} \partial_z \rho = 0, \tag{13}$$

where $\Gamma(\alpha)$ is the common Euler function expressed by

$$\Gamma(\alpha) = \int_0^\infty x^{\alpha-1} e^{-x} dx.$$

Particulary, for $v(x) = $ const we come from (13) to the simplest transport equation

$$\partial_t \rho + v \frac{K}{F} \partial_z \rho = 0,$$

but for $v(x) = x$ we obtain the nonlinear equation

$$\partial_t \rho + 3\left(\frac{K}{F}\right)^{3/2} \sqrt{\rho} \partial_z \rho = 0,$$

which cannot have a continuous solution for most of initial values $\rho(z, 0)$. From (10) we can easily see that usually the mass density ρ loses its continuity in a finite time since this equation has the nonlinearity that causes blow-ups of solutions.

This observation perfectly corresponds to existence results of [29] where the conditions ensuring the continuity of ρ and existence of continuous time-dependent solution are derived.

In conclusion, I would like to express my deep gratitude to Professor V.I. Agoshkov who read this note and made useful critical remarks. Also, I am thankful to Dr. M. Grinfeld and Dr. V.P. Shutyaev for interesting discussions.

REFERENCES

1. L.E. Sternin, A.A. Shraiber, *Multi-phase flows of gas with particles*, Mashinostroenie, Moscow (1994) (in Russian).
2. V.M. Voloschuk, Yu.S. Sedunov, *Coagulation processes in disperse systems*, Hydrometeoizdat, Leningrad (1975) (in Russian).
3. P.B. Dubovskii, Mathematical theory of coagulation, *Lecture Notes Series* 23, Seoul, Seoul National University, Research Institute of Mathematics, Global Analysis Research Center: 169 (1994).
4. C. Bardos, F. Golse and D. Levermore, Fluid dynamic limits of kinetic equations. I. Formal derivations, *J. Stat. Phys.* 63, 1/2: 323 (1991).
5. K. Asano and S. Ukai, On the fluid dynamical limit of the Boltzmann equation, *Recent topics in nonlinear PDE*, North-Holland Math. Stud. 98: 1 (1984).
6. R. Caflisch, The fluid dynamic limit of the nonlinear Boltzmann equation, *Comm. Pure Appl. Math.* 33: 651 (1980).

7. R. Caflisch, Gas dynamics and Boltzmann equation, in *Nonequilibrium phenomena I. The Boltzmann equation*, J.L. Lebowitz and E.W. Montroll (eds), Amsterdam, North-Holland Publishing Co.: 204 (1983).
8. R. Ellis and M. Pinsky, The first and second fluid approximation to the linearized Boltzmann equation, *J. Math. Pures Appl.* 54: 125 (1975).
9. T. Nishida, Fluid dynamical limit of the nonlinear Boltzmann equation to the level of the compressible Euler equation, *Comm. Math. Phys.* 61: 119 (1978).
10. S. Kawashima, A. Matsumura and T. Nishida, On the fluid dynamical approximation to the Boltzmann equation at the level of the Navier–Stokes equation, *Comm. Math. Phys.* 70: 97 (1979).
11. H. Grad, Singular and nonuniform limits of solutions of the Boltzmann equations, *SIAM–AMS Proceedings I, Transport Theory., R.I.*: 296 (1969).
12. A.A. Arsen'ev, *Lectures on kinetic equations*, Moscow, Nauka (1992).
13. Z. Xin, The fluid-dynamic limit of the Broadwell model of the nonlinear Boltzmann equation in the presence of shocks, *Comm. Pure Appl. Math.* 44; 679 (1991).
14. R. Caflisch and G.C. Papanicolaou, The fluid dynamic limit of the nonlinear Boltzmann equation, *Comm. Pure Appl. Math.* 32: 589 (1979).
15. H. Fan, Self-similar solutions for a modified Broadwell model and its fluid-dynamic limits, *SIAM J. Math. Anal.* 28, 4: 831 (1997).
16. J.-G. Liu, Z. Xin, Boundary-layer behavior in the fluid-dynamic limit for a nonlinear Boltzmann equation, *Arch. Ration. Mech. Anal.* 135, 1: 61 (1996).
17. G. Russo. and P. Smereka, Kinetic theory for bubbly flow. II: Fluid dynamic limit, *SIAM J. Appl. Math.* 56, 2: 358 (1996).
18. A.E. Tzavaras, Wave structure induced by fluid-dynamic limits in the Broadwell model, *Arch. Ration. Mech. Anal.* 127, 4: 361 (1994).
19. M. Slemrod and A.E. Tzavaras, Self-similar fluid-dynamic limits for the Broadwell system, *Arch. Ration. Mech. Anal.* 122, 4: 353 (1993).
20. M. Slemrod, Fluid dynamic limit for a modified Broadwell system, *Matematiche* 46, 1: 439 (1991).
21. W.E. Fitzgibbon, The fluid dynamical limit of the Carleman equation with reflecting boundary, *Nonlinear Anal.* 6: 695 (1982).
22. T. Platkowski, R. Illner, Discrete velocity models of the Boltzmann equation: A survey on the mathematical aspects of the theory, *SIAM Rev.* 30, 2: 213 (1988).
23. S. Ukai, Solutions of the Boltzmann equation, in *Patterns and waves. Qualitative analysis of nonlinear differential equations*, Stud. Math. Appl. 18: 37 (1986).
24. V. Comincioli, G. Naldi, G. Toscani, Nonlinear diffusion and fluid dynamical limit from discrete velocity models, *Commun. Appl. Nonlinear Anal.* 2, 4: 1 (1995).
25. V. Gorunovich, The fluid-dynamical limit for the BBGKY hierarchy of a discrete velocity model, in *On three levels. Micro-, meso-, and macro- approaches in physics, Proceedings of a NATO Advanced Research Workshop*, M. Fannes *et al* (eds.), Leuven, Belgium, (1993), NATO ASI Ser., Ser. B, Phys. 324: 315 Plenum Press, NY (1994).
26. K. Asano, Fluid dynamical limit of the Boltzmann equation. I., *Transp. Theory Stat. Phys.* 24, 1–3: 329 (1995).
27. M. Aizenman and T.A. Bak, Convergence to equilibrium in a system of reacting polymers, *Comm. Math. Phys.* 65: 203 (1979).
28. I.W. Stewart and P.B. Dubovskii, Approach to equilibrium for the coagulation–fragmentation equation via a Lyapunov functional, Math. Meth. Appl. Sci. 19: 171 (1996).
29. P.B. Dubovski., Initial boundary value problems for Boltzmann equation of the coagulation–fragmentation kinetics, Sbornik: Mathematics, submitted.

NECESSARY AND SUFFICIENT CONDITIONS FOR SOLVABILITY OF THE INITIAL–BOUNDARY VALUE TRANSPORT PROBLEM*

Victor P. Shutyaev

Institute of Numerical Mathematics, Russian Academy of Sciences,
117924 Moscow, Russia

INTRODUCTION

The problems of solvability of transport equations have attracted the attention of many researchers (see the well-known papers by V.S. Vladimirov [1], T.A. Germogenova [2,3], V.I. Agoshkov [4,5] for stationary problems, and the papers by A. Douglis [6], Yu. Kuznetsov and S.F. Morozov [7], U.M. Sultangazin [8], V.M. Novikov and S.B. Shikhov [9], W. Greenberg, Van der Mee, and V. Protopopescu [10], and by the author [11] – for nonstationary problems).

When studying the boundary value transport problems, it turned out to be important to choose properly the spaces of boundary value functions which structure is closely connected with the transport operator in the equation. The properties of boundary values (traces) of solutions to the transport equations have been studied in the papers by V.S. Vladimirov [1], T.A. Germogenova [2,3], C. Bardos [12], V.I. Agoshkov [4,13], M. Cessenat [14], S. Ukai [15], S. Mischler [17].

The general approach to constructing spaces of boundary values was formulated in [4,10]. In [4], with the use of properly chosen spaces of traces, the necessary and sufficient conditions for solvability of stationary boundary value transport problems were given. Some related results for nonstationary transport problems were presented in [10,15,16,17].

In this paper, with the use of the technique developed in [10,4,16] and the results of [11], the space of boundary values for the time-dependent transport problem in a slab is constructed. The necessary and sufficient conditions for solvability of the initial-boundary value problem are given.

STATEMENT OF THE PROBLEM AND FUNCTIONAL SPACES

Consider the time-dependent initial-boundary value transport problem in a slab [18]:

---

* The work was partially supported by the Russian Foundation for the Basic Research (grant no. 97–01–00333).

$$\frac{\partial \varphi}{\partial t} + \mu \frac{\partial \varphi}{\partial z} + \sigma(z,\mu,t)\varphi = \int_{-1}^{1} K(z,\mu,\mu',t)\varphi(z,\mu',t)d\mu' + f(z,\mu,t), \qquad (1)$$

$z \in (-H,H), \mu \in (-1,1), t \in (0,T)$

$\varphi(-H,\mu,t) = v_1(\mu,t), 0 < \mu < 1, t \in (0,T),$

$\varphi(H,\mu,t) = v_2(\mu,t), -1 < \mu < 0, t \in (0,T),$ \qquad (2)

$\varphi(z,\mu,0) = v_3(z,\mu), z \in (-H,H), \mu \in (0,T),$ \qquad (3)

where $\varphi(z,\mu,t)$ is an unknown distribution function, $\sigma(z,\mu,t)$, $K(z,\mu,\mu',t)$, $f(z,\mu,t)$, $v_1(\mu,t)$, $v_2(\mu,t)$, $v_3(z,\mu)$ are prescribed functions, $T, H < \infty$.

Introduce the notations:

$D = \{(z,\mu,t) : z \in (-H,H), \mu \in (-1,1), t \in (0,T)\},$

$\partial D_- = \{(z,\mu,t) : \Gamma_- \cup (t=0)\}, \partial D_+ = \{(z,\mu,t) : \Gamma_+ \cup (t=T)\},$

where

$\Gamma_- = \{(z,\mu,t) : (\mu \in (0,1), z=-H) \cup (\mu \in (-1,0), z=H), t \in (0,T)\},$

$\Gamma_+ = \{(z,\mu,t) : (\mu \in (-1,0), z=0) \cup (\mu \in (0,1), z=H), t \in (0,T)\}.$

Hhere ∂D_- is the "incoming" part of the boundary of D, i.e. the part where the initial and boundary conditions are set.

Introduce the spaces $L_p = L_p(D)$, $H_p^1 = H_p^1(D)$ of real-valued functions with the norms

$$\|u\|_{L_p} = \left(\int_D |u|^p dD\right)^{1/p} = \left(\int_0^T dt \int_{-1}^1 d\mu \int_{-H}^H |u(z,\mu,t)|^p dz\right)^{1/p},$$

$$\|u\|_{H_p^1} = \|u\|_{L_p} + \left\|\frac{\partial u}{\partial t} + \mu \frac{\partial u}{\partial z}\right\|_{L_p}, 1 \leq p < \infty.$$

Let $\varphi_{(\Gamma)} = (v_1, v_2, v_3)$ be a vector-function with the components from the initial-boundary conditions (2)–(3), given on ∂D_-. We define the space of "boundary values" $L_{p,-}$ as a space of vector-functions $v = (v_1(\mu,t), v_2(\mu,t), v_3(z,\mu))$ with the norm

$$\|v\|_{L_{p,-}} = \left[\int_0^T dt \int_0^1 (T-t)|v_1(\mu,t)|^p \mu d\mu + \int_0^T dt \int_{-1}^0 (T-t)|v_2(\mu,t)|^p |\mu| d\mu \right.$$

$$\left. + \int_{-1}^1 d\mu \int_{-H}^H \lambda(z,\mu)|v_3(z,\mu)|^p dz\right]^{1/p}, \qquad (4)$$

where the function $\lambda(z, \mu)$ is defined by

$$\lambda(z,\mu) = \begin{cases} T, T|\mu| \le H - l_\mu z \\ \dfrac{H - l_\mu z}{\mu}, T|\mu| \ge H - l_\mu z, \end{cases} \quad (5)$$

and

$$l_\mu = \operatorname{sign}\mu = \begin{cases} 1, \mu > 0 \\ -1, \mu < 0. \end{cases}$$

WEAK FORMULATION OF THE PROBLEM

Let $f \in L_p(D)$, $\varphi_{(\Gamma)} \in L_{p,-}$, $\sigma \in L_\infty(D)$, $\sigma \ge 0$. Consider the problem (1)–(3) when $K \equiv 0$. The function $\varphi \in H_p^1(D)$ is said to be a weak solution of this problem if

$$\left\| \frac{\partial \varphi}{\partial t} + \mu \frac{\partial \varphi}{\partial z} + \sigma\varphi - f \right\|_{L_p} = 0, \left\| \varphi \big|_{\partial D_-} - \varphi(\Gamma) \right\|_{L_{p,-}} = 0. \quad (6)$$

The following theorem holds.

Theorem 1. The problem (1)–(3) for $K \equiv 0$ has a weak solution $\varphi \in H_p^1(D)$ if $\varphi_{(\Gamma)} \in L_{p,-}$, $f \in L_p(D)$. Under these constraints there exists a unique function $\varphi \in H_p^1(D)$ such that the equations (6) are satisfied and the estimates

$$c_1 \left(\left\| \varphi(\Gamma) \right\|_{L_{p,-}} + \left\| f \right\|_{L_p} \right) \le \left\| \varphi \right\|_{H_p^1} \le c_2 \left(\left\| \varphi(\Gamma) \right\|_{L_{p,-}} + \left\| f \right\|_{L_p} \right) \quad (7)$$

hold, where the constants c_1, c_2 do not depend on f, φ, $\varphi_{(\Gamma)}$.

Proof. The equations of characteristic curves corresponding to (1) are as follows:

$$\frac{dt}{d\tau} = 1, \frac{dz}{d\tau} = \mu, \frac{d\mu}{d\tau} = 0.$$

Hence, the characteristic curves are the lines $t = \tau + c_1$, $z = \mu\tau + c_2$, $\mu = c_3$, $c_i = \operatorname{const}$, $i = 1,2,3$. Consider the characteristic curves entering D at the points of ∂D_- and going out of D at the points of ∂D_+. Let $(z, \mu, t) \in \partial D_-$, then by $l(z, \mu, t)$ we denote the value of τ, for which the characteristic curve, going through the point (z, μ, t), goes out of D (the characteristic curves are assumed to enter D when $\tau = 0$). The function $l(z, \mu, t)$ is said to be the "length" of the characteristic curve in D. Thus, each point $(z, \mu, t) \in D$ may be defined by a point $(z', \mu', t') \in \partial D_-$ and some value of τ. Then, the domain D may be treated as a Cartesian product

$$D = \{(z', \mu', t', \tau) : (z', \mu', t') \in \partial D_-, 0 < \tau < l(z', \mu', t')\}.$$

If the function $u(z,\mu,t)$ is summable on D, then, as follows from [10], the equality holds

$$\int_D u\,dD = \int_{\partial D_-} \int_0^{l(z',\mu',t')} u(z',\mu',t',\tau)\,d\tau\,dv^-,$$

where

$$dv^- = \begin{cases} d\mu'dz', & t=0 \\ |\mu|d\mu'dt', & (z',\mu',t') \in \Gamma_-. \end{cases}$$

Using the results of [11], it is readily seen that the function $l(z,\mu,t)$ has the form:

$$l(z,\mu,t) = \begin{cases} \lambda(z,\mu), & t=0 \\ (T-t), & (z,\mu,t) \in \Gamma_-, \end{cases} \qquad (8)$$

where the function $\lambda(z,\mu)$ is defined by formula (5). From (8), the function $l(z,\mu,t)$ is upper bounded: $l(z,\mu,t) \le T$, but it vanishes at the points $t=T$, $(z,\mu,t) \in \Gamma_-$ and $t=0$, $(z,\mu,t) \in \Gamma_+$.

According to [10, 4], the function $l(z,\mu,t)$ should be the weight function when constructing the boundary norm:

$$\|v\| = \left(\int_{\partial D_-} l(z,\mu,t)|v|^p\,dv^- \right)^{1/p}.$$

This is why the norm of $L_{p,-}$ is defined in the form (4).

Following [4, 10] and taking into account the boundedness of $l(z,\mu,t)$, it is easy to show that the function $u \in H^1_p(D)$, $1 \le p < \infty$, has the trace $v = u|_{\partial D_-} \in L_{p,-}$, and

$$\left\| u|_{\partial D_-} \right\|_{L_{p,-}} \le c_1 \|u\|_{H^1_p}, c_1 = \text{const} > 0. \qquad (9)$$

Conversly, if $v \in L_{p,-}$, then there exists a function $u \in H^1_p(\Omega)$, such that $v = u|_{\partial D_-}$ and

$$\|u\|_{H^1_p} \le c_2 \left\| u|_{\partial D_-} \right\|_{L_{p,-}}, c_2 = \text{const} > 0, \qquad (10)$$

where the constant c_1, c_2 do not depend on u, v.

Using these results, it is readily seen that for the space $H^1_p(D)$ the norm $\|\cdot\|_{H^1_p}$ is equivalent to the norm

$$[u]_1 = \left\| u|_{\partial D_-} \right\|_{L_{p,-}} + \left\| \frac{\partial u}{\partial t} + \mu \frac{\partial u}{\partial z} + \sigma u \right\|_{L_p}.$$

Then, for a sequence of regular functions f^n, $\varphi^n_{(\Gamma)}$, $n = 1,2,\ldots$, convergent to f (in $L_p(D)$) and to $\varphi_{(\Gamma)}$ (in $L_{p,-}$), respectively, consider the problem

$$\begin{cases} \dfrac{\partial \varphi^n}{\partial t} + \mu \dfrac{\partial \varphi^n}{\partial z} + \sigma \varphi^n = f^n, (z,\mu,t) \in D \\ \varphi^n \big|_{\partial D_-} = \varphi^n_{(\Gamma)}. \end{cases} \qquad (11)$$

For every n, the problem (11) has a unique solution $\varphi \in H^1_p, 1 \le p < \infty$. Then, using the norm equivalency, we find

$$c_1 [\varphi]_1 = c_1 \left(\left\| \varphi^n_{(\Gamma)} \right\|_{L_{p,-}} + \left\| f^n \right\|_{L_p} \right) \le \left\| \varphi^n \right\|_{H^1_p} \le c_2 [\varphi^n]_1 =$$

$$= c_2 \left(\left\| \varphi^n_{(\Gamma)} \right\|_{L_{p,-}} + \left\| f^n \right\|_{L_p} \right), c_1, c_2 = \text{const} > 0, \qquad (12)$$

and

$$\left\| \varphi^n - \varphi^m \right\|_{H^1_p} \le c_2 [\varphi^n - \varphi^m]_1 = c_2 (\left\| \varphi^n_{(\Gamma)} - \varphi^m_{(\Gamma)} \right\|_{L_{p,-}} + \left\| f^n - f^m \right\|_{L_p}) \to 0,$$

$n, m \to \infty,$

i.e. the sequence φ^n is convergent in $H^1_p(D)$. Let $\varphi^n \to \varphi \in H^1_p$ as $n \to \infty$. Going to the limit in (11), (12) as $n \to \infty$, we get (6), (7). This proves the theorem.

Using the representation (5) for $\lambda(z,\mu)$, the norm (4) may be written in the following more detailed form:

$$\begin{aligned} \|v\|^p_{L_{p,-}} &= \int_0^T dt \int_0^1 (T-t) |v_1(\mu,t)|^p \mu d\mu + \int_0^T dt \int_{-1}^0 (T-t)|v_2(\mu,t)|^p |\mu| d\mu + \\ &+ T \int_0^1 d\mu \int_{-H}^{H-\mu T} |v_3(z,\mu)|^p dz + \int_0^1 \dfrac{d\mu}{\mu} \int_{H-\mu T}^{H} (H-z)|v_3(z,\mu)|^p dz + \\ &+ \int_{-1}^0 \dfrac{d\mu}{|\mu|} \int_{-H}^{-H-\mu T} (H+z)|v_3(z,\mu)|^p dz + T \int_{-1}^0 d\mu \int_{-H-\mu T}^{H} |v_3(z,\mu)|^p dz, \end{aligned} \qquad (13)$$

hence, the functions $v_1(\mu,t)$, $v_2(\mu,t)$ may have singularities at the point $t=T$, and the function $v_3(z,\mu)$ may be singular at the ponts of Γ_+, i.e. when $z=H$, $\mu > 0$ and $z=-H$, $\mu < 0$.

As an example, consider the case when

$$v_1(\mu,t) = v_2(\mu,t) = (T-t)^{-(1+\alpha)/p}, 0 < \alpha < 1,$$

$$v_3(z,\mu) = \begin{cases} (H-z)^{-(1+\alpha)/p}, \mu > 0 \\ (H+z)^{-(1+\alpha)/p}, \mu < 0. \end{cases}$$

Then, it is easily seen that $\varphi_{|\Gamma} = (v_1, v_2, v_3) \in L_{p,-}$, but $\varphi_{|\Gamma} = (v_1, v_2, v_3) \notin L_p(\partial D_-)$.

SOLVABILITY OF THE INITIAL–BOUNDARY VALUE TRANSPORT PROBLEM

To formulate the solvability theorem for the problem (1)–(3) let us impose restrictions on the function K. Let for almost every point $(z, \mu, t) \in D$ the condition

$$\int_{-1}^{1} |K(z, \mu, \mu', t)| d\mu' = \int_{-1}^{1} K(z, \mu, \mu', t) d\mu \le \theta_0, \quad \theta_0 = \text{const} > 0 \tag{14}$$

is satisfied. Moreover, we assume that the function $\sigma(z, \mu, t) = \sigma(z, t)$ does not depend on μ and is bounded.

The function $\varphi \in H_p^1(D)$ is said to be a weak solution to the problem (1)–(3) if

$$\left\| \frac{\partial \varphi}{\partial t} + \mu \frac{\partial \varphi}{\partial z} + \sigma \varphi - \int_{-1}^{1} K \varphi d\mu' - f \right\|_{L_p} = 0, \left\| \varphi|_{\partial D_-} - \varphi(\Gamma) \right\|_{L_{p,-}} = 0. \tag{15}$$

The following theorem holds.

Theorem 2. The problem (1)–(3) has a unique weak solution $\varphi \in H_p^1(D)$ if $f \in L_p(D)$, $\varphi_{|\Gamma} \in L_{p,-}$. Under these constraints, there exists a unique function $\varphi \in H_p^1(D)$ such that the equations (15) are satisfied and the estimates (7) hold.

Proof. Under the condition $\sigma(z, t) \le \sigma_1 = \text{const} < \infty$ and by the substitution $\varphi = \exp(\gamma t) \tilde{\varphi}$, where $\gamma = \text{const} > \sigma_1$, the problem (1)–(3) may be reduced to the problem for the function $\tilde{\varphi}$, with the coefficient $\tilde{\sigma}(z, t) = \gamma + \sigma(z, t) \ge \gamma - \sigma_1 > 0$ instead of $\sigma(z, t)$. Taking γ sufficiently large, under the condition (14) one can obtain

$$\int_{-1}^{1} \left| \frac{K(z, \mu, \mu', t)}{\sigma(z, t)} \right| d\mu' \le k_0, \quad k_0 = \text{const} < 1. \tag{16}$$

Therefore, in further proof, we assume the condition (16) be satisfied and $\sigma(z, t) \ge \sigma_0 > 0$.

Following the technique of [4], we should, first, show that for the space $H_p^1(D)$ the norm $\|\cdot\|_{H_p^1}$ is equivalent to the norm

$$[u]_2 = \|u|_{\partial D_-}\|_{L_{p,-}} + \|Lu - Su\|_{L_p},$$

where $Lu = \dfrac{\partial u}{\partial t} + \mu \dfrac{\partial u}{\partial z} + \sigma u$, $Su = \int_{-1}^{1} K(z, \mu, \mu', t) u(z, \mu', t) d\mu'$.

The inequality $[u]_2 \le c\|u\|_{H_p^1}$ is easily obtained, taking into account the result on traces of H_p^1, formulated when proving Theorem 1. To derive the backward inequality, we consider the function $u \in H_p^1$ and introduce the notation $f = Lu - Su$. Then, by integrating along the characteristic curve we get the formula:

$$u(z',\mu',t',\tau) = \exp(-\int_0^\tau \sigma(z',t',s)ds)u(z',\mu',t',0) +$$
$$\int_0^\tau \exp(-\int_\xi^\tau \sigma(z',t',s)ds)(f + Su)(z',\mu',t',\xi)d\xi, \qquad (17)$$
$$(z',\mu',t') \in \partial D_-, \; 0 < \tau < l(z',\mu',t').$$

Hence,

$$\|u\|_{B_p} \le \|u|_{\partial D_-}\|_{L_{p,-}} + \|(f + Su)/\sigma\|_{B_p}, \qquad (18)$$

where $\|u\|_{B_p} = \|\sigma^{1/p} u\|_{L_p}$.

From (16), it follows that $\|Su/\sigma\|_{B_p} \le k_0 \|u\|_{B_p}$. Then, along with (18) and following [4], we find successively

$$(1-k_0)\|u\|_{B_p} \le \|u|_{\partial D_-}\|_{L_{p,-}} + \|f/\sigma\|_{B_p} \le c[u]_2,$$

$$\|u\|_{L_p} \le c[u]_2,$$

$$\left\|\frac{\partial u}{\partial t} + \mu \frac{\partial u}{\partial z}\right\|_{L_p} \le \|\sigma u - Su\|_{L_p} + \|(L-S)u\|_{L_p} \le c[u]_2,$$

$$\|u\|_{H_p^1} \le c[u]_2.$$

Using the norm equivalence, and the well-known results on properties of the solution to the time-dependent transport problem with regular prescribed functions [12, 11], we come to the conclusion of the theorem similarly to the proof of Theorem 1.

REFERENCES

1. V.S. Vladimirov, Mathematical problems of one-velocity transport theory, *Papers of Math. Steklov Inst.* 61: 3 (1961).
2. T.A. Germogenova, Local properties of solutions of the transport equation, *Sov. Doklady.* 187: 5 (1969).
3. T.A. Germogenova. *Local Properties of Solutions of Transport Equations*, Nauka, Moscow (1986).

4. V.I. Agoshkov. *Generalized Solutions of the Transport Equation and Regularity Properties*, Nauka, Moscow (1988).
5. V.I. Agoshkov. *Boundary Value Problems for Transport Equations*, Birkhauser, Boston (1998).
6. A. Douglis, The solution of multidimensional generalized transport equations and their calculation by difference methods,in: *Numerical Solutions of Partial Diff. Eq. - Proc. of a Symposium*, I.H.Bramble, ed., Maryland Univ., London (1966).
7. Yu.A. Kuznetsov and S.F. Morozov, Integro-differential set of reactor kinetic equations, *Diff. Eqs.* 10: 8 (1974).
8. U.M. Sultangazin. *Spherical Harmonics and Discrete Ordinate Methods in Problems of Kinetic Transport Theory*, Nauka, Alma-Ata (1979).
9. V.M. Novikov and S.B. Shikhov. *The Theory of Parametric Influence on the Neutron Transport*, Energoatomizdat, Moscow (1982).
10. W. Greenberg, C. Van der Mee, and V. Protopopescu. *Boundary Value Problems in Abstract Kinetic Theory*, Birkhauser Verlag, Basel (1987).
11. V.P. Shutyaev. *Some regularity properties of the solution of the time-dependent transport-problem in a slab*, Preprint No.81, Dept. Numer. Math., USSR Academy of Science, Moscow (1985).
12. C. Bardos, Problemes aux limites pour les equations aux derivees partielles du premier ordre a coefficients reels; theoremes d'approximation; application a lequation de transport, *Ann. Scient. Ec. Norm. Sup.*, 4: 3 (1970).
13. V.I. Agoshkov, On existence of traces of functions in spaces used in transport theory, *Soviet Doklady*. 288:2 (1986).
14. M. Cessenat, Theoremes de trace L^p pour les espaces de fonctions de la neutroniques, *Note C.R.Acad. Sci. Paris*, 299: 16 (1984).
15. S. Ukai, Solutions of the Boltzmann equations, in: *Paterns and Waves-Qualitative Analysis of Nonlinear Differential Equations.* Stud. Math. Appl, North-Holland, Amsterdam (1986).
16. V.I. Agoshkov, *Necessary and sufficient conditions for sovability of some first-order hyperbolic problems*, Preprint No.248, Dept. Numer. Math., USSR Academy of Science, Moscow (1990).
17. S. Mischler, *Equation de Vlasov avec Regularite Sobolev du Champ: Theoremes de Trace and Applications*. Preprint No.13, Universite de Versailles, Versailles (1997).
18. G.I. Marchuk and V.I. Lebedev, *Numerical Methods in the Theory of Neutron Transport*, Harwood Academic Publisher, New York (1986).

SOLITARY WAVES IN TWO-COMPONENT RESONANTLY ABSORBING MEDIA[*]

Yurii V. Kistenev[1], and Alexander V. Shapovalov[2]

[1]The Institute of Atmospheric Optics
SB of the Russian Academy of Sciences
634055 Tomsk, Russia

[2]Tomsk State University
634050 Tomsk, Russia

INTRODUCTION

Loss-free pulse propagation through a resonantly absorbing medium is one of the exiting phenomena of nonlinear optics. Solitons and solitary waves (SWs) are two kinds of optical pulses with the above property. Particle-like behavior is inherent to solitons which keep the shape of wave stable after collisions. This is similar to the elastic colliding of particles.

By solitons in the strict sense is meant a special class of solutions of non-linear field propagation equations which are integrated via the Inverse Scattering Transform (IST) method [1,2]. Fairly narrow set of non-linear equations can be solved by the IST theory, including the equations describing the phenomenon of self-induced transparency (SIT), the nonlinear Schroedinger equation (NLSE) and some others. Solitary waves (or quasi-solitons, soliton-like solutions, etc.) appear in more wide class of non-linear optical models which are not integrated in the frame of the IST. They are localized (in a sense) solutions of the field equations and can change their characteristics under the pulse propagation, conserving the localization.

Inelastic properties of solitary wave coupling appear in nearly integrable models differing from integrable ones by small terms due to action of the last ones. Peculiarities of solitons in such models are studied in the frame of the soliton perturbation theory [3,4,5]. Systems being far from integrable ones and having soliton-like solutions can show deep inelasticity of SW collisions. In such models an annihilation of SWs is observed.

Although solitary waves are not identical to solitons, their properties can serve as a methodology to study solitary waves.

---

[*] The work is supported by RFFR grant 98-02-16195.

Optical solitons are caused by various physical mechanisms. In non-resonant optical media SWs appear due to competition between dispersion and non-linear self-compression of the pulse, propagation of which is described in terms of the NLSE. Another mechanism creates SWs in a resonantly absorbing medium with dispersion. Here, a localized pulse arises as a result of coherent energy exchange between the pulse and the medium. The well-known 2-π-SIT-solitons are an example of such solitons.

Soliton properties of non-linear systems are manifested, among other things, in the form of spontaneous soliton formation (SSF) phenomenon which implies that an initial pulse of a non-soliton form transforms into the soliton-like pulse (or pulses) in the course of the pulse propagation process.

The SSF was studied in numerous works, in particular, for the NLSE in Refs.[6,7] and for one-component resonant media in Refs.[8,9]. In Ref.[10] strong wave turbulence is studied in the frame of 3D-NLSE and it is shown that system states decay into a soliton and weak non-linear waves in the course of the system evolution under special conditions. In Refs.[6,7] theoretical estimations are given for a number of NLS-solitons which are formed in the transition process of initial pulses of non-soliton form. The obtained estimations agree with results of numerical simulations. In Ref.[8] the SSF is studied numerically for a one-component resonant two-level medium of large optical density. The analysis is based on the correspondent Maxwell-Bloch system of equations.

The SSF was considered in resonantly absorbing molecular media in Ref.[9]. Here it was assumed that the structure of rotational sublevels corresponds to the model of a rigid rotator with resonant transitions $J \to J \pm 1$. It was shown that the SSF in molecular media is qualitatively similar to analogus process in two-component media. The peculiar feature of pulse interaction with a molecular medium is in the fact that in some cases the difference is observed in speeds of propagation of SWs.

Let us recall some properties of the SSF depending on the pulse duration, shape, energy and the pulse square (S). If the duration of an initial pulse (τ_p) exceeds the phase memory time (T_2) of the medium, the SSF is accompanied by a decreasing of the pulse duration until it does not become much smaller than T_2. In resonant media the inequality

$$\mu = \frac{T_2}{\tau_p} \gg 1 \qquad (1)$$

is the necessary condition for the solitons existence caused by completely coherent interaction. The decomposition of an initial pulse into a set of soliton-like pulses also takes place when the square of an initial pulse exceeds certain threshold.

There are known only a few essential results concerning the soliton regime of pulse propagation for a media with more than one component. For example, the existence was shown [11,12] of the SIT-NLSE-solitons in a resonantly absorbing medium with an additional Kerr non-linearity. The soliton-like solutions for a two-component resonantly absorbing-amplifying medium was found [13] for the zero detunings and with $d_2/d_1 = 2$.

The aim of the present work is to show that the solitary wave solutions can exist in two-component resonantly absorbing media with arbitrary detunings and dipole momenta.

We use the model of a two-level medium with arbitrary dipole momenta and detunings for every of medium components. The correspondent propagation problem is described in the frame of standard Maxwell-Bloch equations [14] modified to fit the two-component medium with both components coupling by means of the field:

$$\frac{\partial V}{\partial \tau} = -i\left(P_1 + \frac{N_2 d_2}{N_1 d_1} P_2\right), \quad (2a)$$

$$\frac{\partial P_j}{\partial \eta} = -\gamma_j P_j + i\frac{d_j}{d_1} w_j V, \quad (2b)$$

$$\frac{\partial w_j}{\partial \eta} = -\frac{d_j}{d_1} Im(P_j V^*) - \frac{w_j + 1}{Q}. \quad (2c)$$

Here, $j = 1,2$, $V(\tau,\eta) = 2T_2 d_1 E(\tau,\eta)/\hbar$, T_2 is a phase memory time of the medium, d_j is a dipole momentum of the resonant transition for the j-th medium component with its density N_j, $E(\tau,\eta)$ is a complex amplitude of the optical pulse, P_j is a polarization, w_j is a difference between populations of the levels of resonant transition, $\gamma_j = 1 - iT_2 \Delta_j$, Δj is a detuning from the resonance, $Q = T_1/T_2$, T_1 is a time of population relaxation, τ is an optical density of the first medium component ($\tau = 2\pi k N_1 d_1^2 \hbar^{-1} T_2 z$, $k = \omega/c$, ω is a pulse frequency), $\eta = (t - z/c)/T_2$, t and z are temporal and spatial coordinates, respectively.

QUALITATIVE ANALYSIS

Solitary waves in non-integrable models can be revealed numerically. Before doing numerical simulation, consider some theoretical arguments in favor of SWs existence for the model under consideration. This allow one both to specify conditions for SWs construction by numerical methods and to analyze results of numerical calculations.

The above necessary condition (1) for the soliton-like regime of pulse propagation in an absorbing medium requires $T_2 \gg \tau_p$. Therefore, we can restrict ourselves for the qualitative analysis by a special case of Eqs.(2) taking into account this inequality. The values P_j, w_j alter essentially on the temporal interval equal to the pulse duration τ_p that offers the following estimations of the derivatives in Eqs. (2b), (2c): $\partial P_j/\partial \eta \sim P_j T_2/\tau_p$, $\partial w_j/\partial \eta \sim w_j T_2/\tau_p$. If $\Delta_j = 0$, the conditions $\mu \gg 1$ and $T_1 \sim T_2$ allow one to simplify Eqs.(2b), (2c) by dropping the relaxation term $\gamma_j P_j$ in (2b) and the last term in (2c) with. Then, we can reduce Eqs.(2) in a standard way (see, for example, Ref. 14) to the form

$$\frac{\partial^2 S}{\partial \tau \partial \eta} = sin(S) + \frac{N_2 d_2}{N_1 d_1} sin\left(\frac{d_2}{d_1} S\right). \quad (3)$$

Here,

$$S(\tau,\eta) = \int_{-\infty}^{\eta} V(\tau,q) dq.$$

Eq.(3) becomes the well-known sine-Gordon equation (SG) when the second term in the right-hand side of (3) is omitted and does the double-SG (DSG) equation [2] if $d_2/d_1 = 2$.

The SG equation is known to have soliton solutions in terms of the IST theory. The DSG equation, being non-integrable by the IST method, nevertheless has solitary wave solutions.

To analyze existence conditions of the solitary waves for Eq.(3), let us use the techniques developed in the classical field theory for similar purposes. The Eq. (3) from this standpoint is the special case of the generalized Klein-Gordon equation (KGE)

$$\frac{\partial^2 S}{\partial x^2} - \frac{\partial^2 S}{\partial y^2} = -\frac{\partial U(S)}{\partial S} \qquad (4)$$

Eq.(4) follows from (3) when $x = \tau+\eta, y=\tau-\eta$, $-\partial U(S)/\partial S = \sin S + a \sin(\beta S)$, $a = \frac{N_2 d_2}{N_1 d_1}$, $\beta = \frac{d_2}{d_1}$. Here, x plays the role of the evolution variable. The Eq.(4) can be considered as a classical Hamilton field theory[15] with the Hamiltonian density

$$H = \frac{1}{2}\left(\frac{\partial S}{\partial x}\right)^2 + \frac{1}{2}\left(\frac{\partial S}{\partial y}\right)^2 + U(S)$$

and the field S energy

$$W = \int_{-\infty}^{\infty} H dy .$$

The function $U(S)$ plays the role of potential energy density and it is a non-negative function vanishing for definite values of the variable S.

Let the global minima of $U(S)$ are placed at the points S_i, $i = 1,2,...$. Then the energy functional W reaches the minimum for the constant field S taking one of the values S_i which annulling the potential function $U(S)$. A solitary wave must have a finite energy and localized energy density[15]. In this case S has to approach to one of the values S_i at $x \to \pm\infty$. So, we come to the following evident necessary condition of soliton existence:

Notice 1. For the solitary-like solutions to exist for Eq.(3), the energy potential $U(S)$ have to possess several global minima.

For Eq.(3) we have

$$U(S) = c + \cos S + \frac{a}{\beta}\cos(\beta S) \qquad (5)$$

where the constant c is specified from the condition $U(S) \geq 0$. The Fig. 1 depicts the dependence of $U(S)$ on S for $N_1 = N_2$ and various values of d_2/d_1. The presented curves show the existence of global minima of $U(S_i) = 0$ with $S_i > 0$. According to our computation of the function $U(S)$, this takes place when the ratio d_2/d_1 is a rational.

Thus, to construct SWs numerically in the considered model we have to take d_2/d_1 to be a rational and do put the square S of the initial pulse to be equal to one of the values S_i.

Since the potential function $U(S)$ possess infinitely many degenerative vacua, a topological number[15] can be introduced to characterize the SWs of the Eq.(3). These solitary waves belong to the class of topological SWs in the field theoretic terms[15]. The topological number accounts for the stability of a solitary wave with respect to a changing of its topological structure which is determined by asymptotic values of the field S at the space infinities, where S must approach to one of the vacuum values S_i.

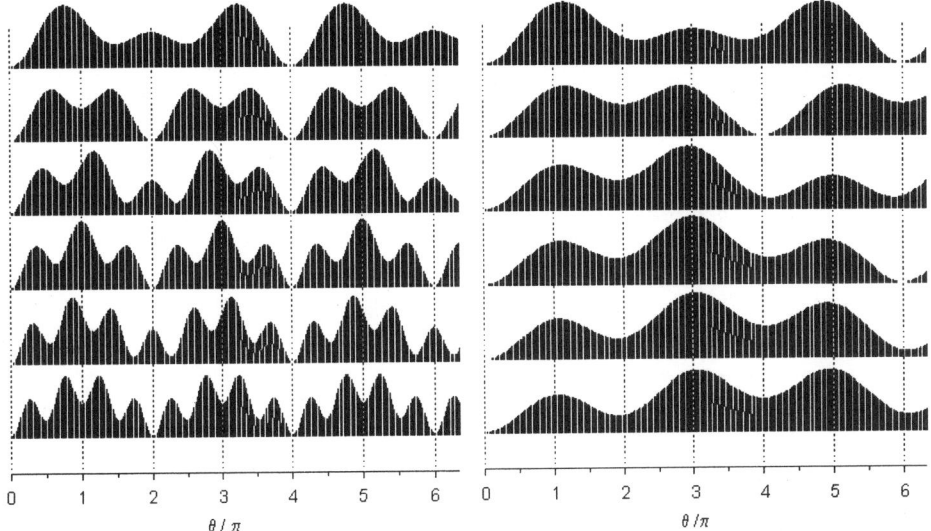

Figure 1a. The dependence of $U(S)$ on S for $N_1 = N_2$. The ratio of d_2/d_1 equals to 1.5, 2, 2.5, 3, 3.5, 4 (from top curve to bottom).

Figure 1b. The dependence of $U(S)$ on S for $N_1 = N_2$. The ratio d_2/d_1 equals to 1/1.5, 1/2, 1/2.5, 1/3, 1/3.5, 1/4 (from top curve to bottom)

The topological number is conserved in the course of SW evolution and admits one to classify both solitary waves and their collisions.

Apart from the points S_i of the true vacua, the potential function $U(S)$ have an infinite set of the points $S_i^{(f)}$ of the 'pseudo'-vacuum. By definition $S_i^{(f)}$ are the vacuum points with non-minimal energy. In other words, $S_i^{(f)}$ are the points of local minima of $U(S)$ with $U(S_i^{(f)}) \neq 0$. The pulses with initial square equal to $S_i^{(f)}$ are not absolutely stable and decay during the propagation.

In this way, optical pulse propagation in two-component resonance media would be expected to possess a rich set of solitary waves and their interactions.

NUMERICAL METHOD

Let us note some basic points of the numerical simulation method used for the solution of the model system of equations (2).

The Bloch equations (2b), (2c) can be represented in the form

$$\mu \frac{\partial X(q)}{\partial q} = A(q)X(q) + R \quad , \tag{6}$$

where $q = \mu \eta$, $XT = (P_j, P_j^*, w_j)$; A is a matrix of coefficients for X, R is a relaxation matrix including relaxation terms in (2). This system is singularly perturbed for $\mu < 1$. The presence of a boundary layer for $0 < q < \mu$ considerably lowers the stability of standard difference methods when the full system of equations (2) is solved. This requires either the use of a non-uniform adaptive grid for the time variable or the use of numerical algorithms for solving Eq. (6) which possess the property of uniform convergence16. Here we have employed the following algorithm. The system (6) is approximated not by a system of algebraic equations, as is customary, but does by a system of ordinary differential equations with constant coefficients

$$\mu \frac{\partial X(q)}{\partial q} = A_k X(q) + R \quad , \tag{7}$$

where $A_k = A(V_k)$, V_k is an approximation of the pulse shape in the k-th node of the grid $q=k \Delta q$; Δq is the grid pitch. For example, for a step-function approximation $V_k = V(k \Delta q)$. System (7) is solved using the Laplace transformation. The inverse Laplace transform is found using the theory of residues. The poles of the inverse matrix A_{-1} are found numerically or analytically if the equation det $A = 0$ degenerates into a quadratic equation. The latter takes place if $Q=1$ or $\Delta_j = 0$. Discrete values of X are found as follows: $X_k = X(q = k \Delta q)$. A numerical solution of Eq. (2a) is found by a second order predictor-corrector method.

RESULTS OF NUMERICAL SIMULATION

Here, we present the results of numerical simulation of Eqs.(2). The initial pulse shape is taken in the form

$$V(0,\eta) = \begin{cases} V_0 \sin^p(\pi \mu \eta), & \mu \eta \in [0,1]; \\ 0, & \mu \eta \notin [0,1]; \end{cases}$$

which is varied from a quasi-rectangular (with $p \ll 1$) to a quasi-gaussian (with $p \gg 1$). The following results correspond to the case of $p = 4$ since the SSF is perturbed by the coherent transition processes (nutation and free induction decay). Therefore, the SSF occurs only for the pulses with a smooth envelope and in the absence of perturbations.

Our computations show that variations of detunings do not effect essentially to the process of the SSF if the pulse duration is comparable with T_2.

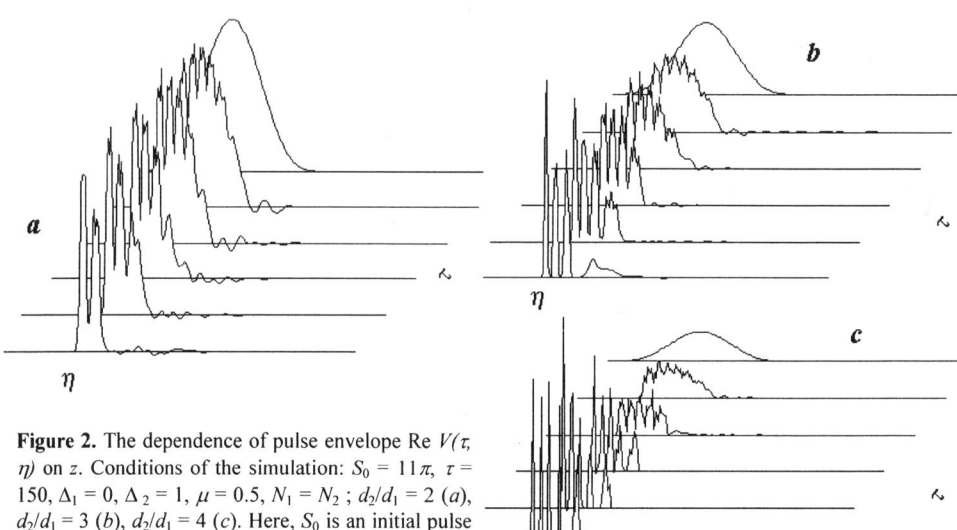

Figure 2. The dependence of pulse envelope Re $V(\tau, \eta)$ on z. Conditions of the simulation: $S_0 = 11\pi$, $\tau = 150$, $\Delta_1 = 0$, $\Delta_2 = 1$, $\mu = 0.5$, $N_1 = N_2$; $d_2/d_1 = 2$ (a), $d_2/d_1 = 3$ (b), $d_2/d_1 = 4$ (c). Here, S_0 is an initial pulse square.

Our computations show that variations of detunings do not effect essentially to the process of the SSF if the pulse duration is comparable with T_2.

The Fig. 2 depicts the results of calculations for transformation of the optical pulse envelope $Re\ V(\tau, \eta)$ versus τ with $\mu = T_2/\tau_p = 1$. We can see that the formation of a soliton-like pulse (the SSF effect) is observed in two-component absorbing media. However, in contrast with one-component media, only the totality of pulses possesses by soliton properties. The separate pulse shape, square and energy are changing but, in spite of change of parameters of separate pulses, their total square satisfies to the above condition of soliton existence according to Notice 1 and the total energy conserves. Our computations show that variations of detunings do not effect essentially to the process of the SSF if the pulse duration is comparable with T_2.

The Fig. 3 depicts the pulse envelope $Re\ V(\tau, \eta)$ when $\mu = 10$. Here, the SSF courses much faster than in the above case due to the conditions of coherent interaction are satisfied from the very beginning.

The Fig. 3b illustrates an ability of the soliton decomposition in two- component media similarly to the case of one-component. As opposite to the latter case, every forming solitary wave consists of several pulses as was mentioned above.

The Figs. 2, 3 also show that 2π-soliton of the SIT can not be realized in two-component absorbing media as a single pulse soliton. The number of pulses in the soliton-like totality is shown by computation to equal to the ratio of d_2/d_1.

In addition, let us note that the results presented here correspond to the case of $d_2/d_1 > 1$. Evidently, the case of $d_2/d_1 < 1$ can be reduced to the previous one by a renormalization of the pulse square S.

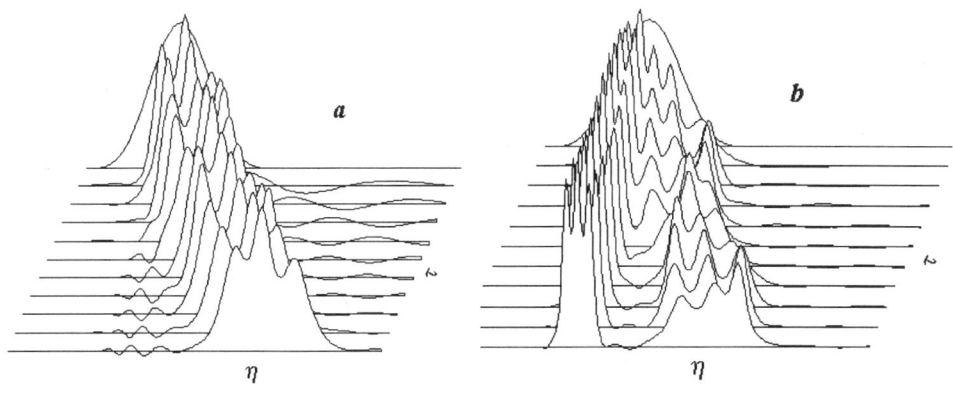

Figure 3. The dependence of pulse envelope $Re\ V(\tau, \eta)$ on z. Conditions of the simulation: $\tau = 20$, $\Delta_{1,2} = 0$, $\mu = 10$, $N_1 = N_2$, $d_2/d_1 = 3$; $S_0 = 2.2\pi\ (a)$, $S_0 = 3.9\pi\ (b)$.

CONCLUSION

In the work we have shown that the soliton-like pulses can exist in two-component resonantly absorbing media when the ratio of the dipole momenta of the medium components is a rational. The initial pulse must have smooth envelope and specified values of the pulse square S. The first stage of propagation of such a pulse is accompanied by the pulse transformation into a soliton-like formation consisting of a set of pulses. This pulse totality and not a separate one possesses by soliton properties. The number of pulses entering this collection is defined by the ratio of d_2/d_1.

The existence of soliton-like solutions for the model under consideration is verified indirectly by arguments originated from the field theoretical methods.

The set of soliton-like solutions of the considered model (2) includes additional SWs connected with the above 'pseudo-vacua' of the potential energy function $U(S)$. These 'pseudo'-SWs when propagated have a definite 'time of live' and transform into another SWs with less pulse square if they are exist or small amplitude waves ('radiation').

So, solitary waves in two-component resonantly absorbing media are rather various and have more reach behavior in comparison to single-component ones.

REFERENCES

1. V.E. Zakharov, S.V. Manakov, S.P. Novikov, and L.P. Pitaevskii. Solitons: Inverse Scattering Method, Nauka, Moscow, (1980).
2. R.K. Dodd, J.C. Eilbeck, J.D. Gibbon, and H.C. Morris. Solitons and Nonlinear Wave Equations, Academic Press, New York, (1982)
3. D.J. Kaup, A perturbation theory for inverse scattering transforms, *SIAM J.Appl. Math.*, 31: 121 (1976).
4. V.I. Karpman and E.M. Maslov, Perturbation theory for solitons, *Sov. Phys. JETP*, 46: 281 (1977).
5. Yu.S. Kivshar and B.A. Malomed, Dynamics of solitons in nearly integrable systems, *Rev. Mod. Phys.*, 61: 4 (1989).
6. V.A. Donchenko, M.V. Kabanov, E.V. Lugin, et al, On a formation of optical pulses in absorption line of weakly nonlinear medium, *Atmos. Oceanic Opt.*, 1: 67 (1988).
7. A.V. Shapovalov and S.N. Yurchenko, Application of complex VKB method in investigation of evolution of initial pulse according to NSE, *Izvestiya. Vysh. Uchebn. Zaved., Fizika*, 6: 19 (1992).
8. Yu.V. Kistenev, The processes of soliton formation in optically dense media, *Izvestiya. Vysh. Uchebn. Zaved., Fizika*, 10: 95 (1994).
9. Yu.V. Kistenev and I.A. Shevchuk, Spontaneous soliton formation in a region of resonant absorption of molecular media, *Atmos. Oceanic Opt.*, 8: 975 (1995).
10. V.E. Zakharov, A.N. Pushkarev, V.F. Shvetz, et al, On a soliton turbulence, *Zh.Exp.Teor.Fiz.*, 48: 79 (1988).
11. A.I. Maymistov and E.A. Manykin, To propagation of ultra-short optical pulses in resonant nonlinear waveguides, *Sov. Phys. JETF*, 85: 1177 (1983).
12. M. Nakazawa and H. Kubota, Coexistence of self-induced transparency soliton and Schroedinger soliton, *Phys. Rev. Let.*, 66: 2625 (1991).
13. A.V. Andreev and P.V. Polevoy, Dynamics of amplification and propagation of pulses in two-component media, *Sov.Phys. JETF*, 106: 1343 (1994).
14. M. Shubert and B. Vilgelmi, Introduction in Nonlinear Optics, Mir, Moscow (1979).
15. R. Rajaraman, Solitons and Instantons, North-Holland, Amsterdam (1982).
16. Yu.P. Boglaev, About numerical methods of solution of singularly perturbed tasks, *Sov. Differentialnye Uravnenya*, 21: 1804 (1985).

NON-LINEAR VECTOR MODEL OF THE INTERNAL DNA DYNAMICS

Ludmila V. Yakushevich

Institute of Cell Biophysics of the Russian Academy of Sciences,
142292 Pushchino, Moscow region, Russia

INTRODUCTION

Mathematical modeling of DNA structure and dynamics is an important and promising part of modern bioscience. Nonlinear mathematical models of DNA dynamics are of most interest because they permit us to imitate large amplitudes internal motions such as local unwinding of the double helix or conformation transitions which play an important role in the functioning of the molecule.

Studies of nonlinear mathematical models of DNA dynamics were started in 1980 when the article of five American authors: Englander, Kallenbach, Heeger, Krumhansl and Litwin was published[1]. The article was titled "Nature of the open state in long polynucleotide double helices: possibility of soliton excitations", and it was the first time when the nonlinear conformational excitations (or DNA solitons) imitating the local opening of base pairs has been introduced. In the article the first nonlinear hamiltonian of DNA has been presented and this gave a powerful impulse for theoretical investigations. A large group of authors, including Yomosa[2,3], Takeno and Homma[4,5], Krumhansl and co-authors[6,7], Fedyanin and co-authors[8,9,10], Yakushevich[11,12,13], Zhang[14], Prohofsky[15], Muto and co-authors[16,17,18], van Zandt[19], Peyrard[20,21], Dauxois[22], Gaeta[23,24], Salerno[25], Bogolubskaya and Bogolubsky[26], Hai[27], Gonzalez and Martin-Landrove[28] made contributions to the development of this field by improving the model hamiltonian proposed by Englander and co-authors or suggesting some new models, by investigating corresponding nonlinear differential equations and their soliton-like solutions, by consideration of DNA solitons and calculation of corresponding correlation functions. The results obtained by them formed a theoretical basis of the nonlinear DNA dynamics.

Experimental basis of the nonlinear DNA dynamics was formed by the results of experimental investigations on the DNA dynamics and interpretations some of them in the frameworks of the nonlinear concept. The most important results were obtained by Englander and co-authors[1] on hydrogen-tritium exchange in DNA, by Webb and Booth[29], Swicord and co-authors[31,32] on resonant microwave absorption (interpretations were made by Muto and co-authors[16] and by Zhang[33]), and by Baverstock and Cundall[34] on neutron scattering by DNA. All these results, however, admitted alternative interpretations (see the discussion in the review[35]), and only after publication of the work of Selvin and co-authors[36] where the torsional rigidity of positively and negatively supercoiled DNA was measured, the reliable experimental basis for theoretical predictions was given.

Besides theoretical results and experimental data an important contribution to formation of the nonlinear DNA science was made by numerous applications where the nonlinear concept was used to explain the dynamical mechanisms of DNA functioning such as the mechanisms of transitions between different DNA forms[37,38,39], of long-range effects[40,41,42], of regulation of transcription[43], of DNA denaturation[20], of protein synthesis (namely, insulin production)[44], of carcinogenesis[45].

This period of investigations of the nonlinear properties of DNA can be considered as a first step to the solution of the problem, which was characterized by many interesting simplified models, suggestions and hypothesizes. But now this period is finishing and a new more deep study of the subject both theoretically and experimentally is required to receive new and more reliable results.

It seems that one of the most promising directions in theoretical studies of the problem is that associated with the improvement of simplified models. This can be done by construction of the so-called combined models. In this approach, simple models which were proposed earlier by different authors to imitate only one of many types of internal motions in DNA, are used to construct new models describing two or more types of internal motions and interactions between them.

In this article we present the method of constructing combined models, and we apply it to construct a simple combined model describing two types of internal motions in DNA: torsional and transverse ones, and interactions between them. When constructing we used two basic models: the model of Peyrard and Bishop[20] and the model proposed in one of our works[12]. Both models were analyzed in details by Gaeta et al.[46]

BASIC MODELS

The first basic model has been proposed by Peyrard and Bishop[20] to imitate the process of DNA melting. In the model it is assumed that transverse motions make an important contribution to the process of formation of local melting regions in DNA. Especially these motions play an important role in local separation of the DNA strands. So, to model the process of separation, it is enough (in the first approximation) to restrict ourselves by consideration of only two degrees of freedom $u_n(t)$ and $v_n(t)$ which correspond to the displacements of the bases from their equilibrium positions along the direction of the hydrogen bonds that connect the two bases in a pair. Corresponding model hamiltonian has the form

$$H_1 = \sum_n \frac{m}{2}\left[\left(\frac{du_n}{dt}\right)^2 + \left(\frac{dv_n}{dt}\right)^2\right] + \sum_n \frac{K}{2}\left[(u_n - u_{n-1})^2 + (v_n - v_{n-1})^2\right] + \sum_n V(u_n - v_n) \tag{1}$$

with $V(u_n - v_n) = D\{\exp[-a(u_n - v_n)] - 1\}^2$. Here m is a common mass of bases, K is the coupling constant along each strand, and V is the potential for the hydrogen bonds, which is approximated by a Morse potential.

By using new variables $x_n = (u_n + v_n)^{1/2}/2$; $y = (u_n - v_n)^{1/2}/2$ hamiltonian (1) transforms to that consisting of two independent parts

$$H_1 = H(x) + H(y), \tag{2}$$

where

$$H(x) = \sum_n \frac{m}{2}\left(\frac{dx_n}{dt}\right)^2 + \sum_n \frac{K}{2}(x_n - x_{n-1})^2, \qquad (3)$$

$$H(y) = \sum_n \frac{m}{2}\left(\frac{dy_n}{dt}\right)^2 + \sum_n \frac{K}{2}(y_n - y_{n-1})^2 + \sum_n D\left[\exp(-ay_n)^2 - 1\right]^2 \qquad (4)$$

The problem with $H(x)$ is a usual harmonic (or linear) problem. The problem with hamiltonian $H(y)$ is an anharmonic (or nonlinear) problem. And it is necessary to note that just the variable y_n describes the process of separation.

The second basic model has been proposed in one of our works[12] to describe the process of formation of local open states in DNA. In this model it is assumed that the torsional motions of bases around the sugar phosphate chains make the main contribution to the process of local opening of the double helix. So, this model includes two degrees of freedom $\varphi_{n,1}$ and $\varphi_{n,2}$ which correspond to the angular displacements of the bases from their equilibrium positions. Corresponding model hamiltonian has the form

$$H_2 = \sum_n \left[\frac{mR_0^2}{2}\left(\varphi_{n,1}^2 + \varphi_{n,2}^2\right)\right] + \sum_n \left\{\frac{KR_0^2}{2}(\varphi_{n,1} - \varphi_{n-1,1})^2 + \frac{KR_0^2}{2}(\varphi_{n,2} - \varphi_{n-1,2})^2 + \right.$$
$$\left. + KR_0^2\left[2(1-\cos\varphi_{n,1}) + 2(1-\cos\varphi_{n,2}) - (1-\cos(\varphi_{n,1}+\varphi_{n,2}))\right]\right\}. \qquad (5)$$

Here R_0 is the length of pendula imitating DNA bases; k is the constant of rigidity of springs imitating interactions of bases in pairs.

Both these models are very simplified and each of them describes independently one type of the internal motions. In the reality, however, these two types of internal motions are not independent and they both make an important contribution to the process of local unwinding of the double helix. So, to improve mathematical description of the process it is necessary to combine these two models.

The problem which is now appeared is the following: how to construct the combined model? In other words, we should find the method of construction of the terms of the model hamiltonian, which describe interactions between these two types of internal motions.

In the next section we describe a simple method of construction of corresponding terms.

METHOD OF CONSTRUCTION OF COMBINED MODELS

Our efforts to find the method were stimulated by discussions with Barbi and Ruffo[47] who tried to improve the model of Peyrard and Bishop (1). Just these discussions gives us an idea to try to generalize the hamiltonian of Peyrard and Bishop. And only later we understood that this was the first step towards the construction of the combined model.

Generalization of the model of Peyrard and Bishop

To generalize the model let us rewrite model hamiltonian (1) in the vector form

$$H_1 = \sum_n \frac{m}{2}\left[\left(\frac{dU_{n,1}}{dt}\right)^2 + \left(\frac{dU_{n,2}}{dt}\right)^2\right]$$

$$+ \sum_n \frac{K}{2}\left[(U_{n,1} - U_{n-1,1})^2 + (U_{n,2} - U_{n-1,2})^2\right] + \sum_n V(|U_{n,1} - U_{n,2}|) \qquad (6)$$

with $V(|U_{n,1} - U_{n,2}|) = D\{\exp[-a|U_{n,1} - U_{n,2}|] - 1\}^2$. Here $U_{n,i}(t)$ is the displacement vector of the n-th nucleotide of the i-th chain. In the case of the model of Peyrard and Bishop vector $U_{n,i}(t)$ is determined by the following formulas

$$U_{n,1} = \{u_n; 0; 0\}; \qquad U_{n,2} = \{v_n; 0; 0\}. \qquad (7)$$

Inserting (7) into (6), we can easily obtain hamiltonian (2).
The other thing we should do now is to try to generalized the model hamiltonian (5) in the same way.

Generalization of the model (5)

To generalize hamiltonian (5) let us rewrite it in the vector form

$$H_2 = \sum_n \frac{m}{2}\left[\left(\frac{dU_{n,1}}{dt}\right)^2 + \left(\frac{dU_{n,2}}{dt}\right)^2\right]$$
$$+ \sum_n \frac{K}{2}\left[(U_{n,1} - U_{n-1,1})^2 + (U_{n,2} - U_{n-1,2})^2\right] + \sum_n \frac{k}{2}|U_{n,1} - U_{n,2}|^2, \qquad (8)$$

Here $U_{n,i}(t)$ is the displacement vector of the *n*-th nucleotide of the *i*-th chain and it is determined by the following formulas

$$U_{n,1} = \{R_0(1 - \cos\Theta_{n,1}); \; -R_0\sin\Theta_{n,1}; \; 0\};$$
$$U_{n,2} = \{-R_0(1 - \cos\Theta_{n,2}); \; R_0\sin\Theta_{n,2}; \; 0\}; \qquad (9)$$

where $\Theta_{n,1} \equiv \varphi_{n,1}$, $\Theta_{n,2} \equiv -\varphi_{n,2}$.

Discussion of the results of the generalizations

Let us note that hamiltonian (8) contains the term

$$\sum_n \frac{k}{2}|U_{n,1} - U_{n,2}|^2, \qquad (10)$$

which can be considered as some approximation of the potential

$$V(|U_{n,1} - U_{n,2}|) = D\{\exp[-a|U_{n,1} - U_{n,2}|] - 1\}^2. \qquad (11)$$

Indeed, let us expand (11)

$$V(|U_{n,1} - U_{n,2}|) = D[a|U_{n,1} - U_{n,2}|]^2 + ... = \frac{k}{2}|U_{n,1} - U_{n,2}|^2 \qquad (12)$$

If we suggest now that $k = 2Da^2$, we obtain that (10) is the first term of expansion (12).
So, we can conclude that expression (6) is more general than expression (8) or in other words we can conclude that (8) is an approximative version of (6).

We could make one more important conclusion: hamiltonian (6) (or its approximate variant (8)) is valid for modelling both internal DNA motions: transverse motions (when torsional motions are absent) and torsional motions (when transverse motions are absent). It remains only to write correctly vectors of displacements $U_{n,i}$ and insert them into the hamiltonian. In other words we can conclude that the form of model hamiltonian is invariant and the vectors $U_{n,i}$ are values which changes from one simple basic model to another.

APPLICATION OF THE METHOD

A remarkable consequence of the result obtained in the previous section is that it gives us a possibility to construct easily a combined model. Let us illustrate now how this method can be applied to construct the model which describes two types of internal motions: transverse motions of bases imitated by the model of Peyrard and Bishop[20], and torsional motions imitated by the model proposed in one of our models[12].

Assuming that the form of model hamiltonian is invariant and that only the vectors $U_{n,i}$ are values which changes from one model to another we concentrate our efforts on the writing of the formulas for vectors of displacements. In the case of combined model the vectors are

$$U_{n,1} = \{u_n + R_0(1 - \cos\Theta_{n,1}); \ -R_0\sin\Theta_{n,1}; \ 0\};$$
$$U_{n,2} = \{v_n - R_0(1 - \cos\Theta_{n,2}); \ R_0\sin\Theta_{n,2}; \ 0\};$$
(13)

To receive the model hamiltonian and the terms describing interactions between transverse and torsional motions it is enough only to insert (13) into invariant form (6). The resulting hamiltonian will have then the form

$$H = H_1 + H_2 + H(interact.). \tag{14}$$

To calculate H, let us calculate step by step the following values

$$U_{n,1} = \{u_n + R_0(1 - \cos\Theta_{n,1}); \ -R_0\sin\Theta_{n,1}; \ 0\}; \tag{15}$$

$$\frac{dU_{n,1}}{dt} = \left\{\left(\frac{du_n}{dt}\right) + R_0\left(\frac{d\Theta_{n,1}}{dt}\right)\sin\Theta_{n,1}; \ -R_0\left(\frac{d\Theta_{n,1}}{dt}\right)\cos\Theta_{n,1}; \ 0\right\}; \tag{16}$$

$$\left(\frac{dU_{n,1}}{dt}\right)^2 = \left(\frac{du_n}{dt}\right)^2 + 2R_0\left(\frac{du_n}{dt}\right)\left(\frac{d\Theta_{n,1}}{dt}\right)\sin\Theta_{n,1} + R_0^2\left(\frac{d\Theta_{n,1}}{dt}\right)^2\left(\sin^2\Theta_{n,1} + \cos^2\Theta_{n,1}\right) =$$

$$= \left(\frac{du_n}{dt}\right)^2 + \underline{2R_0\left(\frac{du_n}{dt}\right)\left(\frac{d\Theta_{n,1}}{dt}\right)\sin\Theta_{n,1}} + R_0^2\left(\frac{d\Theta_{n,1}}{dt}\right)^2 \tag{17}$$

Here and in further formulas underlined terms are those which contribute to $H(interact.)$ And let us continue calculations

$$U_{n,2} = \{v_n - R_0(1 - \cos\Theta_{n,2}); \ R_0\sin\Theta_{n,2}; \ 0\}; \tag{18}$$

$$\frac{dU_{n,2}}{dt} = \left\{ \left(\frac{dv_n}{dt}\right) - R_0\left(\frac{d\Theta_{n,2}}{dt}\right)\sin\Theta_{n,2}; \; R_0\left(\frac{d\Theta_{n,2}}{dt}\right)\cos\Theta_{n,2}; \; 0 \right\}; \quad (19)$$

$$\left(\frac{dU_{n,2}}{dt}\right)^2 = \left(\frac{dv_n}{dt}\right)^2 - 2R_0\left(\frac{dv_n}{dt}\right)\left(\frac{d\Theta_{n,2}}{dt}\right)\sin\Theta_{n,2} + R_0^2\left(\frac{d\Theta_{n,2}}{dt}\right)^2\left(\sin^2\Theta_{n,2} + \cos^2\Theta_{n,2}\right) =$$
$$= \left(\frac{dv_n}{dt}\right)^2 + 2R_0\left(\frac{dv_n}{dt}\right)\left(\frac{d\Theta_{n,2}}{dt}\right)\sin\Theta_{n,2} + R_0^2\left(\frac{d\Theta_{n,2}}{dt}\right)^2. \quad (20)$$

So, the kinetic part of the hamiltonian (14) is equal to

$$T = \sum_n \frac{m}{2}\left[\left(\frac{dU_{n,1}}{dt}\right)^2 + \left(\frac{dU_{n,2}}{dt}\right)^2\right] =$$
$$= \sum_n \left\{ \frac{m}{2}\left[\left(\frac{du_n}{dt}\right)^2 + \left(\frac{dv_n}{dt}\right)^2\right] + \frac{m}{2}\left[\left(\frac{d\Theta_{n,1}}{dt}\right)^2 + \left(\frac{d\Theta_{n,2}}{dt}\right)^2\right]\right\}. \quad (21)$$

It follows from (21) that there are not cross terms in the kinetic part of hamiltonian (14).
And now let us continue calculations to find interaction terms in the potential part of the hamiltonian (14)

$$|U_{n,1} - U_{n-1,1}|^2 = \{u_n - u_{n-1} + R_0(-\cos\Theta_{n,1} + \cos\Theta_{n-1,1}); \; R_0(-\sin\Theta_{n,1} + \sin\Theta_{n-1,1}); \; 0\}^2 =$$

$$= (u_n - u_{n-1})^2 + 2R_0(u_n - u_{n-1})(-\cos\Theta_{n,1} + \cos\Theta_{n-1,1}) +[$$

$$+ R_0^2\left\{\left(\cos^2\Theta_{n,1} + \cos^2\Theta_{n-1,1} - 2\cos\Theta_{n,1}\cos\Theta_{n-1,1}\right) +[\right.$$

$$+ \left(\sin^2\Theta_{n,1} + \sin^2\Theta_{n-1,1} - 2\sin\Theta_{n,1}\sin\Theta_{n-1,1}\right)\Big\} =[$$

$$= (u_n - u_{n-1})^2 + 2R_0(u_n - u_{n-1})(-\cos\Theta_{n,1} + \cos\Theta_{n-1,1}) + 2R_0^2\left[1 - \cos(\Theta_{n,1} - \Theta_{n-1,1})\right]. \quad (22)$$

$$|U_{n,2} - U_{n-1,2}|^2 = \{v_n - v_{n-1} + R_0(-\cos\Theta_{n,2} + \cos\Theta_{n-1,2}); \; R_0(-\sin\Theta_{n,2} + \sin\Theta_{n-1,2}); \; 0\}^2 =$$

$$= (v_n - v_{n-1})^2 + 2R_0(v_n - v_{n-1})(-\cos\Theta_{n,2} + \cos\Theta_{n-1,2}) +[$$

$$+ R_0^2\left\{\left(\cos^2\Theta_{n,2} + \cos^2\Theta_{n-1,2} - 2\cos\Theta_{n,2}\cos\Theta_{n-1,2}\right) +[\right.$$

$$+ \left(\sin^2\Theta_{n,2} + \sin^2\Theta_{n-1,2} - 2\sin\Theta_{n,2}\sin\Theta_{n-1,2}\right)\Big\} =[$$

$$= (v_n - v_{n-1})^2 + 2R_0(v_n - v_{n-1})(-\cos\Theta_{n,2} + \cos\Theta_{n-1,2}) + 2R_0^2\left[1 - \cos(\Theta_{n,2} - \Theta_{n-1,2})\right]. \quad (23)$$

$$|U_{n,1} - U_{n,2}|^2 = \{u_n - v_n + R_0(1 - \cos\Theta_{n,1} + 1 - \cos\Theta_{n,2}); \; R_0(-\sin\Theta_{n,1} + \sin\Theta_{n,2}); \; 0\}^2 =$$

$$= (u_n - v_n)^2 + 2R_0(u_n - v_n)(1 - \cos\Theta_{n,1} + 1 - \cos\Theta_{n,2}) +[$$

$$+ R_0^2\{(\cos^2\Theta_{n,1} + \cos^2\Theta_{n,2} + 4 - 4\cos\Theta_{n,1} - 4\cos\Theta_{n,2} + \cos\Theta_{n,1}\cos\Theta_{n,2}) +$$

$$+ (\sin^2\Theta_{n,1} + \sin^2\Theta_{n,2} + 2\sin\Theta_{n,1}\sin\Theta_{n,2})\} =$$

$$= (u_n - v_n)^2 + 2R_0(u_n - v_n)(2 - \cos\Theta_{n,1} - \cos\Theta_{n,2}) +$$

$$+ R_0^2[6 - 4\cos\Theta_{n,1} - 4\cos\Theta_{n,2} + 2\cos\Theta_{n,1}\cos\Theta_{n,2} + 2\sin\Theta_{n,1}\sin\Theta_{n,2}] =$$

$$= (u_n - v_n)^2 + 2R_0(u_n - v_n)(2 - \cos\Theta_{n,1} - \cos\Theta_{n,2}) +$$

$$+ 2R_0^2[3 - 2\cos\Theta_{n,1} - 2\cos\Theta_{n,2} + 2\cos(\Theta_{n,1} - \Theta_{n,2})] =$$

$$= (u_n - v_n)^2 + 2R_0(u_n - v_n)(2 - \cos\Theta_{n,1} - \cos\Theta_{n,2}) +$$

$$+ 2R_0^2[2(1 - \cos\Theta_{n,1}) + 2(1 - \cos\Theta_{n,2}) - (1 - 2\cos(\Theta_{n,1} - \Theta_{n,2}))]. \quad (24)$$

Taking into account formulas (6), (12) and (22)-(24), we find the potential part of the model hamiltonian in the form

$$V = \sum_n \left\{ \frac{K}{2}\left[(U_{n,1} - U_{n-1,1})^2 + (U_{n,2} - U_{n-1,2})^2\right] + V(|U_{n,1} - U_{n,2}|) \right\} =$$

$$= \sum_n \left\{ \frac{K}{2}\left[(u_{n,1} - u_{n-1,1})^2 + 2R_0(u_{n,1} - u_{n-1,1})(-\cos\Theta_{n,1} + \cos\Theta_{n-1,1}) + \right.\right.$$

$$+ 2R_0^2[1 - \cos(\Theta_{n,1} - \Theta_{n-1,1})]] +$$

$$+ \frac{K}{2}\left[(v_{n,1} - v_{n-1,1})^2 + 2R_0(v_{n,1} - v_{n-1,1})(\cos\Theta_{n,2} - \cos\Theta_{n-1,2}) + \right.$$

$$+ 2R_0^2[1 - \cos(\Theta_{n,2} - \Theta_{n-1,2})]] +$$

$$+ 2Da^2\left[(u_{n,1} - v_{n,1})^2 + 2R_0(u_{n,1} - v_{n,1})(2 - \cos\Theta_{n,1} - \cos\Theta_{n,2}) + \right.$$

$$+ 2R_0^2\{2(1 - \cos\Theta_{n,1}) + 2(1 - \cos\Theta_{n,2}) - [1 - \cos(\Theta_{n,2} - \Theta_{n-1,2})]\}\right\} \quad (25)$$

And the model hamiltonian (6) is equal now to

$$H = T + V =$$

$$= \sum_n \frac{m}{2}\left[\left(\frac{dU_{n,1}}{dt}\right)^2 + \left(\frac{dU_{n,2}}{dt}\right)^2\right] = \sum_n \frac{m}{2}\left\{\left(\frac{du_n}{dt}\right)^2 + \left(\frac{dv_n}{dt}\right)^2 + R_0^2\left[\left(\frac{d\Theta_{n,1}}{dt}\right)^2 + \left(\frac{d\Theta_{n,2}}{dt}\right)^2\right]\right\} +$$

$$+ \sum_n \left\{ \frac{K}{2} \left[(u_n - u_{n-1})^2 + 2R_0(u_n - u_{n-1})(-\cos\Theta_{n,1} + \cos\Theta_{n-1,1}) + 2R_0^2[1 - \cos(\Theta_{n,1} - \Theta_{n-1,1})] \right] + \right.$$

$$+ \frac{K}{2} \left[(v_n - v_{n-1})^2 + 2R_0(v_n - v_{n-1})(\cos\Theta_{n,2} - \cos\Theta_{n-1,2}) + 2R_0^2[1 - \cos(\Theta_{n,2} - \Theta_{n-1,2})] \right] +$$

$$+ \frac{k}{2} \left[(u_n - v_n)^2 + 2R_0(u_n - v_n)(2 - \cos\Theta_{n,1} - \cos\Theta_{n,2}) + 2R_0^2[1 - \cos(\Theta_{n,2} - \Theta_{n-1,2})] \right] +$$

$$\left. + 2R_0^2 \left[2(1 - \cos\Theta_{n,1}) + 2(1 - \cos\Theta_{n,2}) - (1 - \cos(\Theta_{n,2} - \Theta_{n-1,2})) \right] \right\} \tag{26}$$

or

$$H = H_1 + H_2 + H(\text{interact.}); \tag{27}$$

where

$$H_1 = \sum_n \left\{ \frac{m}{2} \left[\left(\frac{du_n}{dt} \right)^2 + \left(\frac{dv_n}{dt} \right)^2 \right] + \frac{K}{2} \left[(u_n - u_{n-1})^2 + (v_n - v_{n-1})^2 \right] \right\} \tag{28}$$

$$H_2 = \sum_n \frac{m}{2} R_0^2 \left[\left(\frac{d\Theta_{n,1}}{dt} \right)^2 + \left(\frac{d\Theta_{n,2}}{dt} \right)^2 \right] +$$

$$+ \sum_n KR_0^2 \left[1 - \cos(\Theta_{n,1} - \Theta_{n-1,1}) + 1 - \cos(\Theta_{n,2} - \Theta_{n-1,2}) \right] +$$

$$+ \sum_n kR_0^2 \left[2(1 - \cos\Theta_{n,1}) + 2(1 - \cos\Theta_{n,2}) - (1 - \cos(\Theta_{n,1} - \Theta_{n,2})) \right] \tag{29}$$

$$H(\text{interact}) = \sum_n \left\{ K \left[R_0(u_n - u_{n-1})(-\cos\Theta_{n,1} + \cos\Theta_{n-1,1}) + \right. \right.$$

$$\left. \left. + R_0(v_n - v_{n-1})(\cos\Theta_{n,2} - \cos\Theta_{n-1,2}) \right] + kR_0(u_n - v_n)(2 - \cos\Theta_{n,1} - \cos\Theta_{n,2}) \right\}. \tag{30}$$

Thus we are success in finding the term *H(interact)* describing the interaction between the torsional and transverse motions in DNA.

The work was supported by the Russian Foundation for Basic Research (Grant No 97-04-48417).

REFERENCES

1. S.W. Englander, N.R. Kallenbach, A.J. Heeger, J.A. Krumhansl and A. Litwin, Nature of the open state in long polynucleotide double helices: possibility of soliton excitations, *Proc. Natl. Acad. Sci. USA* 77: 7222 (1980).
2. S. Yomosa, Soliton excitations in deoxyribonucleic acid (DNA) double helices, *Phys. Rev.*, A-27: 2120 (1983).
3. S. Yomosa, Solitary excitations in deoxyribonucleic acid (DNA) double helices, *Phys. Rev.*, A-30: 474 (1984).
4. S. Takeno and S. Homma, Topological solitons and modulated structure of bases in DNA double helices, *Prog. Theor. Phys.*, 70: 308 (1983).

5. S. Homma and S. Takeno, A coupled base-rotator model for structure and dynamics of DNA. *Prog. Theor. Phys.*, 72: 679 (1984).
6. J.A. Krumhansl and D.M. Alexander, Nonlinear dynamics and conformational excitations in biomolecular materials, in: *Structure and Dynamics: Nucleic Acids and Proteins*, E. Clementi and R.H. Sarma, eds., Adenine Press, New York, p. 61 (1983).
7. J.A. Krumhansl, G.M. Wysin, D.M. Alexander, A. Garcia, P.S. Lomdahl and S.P. Layne, Further theoretical studies of nonlinear conformational motions in double-helix DNA, in: *Structure and Motion: Membranes, Nucleic Acids and Proteins*, E. Clementi, G. Corongiu, M.H. Sarma and R.H. Sarma, eds., Adenine Press, New York, p. 407 (1985).
8. V.K. Fedyanin and L.V. Yakushevich, Scattering of neutrons and light by DNA solitons, *Stud. biophys.*, 103: 171 (1984).
9. V.K. Fedyanin, I. Gochev and V. Lisy, Nonlinear dynamics of bases in continual model of DNA double helices, *Stud. biophys.*, 116: 59 (1986).
10. V.K. Fedyanin and V. Lisy, Soliton conformational excitations in DNA, *Stud. biophys.*, 116: 65 (1986).
11. L.V. Yakushevich, The effects of damping, external fields and inhomogeneity on the nonlinear dynamics of biopolymers, *Stud. biophys.*, 121: 201 (1987).
12. L.V. Yakushevich, Nonlinear DNA dynamics: a new model, *Phys. Lett.*, A-136: 413 (1989).
13. L.V. Yakushevich, Investigation of a system of nonlinear equations simulating DNA torsional dynamics, *Stud. biophys.*, 140: 163 (1991).
14. Ch.-T. Zhang, Soliton excitations in deoxyribonucleic acid (DNA) double helices. *Phys. Rev.*, A-35: 886 (1987).
15. E.W. Prohofsky, Solitons hiding in DNA and their possible significance in RNA transcription, *Phys. Rev.*, A-38: 1538 (1988).
16. V. Muto, J. Holding, P.L. Christiansen and A.C. Scott, Solitons in DNA, *J. Biomol. Struct. Dyn.*, 5: 873 (1988).
17. V. Muto, A.S. Scott and P.L. Christiansen, Thermally generated solitons in a Toda lattice model of DNA. *Phys. Lett.*, A-136: 33 (1989).
18. V. Muto, P.S. Lomdahl and P.L. Christiansen, Two-dimensional discrete model for DNA dynamics: longitudinal wave propagation and denaturation, *Phys. Rev.*, A-42: 7452 (1990).
19. L.L. Van Zandt, DNA soliton realistic parameters, *Phys. Rev.*, A-40: 6134 (1989).
20. M. Peyrard and A.R. Bishop, Statistical mechanics of a nonlinear model for DNA denaturation. *Phys. Rev. Lett.*, 62: 2755 (1989).
21. T. Dauxois, M. Peyrard and C.R. Willis, Localized breather-like solutions in a discrete Klein-Gordon model and application to DNA, *Phys.*, D-57: 267 (1992).
22. T. Dauxois, Dynamics of breathers modes in a nonlinear helicoidal model of DNA, *Phys. Lett.*, A-159: 390 (1991).
23. G. Gaeta, On a model of DNA torsion dynamics, *Phys. Lett.*, A-143: 227 (1990).
24. G. Gaeta, Solitons in planar and helicoidal Yakushevich model of DNA dynamics, *Phys. Lett.*, A-168: 383 (1992).
25. M. Salerno, Discrete model for DNA-promoter dynamics, *Phys. Rev.*, A-44: 5292 (1991).
26. A.A. Bogolubskaya and I.L. Bogolubsky, Two-component localized solutions in a nonlinear DNA model, *Phys. Lett.*, A-192: 239 (1994).
27. W. Hai, Kink couples in deoxyribonucleic acid (DNA) double helices, *Phys. Lett.*, A-186: 309 (1994).
28. J.A. Gonzalez and M. Martin-Landrove, Solitons in a nonlinear DNA model, *Phys. Lett.*, A-191: 409 (1994).

29. S.J. Webb and A.D. Booth, Absorption of microwave by microorganisms, *Nature*, 222: 1199 (1969).
30. M.L. Swicord and C.C. Davis, Microwave absorption of DNA between 8 and 12 GHz, *Biopolymers*, 21: 2453 (1982).
31. M.L. Swicord and C.C. Davis, An optical method of investigating the microwave absorption characteristics of DNA and other biomolecules in solution, *Bioelectromagnetics*, 4: 21 (1983).
32. G.S. Edwards, C.C. Davis, J.D. Saffer and M.L. Swicord, Resonant absorption of selected DNA molecules, *Phys. Rev. Lett.*, 53: 1284 (1984).
33. Ch.T. Zhang, Harmonic and subharmonic resonances of microwave absorption in DNA, *Phys. Rev.*, A-40: 2148 (1989).
34. K.F. Baverstock and R.D. Cundal, Are solitons responsible for energy transfer in oriented DNA? *Int. J. Radiat. Biol.*, 55: 152 (1989).
35. L.V. Yakushevich, Nonlinear dynamics of biopolymers: theoretical models, experimental data, *Quart. Rev. Biophys.*, 26: 201 (1993).
36. P.R. Selvin, D.N. Cook, N.G. Pon, W.R. Bauer, M.P. Klein and J.E. Hearst, Torsional rigidity of positively and negatively supercoiled DNA, *Science*, 255: 82 (1992).
37. A. Khan, D. Bhaumic and B. Dutta-Roy, The possible role of solitonic process during A to B conformational changes in DNA, *Bull. Math. Biol.*, 47: 783 (1985).
38. Z. Zhang and W. Olson, A model of the B-Z transition of DNA involving solitary excitations, in: *Proceedings, 6th Annual Conference on Nonlinearity of Condensing Matter, Los Alamos, New Mexico, 5-9 May 1986*, A.R. Bishop, D.K. Campbell, P. Kumar and S.E. Trullinger, eds., Springer, Berlin, p. 265 (1987).
39. H.M. Sobell, Kink-antikink bound states in DNA structure, in: *Biological Macromolecules and Assemblies*, F.A. Jurnak and A. McPherson, eds., John Wiley and Sons, New York, p. 172 (1984).
40. L.V. Yakushevich, DNA dynamics, *Mol. Biol., (Russian J.)* 23: 652 (1989).
41. S.N. Volkov, Conformational transition. Dynamics and mechanism of long-range effects in DNA, *J. Theor. Biol.*, 143: 485 (1990).
42. L.V. Yakushevich, Non-linear DNA dynamics and problems of gene regulation, *Nanobiology*, 1: 343 (1992).
43. R.V. Polozov and L.V. Yakushevich, Nonlinear waves in DNA and regulation of transcription, *J. Theor. Biol.*, 130: 423 (1988).
44. E. Balanovski and P. Beaconsfield, Solitonlike excitations in biological systems, *Phys. Rev.*, 32: 3059 (1985).
45. J.J. Ladik, S. Suhai and M. Seel, Electronic structure of biopolymers and possible mechanisms of chemical carcinogenesis, *Int. J. Quant. Chem. QBS Suppl.*, 5: 35 (1978).
46. G. Gaeta, C. Reiss, M. Peyrard and T. Dauxois, Simple models of nonlinear DNA dynamics, *Rev. Nuovo Cimento*, 17: 1 (1994).
47. M. Barbi, S. Cocco, M. Peyrard and S. Ruffo, A twist opening model for DNA, *Report at the First International Conference Nonlinear Phenomena in Biology (22-27 June, 1998, Pushchino, Russia)*.

THE DYNAMICS OF THE ECONOMIC SOCIETY STRUCTURE

Dmitrii S. Chernavskii[1], Olga D. Chernavskaya[1], Andrei V. Scherbakov[2],
Boris A. Suslakov[3], Nikolai I. Starkov[1]

[1]P.N. Lebedev Physical Institute Russian Academie of Scienties,
117934 Moscow, Russia

[2]"Kurs" LTD,
105005 Moscow, Russia

[3]Socio-Technological Institute,
109380 Moscow, Russia

INTRODUCTION

Recently – within so called post-soviet period – countries of East Europe have met with a set of considerable problems. Attempts to solve these problems by methods of practiced in developed market economy have not lead to desirable results. One of the causes – and, from our point of view, the most important one – is the following: reforms are proceeded without taking into account peculiarities of the economic structures of the societies.

The Economic Structure of the Society (referred bellow on as ESS) is treated here as the distribution $\rho(x)$ of the society elements (i.e. household) by their money accumulation x. Few examples of such distribution are shown at the Fig. 1, where the ordinate axis corresponds to the number of households with money accumulations from x to $x +\Delta\ x$; the abscissa represents the volume of money accumulation x in some conventional units. Two kinds of distributions are mostly realistic ones: the unimodal one (containing single only hump and, in this sense, being closed to the so called Gaussian, (or Normal one) and the bimodal one (containing two humps); they are illustrated at the Fig. 1 by the curves (a) and (b) respectively.

Unimodal ESS is inherent to modern developed countries. Here the main part of society can be aggregated into the so called middle class, with only a few of the bulk being concentrated in the tails of distribution (the left tail corresponds to the marginal group, and the right one represents the elitarian group). The members of that middle class occupy a middle position in the status hierarchy, dispose over certain wealth, have relatively high education and cultural levels, high skills, enough professional experience, and enjoy a rather stable standing in the society. The middle class includes a part of technical personnel, middle level managers, small business owners, high-level research workers, skilled blue-collars and officers.

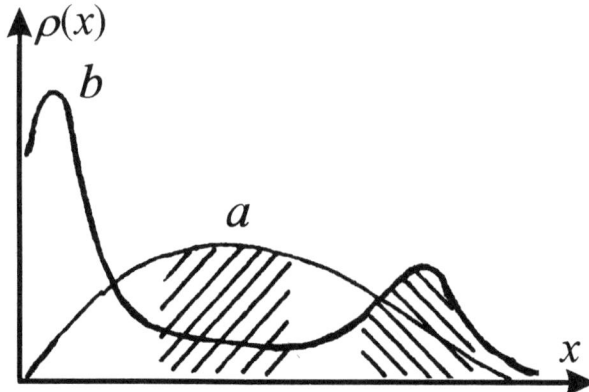

Figure 1. The ρ(x) distributions examples: (a) – unimodal society; (b) – the bymodal one. Determining free market prices distribution intervals are marked.

The bimodal distribution is common mainly to developing countries, where middle layers are practically absent and the main part of society is concentrated at the left tail of distribution. Below we'll consider the bimodal society structure in more details, using example of the ESS during so called stagnation period in USSR.

ESS is important due to the following reasons.

Firstly, ESS modality influences considerably on the behavior and mental peculiarities of the society elements – such as work incentives and business activity, social climate and so on. Commonly the work incentives in unimodal society is higher than that in bimodal ones.

Second, character of the response of society members to any policy impacts (such as law and regulating acts etc.) depends hardly on the ESS. Let's clarify the last statement on specific, but rather essential examples which are a important for the transformation from planed to market economy.

The processes of the pricing in unimodal and bimodal society result in essentially different outcomes. The liberalization of the prices in unimodal society gives rise to development of economy and production. In bimodal society the same measure results in the turning down solvent demand and production. The largest part of society is excluded both from consumption and from production. The currency issue also can result in different outcomes depending on a structure of society. Below we shall discuss these problems more in detail.

Nevertheless, the investigation of ESS and its role is not popular in modern economy. There are several reasons. In the developed countries theoretic economy is based on the assumption of social unimodality, what is really the fact and doesn't require for any discussion. The analysis of bimodal distribution is not of interest here; the dynamics of ESS becomes to be actual problem within reformation periods (such as modern situation in Germany after the joining). It have been discussed in literature[1], but at rather verbal level.

Soviet economic science didn't consider the ESS due to ideological reasons: it have been believed that the poor as well as the reach people shouldn't exist in socialistic countries, and all the economic calculations were proceeded per capita, i.e. referred to some mysterious soul of the population. Besides, it is important to stress that under so called command-administrative conditions any response of society have been controlled by the state apparatus, thus the role of ESS was not too important. Under the absence of hard state policy, within social instability conditions, the role of ESS becomes to be the most essential factor.

There exist two methods of the ESS analysis.
- The panel-technique with subsequent statistical treatment of the results. The results are represented in Lorentz's diagram and Jiny's index[2].
- Mathematical modeling using free external parameters, which are defined partly by the interrogatory, partly by the expert estimations.

The first method is well known but not trustworthy one, because people often try to hide the information on their wealth. Moreover, the current interrogatory doesn't give any information on the ESS dynamics, so in the unstable society the interrogatory data loss its sense.

The second method has been treated in biophysics, ecology and sinergetics. In the economy and sociology this method have been used not so long ago[3–6]. This method enables not only to estimate the current distribution of interest, but also to analyze the dynamics of distribution, i.e. to predict the tendencies of society evolution.

Note, that the attempts (and rather successful ones) to analyze social problems by methods of natural sciences (such as theory of complex developing systems, sinergetics and so on) have been performed already[3,4]. It concerned the problem of public opinion formation, problem of prediction of definite ware popularity, etc.

However, the problem of ESS investigation is methodically different[5,6,7]: it permits to describe the society members behavior in some self-consistent field, created by the hole society. In other words, here the economic situation in the society is considered not from the macroeconomy position (i.e. at the state level), but from the point of view of one family. Such formulation of the problem enables to use the mathematical apparatus of the self-conjugated field (with accidental factors being taken into account), which is well-worked in physics and leads to equations of Langeven and Fokker-Planck type[8]. Below it will be shown that, depending on the self-conjugated field parameters, the method in question allows to describe rather broad spectrum of possible ESS types and its dynamics.

The mathematical model of ESS have been suggested in our papers[5,6,7]. In the work presented we'll analyze the ESS model and demonstrate its abilities on a set of examples. The conclusions from the model seem to show the actuality of the treatment in question.

MATHEMATICAL MODEL OF ESS

The model is constructed in the following way.

The 1-st step – determination of the family economic behavior, i.e. its possible income sources and expenses. Let's formulate the main features of this behavior.

i) There are two income sources:
a) stable income per definite time, i.e. salary; below it will be signed as P.
b) the income connected with some business operations, where there is the possibility to lose all input money (i.e. income risk). We assume that this income is proportional to the input money; the difference between the input and the income is represented in the form Ax, where A – the "activity coefficient" (or: "effective coefficient") is the profit percent during time unity.

In order to get pure income it is necessary to extract taxis. In the simplest case the taxis function $T(x)$ can be accepted in the form: $T(x)=k(P+Ax)$, where k is constant taxis coefficient.

ii) There exist some random factors, which also can be divided into two groups:

a) factors undepending on the value of money accumulations (such as: sickness, possibility to get/loss good work, unexpected inheritance, etc.). It will be represented as a function of time t in the form $g\xi(t)$, where $\xi(t)$ is accidental function of the unit amplitude, g is the characteristic value of these factors (the "noise amplitude").

b) factors connected with the business risk and influenced on the business income only. It will be represented in the form of factor of the business income: $Ax[1 + a\xi(t)]$, where a reflects the risk measure; note that the whole expression can be positive as well as negative (if the business operation was not successful).

Thus, the total income of the family can be represented in the form:

$$P + g\xi(t) + Ax[1 + a\xi(t)] - k(P + Ax) \qquad (1)$$

iii) There are two kinds of expenses:
a) everyday necessary expenses: it depends on the value of money, and it is limited by natural reasons (even for reach families). We represent such expenses in the form:

$$C\frac{x}{x + x_0}, \qquad (2)$$

where C is expenses fully satisfying everyday needs, x_0 is money accumulation of family which can spend everyday only the half of the money sufficient for full satisfaction. Below these very units will be used as basic ones for income/input and money mass respectively; the unit of time will be month.

b) elite expense (automobile, house, etc) – it has threshold character, i.e. those wares are accessible for that families whose money mass x is more than some value x_1. The desired function is:

$$B\frac{(x - x_1)\cdot\theta(x - x_1)}{(x - x_1) + (x - x_2)}, \qquad \theta(x - x_1) = \begin{cases} 1 \to if \cdot x \geq x_1 \\ 0 \to if \cdot x_1 \triangleleft x_1 \end{cases}, \qquad (3)$$

where $\theta(x)$ is threshold function, x_1 is the minimal price of elite ware, x_2 is its "middle" price, B corresponds to maximal elite expenses (analogous to "C" value in everyday expenses). Later on these value will be measured in its basic units.

Note that there is possible once more channel of expenses: large infestations into production sphere. Here, however, we wouldn't take it into account because, due to some specific reasons, in Russia (as well as in USSR before) this channel doesn't work, i.e. rich people doesn't input their money into production sphere, preferring to export money into western banks or foreign currency.

Let's stress that the analytic form of the income and expenses x dependence doesn't play essential role; it is important to choose correctly only asymptotic behavior (controlled by the coefficients C, B) and threshold character (x_1) and angles of the curves (determined here by x_0, x_2).

Then the balance equation for the single family money mass x has the form of Langeven type:

$$\frac{dx}{dt} = P + Ax(1 + a\xi(t)) - C\frac{x}{x + x_0} - k(P + Ax) - B\frac{(x - x_1)\cdot\theta(x - x_1)}{(x - x_1) + (x - x_2)} + g\xi(t); \qquad (4)$$

Note that some parameters of the equation (4) such as k, C, B, x_2 are fixed by the laws or traditions of the society. The value of salary P is different for different families: there are groups of families with low salary, middle salary and high salary. The same is true for the business activity coefficient: there are people, who are afraid of risk they put their money into reliable banks and are satisfied with low income percent; there are people ready for venture to get higher profit. So the coefficient A depends partly on the mentality. Thus, it is possible to choose a set of groups of families so that within each group all the parameters being fixed. Below such groups will be called the layers.

The 2-nd step: analysis of the distribution $\rho_{i,j}(x)$ within the given layer – it obeys the following Fokker-Planck equation:

$$\frac{\partial \rho_{i,j}(x,t)}{\partial t} = \frac{\partial}{\partial x}\left[\frac{\partial U_{i,j}(x)}{\partial x} \cdot \rho_{i,j}(x,t)\right] + \frac{1}{2}\frac{\partial}{\partial x}\left[G_{i,j}^2 \cdot \frac{\partial}{\partial x}\rho_{i,j}(x,t)\right], \quad (5)$$

where the functions $U_{i,j}(x)$ and $G_{i,j}(x)$ are determined by the right side of the equation (4):

$$U_{i,j}(x) = -\int_0^x \left[P_{i,j} + A_{i,j}x - C\frac{x}{x+x_0} - B\frac{(x-x_1)\cdot\theta(x-x_1)}{(x-x_1)+(x-x_2)} - k(P_{i,j} + A_{i,j}x)\right]dx; \quad (6)$$

$$G_{i,j}(x) = A_{i,j}x + g_{i,j}.$$

The $U_{i,j}(x)$ function is usually called "the potential" although it is not connected with any energy; $G_{i,j}(x)$ is the function, which play the role of temperature.

The 3-rd step: In order to get $\rho(x)$ for the hole society it is necessary to calculate $\rho_{i,j}(x)$ for each layer and then summarize the results, taking into account its statistical weights in this society:

$$\rho(x,t) = \sum_{i,j} v_{i,j} \cdot \rho_{i,j}(x,t); \quad (7)$$

The weight coefficients $v_{i,j}$ are defined by the known statistical data on the distribution of families by their salary and business activity; here $v_{i,j}$ play the role of parameters.

There are two additional conditions the whole ESS have to satisfy to:

$$\int_0^\infty \rho(x,t)dx = N, \qquad \int_0^\infty x\rho(x,t)dx = M, \quad (8)$$

where N is full number of families, M – full money mass in the society.

The combined model, i.e. equations (4), (5), (8) and conditions (8) allows to receive the following results.

1. It is possible to investigate the stationary ESS. The stationary solution of the Fokker-Planck equation (2) is:

$$\rho_{i,j}(x) = v_{i,j} \cdot N \cdot \exp\left[\frac{2U_{i,j}(x)}{G_{i,j}^2}\right], \quad (9)$$

where N is determined by the normalized condition (8).

It is clear from the eq. (9) that all maxima of the $\rho_{i,j}(x)$ function, its number and positions are determined by the potential $U_{i,j}(x)$ form. One can say that $U_{i,j}(x)$ appears to be the function of the social niches. The conception of social niche is used in economy and sociology on rather verbal level; here it has the same meaning, but is defined more correctly. Besides, within this model it is possible to define another concept, also used before only verbally: social barrier; here it means the $U_{i,j}(x)$ maximum separating two neighboring potential niches.

Thus, all properties of the stationary ESS are determined by the $U(x)$. $G(x)$ function defines the width of the $\rho(x)$ distribution; in some sense it plays the role of temperature.

2. It is possible to analyze the stability of the ESS.
The society is stable if each layer occupy its own social niche. The instability occurs if some social niche disappears (and/or new niche appears).

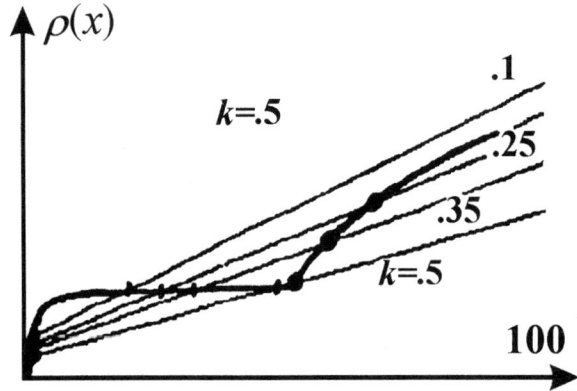

Figure 2. Bifurcation diagram: bold line corresponds to the expenses, thin line – the income without taxis; k is the taxis coefficient.

For qualitative analysis it is sufficient to use simple graphic technique: at the Fig. 2 there is presented the combination of the expense function (bold line) and the pure income function (after taxis excluded) under different values of the taxis parameter k (all others are fixed). The points of intersection of these function correspond to the extreme points of the potential: maxima (dotes) and minima (crosses). After parameters change these points may coincide (and then disappear), or new couple the intersections may occur. These situations are known in literature as bifurcations, corresponding set of parameters is called bifurcating, or critical ones.

Thus the diagram at Fig. 2 allows to predict the social niche function $U(x)$ evolution with parameters changing and the degree of the parameters vicinity to their critical values.

3. In the case of nonstable and nonstationary ESS this model permits to describe the exchanging society dynamics. Differently speaking, one can say how rapidly and in what direction the society would change.

Let's point out one important feature of the model. It contains two kind of processes: rapid one (with the characteristic time approximately equal to month) and slow one (with characteristic time by the order of years). The rapid processes start just after any social niche disappears, with the group occupied this niche being pull away without potential barriers restrictions. Usually it is the very process which is realized in crisis situations.

Slow processes are commonly connected with the population of new social niche, separated from the rest of society by sufficiently high barrier.

Below we'll demonstrate abilities of the model at several examples.

SEVERAL EXAMPLES OF THE ESS

Stationary $\rho(x)$ for the "stagnation period" in USSR

The following set of parameters have been used: $x_1=50x_0$, $x_2=110x_0$, $B=4$; basic units were accepted to be equal to: $x_0=400$ rubles, $C=400$ rubles per month.

Table 1. The ESS of former USSR.

$A \setminus P$	0.5	2
0.003 (3% p.y.)	Low-wage, unectiv peasants, serviser, workers $v_{1,1}=0.7$	High-salary, unactive servisers, research workers, artists, officials $v_{1,2}=0.09$
0.03 (30% p.y.)	Low-wage but active commerce, supply workers $v_{2,1}=0.2$	High-salary, active goverment officials $v_{2,2}=0.01$

There were extracted four layers with different values of distributed parameters (presented at the Table 1 with corresponding statistical weights and social belongings). So, the values $P=0.5$ and $P=2$ correspond to low salary and high salary people respectively. The value $A=3\cdot 10^{-3}$ (or 3% per year) corresponds to low business activity (*t* was common income percent in Soviet banks); the other case $A=3\cdot 10^{-2}$ (or 30% per year: the most income percent legalized in civilized countries) refers to high business activity.

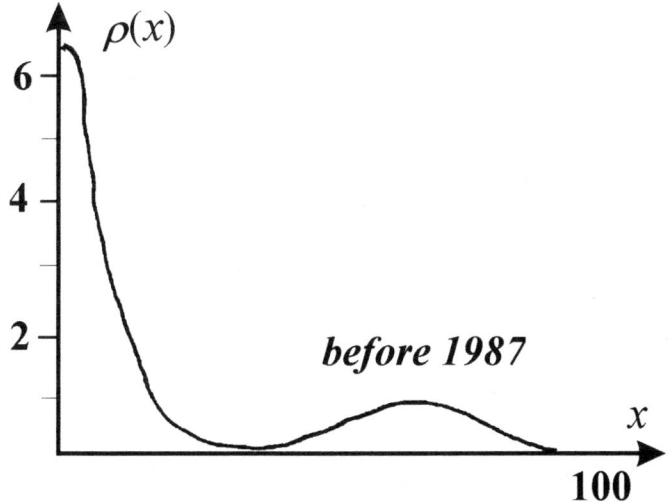

Figure 3. The ESS for USSR within the stagnation period.

Stationary ESS structure is presented at Fig. 3 (solid line). Here $\rho(x)$ has bimodal form, with the maxima being separated considerably: $D=(x_{m,2}-x_{m,1})/x_{m,1}=70$. These layers differ by their social composition and tasks: the "reach" group consists of commerce workers, scientific and art people, officers. Below we'll call this group "active layer" because they were rather poor than reach by compare with international standard. Note, that the "middle class" (with money mass between $x_{m,1}$ and $x_{m,2}$) here is practically absent.

Thus, the ESS in USSR during the "stagnation period" was bimodal but stable one.

The ESS for developed countries

It differs from that in USSR because of parameters diversity. First, forth majority of working people their income exceeds sufficiently their everyday expenses; besides, the percent of active people in the society here is higher than that in USSR.

Secondly, the interval of elite ware prices is broader: the minimal price x_1 (measured in corresponding x_0 units) here is less than that in the previous example while middle price x_2 is higher. As a result, for active people group their income and expenses curves approximately coincides. It means that the potential barrier separated the "poor" and the "reaches" is low so that, roughly speaking, there is only one broad niche for active layer, where exact position of each family (i.e. its real wealth) may fluctuate with time. The same is true for the working people group, analogous to the high salary group in USSR.

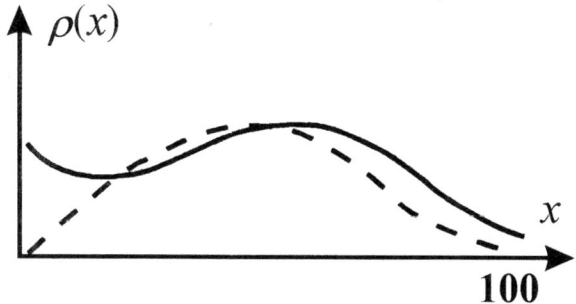

Figure 4. The ESS structure for developed countries. Solid line corresponds to the model, dashed – date from[7].

The stationary society structure is practically unimodal (see Fig. 4), i.e. there is middle class representing almost all social layers of society. This picture has to be treated as illustration, because in different developed countries those distributions are different. At the same Fig. 4 (dashed line) the ESS of Japan is shown (data have been extracted from[9]); this two curves are qualitatively agree with each other, the main discrepancy is at the small x region. We had discussed this picture only to demonstrate the ability of the model in question to describe, depending on the parameter sets, the unimodal society as well as bimodal one. The most important parameters are the prices of the wares (everyday and elitare ones) and the values of its productions; it is possible to estimate qualitatively the ESS character without any others parameters.

THE DYNAMICS OF ESS IN MODERN RUSSIA

Modern ESS in Russia differs essentially from that of the "stagnation period" due to abrupt change of all parameters.

The price liberalization almost immediately has leaded to the 70–100 times jump of the prices, with elite wares prices have been enlarged even more. Only the "rich" part of the people appears to be able to buy the industrial products; the solvent demand was reduced. Correspondingly, the value of production starts to fall down. This immediate result of the liberalization was obvious consequence of bimodal ESS which presented in the country before. This result has been predicted in 1991 y.[5] before the liberalization.

Further, the salary distribution has been altered. For the majority of high-salary people their real salary have decreased (to be less than everyday necessary expenses). As a result the part of high salary people have been reduced, with their niche have been removed to large x value. The taxis remained unprogressive but taxis coefficient "k" really becomes increased and appears to be inadequate to real ESS in Russia. Activity distribution has been changed (so that the averaged activity has decreased), with real taxis being increased. Thus, ESS in Russia, being bimodal, becomes unstable.

The privatization of natural monopolies has leaded to formation of tiny group of very rich people. According to our estimations[7,10–13] this group contains very small part of society – 0.2 % (about 10^5 families), but processes main part of money (including rubels as well as dollars) in Russia. This group having super-incomes, can not disburse the money in Russia and does not invest the money in russian industry (or safety, science, education and so on) because it is not profitable. As a result, the main part of this super-profit money is exported into western Eauropen and American banks, thus it appears to be lost for the country. The value of money autflux from Russia according to our estimations[8–11] (taking into account accumulation of money, incomes and quantity of this group) is about 20 billions dollars per year. The same values of autflux has been obtained some later other authrs[14,15]. This group has special title: "running tail" (it correspond to the place of this group in the distribution) or "compradores" (the term is adopted from Latin American history) or "new Russians" (popular term in modern Russia).

In the last 6 years the rate of inflation essentially advanced rate of currency issue. Last was supported on a low level from ideological ("monetarist") reasons. In an outcome the deficit of a money supply $M2$ was derivated which now is less necessary in several times[13,17]. This hinders mutual offset, results in non-payment and in the issue aggravates a polarization of society.

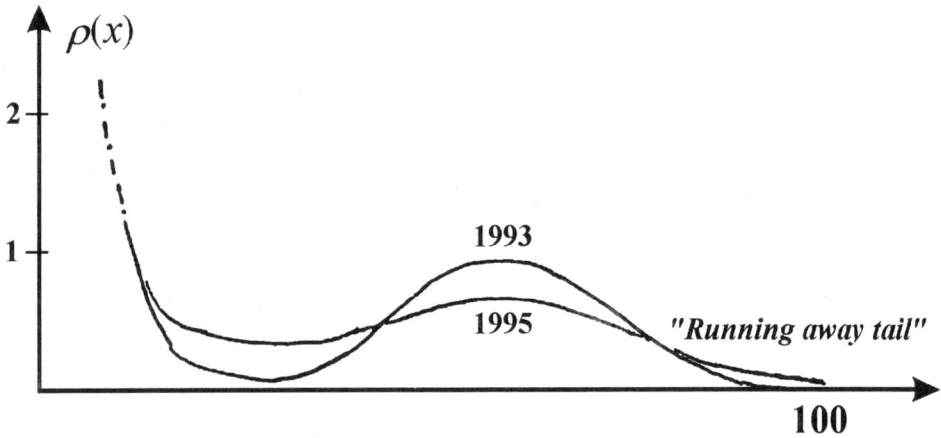

Figure 5. The dynamics of "rich" part of Russia.

The modern ESS in Russia contains seven groups which are shown in the Table 2 and at Fig. 5.

The social and functional composition of the groups:
1. The low – wages workers and employers.
1a The revagning members of the 2-th, 4-th and 5-th groups.
2. The midle – wages workers and employers serving group 5.

3. The hight – wages workers and employers serving group 6.
4. The petti buisnesmen.
5. The buisnesmen working on internal market of Russia.
6. "New russians" (compradores).

Table 2. The ESS of Russia in last years.

	1993		1995		1997	
Number of group	Number of families in group in %	Quantity of money in group in %	Number of families in group in %	Quantity of money in group in %	Number of families in group in %	Quantity of money in group in %
1	70	10	70	10	70	6
1a	–	–	10	3	11	2
2	20	27	12	7	12	4
3	4	5	4	6	3	4
4	3	8	2	3	2	2
5	3	8	2	3	2	2
6	0.2	42	0.2	68	0.2	80

It is important note that the 6-th group has not the social niche. On bifurcation diagram (see Fig. 2) this group corresponds to the line of incomes which does not intersect the expenses line and displaced above the last one. The 5-th group also has not social niche because taxis coefficient is too big. It corresponds to the incomes line which displaced below expenses one.

As a result, modern Russian ESS remains to be bimodal, but becomes unstable and nonstationar. The economic polarization of the society increases. Social composition of the rich group has been modified: now it consists of the people who accumulate money and don't take part in any producing process. Thus this group can't carry out social functions of the former active layer – trade, supply, safety, science, education, and so on.

Analogous society structure including six grour was derived later by Ayvasyan[18]. The difference is that he considers the profit distribution but we accumulation one.

THE ROLE ESS AT PRICING[16]

In this section we shall consider the mechanism of formation of market prices in case, when the main factor is the aim to obtaining a maximum of the profit. Let us show, that the aim of the maximum profit results in to opposite outcomes in societies of a different type (unimodal and bimodal). Accordingly role of state regulation also should be various in these cases.

We shell use such concepts as a function of consumption (demand), elasticity on the price and elasticity under the incomes (or accumulation). Let us remind briefly their sense and properties.

Function of consumption (or, that the same, a function of demand) $Q_i(x, p_1, p_2, ... p_n)$ represents quantity of the goods a type of i which is consumed by family per unit of time. It depends on the price of the goods p_i, incomes of family (more exact from quantity of means intended on consumption, which we shall designate x) and factor of the consumer cost k_i. Last reflects the contribution of the given goods a type of i in a set of the consumed goods, whose prices $(p_1, p_2, p_3 ... p_n)$ are generally various. Further for simplicity we shall consider a set of the goods as the unified universal goods of the total price p. Thus we shall not consider a problem of preference of one goods to other.

The dependence of a function of demand on accumulation x has the following properties:

i) The demand will increase with increase of accumulation.
ii) At small values x functions of consumption is small, since if money is absent, to purchase there is nothing.
iii) the function of demand is sated, that is when x increases, $Q(x, p)$ aims at a constant Q_m:

$$Q_m(x, p_m)_{x \to \infty} \to Q_m = Const. \tag{10}$$

The function of demand is a homogeneous function of a zero degree[3], that is the demand should not vary, if all incomes, accumulation and price are changed in an identical number of times (for example, at denomination). From here follows, that the function of demand should depend on the relation of accumulation to the price, that is from variable

$$t_m = \frac{x}{p_m}. \tag{11}$$

We shall consider two types of functions $Q(t)$.
The function of the first type is everywhere convex, that is its second derivative is negative at all values t:

$$\frac{\partial^2 Q}{\partial t^2} \leq 0 \tag{12}$$

at $0 < t < \infty$.

This type of dependence is characteristic of the essential commodities and the necessary of life services such as, food, municipal services, transport and so on. These goods the people are enforced to acquire under any conditions. If the price of them grows, the demand decreases, but does not drop up to zero.

The function of consumption of the second type has a cusp, where the second derivative is equal to zero. (The function of a threshold type is an example of one). In a limiting case it can be represented by a step-function:

$$Q(t) = \frac{Q_m \, at \cdot t \geq 1}{0 \cdot at \cdot t \cdot < 1} = \theta(t-1) Q_m. \tag{13}$$

This type of dependence is characteristic of industrial goods (durable goods) and the goods for elite (luxury goods), which are acquired only if there is sufficient accumulation.

The profit from production and realization of the goods depends on the price, function of demand and distribution of families on accumulation (that is from ESS) $\rho(x)$. The latter determines volume of solvent demand at the given price. The profit Π is equal to a difference of the gross income R_T and total costs C_T

$$\Pi = R_T - C_T. \tag{14}$$

The gross income depends on the price p:
$R_T = pQ_T$,

where Q_T is full volume produced (and realized) product. It is equal:

$$Q_T = \int_0^\infty Q(x, \kappa p) \cdot \rho(x) dx. \qquad (15)$$

The total costs C_T consist of constants (C_F) and variable (C_V) of costs: $C_T = C_F + C_V$. First include costs of maintenance of production in working order. The variable costs include the salary, costs of raw (supply), components and their transport. As a first approximation they are proportional to quantity of made production Q_T: $C_V = sQ_T$ where s – variable specific costs. The full specific costs s_T are equal: $s_T = C_T/Q_T = s + C_F/Q_T$. At constants values s and C_F the size s_T monotonically drops when Q_T increases and aims at an asymptotic limit $s_T = s$. A sense of a factor s is clear from here. With allowance for said the profit Π is equal:

$$\Pi = pQ_T - sQ_T - C_F. \qquad (16)$$

Let's separately consider formation of the prices for functions of consumption of the first and second type.

1. Formation of prices on essential commodities and necessaries of life service (function $Q(t)$ is first type).

The maximum of the profit is reached at the price p_{opt}, which satisfies to a condition:

$$\frac{d\Pi}{dp} = \int_0^\infty Q(x, p)\rho(x)dx + p\int_0^\infty \frac{dQ}{dp}\rho(x)dx - s\int_0^\infty \frac{dQ}{dp}\rho(x)dx = 0. \qquad (17)$$

The analysis of expression (17) is adduced in appendix 1. It is shown, that profit, as the function of the price has not an extremum and monotonically rises when last increases. This statement is true at any structure of society; the details of relation are various only. In unimodal society the profit fast will increase before as the price will reach value $p = x_m$, where x_m – accumulation appropriate to a maximum of distribution $\rho(x)$, then growth is decelerated and the profit also aims at an approximate limit.

In bimodal society the picture is qualitative same, but the slowdown of growth of the profit happens near to value of the price $p = x_{m,2}$ where $x_{m,2}$ – accumulation appropriate to the right hump of distribution, then the growth continues but already slower.

In both cases the maximum profit is reached at an extreme stiff price. It means, that it is expedient to make not enough production, but to sell it on a very stiff price.

Certainly, such strategy is not acceptable to society as a whole at anyone of structure.

On the first sight this conclusion seems paradoxical, but at detailed study it become clear, that in all countries, including developed, the measures of state regulation having not market character are used. In absence of such measures the price of essential commodities and the necessaries of life services are constantly increased, that leads to inflation of price.

The particular measures depend on character of the goods (or services) and on a structure of society. So, the services in organization of communication and transport in majority of developed countries are implemented by the state and the prices on them are fixed. In power and production of foods depending on a situation are used or economic measures (grant, preferential and differentiated taxation), or administrative (limitation of the prices, antitrust laws and others). It is obvious to use only antitrust laws (with the purpose of artificial creation of a competition) is insufficiently.

2. Formation of the prices of manufactured durable goods (function of demand of the second type, which has a cusp). For simplicity we shall use a threshold function (13). Profit,

as the function of the price is given by expression (16), which with allowance for (13) and (15) can be represented in the form:

$$\Pi(p) = (p-s)Q_m \int_p^\infty \rho(x)dx - C_F. \quad (18)$$

The condition of a maximum of the profit has a kind:

$$\frac{d\Pi}{dp} = Q_m \left[\int \rho(x)dx - p\rho(p) + s\rho(p) \right] = 0. \quad (19)$$

From (19) follows, that the price appropriate to a maximum of the profit p_{opt} can be fixed up (and be stabilized) without interference of the state. However, this price essentially depends on a structure of society. The situation in unimodal and bimodal society differ qualitatively.

i) The unimodal company is characterized in two parameters: by a position of a maximum of distribution x_m and dispersion (width) σ. In case, when $\sigma < x_m$, the price p_{opt}, is close to value $p_{opt} \cong x_m$, that is the large part of society is solvent. With allowance for of spread of the prices and qualities of the goods such situation suits all society and the additional state regulation is not required.

In case $\sigma \gg x_m$ the distribution $\rho(x)$ practically is a monotonically dropping function. This case we shall consider separately.

ii) The bimodal society is characterized by four parameters: two positions of a maximum $x_{m,1}$ and $x_{m,2}$ and two dispersions σ_1 and σ_2. It essentially differs from unimodal if $x_{m,2} \gg x_{m,1}$, $\sigma_1 \cong x_{m,1}$ and $\sigma_2 < x_{m,2}$. This case we shall just consider. Profit, as the function of the price has two maxima. First is close to value $p_{opt,1} \cong x_{m,1}$ (or $p_{opt,1} \cong \sigma_1$) and second to value $p_{opt,2} \cong x_{m,2}$. However, in unified society there can not be two, hardly distinguished from each other prices for the same goods. Therefore that price is set in which ensures the greatest profit.

In case of sale on the first price the profit is equal:

$$\Pi_1 = (x_{m,1} - s) Q_m \int_{x_{m,1}}^\infty \rho(x)dx - C_F \approx (x_{m,1} - s) N - C_F. \quad (20)$$

By sale on the second price the profit is equal:

$$\Pi_2 = (x_{m,1} - s) Q_m \int_{x_{m,2}}^\infty \rho(x)dx - C_F \approx (x_{m,2} - s)\rho_2 - C_F. \quad (21)$$

Where ρ_2 – number of families entering into the second hump. In case, when:

$$\rho_2 \, x_{m,2} \gg N \, x_{m,1} \quad (22)$$

most expedient appears the second price, which is set in society. Thus the solvent market actuates only with the right of hump. Production is adapted for effective demand and up to a level satisfying only a rich part of society is reduced. Thus people which are included in a left-hand hump is excluded as from a consumption sphere so from production one. To avoid such situation the rigid state regulation is necessary.

iii) We shall consider case, when the distribution $\rho(x)$ has acclivous "tail" of a Paretto's type. It signifies, that in a broad interval x from x_{min} up to x_{max} $\rho(x)$ has a form:

$$\rho(x) = \rho \frac{1}{x^n}. \qquad (23)$$

The limitation $x < x_{max}$ means, that at very large accumulation $x \approx x_{max} \approx M$ the noneconomic factors begin to play a role. It results in limitation of accumulation. It is important to note, that x_{max} is much more of characteristic accumulation of a main part of society.

If $n > 1$, "tail" does not play an essential role, as integrals in (20) and (21) converges on a upper limit and x_{max} is not included in outcomes of calculations. In case $n \leq 1$ upper limit plays an essential role. In appendix 2 it is shown, that in this case the price p_{opt} is established at a level, accessible only for the richest part of society. It is true, if not too small number of families enter in considered part of distribution, that is under condition:

$$x_{m,1} \cdot N \leq \rho \cdot x_{max} / e, \qquad (24)$$

where $e = 2.72...$ – number of the Euler.

Thus, even at unimodal distribution (or slowly dropping) the availability of a slanting part of a type Paretto is equivalent to existence second hump hardly shifted to the right. In this situation the state price control also is necessary.

ROLES ESS AT MONEY ISSUE

It is accepted to think, that the currency issue always gives rise to an increase of price, that is to inflation. This is one of dogmas of monetary economy. The arguments usually are following: the price in the market is increased if the effective demand exceeds the supply and is lowered otherwise. After a currency issue, generally speaking, mass of effective demand is increased and if it becomes more than mass of the goods, the prices of goods are increased. Actually outcome of issue depends on a structure of society, a ratio of actual trade turnover and cash money supply and from to what layer of society the issue is addressed.

In an one-hump society the issue always falls in a middle layer. In these conditions the additional issue of money really gives rise to an increase in price. (that is in this case the monetarism dogma is justified).

In bimodal society a situation is other. Here the issue is not so necessary accompanied by inflation, even such issue is possible which will cause to decrease of the prices. Let us assume that the issue is addressed to the people who not by poorest and not by richest, and such, that after obtaining money will take a place between "humps". Then the fall between "humps" will be filled and the form of society structure will be nearer to unimodal. Besides the top of a "rich" hump will be displaced to the left and the prices for consumer goods should decrease.

This statement contradicts monetaruism dogma and it seems paradoxical; however, its mechanism is simple. Due to issue the effective demand is increased and it becomes expedient to produce and to sell more goods, though at the lower price. Let's underline, this effect is possible only in bimodal society, in particular in modern Russia. It is possible to name such issue "inflationless" (or even antiinflationary).

This effect is adduced on Fig. 6. The illustration is based on evaluations effected in the 1995 by the account ESS of Russia[13]. From Figure it follows, that at volume of such issue of the order M2 = 40 – 50 billions roubles ESS should be changed, the solvent demand should be increased (approximately in 1,5 – 2 times), accordingly the production must increase, and optimum prices should fall. However, for reaching it the fulfillment of two conditions is necessary:

1) Availability of the goods in a reserve and/or a capability of their fast production.

2) Prevention from transfer of the addressed money in a tail of distribution. Such overflow is possible, though is criminal. In this case issue will not give desirable outcomes.

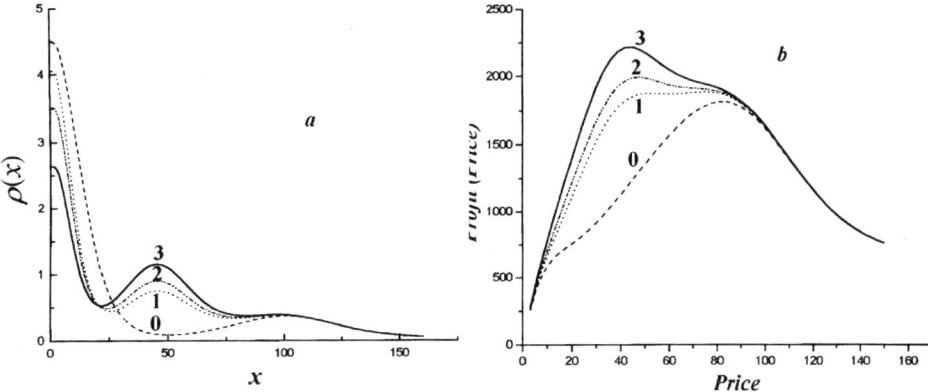

Figure 6. The accumulation dencity (a) and the dependence of profit from the value of price (b) at different money issue: 1 – 30 billion rubels, 2 – 40 billion rubels, 3 – 50 billion rubels, 0 – the money issue is absent.

Necessary for fulfilment of these conditions of a measure are discussed in[13] in more detail.

Unfortunately, in 1995 this measure was not realized. Per 1998 the situation was changed to worse but the issue (now of the order 100 – 150 billions of denominated roubles) is necessary as before.

CONCLUSION

It has been shown that, depending on the parameters, the model in question enables to describe broad spectrum of possible situations: unimodal as well as bimodal societies, the dynamics (i.e. tendencies of evolution) of unstable societies, etc. However, this model is not completely closed one. In fact, the model parameters depend on the macro-economic situation and, simultaneously, do influence on that. To describe this fit-back connection it is necessary to consider more complex model, taking in to account mutual influence of the ESS and the productivity within given society; now it is in progress.

The most actual problem for us – how it is possible to escape the growing crisis in Russia – is not discussed in the work presented. Within our model this problem can be reduced to the following one: what is necessary to achieve normal and stable ESS. The first aim it is necessery to create social niche for the active layer; the second one means creation of the uniform social niche for the whole society. Now only the first aim seems to be actual and realistic one.

APPENDIX 1

Let us show, that the derivative of a function of the profit on the price is positive at any form of distribution $\rho(x)$ if functions of demand everywhere is convex.

We begin from a more simple macroeconomic version where the condition of a maximum of the profit has the form (15). Let us show, that integrand in (15) everywhere positively.

The convexity condition means, that:

$$\frac{d^2Q(x,p)}{dx^2} < 0, \quad \frac{d^2Q(t)}{dt^2} = \frac{d^2Q}{dx^2} \cdot (p)^2 < 0, \quad \frac{dQ(t)}{dt} > 0 \tag{1.1}$$

at $0 < t < \infty$.

Let us consider first two members of a derivative of the profit on the price.

$$\left\{ \int_0^\infty Q(x,p)\rho(x)dx + p\int_0^\infty \frac{dQ}{dp}\rho(x)dx \right\} = \left\{ \cdot \int_0^\infty Q(t)\rho(pt)pdt + p\int_0^\infty \frac{dQ}{dt}\frac{dt}{dp}pdt \cdot \right\} =$$

$$= \int_0^\infty pQ(t)\rho(pt) \cdot \left(1 - \frac{dQ}{dt}\frac{x}{Qp}\right)dt = \cdot p \int_0^\infty Q(t)\rho(pt)(1 - E_t)dt, \tag{1.2}$$

where: $E_t = \frac{t}{Q}\frac{dQ}{dt}$ is elasticity of a function of demand on a variable t.

We research properties of a function of elasticity E_t. It reaches maximum value $E_{t,vax}$ at

$$\frac{dE_t}{dt} = \frac{d}{dt}\left(\frac{dQ}{dt} \cdot \frac{t}{Q}\right) = \frac{d^2Q}{dt^2}\frac{t}{Q} + \frac{dQ}{dt}\frac{1}{Q}(1 - E_t) = 0. \tag{1.3}$$

From here:

$$E_{t,max} = 1 + \frac{d^2Q}{dt^2} \cdot t \left(\frac{dQ}{dt}\right)^{-1}. \tag{1.4}$$

From (1.4) follows, that at observance (1.1) the values $E_{t,vax}$ are less than unit and, therefore, $(1 - E_t) > 0$ in all an interval of an integration. The remaining multiplicands of integrand in expression (1.2) are positive by definition and, therefore, the integrand is positively on the whole.

The full derivative of the profit Π on the price p (at s is constant) is equal:

$$\frac{d\pi}{dt} = \int_0^\infty Q(x,p)\rho(x)dx + \int_0^\infty p\frac{dQ}{dp}\rho(x)dx - s\int_0^\infty \frac{dQ}{dp}\rho(x)dx =$$

$$= \int_0^\infty Q(t)\rho(pt)p \cdot dt + \int_0^\infty \frac{dQ}{dt} \cdot \frac{dt}{dx}\rho(pt)p \cdot dt - s\int_0^\infty \frac{dQ}{dt} \cdot \frac{dt}{dp}\rho(pt)p \cdot dt = \tag{1.5}$$

$$= \int_0^\infty pQ(t)[1 - E_t]\rho(pt) \cdot dt + s\int_0^\infty \frac{dQ}{dt}\rho(pt)t \cdot dt$$

The first member of expression (1.5), as shown is higher, is always positive. At observance of conditions (1.1) the second member is proportional costs s and therefore is positive too. Thus the derivative of the profit on the price is always positive and maximum of the profit is not accessible, as was to be proved.

APPENDIX 2

Let us consider a limiting case of distribution Паретто, when a parameter n = 1 and in an interval $x_{min} < x < x_{max}$ a function $\rho(x)$ has a form:

$$N_t = \bar{\rho} \int \frac{dx}{x} = \rho \ln \frac{x_{max}}{x_{min}}. \tag{2.1}$$

For a determinacy we shall suppose, that in the region of $x < x_{min}$ $\rho(x)$ has one maximum or monotonically drops, but is steeper, than in (4.1). Parameter $\bar{\rho}$ reflects number of families N_t, located in the field of $x_{min} < x < x_{max}$ so, that:

$$N_t = \bar{\rho} \int \frac{dx}{x} = \bar{\rho} \ln \frac{x_{max}}{\cdot x_{min}}, \tag{2.2}$$

The condition of a maximum of the profit (13) in this case has a form:

$$\ln \frac{x_{max}}{p} - 1 = 0; \quad \text{or} \quad p_{opt} = x_{max}/2.72; \tag{2.3}$$

In this case the profit is equal

$$\Pi = (1-\lambda)Q_m x_{max} \frac{\bar{\rho}}{2.72}, \tag{2.4}$$

Price p_{opt} in equation (3.3) is the most expedient, if number of families N_t is rather great, that is under condition of:

$$x_{m,1} \cdot N < x_{max} \cdot N_t \frac{1}{2.72 \cdot \ln x_{max}/x_{min}}, \tag{2.5}$$

where $x_{m,1}$ is characteristic accumulation of a main part of society.

In case $N_t \ll N$ the condition (4.5) is not observed, whereat the "tail" of a type Paretto does not influence on pricing.

REFERENCES.

1. R. Kravchik, *The degradation and thrive of Polish economy*, M, 1991 (in Russian)
2. C.P. McKonnell, S.L. Brue S.L., *Economics*, McGrow-Hill Publishing Company, 1990.
3. *The Economy as an Evolving Complex System*, Univ. Santa-Fe, Ed. Ph.W. Anderson, K.J. Arrow, D. Pines.

4. W. Weidlich, Physics and Social Science – the Approach to Synergetics, *Phys.Rep.*, (1991).
5. D.S. Chernavskii, A.H. Rakhimov, Mathematical model of the Economic Structure of Society, Preprint of the P.N. Lebedev Physical Inst., 15, (1991) (in Russian); D.S. Chernavskii, On the Economic Structure of Society, Radical. *Journal of Business World*, 17 (1992).
6. D.S. Chernavskii, A.S. Popkov, A.X. Rakhimov, The Mathematical Model of the Society Economic Structure, *Mathematical Economy*, 30, 2: 98 (1993)
7. D.S. Chernavskii, G.G. Pirogov, O.D. Chernavskaia, A.V. Scherbakov, B.A. Suslakov, The dynamics of economic structure of society, *Applied Nonlinear dynamics*, 4, 3: 67 (1996).
8. E.M. Lifshitz, L.P. Pitaevsky, *Physical Kinetics*, p. 118, Science, Moscow (1979) (in Russian).
9. Rodo Tokay Eran, Tokyo, p. 185 (1988).
10. D.S. Chernavskii, B.A. Suslakov, O.D. Chernavskaia, G.G. Pirogov, N.I. Starkov, On the socio-economical structure of socity, *Zakonodatelstvo i economika*, 7/8: 8 (1995).
11. D.S. Chernavskii, B.A. Suslakov, A.V. Scherbakov, How the russian capital becomes the "forigen" one, *Buisness-class*, 1(10): 35 (1996).
12. D.S. Chernavskii, A.V. Scherbakov, What is happened in Russia?, *Intellectual World*, 13 (1997).
13. D.S. Chernavskii, A.V. Scherbakov, Russian economics; stabilisation or stagnation?, *Politia*, 3(5): 49 (1997).
14. L. Makarevich, Bank crisis is spreaded with the economical collaps, *Finansovye isvestia*, 102, October 31 (1996).
15. A.S. Bulatov, In collective textbook *"Economics"*, , Moscow, ed. A.S. Bulatov (1997)
16. D.S. Chernavskii, B.A. Suslakov, A.V. Scherbakov, N.I. Starkov, The price issue at a profit maximization, *Economics and mathematical methods*, 34, 2: 44 (1998) (in Russian).
17. S. Yu. Glasyev, *Voprosi economiki*, 2: 90 (1997);
    S. Yu. Glasyev, *Voprosi economiki*, 1: 6 (1998);
    S. Yu. Glasyev, *Voprosi economiki*, 2: 37 (1998) (in Russian).
    N.P. Shmelev, *Voprosi economiki*, 3: 26 (1997); 1: 4 (1998); (in Russian).
18. S.A. Ayvasyn, *Economics and mathematical methods*, 3, 4: 74 (1997) (in Russian).

MATHEMATICAL MODEL FOR HEAT AND MASS TRANSFER IN THE SYSTEMS WITH THE NONLINEAR PROPERTIES INDUCED BY THE ELECTROMAGNETIC RADIATION

Ludmila A. Uvarova

Moscow State University of Technology "STANKIN",
101472 Moscow, Russia

INTRODUCTION

The problems of the heat and mass transfer arise from the solution of many physical and technical problems. The spreading of the electromagnetic radiation in real media, always practically absorbed, even if the absorption coefficient is very small, induced the heat and mass transfer and the phase transition under certain conditions. There are the following such problems: the problem of electromagnetic radiation interaction with dispersion systems[1], with biosystems[2], the problems of the electromagnetic radiation and heat mass transfer spreading in the parallel–sided structures (waveguides)[3] and others. The transport processes are nonlinear in general case. Therefore it would be interesting to determine exact solutions describing these king of processes. This work is devoted to the founding and the analysis of such class of solutions.

FORMULATION OF THE MATHEMATICAL MODEL

Let us introduce the mathematical model of the heat and mass transfer induced by the electromagnetic radiation. We shell assume that the problem is enough general, so it may be formulated two principal restrictions: 1) first of all, we suppose that the process of transfer proceeds quasi stationary : all typical relaxation times are enough small and the fields of the process phenomenological characteristics have time to concord with each new configuration; 2) secondly, we shall suppose that the approach of continuous medium may be used. Then the following mathematical model of the heat and mass transfer is proposed:

$$\Delta \mathbf{E}_i + k^2 \varepsilon_i \mathbf{E}_i (E_{i1}, E_{i2}, E_{i3}, T) = \nabla(\nabla \cdot \mathbf{E}_i) \tag{1}$$

$$\Delta \mathbf{H}_i + k^2 \varepsilon_i \mathbf{E}_i (E_{i1}, E_{i2}, E_{i3}, T) = \nabla k_{1i} \times \mathbf{E}_i \tag{2}$$

$$\nabla \cdot (k_{1i} \cdot \mathbf{E}_i) = 0 \tag{3}$$

$$\nabla \cdot \mathbf{H}_i = 0 \tag{4}$$

$$\nabla(\lambda_i \nabla T_i) + q_i(T_i) = 0 \tag{5}$$

$$q_i = \gamma_i(T_i)|\mathbf{E}_i|^2 \tag{6}$$

$$\varepsilon_i = f_i(\omega|\mathbf{E}_i|, T) \tag{7}$$

$$\nabla \cdot \mathbf{I}_{ji} = 0 \tag{8}$$

$$\mathbf{I}_{ji} = \mathbf{I}_{ji}(C_{1i}, C_{2i}, \ldots C_{ni}, D_{ji}, \nabla C_{ji}) \tag{9}$$

$$\lambda_i = \lambda_i(c_{1i}, C_{2i}, \ldots \ldots C_{ni}, T_i) \tag{10}$$

$$D_{ji} = D_{ji}(C_{1i}, C_{2i}, \ldots \ldots C_{ni}, T_i) \tag{11'}$$

$$F_i(T_i, T_{i-1}, q_i, q_{i-1}, \{I_{ji}\}, \{T_{ji-1}\}) = 0 \tag{12}$$

$$r_i(t, T_i, T_{i-1}, \{I_{ji-1}\}) = 0 \tag{13}$$

Here the following notations are used: i is the medium number, $i=1 \div m$; j is the component number, $j = 1 \div N_i$; $\mathbf{E}$ is amplitude of the electric field strength; $\mathbf{H}$ is the amplitude of the magnetic field strength; $k = \omega/c$, ω is the electromagnetic wave frequency, c is the velocity of light; T is the temperature; λ is the thermal conductivity coefficient; q is the density of the heat sours, $\gamma(T)$ is the coefficient, depending on the medium optical properties, $k_{1i} = k\varepsilon_i$, $\varepsilon = \varepsilon' + i\varepsilon''$, $i = \sqrt{-1}$, ε is the complex permittivity, I_j is the flow of matter density of the j-component, C_{ji} is the relative concentration, j is the component in in the i–medium, D_j is the effective diffusion coefficient of j–component. Eqs (1)–(4) are written for the amplitudes of the electric and obtained form Maxwell equations by supposing that $\mathbf{E} \sim e^{-iwt}$, $\mathbf{H} \sim e^{-iwt}$, where t is the heat balance on the boundary of the medium partition (i) and ($i-1$). The equation (13) is the law of the motion of the medium partition boundary (i) and ($i-1$). Such law may be set or defined at the solution of the problem. In particular case of the specified boundaries the equation (13) turns into the condition $r_i = r_{i0}$. It is necessary to add the conditions on the boundaries and others fundamental conditions. The boundary conditions can be written in the following way:

$$E_{ji-1} = E_{ji}, H_{ji-1} = H_{ji}, \tag{14}$$

$$T_{i-1} = T_i, \quad -\lambda_{i-1}\nabla T_{i-1} = -\lambda_i \nabla T_i \tag{15}$$

$$\phi_i(C_i, C_{is}) = 0 \tag{16}$$

In particularity

$$C_i = C_{is} \tag{16'}$$

The E_j, H_j are the tangential components of the electric and magnetic vectros, C_{is} are setting or obtaining by (12) equation meanings of the relative concentrations on the bounderies of the medium partition. It is necessary also to add to the conditions (14)–(16) the conditions on the external boundary, for example at infinite for the case of unlimited space. As the addition conditions it may be considered the dependences of C_{is} on the medium composition $(i-1)$, laws on the basis of which it can be defined the dependences $r_i(T)$ and others.

SOME SOLUTION CLASSES

System of equations (1)–(4) permits the solutions for which it is valid the following equation.

$$\varepsilon_i(E_{i1}, E_{i2}, E_{i3}, T) = 0 \tag{17}$$

It can be distinguished in orthogonal coordinate system the following two solution classes:

1. Vector E is expressed in terms of the potential φ: $E_i = \vec{\nabla}\varphi_i$ and the function φ_i is determined from the following equation:

$$f_i\left(\omega \left| \frac{1}{h_1}\frac{\partial \varphi_i}{\partial x_1}, \frac{1}{h_2}\frac{\partial \varphi_i}{\partial x_2}, \frac{1}{h_3}\frac{\partial \varphi_i}{\partial x_3}, T_i\right.\right) = 0, \tag{18}$$

where h_j are the metric coefficients, $j = 1, 2, 3$. In many cases the equation (17) depends on $|E_i|$:

$$\varepsilon_i\left(|E_i|^{2\upsilon}, T\right) = 0, \upsilon \in Z \tag{19}$$

In such case it can be determined the exactly or approximate dependence:

$$|E_i|^2 = W(T_i) \tag{20}$$

So if it not necessary to take into account the phase for the electric vector[4], the fulfilment of the condition (17) will permit to solve independently Maxwell equations and heat and mass transfer equations. We shall discuss this question later.

2. The H and E vectors depend on two coordinates:

$$E_i = E_i(x_k, x_e), \qquad H_i = H_i(x_k, x_e).$$

Then it is took place the following solution:

$$E_{ik} = \frac{1}{hk}\frac{\partial V_i}{\partial x_k}, \quad E_{il} = \frac{1}{hl}\frac{\partial V_i}{\partial x_l}, \quad E_{im} = \frac{1}{hm}\psi_i, \quad k \neq l \neq m \tag{21}$$

where ψ_i and v_i functions are determined from following equations:

$$\frac{\partial}{\partial x_k}\left(\frac{h_e}{h_k h_m}\frac{\partial \psi_i}{\partial x_k}\right)+\frac{\partial}{\partial x_e}\left(\frac{h_k}{h_e h_m}\frac{\partial \psi_e}{\partial x_e}\right)=0 \qquad (22)$$

$$f_i\left(\omega\left|\frac{1}{h_k}\frac{\partial V_i}{\partial x_k},\frac{1}{h_e}\frac{\partial V_i}{\partial x_e},\frac{1}{h_m}\psi_i,T_i\right.\right)=0 \qquad (23)$$

It may be noted that in the case of quadratic dependence:

$$f_i(|\mathbf{E}|)=\varepsilon_{i0}-\alpha|\mathbf{E}_i|^2+i\frac{4\pi\sigma_i}{\omega}, \qquad (24)$$

where ε_{i0} is the permittivity in the lack of the field, α is the parameter of the nonlinearity, magnitude σ is the conductivity. The equation (23) is Hamiltonian equation of the flat motion of the point:

$$H_i+W_i=const_i \qquad (25)$$

In the (25) T_0 is the system temperature in the lack of the radiation distribution. Mentioned solutions can be obtained directly from the equation system (1), (3), (17) taking into account the vector ratios in orthogonal coordinates. In this case the solutions for $\mathbf{H}_i$ are the harmonic functions.

It is important to note that obtained solutions satisfy boundary terms (14) under certain conditions for the problem parameters (typical dimensions of *i*-regions, the radiation frequency and others). So for the optically large spherical particle ($KR >> 1$, R is the particle radius), suspending in continual media, the following ratio takes place which connects the problem parameters[5]:

$$\frac{\varepsilon_0}{\alpha}=\frac{9E_0^2}{(KR)^2}, \qquad (26)$$

where R is the macro-particle radius, E_0 is the amplitude of the falling wave electrical vector.

Let us give as an example the equations for two cylinders being in continual medium. It will be supposed that written above Kerr's law is valid for defocus medium. For the first class of the problems we obtain the following equation:

$$\frac{1}{a^2(ch\tau-\cos\sigma)^2}\left(\left(\frac{\partial\varphi_i}{\partial\sigma}\right)^2+\left(\frac{\partial\varphi_i}{\partial\tau}\right)^2\right)+\left(\frac{\partial\varphi_i}{\partial z}\right)^2=\varepsilon_{i0}-i\frac{4\pi\sigma_i}{\omega}, \qquad (27)$$

where τ, σ, z are bicylindrical coordinates, $2a$ is the distance between centers of cylinders. For the second mass of the problems we obtain the following equations:

$$\frac{\partial^2\varphi_i}{\partial\tau^2}+\frac{\partial^2\psi_i}{\partial\sigma^2}=0 \qquad (28)$$

$$\frac{1}{a^2(ch\tau - \cos\sigma)}\left(\frac{\partial V_i}{\partial r} + \frac{\partial V_i}{\partial \sigma}\right) + \psi_i = \frac{\varepsilon_{i0} - i\frac{\sigma_i}{\omega}}{\alpha} \qquad (29)$$

So if the condition (17) is fulfilled it will be separated the electromagnetic radiation and heat transfer equations. Taking into account either dependence of the optical characteristics on the temperature we obtain in Kerr's law case the following equations:

$$|\nabla \Phi|^2 = v^2(x_1, x_2, x_3), \qquad (30)$$

where

$$v^2(x_1, x_2, x_3) = a'(T(x_1, x_2, x_3))\varepsilon_0(T(x_1, x_2, x_3))$$

$$\left(1 + \frac{16\pi^2 \sigma^2(T(x_1, x_2, x_3))}{\omega^2 \varepsilon_0^2(T(x_1, x_2, x_3))}\right)|\alpha_i|^{-2} \qquad (31)$$

The equation (26) coincides with the eikonal equation in heterogeneity medium[6] in the case of the real function ϕ. Determining the electrical vector components by means of the equation (22) it will be apparently that this equation will transform into (23) equation in which the constant and the function W are equal correspondingly:

$$const_i = 0.5(\gamma_0)^2(T_0), \qquad W_i = 0.5\left\{[(\gamma_i(T_0))^2 - (\gamma_i(T_i))^2] + \frac{\varphi_i^2}{h_m}\right\}$$

where T_0 is the temperature of the undisturbance medium, $h_m = 1$ for the bicylindrical coordinate system. From the expression for W_i it follows that the potential function is obtained not only by one of the electrical vector component but also by the function dependence of the optical properties on the temperature. As follows from the expression for W_i for lack of the relation between boundary conditions of electrodynamical and heat problems $v^2(T)$ dependence leads to the changing only two electrical vector components E_k and E_l while the component E_m is not depending. Present result signifies the possibility of the control over system. Really such solution takes a place for systems, which contains defocus transparent medium with assumed temperature gradient. In this case the field of temperature is, and magnitude of T_i serves the control parameter. The heat equation (6) in the case $\lambda = \lambda(T)$ by the help of Kirchgoff's substitution $\lambda \nabla T = \nabla \phi$ can rewrite

$$\Delta \phi_i + q_i(T_i(\phi_i)) = 0.$$

Since in the considered case a magnitude of $|\mathbf{E}|^2$ is determined only by the dependence of dielectric permittivity on the field then heat equation can solve without preliminary founding the electrical vector components. Then we receive the solution for temperature if will assume concrete dependence $q_i(T_i)$. So, for one-dimensional case we can write:

$$x - x_0 = \int_{T_0}^{T_i} \frac{\lambda_i(T_i)dT_i}{\left(C_i - 0.5\int_{T_0}^{\vartheta} q_i(\vartheta_i)\lambda_i(\vartheta)d\vartheta\right)^{0.5}}$$

where C_i, x_0 are integrated constants.

In particular for quadratic dependence the heat source on the temperature $q_i = q_{i0} + q_{i1}T_i + q_{i2}T_i^2$, the solution for T_i was presented in form of cnoidal wave[7]

$$T_i = T_{i1} - (T_{i1} - T_{i2})\sinh^2\left(\frac{x}{d_i}, \kappa_i\right), \quad \kappa_i = \frac{T_{i1} - T_{i2}}{T_{i1} - T_{i3}}, \quad d_i = \left(\frac{6\lambda_{i0}}{q_{i2}(T_{i1} - T_{i3})}\right)^{0.5} \tag{32}$$

where $\lambda_i = \lambda_{i0}$, $\sinh(\varphi)$ is Jacobi function, T_{ij} are the roots of the following cube equation:

$$T_i^3 + \frac{3}{2}\frac{q_{i1}}{q_{i2}}T_i^2 + \frac{3q_{i0}}{q_{i2}}T_i - \frac{3c_i^2\lambda_{i0}}{2q_{i2}} = 0 \tag{33}$$

Accordingly, square of the amplitude of the electrical vector is expressed by the function $\sinh\left(\frac{x}{d_i}, \kappa_i\right)$. In particular, we receive the solution for the temperature, having form of solution, under condition $T_{i2} = T_{i3}$ (othewise $\kappa_i = 1$). The quantity $\mathbf{E}_i^2$ is determined with the following formula:

$$\mathbf{E}_i^2 = \beta_{i0} + \beta_{i1}\cosh^{-2}\left(\frac{x}{d_i}\right) + \beta_{i2}\cosh^{-4}\left(\frac{x}{d_i}\right), \tag{34}$$

where magnitudes β_{ij} are constants, expressed by way of $q_{ij}, T_{i1}, T_{i2}, j = 0 \div 2$

The composition of multicomponent mixture may be determined by Stefan–Maxwell's equations:

$$\frac{1}{n}\left[\sum_{j=1}^{w_i}\frac{1}{D_{kj}^{(i)}}\left(\vec{G}_{ji}C_{ki} - \vec{G}_{ki}C_{ji}\right) - \vec{G}_{ki}\sum_{j=N_i+1}^{M_i}\frac{C_{ji}}{D_{kj}^{(i)}}\right] = \nabla C_{ki}, \qquad j \neq k \tag{35}$$

where n is summary concentration, D_{ij} is the binary diffusion coefficient, N_i is the number of components, which molecules cross the phase transition boundaries, $(M_i - N_i)$ is the number of inert components. We suppose that the diffusion coefficients depend on the temperature equally:

$$nD_{kj}^{(i)} = D_{kj}^{(i_0)}n_0\delta_i(T_i) \tag{36}$$

Let's, there is the symmetry in the problem geometry too. We receive the following solution:

$$C_{ki} = \sum_{j=1}^{N_i} B_{ji}\gamma_{ki}(\zeta_{ji})e^{\xi_i \zeta_{ji}} + Z_{ki}\sum_{j=N_i+1}^{M_i} C_{j(i-1)s}\left(\frac{C_{jis}}{C_{j(i-1)s}}\right)^{\frac{\xi_i}{\xi_{is}}} \tag{37}$$

In expression (37) variable is the function of r-coordinate, on which concentrations depend:

$$\xi_i = \int_r^{r_{i-1}} \frac{dr}{h_r \delta_i(T_i(r)) X(r)}, \qquad G_{ji} = \frac{A_{ji}}{X(r)} \tag{38}$$

In one-dimensional case substitution of (38) permits to reduce system (35) to the system of linear differential equations with constant coefficients. Correspondingly values ζ_{ji} and γ_{ki} are determined from the following equations:

$$\det\left|\left(\frac{1}{n_0}\sum_{j=1}^{N_i} \frac{A_{ji}}{D_{kj}^{(i_0)}} + \zeta_i\right)\delta_{kj} - \frac{A_{ki}(1-\delta_{nj})}{n_0 D_{kj}^{(i_0)}}\right| = 0 \qquad j \neq k \tag{39}$$

$$\gamma_{ki}\left(\zeta_{ie} + \frac{1}{n_0}\sum_{j=1}^{N_i}\frac{A_{ji}}{D_{kj}^{(i_0)}}\right) - \frac{A_{ki}}{n_0}\sum_{j=1}^{N_i}\frac{\gamma_{ji}}{D_{kj}^{(i_0)}} = \frac{A_{ki}}{D_{1k}^{(i_0)}}, \qquad j \neq k \tag{40}$$

where δ_{kj} is Kronecker's delta. Constants B_{ji} and Z_{ki} are defined with the help boundary conditions (16). Using the expression (37) it may be determined derivative of the relative concentration, which substitution under condition $r = r_i$ in (35) permit to define A_{ji} fluxes. Another way is the resolution of the system (37) relative to the flux density $\bar{G}_{ki}$. For that it is necessary to put in the effective diffusion coefficients D_{ki}, depending on the mixture composition[8]. It Onzager's ratios $D_{kj} = D_{jk}$ are valid, then the matrix with the help of which it is received the characteristics equation (39), is equivalent to symmetric matrix. In this case, the eigen-values of the system are material. As far as $x(r)$ is defined by the geometry of the system (for example $x = 1$ for the parallel-sided structure, $x = r^2$ for the spherical particle and etc) then the functional dependence of concentrations on coordinates is determined with the function $\delta_i(T_i(r))$ (if it will be monotonic or not and so on). As mentioned above the dependence of optical medium characteristics on the temperature can reduce to the solution in the form of temperature waves. It, in one's turns influences on concentration fields. In the general case it is necessary to take into account the dependence of the diffusion coefficients on concentrations. For the gaseous mediums such dependence has the following form:

$$D_{kj}^0 = [D_{kj}^0](1 + w_{kj} C_k), \tag{41}$$

where $[D^0{}_{kj}]$ is the binary diffusion coefficient determined by Chapman–Enskog's method in null approximation, w_{kj} is the coefficient depending on molecular features. Substitution (41) into (35) and following linearization of the system near by its stationary point permits to receipt the conditions under which the stable or unstable focuses take place on the corresponding phase plane.

The motion time of the medium partition boundary may be determined by the help of Raik's axiom:

where value v is velocity for some extensive value l. The law of conservation of mass M_i is represented by the equation:

$$\pm \frac{dM_i}{dt} = \int_{S_i} \sum_{j=1}^{N_i} \vec{G}_{ji} m_j d\vec{S}_i, \qquad (43)$$

where m_j is molecular mass j-th sort, $\vec{S}_i$ is the surface. Considering $M_i = V_i \rho_i'$ (ρ_i is substance density of medium, $\rho_i \approx$ const, V_i is the volume) we receive:

$$l = V_i, \quad V = \frac{1}{\rho_i} \sum_{j=1}^{N_i} m_i \int_{S_i} \vec{G}_{ji} d\vec{S}_i \qquad (44)$$

Or, according to supposition used on derivation of (37):

$$l = r_i, \quad V = \frac{1}{\rho_i S_i} \sum_{j=1}^{N_i} m_j A_{ji}$$

CONCLUSION

In this work some classes of exact solutions of the equations, describing the heat mass transfer, taking place under electromagnetic radiation action systems with nonlinear properties, were obtained. It is shown that in system with nonlinear optical and heatphysical properties there are effects, not having analogies for linear systems. These effects are following; 1) solutions for the electrical vector components proportional $1/\alpha$ (α is parameter of non–linearity); 2) anisotropic influence of the temperature field on the electrical vector components; 3) arising of temperature cnoidal waves and there dependence on concentration fields and others. Mentioned effects can be used for projection of arrangements which physical mechanisms of working are the transfer processes.

This work is performed and supported by Federal Program «Integration» (project No 43–2.1).

REFERENCES

1. K. Boren, D. Hafmen. Absorption and dispersion of light by small particles, Mir, Moscow (1986).
2. V.D. Lakhno, L.A. Uvarova, The effects of local heating of proteins in electron transfer, *Russ. J. of Physical Chemistry*, 69: 1387 (1995).
3. N.N. Rozanov. Optical Bistability and Hysteresis in distributed nonlinear systems, Nauka, Moscow (1997).
4. I.R. Shan, Principles of nonlinear optics, Nauka, Moscow (1989).
5. L.A. Uvarova, Models of the permittivity dependence on the field for the optical nonlinear evaporating drop, *Mathematical Methods in Chemistry*, Tver State University, Tver (1994).
6. M. Born, E. Wolfe. The principle of optics, Nauka, Moscow (1973).
7. L.A. Uvarova, V.K. Fedyanin, Yu.Z. Bondarev, Specific features of heat transfer in optically nonlinear condensed systems, *Russ. J. of Physical Chemistry*, 69: 1462 (1995).
8. D. Hurshfelder, Ch. Kertiss, G. Berd. Gases and liquids molecular theory, Foreign Literature, Moscow (1961).

MATHEMATICAL MODEL NONLINEAR HEAT TRANSFER IN AN INHOMOGENEOUS DISPERSIBLE SYSTEM

Marina A. Smirnova

Tver State Technical University,
170026 Tver, Russian

INTRODUCTION

Now one from actual problems of modern physics of dispersible mediums is the problem of heat transfer, taking place under an operation of heat sources (for example, electromagnetic nature). Such tasks immediately are connected to a problem of control by natural appearances and technological processes, in personally, with application of a laser engineering. The progress in a modern laser engineering, and also the successes in direction of deriving of substances with specific properties, reduce in necessity of study of singularities nonlinear heat transfer in dispersible systems. In personally, recently has a place active study of processes of transposition in systems containing clusters. Really, the clusters have a direct ratio to processes of association of rigid particles – formation of clusters in clouds, coagulation of particles in smokes, formation of structures for want of relaxation a metal pair etc. Per last years the heavily research of such structures by methods of computing physics [1-3] was carried out.

MATHEMATICAL MODEL

Actual the task about heat transfer in a system consists of N particles stipulated by heat sources in each particle is represented. In common case these heat sources are functions of temperature

$$q_i = q_{i0} f_i(T) \tag{1}$$

Considering the process of heat transfer for a dispersible medium, it is possible to offer the following mathematical model

$$C_i \frac{\partial T_i}{\partial t} + \nabla \chi_i \nabla T_i = -q_i(T_i),$$

$$C_e \rho_e \frac{\partial T_e}{\partial t} + \nabla \chi_e \nabla T_e = 0, \qquad (2)$$

$$-\chi_i \frac{\partial T_i}{\partial n}(S_i,t) = -\chi_i \frac{\partial T_e}{\partial n}(S_i,t),$$

$$T_i(S_i,t) = T_e(S_i,t), \; i = 1 \div N, \; T_e(\infty,t) = T_\infty,$$

where S_i is surface of i-particle, T_e is temperature outside of a particle, C_i is thermal capacity of i- particle, C_{ie} is thermal capacity of an environment, t is time, T_∞ is temperature of an environment, not perturbed presence of particles, ρ_i is denseness of substance of a particle, ρ_e is denseness of substance of an environment, $\vec{n}$ is the normal to a surface.

If the characteristic relaxation time $t_p = \max_{1 \leq i \leq N} \{t_{pi}\} \; t_{pi} = C_i \rho_i L_i^2 / \chi_i$, L_i is effective size of a particles equal to R_i, for example, i.e. radius of i-particle for case of spherical particles) is much greater of time of the process, the task can be decided in a quasistationary approximation.

If the exterior medium gas, and Knuden's number is $K_n = \frac{\lambda}{R_{min}}$, $R_{min} = \min_{1 \leq i \leq N}\{R_i\}$, λ is average length of free run of gas molecules) noticeably differs from zero (more precisely, it is possible to tell, that $0.1 \leq K_n \leq 0.3$, so–called moderately it is necessary to alter large drops on a Yalamov–Deryagin–Galoyan's classification[4]), conditions on the boundary of a particle, introducing in them a saltus of temperature

$$T_e(S_i) = T_S + K_T \frac{\partial T_e}{\partial n}\bigg|_{S_i}, \qquad (3)$$

where K_T is saltus of temperature, T_S is temperature of a surface of a particle. Thus is obtained boundary conditions of the third kind.

If is considered heat transfer in limited volume or the account will be carried out in some part of a continual medium, in which the particles were concentrated, on exterior actual or approximately specific boundary can put boundary conditions of the third kind including the heat exchange factor α.

In a system consisting of several particles, heat transfer depends both on a configuration of a system, and from interaction between particles. The situation is complicated also by that heat transfer is in common case nonlinear.

The large number of works is now devoted to a problem of heat transfer in collectives of particles. The transposition of an electromagnetic energy, heat transfer and mass transfer [5] is considered. In the present work we shall stay on heat transfer and on methods used for its exposition.

The "simple" case is a reviewing of collective from two particles. If it is possible to use particles spherical, bispherical a frame. Then, if the heat conduction equation is linear (i.e. represents the equation of the Poisson), it is possible to receive analytical it a solution, which has an aspect [6]:

$$T_e = \sqrt{2(ch\xi - \cos\eta)} \sum_{n=0}^{\infty} \left(A_n e^{\left(n+\frac{1}{2}\right)\xi} + B_n e^{-\left(n+\frac{1}{2}\right)\xi} \right) P_n(\cos\eta) + T_{e\infty},$$

$$T_1 = \frac{a^2}{2\chi_1}\sqrt{ch\xi - \cos\eta}\sum_{n=0}^{\infty}\left[e^{-\left(n+\frac{1}{2}\right)\xi}\int_{\xi_1}^{\xi}e^{\left(n+\frac{1}{2}\right)\xi}\int_0^{\pi}\frac{q_1 P_n(\cos\eta)\sin\eta}{(ch\xi - \cos\eta)^{5/2}}d\eta\,d\xi - \right.$$

$$-e^{\left(n+\frac{1}{2}\right)\xi}\int_{\xi_1}^{\xi}e^{-\left(n+\frac{1}{2}\right)\xi}\int_0^{\pi}\frac{q_1 P_n(\cos\eta)\sin\eta}{(ch\xi - \cos\eta)^{5/2}}d\eta\,d\xi +\quad (4)$$

$$\left. +e^{\left(n+\frac{1}{2}\right)\xi}\int_{\xi_1}^{\infty}e^{-\left(n+\frac{1}{2}\right)\xi}\int_0^{\pi}\frac{q_1 P_n(\cos\eta)\sin\eta}{(ch\xi - \cos\eta)^{5/2}}d\eta\,d\xi + c_{1n}e^{-\left(n+\frac{1}{2}\right)\xi}\right]P_n(\cos\eta) + T_\infty,$$

$$T_2 = \frac{a^2}{2\chi_2}\sqrt{ch\xi - \cos\eta}\sum_{n=0}^{\infty}\left[e^{\left(n+\frac{1}{2}\right)\xi}\int_{\xi}^{\xi_2}e^{-\left(n+\frac{1}{2}\right)\xi}\int_0^{\pi}\frac{q_2 P_n(\cos\eta)\sin\eta}{(ch\xi - \cos\eta)^{5/2}}d\eta\,d\xi - \right.$$

$$-e^{-\left(n+\frac{1}{2}\right)\xi}\int_{\xi}^{\xi_2}e^{\left(n+\frac{1}{2}\right)\xi}\int_0^{\pi}\frac{q_2 P_n(\cos\eta)\sin\eta}{(ch\xi - \cos\eta)^{5/2}}d\eta\,d\xi +$$

$$\left. +e^{-\left(n+\frac{1}{2}\right)\xi}\int_{-\infty}^{\xi_2}e^{\left(n+\frac{1}{2}\right)\xi}\int_0^{\pi}\frac{q_2 P_n(\cos\eta)\sin\eta}{(ch\xi - \cos\eta)^{5/2}}d\eta\,d\xi + c_{2n}e^{\left(n+\frac{1}{2}\right)\xi}\right]P_n(\cos\eta) + T_\infty,$$

where q_i is denseness of heat sources inside the first and second particle, C_{in} is factors of expansion, ξ_i is equation of surfaces of orbs ($i = 1, 2$), $P_n(\cos\eta)$ is spheroidal harmonic, ξ, η are coordinate.

In common case the accounts are necessary for carrying out numerically, and it is caused by the several reasons: at first, even if the task is linear, the analytical solution will represent the extremely bulky rows; secondly, the task in common case is nonlinear, as a heat source, and heat parameters can depend on temperature; thirdly, the particles can have the various form.

SOLUTION OF THE TASK BY A FINITE ELEMENTS METHOD

In the present work we used for account a finite element method. For want of it the initial mathematical statement should be modified in such a manner that the solution of the Poisson is in one multiply connected area with conditions of the third kind on the exterior boundary. Thus task is formulated as

$$C\rho\frac{\partial T}{\partial t} = \vec{\nabla}\chi\vec{\nabla}T + q(T), \quad (5)$$

$$C = \begin{cases} C_i, \vec{r} \in \theta_i \\ C_e, \vec{r} \notin \theta_i \end{cases}, \quad \chi = \begin{cases} \chi_i, \vec{r} \in \theta_i \\ \chi_e, \vec{r} \notin \bigcup \theta_i \end{cases}, i = 1 \div N \quad q = \begin{cases} q_i, \vec{r} \in \theta_i \\ 0, \vec{r} \notin \bigcup \theta_i \end{cases}$$

where θ_i – point set representing area, in which the particle, $\vec{r} = \{x, y, z\}$ are coordinate.

Let's take a look on , that for want of it is necessary to take into account realizations of account for small particles, condition (3), for want of it temperature near to a particle differs from temperature in an appropriate boundary knot of a particle and pays off under the formulas

$$T_i = T_1^{(i)}\left(a_1^{(i)} + b_1^{(i)}x + c_1^{(i)}y\right)$$

$$T^{(e)}(x,y) = T_1^{(i)} + K_T T_1^{(i)} \left(b_1^{(i)} \cos(\bar{x}, \bar{n}) + c_1^{(i)} \cos(\bar{y}, \bar{n}) \right) \tag{6}$$

that reduces to light correct of some elements of a stiffness matrix.

Certainly actual the research problem heat transfer, stipulated by heating inhomogeneous on the structure of a particle or particle of any form is.

Inhomogeneous on structure of a particle can represent by itself solid particles in which structure the various substances, drops of liquid, condensed on a rigid inhomogeneous particle, imperfections in a semiconducting plane etc. enter.

Heterogeneity of substance of a particle, and also in common case complexity of geometry, reduce in necessity to use methods of mathematical simulation. We developed the program for account by a finite element method of temperature in a single spherical particle with inclusion of any form. The particle was in some continual space, which sizes could vary. In the latter case considered system represented a mathematical model of a particle, weighed in atmosphere. The tasks were considered also, to which the envelope is essential, for example, task of heat transfer in rigid bodies with imperfections.

Thus, the following task was considered to

$$C_p \rho \frac{\partial T}{\partial t} = \nabla(\chi \nabla T) + q_0(T) \tag{7}$$

with boundary conditions of the fourth kind on a particle

$$\begin{cases} -\chi_2 \dfrac{\partial T_2}{\partial \bar{n}} = -\chi_1 \dfrac{\partial T_1}{\partial \bar{n}} \\ T_2 = T_1 \end{cases} \tag{8}$$

and third kind on the exterior boundary

$$\chi_2 \frac{\partial T}{\partial \bar{n}} + \alpha \tilde{T} = q_S, \tag{9}$$

where $\tilde{T} = T - T_\infty$, q_S – heat source on the boundary.

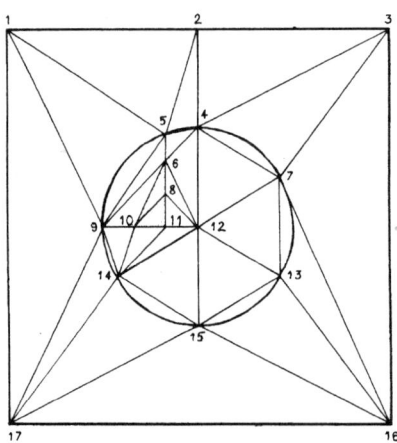

Figure 1. The calculated scheme.

In the present work the following model system was considered: the disk which geometric sizes are comparable to inclusion as prisms were in an environment, with a diameter of the disk. It was supposed, that heat transfer happens owing to a being available heat source. The account was carried out by a finite elements method. The task was decided for disks with radiuses 15 and 30 μ, the heat source varied in limits of 10^8 - 10^{12} W/m^3 (with allowance for of factor of an absorption), the characteristic size of inclusion was about a radius of a particle, and leg of a mesh in which the disk found room made of 3–4 its radiuses. On an axes Z the process was assumed homogeneous.

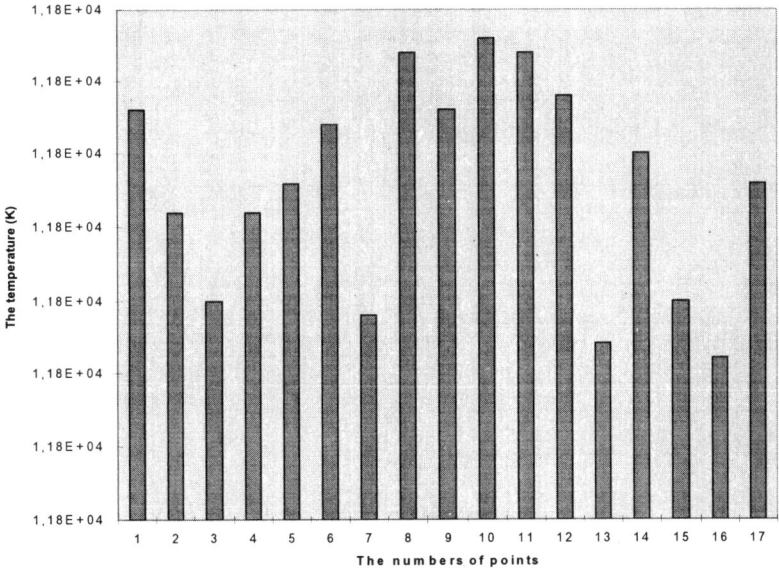

Figure 2. Temperature in a system.

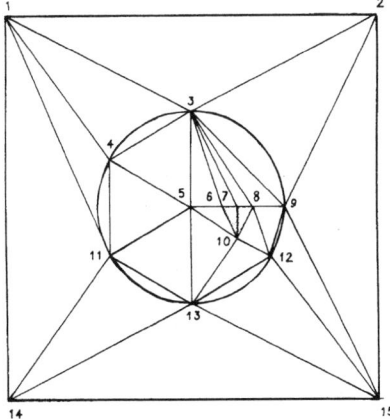

Figure 3. The calculated scheme.

On Fig. 3 and Fig. 4 the characteristic picture for distribution of temperature obtained for a system and partitions represented on Fig. 1 and Fig. 2 accordingly is represented. From obtained outcomes and comparison of the represented diagrams, in particular(personally), follows, that the asymmetric disposition of inclusion essentially influences a field of temperatures.

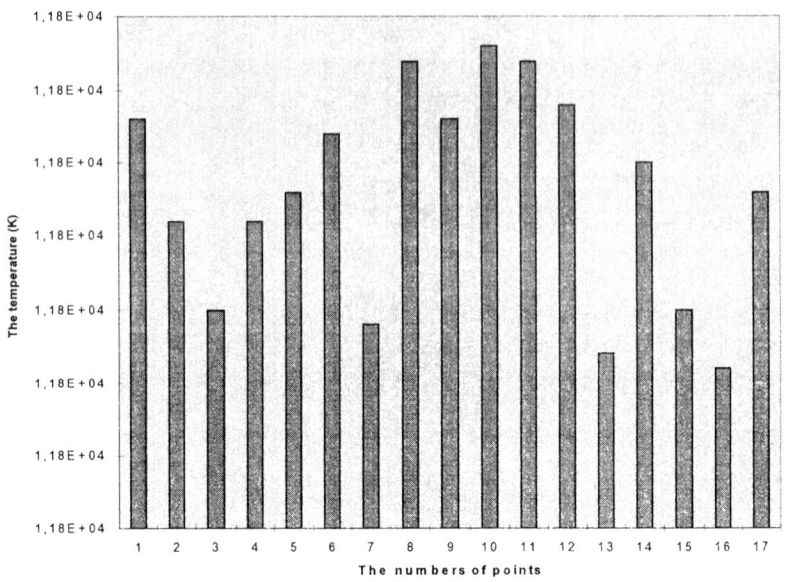

Figure 4. Temperature in a system.

This work is performed and supported by Federal Program «Integration» (project No 43–2.1).

REFERENCES

1. B.N. Smirnov, Fractal clusters, *Successes of physical sciences*, 149, 2: 177 (1986).
2. B.N. Smirnov, Processes in plasma and gases including of clusters, *Successes of physical sciences*, 11: 1169 (1997).
3. I.A. Woyczehovsky, M.T. Medvedev, V.H. Ferlenger, Ionization and fragmentation of clusters sprayed from a surface of metal by accelerated ions, *The technical physics Magazine*, 67, 12: 1 (1997).
4. B.V. Deryagin, Y.I. Yalamov, V.S. Galoyan, The theory of driving moderately of large particles in inhomogeneous gases, The report of Science Academy of USSR., 201: 383 (1971).
5. I.V. Krivenco, Interaction of electromagnetic radiation with a dispersible system. Tver, Tver state engineering university, (1997).
6. Y.I. Yalamov, About singularities of driving of two rigid interacting particles which are heated up with interior heat sources, *VINITI*, 8113 – Б 88.

Numerical Methods and Computer Simulations

UNSTEADY VISCOUS FLOW SIMULATION BASED ON QGD SYSTEM

Boris N. Chetverushkin, and Eugene V. Shilnikov

Institute for Mathematical Modelling Russian Academy of Sciences
125047 Moscow, Russia

INTRODUCTION

One approach to the validation of quasigasdynamic (QGD) equation system is discussed in this paper. There is close connection between kinetically consistent finite difference (KCFD) schemes[1] and QGD system[2]. QGD system may be considered as some kind of differential approximation for KCFD schemes[3]. The new way of obtaining QGD system is demonstrated using the same physical ideas on which the KCFD schemes are based. For this purpose we take advantage of the well known BGK model for one particle distribution function.

The basic assumption used for the construction of as KCFD schemes as QGD system is that one particle distribution function (and the macroscopic gas dynamic parameters too) have small variations on the distances compatible with the average free path length l. Accordingly the QGD system has the inherent correctness from the practical point of view. This correctness of QGD system gives the real opportunity for the simulation of unsteady viscous gas flows in transsonic and supersonic regimes[4]. It's also must be mentioned that the numerical algorithms for QGD system are very convenient for the adaptation on the massively parallel computer systems with distributed memory architecture. This fact gives the opportunity to use high performance parallel computer systems for detailed simulation of viscous gas flows on fine numerical meshes, which permit to study the fine structure of flow. Some results of such calculations are demonstrated in this paper.

QUASIGASDYNAMIC EQUATION SYSTEM

QGD system was originally obtained in[2]. This system is based on the same assumptions as the KCFD schemes[3]. It was shown in[3] that QGD equation system is the differential approximation for the KCFD schemes. Let us pay attention to three knots which are the bedrock of both KCFD schemes and QGD equation system:

a) the kinetic background of these objects is the presentation of the one particle distribution function as Maxwellian function f_{0M} constant (having small variation) on the distances compatible with the average free path length[3];

b) the assumption was essentially used while constructing the KCFD schemes and QGD system that one particle distribution function has Maxwellian form after molecules collisions[1];

c) both KCFD schemes and QGD system may be interpreted as a result of one particle distribution function expansion similar to the Chapmen-Enskog expansion[5]; the fact was essentially exploited that f can be written in the form

$$f = f_{0M} + \Delta f, \quad |\Delta f| \ll f_{0M} \tag{1}$$

and

$$\int \Delta f \, \varphi(\vec{\xi}) \, d\vec{\xi} = 0 \tag{2}$$

where $\varphi(\vec{\xi}) = 1, \vec{\xi}, \vec{\xi}^2/2$ is the collision vector.

These three presumptions permit us to apply the well known BGK model of one particle distribution function in order to obtain the QGD equation system. The BGK equation[6]

$$\frac{\partial f}{\partial t} + \xi_i \frac{\partial f}{\partial x_i} = \nu(f_{0M} - f) \tag{3}$$

is based on the fact that the distribution function near the equilibrium has Maxwellian form after collisions (see knot b for the KCFD and QGD). In (3) ν is the collision frequency ($1/\nu$ is the average time between molecules collisions).

Let's suggest that f has small variations during the time $\tau \approx 1/\nu$. This is equivalent to the small variations at a distance l (see knot a). Let the solution of equation (3) be presented in the following form

$$f = f_{0M} - \frac{1}{\nu} \frac{df_{0M}}{dt} + \frac{1}{\nu^2} \frac{d^2 f_{0M}}{dt^2} + O\left(\frac{1}{\nu^3}\right) \tag{4}$$

Rewrite the equality (4) in the following form

$$f = f_{0M} - \frac{1}{\nu} \frac{\partial f_{0M}}{\partial t} - \frac{1}{\nu} \xi_i \frac{\partial f_{0M}}{\partial x_i} + \frac{1}{\nu^2} \frac{\partial^2 f_{0M}}{\partial t^2} + \frac{2}{\nu^2} \frac{\partial}{\partial t} \xi_i \frac{\partial f_{0M}}{\partial x_i} + \frac{1}{\nu^2} \frac{\partial}{\partial x_k} \xi_k \xi_i \frac{\partial f_{0M}}{\partial x_i} + O\left(\frac{1}{\nu^3}\right) \tag{5}$$

Let's rewrite expression (5) ones more

$$f = f_{0M} - \frac{1}{\nu}\left[\frac{\partial f_{0M}}{\partial t} - \frac{1}{\nu}\frac{\partial^2 f_{0M}}{\partial t^2} + \xi_i \frac{\partial f_{0M}}{\partial x_i} - \frac{2}{\nu}\frac{\partial}{\partial t}\xi_i \frac{\partial f_{0M}}{\partial x_i} + \frac{1}{\nu}\frac{\partial}{\partial x_k} \xi_i \xi_k \frac{\partial f_{0M}}{\partial x_i}\right] + O\left(\frac{1}{\nu^3}\right) \tag{6}$$

The relaxational, convective and dissipative parts of equality (6) can be transformed as follows

$$\frac{\partial f_{0M}^{j+1}}{\partial t} - \frac{1}{\nu}\frac{\partial^2 f_{0M}^{j+1}}{\partial t^2} = \nu\frac{f_{0M}^{j+1} - f_{0M}^{j}}{2} + O\left(\frac{1}{\nu^2}\right) \qquad (7)$$

where $f_{0M}^{j} = f_{0M}(t^j, \vec{x}, \vec{\xi})$, $f_{0M}^{j+1} = f_{0M}(t^{j+1}, \vec{x}, \vec{\xi})$, $\tau = \frac{2}{\nu}$,

$$\xi_i \frac{\partial f_{0M}^{j+1}}{\partial x_i} - \frac{2}{\nu}\frac{\partial}{\partial t}\xi_i\frac{\partial f_{0M}^{j+1}}{\partial x_i} = \xi_i\frac{\partial f_{0M}^{j}}{\partial x_i} + O\left(\frac{1}{\nu^2}\right), \qquad (8)$$

$$\frac{1}{\nu}\frac{\partial}{\partial x_k}\xi_i\xi_k\frac{\partial f_{0M}^{j+1}}{\partial x_i} = \frac{1}{\nu}\frac{\partial}{\partial x_k}\xi_i\xi_k\frac{\partial f_{0M}^{j}}{\partial x_i} + O\left(\frac{1}{\nu^2}\right). \qquad (9)$$

Using (7)–(9) we can rewrite (6) in the following form

$$f^{j+1} = f_{0M}^{j+1} - \frac{1}{\nu}\left[\frac{f_{0M}^{j+1} - f_{0M}^{j}}{\tau} + \xi_i\frac{\partial f_{0M}^{j}}{\partial x_i} - \frac{\tau}{2}\frac{\partial}{\partial x_k}\xi_i\xi_k\frac{\partial f_{0M}^{j}}{\partial x_i}\right] + O\left(\frac{1}{\nu^3}\right).$$

Substituting the second order approximation for Δf into (2) which is the consequence of knot c) we can obtain the following expression

$$\int\left[\frac{f_{0M}^{j+1} - f_{0M}^{j}}{\tau} + \xi_i\frac{\partial f_{0M}^{j}}{\partial x_i} - \frac{\tau}{2}\frac{\partial}{\partial x_k}\xi_i\xi_k\frac{\partial f_{0M}^{j}}{\partial x_i}\right]\varphi(\vec{\xi})d\vec{\xi} = 0 \qquad (10)$$

As a result of integrating in (10) with components of collision vector $\varphi(\vec{\xi})$ the QGD equation system can be obtained. This system written in contravariant velocity components is as follows

$$\frac{\rho^{j+1} - \rho^{j}}{\Delta t} + \nabla_i(\rho u^i) = \nabla_i\tau\nabla_l(\rho u^i u^l) + \nabla_i\tau\nabla^i p \qquad (11)$$

$$\frac{(\rho u^k)^{j+1} - (\rho u^k)^{j}}{\Delta t} + \nabla_i(\rho u^i u^k) + \nabla_i p \\
= \nabla_i\tau\nabla_l(\rho u^i u^l u^k) + \nabla_i\tau\nabla^k(pu^i) + \nabla^k\tau\nabla_i(pu^i) + \nabla_i\tau\nabla^i(pu^k) \qquad (12)$$

$$\frac{E^{j+1} - E^{j}}{\Delta t} + \nabla_i((E+p)u^i) = \nabla_i\tau\nabla_l(u^i u^l(E+2p)) + \nabla_i\tau\nabla^i\left(\frac{p}{\rho}(E+p)\right) \qquad (13)$$

It is the natural result to obtain QGD system in the same form which was encountered in our previous papers because of the use of the similar physical assumptions discussed in a), b), c).

Thus, the QGD system based on presumption of small variations of one particle distribution function on the distances $\approx l$ has the inherent correctness from the practical point of view. This fact gives us the opportunity to use the QGD system as the model for detailed simulation of unsteady viscous gas flows. The description of viscous unsteady separate flows providing a detailed flow structure is very important for modern aerospace investigations. Using such kind of mathematical description one can obtain a set of frequencies relating to pressure oscillations generated by different scale vortices. For such predictions we need high performance parallel computers taking into account the fact that to simulate different scale vortices, fine meshes both on time and space should be used. Some results of such simulation are presented below.

THE TEST PROBLEM DESCRIPTION

The problem under consideration is the simulation of a structure of a shear layer in flows over a 2D rectangular cavity with moving bottom. Supersonic flow near an open cavity is characterized by a complex unsteady flowfields. Under certain freestream conditions such flows may be characterized by regular self-induced pressure oscillations. Frequency, amplitude and harmonic properties depend upon the cavity geometry and external flow conditions. The structure of the flow is of a fundamental physical interest. Moreover, these structures are of interest for different aerospace applications.

Let us consider the following time-constant freestream parameters which were taken in accordance with the experimental data[7] of: freestream Mach number $M_\infty = 1.35$, Reynolds number based on freestream parameters and cavity depth $Re_h = 3.3 \times 10^4$, Prandtl number — $Pr = 0.72$, specific ratio $\gamma = 1.4$ and the thickness of the boundary layer was $\delta/h = 0.041$. The geometrical parameters of cavity are: The ratio of the cavity length l to cavity depth h was $l/h = 2.1$ ($l = 6.3$ mm, $h = 3$ mm). The intensive pressure pulsations in the cavity take place for such parameters.

Let us suppose that the bottom oscillation amplitude are very small (does not exceed the size of the first mesh cell). In this case, the mesh rearrangement inside the computational region is not required, and it is possible to consider that the shape of cells only in the layer adjacent to the bottom are changeable. As a consequence, the deformed cells Ω_{ij} develop a trapezoidal shape. For this case in Cartesian coordinates, the finite difference scheme approximating (11)-(13) in a flux form in dimensionless terms and integrated on space Ω_{ij} can be written as follows (it should be noted that all the gasdynamic parameters are related to the cell mass centres)

$$\frac{\partial M_{ij}}{\Delta t} + \oint_{\partial \Omega_{ij}} \rho W_n dS = \oint_{\partial \Omega_{ij}} \tau \left[n_x \frac{\partial}{\partial x}(\rho u^2 + p) + n_y \frac{\partial}{\partial y}(\rho v^2 + p) \right] dS \qquad (14)$$

$$\frac{\partial (Mu)_{ij}}{\Delta t} + \oint_{\partial \Omega_{ij}} \rho u W_n dS + \oint_{\Omega_{ij}} \frac{\partial p}{\partial x} d\Omega = \oint_{\partial \Omega_{ij}} \tau \left[n_x \frac{\partial}{\partial x}(\rho u^3 + 3pu) + n_y \frac{\partial}{\partial y}(\rho u v^2) \right] dS$$
$$+ \frac{1}{Re} \oint_{\partial \Omega_{ij}} \left[n_x \frac{4}{3} \mu \frac{\partial u}{\partial x} + n_y \mu \frac{\partial u}{\partial y} \right] dS \qquad (15)$$

$$\frac{\partial (Mv)_{ij}}{\Delta t} + \oint_{\partial \Omega_{ij}} \rho v W_n dS + \oint_{\Omega_{ij}} \frac{\partial p}{\partial y} d\Omega = \oint_{\partial \Omega_{ij}} \tau \left[n_y \frac{\partial}{\partial y}(\rho v^3 + 3pv) + n_x \frac{\partial}{\partial x}(\rho u^2 v) \right] dS$$
$$+ \frac{1}{Re} \oint_{\partial \Omega_{ij}} \left[n_y \frac{4}{3} \mu \frac{\partial v}{\partial y} + n_x \mu \frac{\partial v}{\partial x} \right] dS \qquad (16)$$

$$\frac{\partial (ME)_{ij}}{\Delta t} + \oint_{\partial \Omega_{ij}} (p+E)W_n dS = \oint_{\partial \Omega_{ij}} \tau \left[n_x \frac{\partial}{\partial x}\left(u^2\left(E+\frac{5}{2}p\right)\right) + n_y \frac{\partial}{\partial y}\left(v^2\left(E+\frac{5}{2}p\right)\right)\right]dS$$

$$+ \frac{1}{\text{Re}} \oint_{\partial \Omega_{ij}} \mu \left[n_x \left(\frac{4}{3}u\frac{\partial u}{\partial x} + v\frac{\partial v}{\partial x}\right) + n_y \left(u\frac{\partial u}{\partial y} + \frac{4}{3}v\frac{\partial v}{\partial y}\right)\right]dS \quad (17)$$

$$+ \frac{\gamma}{\text{Re}} \oint_{\partial \Omega_{ij}} \frac{\mu}{\text{Pr}} \frac{\partial \varepsilon}{\partial \vec{n}} d\vec{S} + \oint_{\partial \Omega_{ij}} \frac{\tau}{\rho} \frac{\partial}{\partial \vec{n}} \frac{p^2}{\gamma-1} d\vec{S}$$

where ρ – density, p – pressure, u, v – velocity components, E – total energy, ε – internal energy, Ω_{ij} – cell square, $M_{ij} = \rho \Omega_{ij}$ – mass of gas in cell Ω_{ij}, $\vec{n}$ – external normal to boundary $\partial \Omega_{ij}$, $\vec{W} = (U,V)$ – the gas velocity relative to the moving mesh. In this case, the vertical velocity component is determined with the help of linear interpolation on y taken from the lower bound (where it is equal to the wall velocity) to the upper immobile cell bound (where it is equal to 0). The horizontal mesh velocity component is considered to be equal to 0 since the mesh rearrangement is carried out only along one direction.

THE OBTAINED RESULTS

The calculations have been carried out with the usage of various time and space grids. The reduction of time step does not result in any essential change of spectral characteristics of pressure pulsation. The use of more refined space meshes permits to reveal some additional features of flow structure within the cavity as well as to investigate the mechanism of discrete modes generation in pressure pulsation spectrum in details.

The main modes of pressure pulsation have been obtained on rough meshes. Using the mesh of approximately 100 nodes in each direction we predicted the unsteady regime over a cavity with the main frequencies of pressure oscillations corresponding to dimensionless frequencies characterized by Strouhal numbers ($\text{Sh} = lf/u$) $\text{Sh}_2^{calc} = 0.616$, $\text{Sh}_3^{calc} = 1.22$. The values of Strouhal numbers calculated are well agreed with the experimental values $\text{Sh}_1^{exp} = 0.3$, $\text{Sh}_2^{exp} = 0.65$, $\text{Sh}_3^{exp} = 1.2$. The first fluctuation mode is approximately equal to 0.3 which is weakly seen in the spectrum. Even with the help of mesh H2 it is possible to predict the low frequency corresponding to $\text{Sh}_1^{calc} = 0.286$ noticed in experiment. These oscillations result from the interaction between the over-cavity shear layer and large vortices formed inside the cavity and comparable-sized with it. The use of more and more fine meshes makes the predicted flow field more interesting.

To predict a detailed structure of unsteady viscous compressible flows we need to use high performance parallel computer systems. KCFD schemes can be easily adapted to parallel computers with MIMD architecture and give us the real opportunity for the detailed prediction of unsteady flows. These schemes are homogeneous schemes i.e. the one type of algorithm describe as viscous as inviscous parts of the flow. We used the explicit schemes which have soft stability condition. The geometrical parallelism principle have been implemented for constructing their parallel realization. This means that each processor provides calculation in its own subdomain. The explicit form of KCFD schemes allows to minimize the exchange of information between possessors. Having equal number of nodes in each subdomain the homogeneity of algorithm automatically provides load balance of processors. The real efficiency of parallelization for explicit KCFD schemes is close to **100%** and practically do not depend on processors number (see[8,9,10]).

The numerical algorithm was realized on the 12-processor CC-system made by Parsytec based on processors PowerPC 604 with total peak productivity about 3 GFLOPS. The efficiency of parallelization reached at solving the above problems is more than 90%. To adapt the numerical algorithm used, we divide the computational domain onto a set of subdomains at a rate of number equal to the number of processors available. At the same time, all the subdomains contain equal numbers of numerical nodes. The example of domain partition for 12 processors is given on Fig. 1. The equal number of nodes combined with the homogeneity of KCFD schemes naturally provides the processor load balancing.

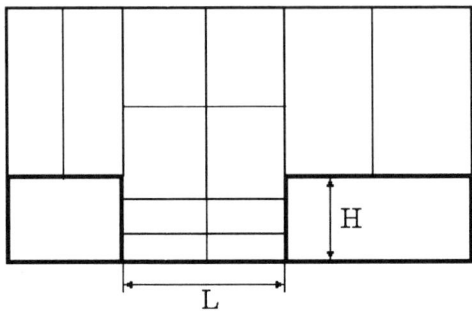

Figure 1. Computational region decomposition.

The use of fine space grid permits to reveal some additional features of flow structure within the cavity and to investigate the mechanism of discrete modes generation in pressure pulsation spectrum in details. We used the 450×2100 nodes) mesh in our calculations. The example of time distribution of static pressure in the point located on the right upper cavity corner is presented on Fig. 2. One can note that the periodic pulsation structure is broken which may be caused by the resolution of additional small vortices interacting with the right cavity wall. The additional frequencies in the pressure pulsation spectrum do not coincide with the main modes obtained on the rough mesh and also observable in the experimental research. This fact is connected with these small vortical structures. The corresponding spectrum is represented on Fig. 3. In addition to the main oscillation modes which are predicted even on coarse meshes pressure pulsation spectrum includes a few new frequencies among which frequency $f = 32400$ Hz with sound pressure level equal to 172.9 dB is the most intensive.

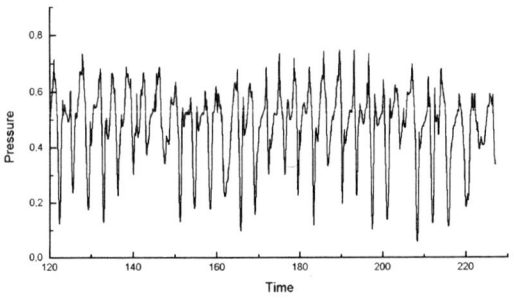

Figure 2. Pressure history for the mesh 1201×451.

As it is customary, spectral characteristics of pressure oscillations are represented as a sound pressure level (*SPL*) in decibels (dB) which is defined as the following

$$\text{SPL} = 20 \log_{10}\left(\frac{\sigma}{\sigma_0} \frac{p_s}{p_\infty}\right),$$

where σ_0 – the acoustic sound reference level of 2×10^{-5} Pa, σ – root-mean-square value of pressure pulsation amplitude, $p_s = 101.325$ kPa – standard pressure, p_∞ – static pressure.

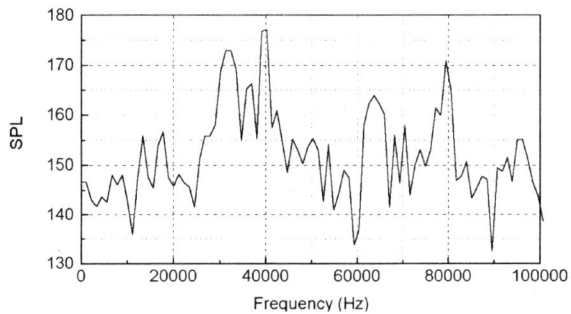

Figure 3. Frequency spectra for mesh 1201×451.

The flow structure in the boundary layer and the share layer are of most interest. Let's consider the fundamental characteristics of these regions. The Fig. 4 represents the vector velocity field for the time $t = 0.002$ s. Every next picture corresponds to the region marked by the rectangle in the previous one. Thus the Fig. 4a corresponds to the whole picture of the flow around cavity. The Fig. 4b and 4c show more and more details of the flow in the vicinity of the cavity left corner. So we can see the effect analogous to the geographical map. The main features of the flow are preserved and are added by new details. One can see the region with the return flow before the corner on Fig. 4b and a very small vortex under this region on Fig. 4c. The more detailed picture of the large-scale vortex generation may be seen on the last figure. During the following time this vortex decreases down to it's total disappearing. The processes of origination and collapse of large scale vortex periodically repeat in the presence of feedback between cavity rear bulkhead and the place of it's origination.

This result is illustrative to the laminar flow. It is in agreement with the experimental data obtained for the such types of cavities. That's just the periodical reappearance of small separated region before front corner of the cavity what shows the mechanism of pressure pulsations generation in laminar flow around the cavity. In this case, the critical difference causing the separation of boundary layer is very small. Therefore, the difference of pressure exceeding the critical value is formed when a compression wave comes to the forward cavity edge which results in a slight separation of boundary layer before a cavity. The separated zone pressure is determined by an acoustic wave which at first reaches its maximum and then weakens. According to these considerations a separated zone is firstly increased and then is reduced and later disappears. A shock wave is formed in front of the separated zone and moves in an agreement with its pulsation. This shock wave is blew down by the flow without a separated zone over a cavity. The specified mechanism of pressure pulsations generation can be also observed experimentally[7].

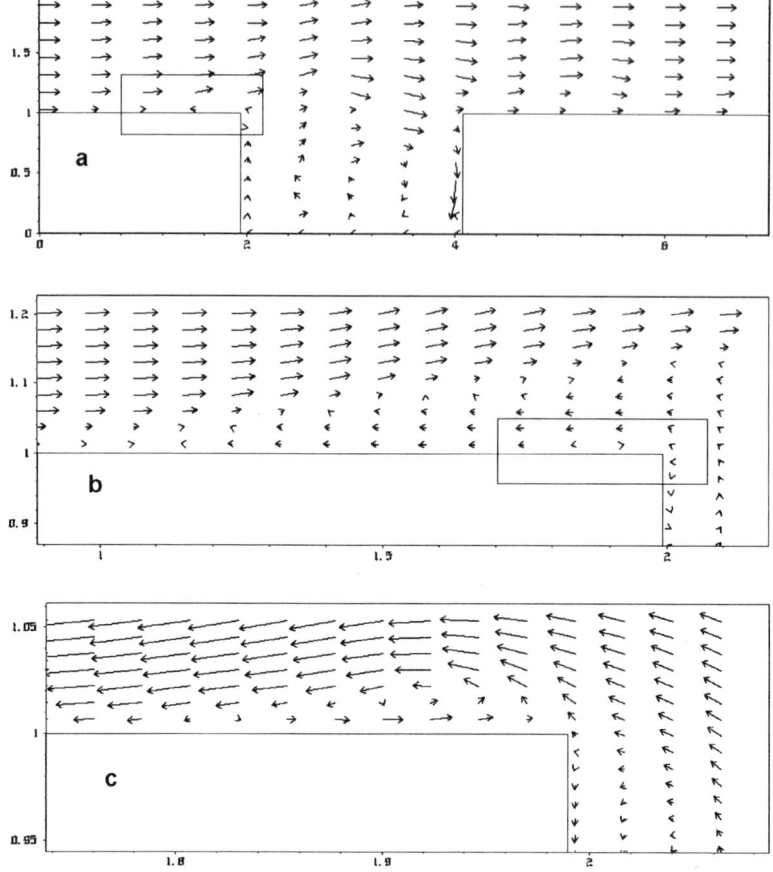

Figure 4. Velocity vector field for the whole region (a), and in the vicinity of left cavity corner (b), (c).

Thus, using of detailed spatial mesh allows to calculate flowfield of separated region of the cavity and visualize middle scale structures. One can hope that more detailed grid will make possible to receive the whole structure of flow in the cavity in transient case. There is also perspective from our point of view to use some kinetical analogue of $K-\varepsilon$ model of turbulence. We intend to combine in our future activity the direct modelling of large-scale and middle-scale flow structures with the description of small-scale turbulent structures basing on $K-\varepsilon$ model.

The influence of low amplitude artificial bottom oscillations of sinusoidal form on the whole flow picture has been examined with the help of the test task. The numerical simulations have shown that the most considerable changes in the flow field are discovered in the vicinity of the bottom. The velocities here have rather noticeable vertical components which change their signs in accordance with the bottom movement. In the regions at a far distance from the bottom, the flow structure appears to be practically unchanged.

The comparison was made of the pressure histories and pulsations spectra in the different points of the flow region for the different frequencies of bottom oscillations. This comparison showed that the resonant bottom oscillations result only in raising the pressure pulsation amplitudes (they amplify primarily the main maximum intensity in spectrum). Besides, there can be noticed some displacement of low frequency mode towards lower frequencies as well as its some growth. However, its amplitude remains to be much less than the main modes. As for the nonresonant frequencies the low frequency modulating modes appear. In addition, the high frequency components arise in the middle of the cavity bottom. In the case of a small displacement of the forcing oscillations from resonance the growth of the modulating mode intensity is quite noticeable. The bottom oscillations with the strongly nonresonant frequency cardinally change the spectrum. The main maximum is amplified practically in the same manner as at the resonance, while a number of additional discrete modes are generated as a result of the forcing frequency overtones.

Thus, the numerical simulations have shown that low amplitude bottom perturbations result in essential change of pressure pulsation field at a rather far distance from the bottom and the difference is well seen even at the upper bound of the cavity.

The main numerical results which have been received were the following:
- it is highly difficult to "excite" the shear layer with formed coherent structures; this can be reached only if the fluctuations of the bottom have sufficiently large amplitude;
- the predictions conducted showed good scope for using QGD system jointly with the specially developed algorithm for the consideration of mobility of streamlined body boundaries in order to investigate the influence of high frequency oscillations on gasdynamic flow;
- the pronounced change of basic flow parameters at a rather long distance from the wall oscillating with a small amplitude was obtained what appeared to be rather unexpected;
- the possibility of using this numerical technique for solving the classical aeroelasticity problem became rather clearly seen; this possibility can be realized in a way of the joint solution of equations describing viscous compressible gas flows (including turbulent flows) and equations simulating oscillations of a streamlined elastic body.

CONCLUSION

QGD equations can be interpreted as a new physical model for the description of viscous gas flows. Its kinetic base is the representation of distribution function in the form of Maxwellian one constant within a domain of size equal to free path length. QGD system can be also derived from the classical kinetic BGK equation. The KCFD schemes give the opportunity of effective use the multiprocessor computer systems for the simulation of unsteady flows.

ACKNOWLEDGEMENTS

The multiprocessor Parsytec CC workstation which have been used to provide all computations has been delivered to the Institute for Mathematical Modelling in the frames of equipment grant of European Economic Community (ESPRIT project No. 21042). The investigations are supported by Russian Foundation for Basic Research (grants No. 97-01-01032 and 96-15-97226) and by Federal Program "Integration" (project No 43–2.1).

REFERENCES

1. M.I. Volchinskaya, A.N. Pavlov, B.N. Chetverushkin, About a scheme for the integration of gasdynamic equations, Keldysh Institute of Applied Mathematics, Preprint No. 113, Moscow (1983), (in Russian).
2. T.G. Elizarova, B.N. Chetverushkin, Using of kinetic models for the computation of gasdynamic flows, in: *Mathematical Modelling. Processes in Nonlinear Media*, Nauka, Moscow (1986), (in Russian).

3. B. Chetverushkin, On improvement of gas flow description via kinetically-consistent difference schemes, in: *Experimentation, Modelling and Computation in Flow, Turbulence and Combustion*, Vol 2, ed. B.N. Chetverushkin, J.A. Desideri et al, Wiley, Chichester (1997).
4. B.N. Chetverushkin, Kinetically consistent finite difference schemes and simulation of unsteady flows, in *Computational Fluid Dynamics 96, Proceedings of III ECCOMAS*, Wiley, Paris (1996).
5. A.V. Lukshin, B.N. Chetverushkin, On the theory of kinetically consistent finite difference schemes, *Matematicheskoe modelirovanie*, 7(11): 109 (1995), (in Russian).
6. M.N. Kogan. *Rarefied Gas Dynamics*, Nauka, Moscow (1967), (in Russian).
7. A. Antonov, V. Kupzov, V. Komarov, *Pressure oscillation in jets and in separated flows*, Moscow 1990, (in Russian).
8. B.N. Chetverushkin, Kinetically Consistent Finite Difference Schemes and Their Application to Transient Flow Prediction, in: *Experimentation, Modelling and Computation in Flow, Turbulence and Combustion*, Vol. 1, J.A. Desideri, B.N. Chetverushkin, Y.A. Kuznetsov, J. Periaux and B. Stoufflet, eds., Wiley, Chichester (1996).
9. I.V. Abalakin, B.N. Chetverushkin, Kinetically consistent difference schemes as a model for description of gasdynamic flows, *Matematicheskoe modelirovanie*, 8(8): 3 (1996), (in Russian).
10. I.V. Abalakin, M.A. Antonov, B.N. Chetverushkin, I.A. Graur, A.V. Jokchova, E.V. Shilnikov, On the opportunity of parallel implementation of the kinetical-consistent finite difference schemes for gas dynamic flow simulation, In: *Parallel Computational Fluid Dynamics: Algorithms and Results Using Advanced Computer*, P.Schiano et. al. eds., Elsevier, Amsterdam (1997).

ECOLOGICAL AFTER-EFFECTS NUMERICAL MODELLING UNDER METHANE COMBUSTION

Boris N. Chetverushkin[1], Mikhail V. Iakobovski[1], Marina A. Kornilina[1], Konstantin Yu. Malikov[2] and Nataliya Yu. Romanukha[1]

[1] Institute for Mathematical Modelling, RAS,
Moscow, Russia
[2] Ural Polytechnic University,
Ekaterinburg, Russia

INTRODUCTION

Increases in atmospheric methane concentrations are highly correlated with increases in population and human-related activities that release methane to the atmosphere. At about 70 % of the total emissions are from anthropogenic sources and only 30 % are from natural sources. Methane is a major component of the natural gas and one of significant anthropogenic sources of the atmospheric methane is it's leakage during natural gas and oil producing, processing and distributing. The catastrophic example is the gas break from the layer. In order to reduce global ecological after-effects the gas gusher is to be ignited. But the methane-air mixtures burning produces some toxic substances as CO, unburned hydrocarbons, NO_x and hence leads to significant ecological problems itself. Mathematical modelling seems to be a good instrument for estimating the picture of the environment contamination under methane combustion. Though combustion is primarily a chemical process followed by the substance's conversion, the analysis of the chemical reactions can't provide good estimation of the transformation's velocities. Combustion processes in many practically important cases are controlled by dynamic factors as diffusion, heat losses and the other. The reaction mechanism selection depends on the pressure and the density ranges under consideration and the objective of the modelling. Hundreds reactions with very strong temperature coupling for tens (more than 30) constituents are necessary to estimate realistic NO_x formation under methane combustion. In the whole, the problem is rather sophisticated due to the large number of species involved and complex flow geometry. Various approaches may be used to minimise the computational costs. It's well known that the main computational costs (80 – 90 %) of the processor time) are presented by solving of the stiff ordinary differential equations, representing the chemical mechanism. That's why it's often desirable to elaborate the models containing complex $2D$ flow geometry and simplified reaction models. In the present study the simplifications are based on the physical properties of the problem under consideration. The temporary scale (gas gushers burn for several days, weeks and sometimes for months) and the very large spatial region occupied by high temperature products of the reactions make possible to neglect the 'prompt' mechanism for *NO* formation in our model. We suppose that the total NO formation is controlled by the modified thermal mechanism (Miller and Bowman, 1989). This restricts the species under consideration to the following 17: CH_4, CH_3, CH_2O, CH_3O, CHO, CO, CO_2, H_2, O_2, H, OH, HO_2, H_2O, N_2, N, NO, NO_2. Even in this case the non - equilibrium chemically reacting flow modelling is very time and labour-consuming problem. The effective numerical methods for simultaneous solution of chemical and gas dynamical problems and highly – efficient modern computers, such as parallel distributed memory computers, are required to solve it (Dorodnitsyn et. al, 1997).

PROBLEM FORMULATION

The system of governing equations corresponding to the non-equilibrium chemically reacting flows is quasi-gasdynamic system, including the conservation equations for species involved into the methane - air reaction mechanism, momentum and energy. In 2D case in Cartesian co-ordinates, the equations may be written in the following form:

$$\frac{\partial \rho_i}{\partial t} + \frac{\partial}{\partial x}(\rho_i u) + \frac{\partial}{\partial y}(\rho_i v) = \frac{\partial}{\partial x} \tau \frac{\partial}{\partial x}(\rho_i u^2 + p_i) + \frac{\partial}{\partial y} \tau \frac{\partial}{\partial y}(\rho_i v^2 + p_i) + F_i,$$

$$\frac{\partial \rho u}{\partial t} + \frac{\partial}{\partial x}(\rho u^2 + p) + \frac{\partial}{\partial y}(\rho u v) = \frac{\partial}{\partial x} \tau \frac{\partial}{\partial x}(\rho u^3 + 3pu) +$$
$$\frac{\partial}{\partial y} \tau \frac{\partial}{\partial y}(\rho u v^2 + pu)$$

$$\frac{\partial \rho v}{\partial t} + \frac{\partial}{\partial x}(\rho u v) + \frac{\partial}{\partial y}(\rho v^2 + p) = \frac{\partial}{\partial x} \tau \frac{\partial}{\partial x}(\rho u^2 v + pv) +$$
$$\frac{\partial}{\partial y} \tau \frac{\partial}{\partial y}(\rho v^3 + 3pv)$$
(1)

$$\frac{\partial E}{\partial t} + \frac{\partial}{\partial x}(u(E+p)) + \frac{\partial}{\partial y}(v(E+p)) = \frac{\partial}{\partial x} \tau \frac{\partial}{\partial x}(u^2(E+2p)) +$$
$$+\frac{\partial}{\partial y} \tau \frac{\partial}{\partial y}(v^2(E+2p)) + \frac{\partial}{\partial x} \tau \frac{\partial \theta}{\partial x} + \frac{\partial}{\partial y} \tau \frac{\partial \theta}{\partial y},$$

The next thermodynamic properties are used to close the system:

$$p = \sum_i p_i = \sum_i \rho_i RT / m_i$$

$$E = \rho(\varepsilon + 0.5(u^2 + v^2)), \quad \rho = \sum_i \rho_i,$$

$$\varepsilon = \sum_i \varepsilon_i \rho_i / \rho, \quad h = \sum_i h_i \rho_i / \rho,$$

$$\varepsilon_i = \int_0^T c_{V_i} dT + h_{0i}, \quad \tau = \frac{\mu}{\rho},$$

$$\theta = \sum_i p_i h_i + 0.5p(u^2 + v^2),$$

where ρ is the density, p is pressure, T is temperature, u and v are velocities along x and y respectively, E is total energy, ε is mean internal energy, h is mean enthalpy of mixture, τ is the averaged molecular collision time, μ is the viscosity, ε_i and h_i are internal energy and enthalpy of the i-th species, h_{0i} is species formation enthalpy, F_i - the species production rate in chemical processes, m_i- molecular weight of the i-th species, c_{pi} and c_{Vi} are the constant pressure and volume heat capacities of the i-th species, R is the universal gas constants.

NUMERICAL PROCEDURE

The significant problem of the considered model formulation is the stiffness of the equations describing the transport characteristics for the species concentrations. We may avoid this difficulty by separating the calculations of flow fields and chemical source terms. It's possible because of the small time scales for chemical reactions compared to gasdynamic time scales. According to the method of summary approximation the problem may be splited into two blocks describing independently gasdynamic and chemical processes. These blocks are to be calculated sequentially. Firstly the system (1) is to be solved in assumption $F_i = 0$. The calculated values of ρi are used to form initial values for 'chemical step'. At this step the species conservation equations are written in the form:

$$\frac{dn_i}{dt} = L_i - n_i Q_i, \qquad (2)$$

where n_i is the concentration of the i-th species $n_i = Mc_i / m_i$, M is mean molecular weight of the mixture, c_i – mass fraction of the i-th species, L_i and Q_i are the formation and loss velocities of the i-th species in reactions. The reactions rates are taken from (Bochkov et al, 1992).

Chemistry strongly depends on temperature. We have no experimentally measured temperature profiles in the flames under consideration, in this case the temperature is to be obtained by solving energy conservation equation. If only chemical processes are considered this equation reduces to the form of mean enthalpy of the mixture remaining constant, which leads to:

$$\frac{dT}{dt} = -\frac{\sum h_i \, dn_i / dt}{c_p \rho}, \qquad (3)$$

where dn_i / dt, is the velocities of species transformation in chemical processes, determined by (2), c_p is the mean specific heat, h_i are calculated as polynomials of T.

The data obtained by solving (2), (3) are used to correct the flow parameters within the gasdynamic step.

The numerical technique chosen to solve system (1) without chemical source terms is constructed on the base of kinetically consistent finite-difference schemes (KCFD) (Chetverushkin, 1996). The used half-implicit schemes are suitable for parallelisation and allow to achieve good efficiency on massively parallel computers. The system of stiff ordinary differential equations (3), (4) is solved by DVODE (Brown, et.al, 1989). Particular attention has been given to achieve a good agreement between gasdynamic and chemical time steps during reacting flow numerical simulation (Chetverushkin et.al, 1998)

COMPUTATIONAL RESULTS

The numerical modelling of combusting methane – air flow propagation was taken as example. In order to illustrate the reaction model counted in numerical simulation the stiff equations (2), (3) were numerically integrated over the time interval $0-100$ sec. The stoichiometric mixture of methane, nitrogen and oxygen was taken to form initial conditions for species CH_4, N_2, O_2 concentrations, the other species concentrations are zero.

The initial temperature was taken 1500 K, the pressure remains constant.

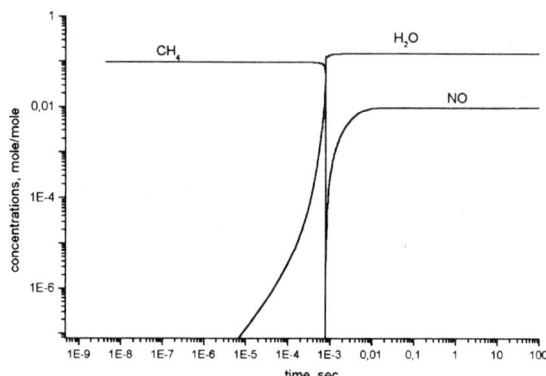

Figure 1. The temporary variations of CH_4, H_2O and NO concentrations in chemical processes

The Fig. 1 demonstrates the behaviour in time (seconds) of CH_4, H_2O and NO concentrations (as molar fractions) during the reaction progress. The NO formation begins in the flame, but the concentration reaches it's equilibrium value at time exceeding $5 \cdot 10^{-2}$ sec. Methane already has been consumed up to the time less than 10^{-3} sec. The equilibrium H_2O concentration is formed simultaneously with methane loss at the time not exceeding $0.78 \cdot 10^{-2}$ sec.

The other figures represent the following numerical experiment. At the initial moment methane under high pressure ($p = 1.8 p_0$, $p_0 = 10^6$ Pa) and the temperature $T = 290$ K flows off the aperture of the diameter 10 cm with velocity along y-axes v=120 m/sec.

The current of methane spreads mixing with the air up to the moment of ignition. The following concentration limits for methane ignition were taken: 5 volume percent as the low limit and 15 volume percent as the upper limit. Setting the temperature to the value 1800 K starts ignition. At the bottom border everywhere except aperture

$$\partial f \; \partial y = 0, \; (f = \rho_i, v, p, T), \; u = 0.$$

At the other borders:

$$\partial f \; \partial n = 0, \; (f = \rho_i, u, v, p, T).$$

The results of numerical modelling – calculated spatial distribution of CH_4 and some intermediate and finite reaction products are shown at the Figs 2–6. The rectangular area of the size 200×200 meters is represented at all figures. Mass fractions of methane and other species concentrations are given by various intensity of black colour. The conformity between colour and numerical value is shown to the right of each figure and the time is given at the bottom. Fig. 2 represents methane distribution in the environment up to the moment of ignition. The other figures represent the species distribution $2.3 \cdot 10^{-5}$ sec and $5.6 \cdot 10^{-5}$ sec later the ignition moment. The burning occurs in the layer, where methane and oxygen concentrations are appropriate for flame reactions progress. During methane combustion the light area with zero methane concentration arises around a formerly formed torch as shown at Fig. 3. The other species concentrations increase from zero values up to shown at Fig. 4, 5 for CH_3O and CO respectively, which are the toxic intermediate products of methane oxidation. Fig.6 demonstrates H_2O (finite product of methane oxidation) concentrations. The flame zone becomes constricted in time. Diffusion processes, supplying oxygen to the reaction zone dictate, the velocity of the flame front propagation.

ACKNOWLEDGEMENTS

The Parsytec CC workstation used to provide all computations was delivered to the Institute for Mathematical Modelling in the framework of the equipment grant of European Economic Community (project No. ESPRIT 21042). The work is supported by Russian Foundation for Basic Research (grants No. 97-01-01032, 96-01-01753) and by Federal Program "Integration" (project No 43–2.1).

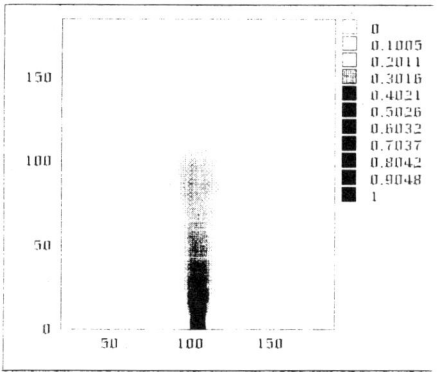
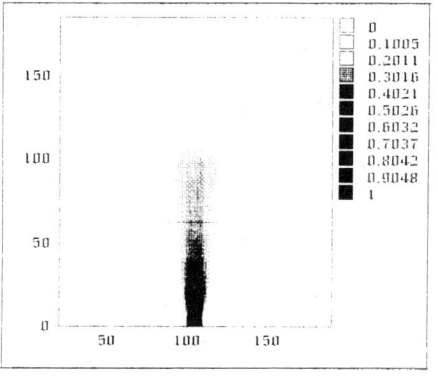

Figure 2. Methane flow up to the ignition moment.

Figure 3. CH_4 concentration (as mass fractions) $5.6 \cdot 10^{-5}$ sec later the ignition.

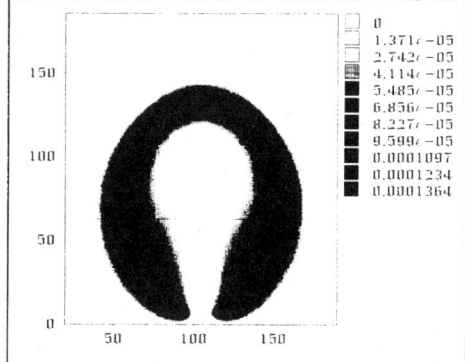
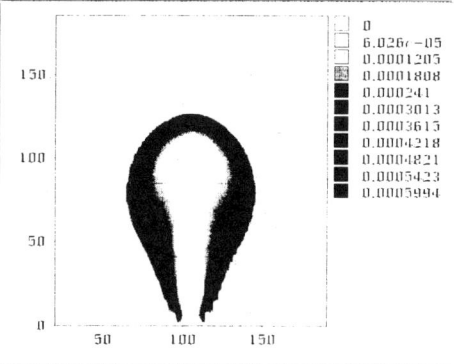

Figure 4. CH_3O distribution $2.3 \cdot 10^{-5}$ and $5.6 \cdot 10^{-5}$ sec later ignition.

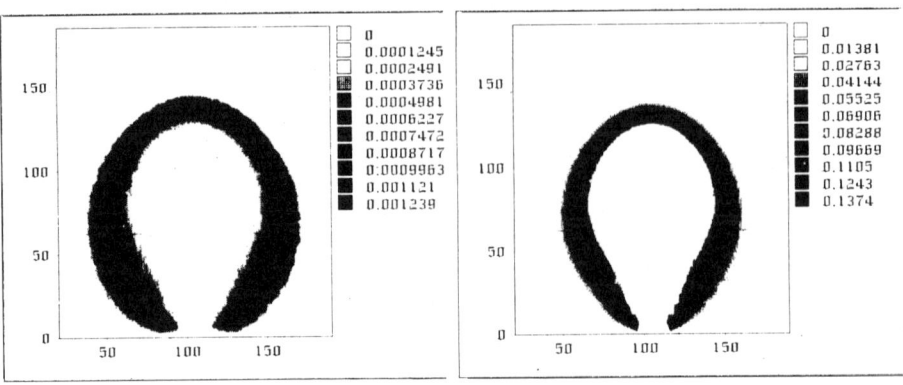

Figure 5. CO CO distribution $2.3 \cdot 10^{-5}$ and $5.6 \cdot 10^{-5}$ *sec* later ignition.

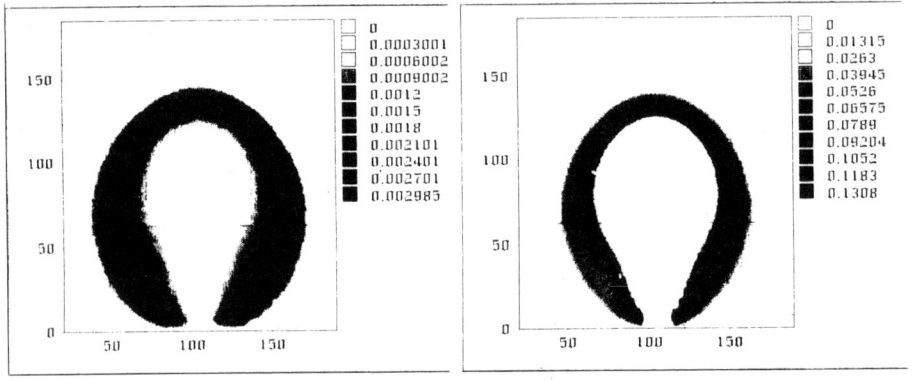

Figure 6. H_2O distribution $2.3 \cdot 10^{-5}$ and $5.6 \cdot 10^{-5}$ *sec* later ignition.

REFERENCES

1. M.V. Bochkov, L.A. Lovachev and B.N. Chetverushkin, Chemical kinetic of NO_x formation under methane combustion in the air, *Mathematical Modelling*, 4, 9: 3 (1992).
2. P.N. Brown, G.D. Byrne, and A.C. Hindmarsh, VODE: a variable coefficient ODE solver, *SIAM J. Sci. Stat. Comput.*, 10: 1038 (1989).
3. B.N. Chetverushkin, Kinetically consistent finite difference schemes and their application to transient flow prediction, in: Experimentations, Modelling and Computation in Flow, Turbulence and Combustion. Edited by J.A. Desideri, B.N. Chetverushkin, et al. John Wiley and Sons Ltd, Chichester, 1: 211 (1996).
4. B.N. Chetverushkin, M.A. Kornilina, N.Yu. Romanukha, Multicomponent gasdynamics flows simulation on distributed memory parallel computers: *Parallel CFD'98* (Taiwan), Eds. P. Schiano et. al, Elsevier, Amsterdam (1998) (to be published).
5. L.W. Dorodnitsyn, M.A. Kornilina, B.N. Chetverushkin, M.V. Iakobovski,), Computer modelling of gas flows containing chemically active components, *Russian Jurnal of Physical Chemistry*, 71, 12: 2059 (1997).
6. A.J. Miller, C.T. Bowman, Mechanism and modeling of nitrogen chemistry in combustion, *Progress. in Energy and Combust Sci.*, 15: 287 (1989).

COMPUTER SIMULATION OF EVAPORATION PROCESS INTO THE VACUUM

L.V. Pletnev[1], N.I. Gamayunov[2], V.M. Zamyatin[3]

[1] Mogilev Machine-Building Institute,
Mogilev, Belarus

[2] Tver State Technical University,
170002 Tver, Russia

[3] Center programsystem (CPS),
170002 Tver, Russia

INTRODUCTION

One of the important problems of a gas dynamics is a question of interaction of gas molecules with the surface of a condensed phase1. Near the surface of a gas phase the molecules also collide with the molecules reflected from the surface. Near the surface there is a Knudsen's layer, in which practically there are no collisions of the molecules with each other[2,3]. The value of this layer is equal to the length of a free molecule run of the gas phase. Thus, there are two independent flows of molecules in the Knudsen's layer: one flow consists of molecules flying to the surface from the gas phase and the other consisting of molecules flying from the surface into the gas phase. The difficulty for obtaining a distribution function of molecule speeds in the Knudsen's layer is that the process of molecule interaction with the surface is non-equilibrium and the function of the distribution of such molecules differs from that of equilibrium.

While calculating the heat and mass transfer processes the molecules that have left the surface are supposed to have the Maxwell function of the distribution in speeds with the temperature of the condensed phase 4. In case of a hydrodynamic mode of the flow there are jumps of the temperature and the density on the surface and in the gas phase[5]. In case of a gasdynamic mode of the flow it is necessary to use the Boltzmann equation or model equations. The above-mentioned assumptions are used for setting boundary conditions in describing the gas medium movement near the surface.

The process of evaporation from the surface of the condensed phase into the vacuum is a more simple case (Overcoming by the molecules of a potential barrier on the surface of the condensed phase). In case of the evaporation of a monoatomic condensed phase there will be no back flow of the atoms. The back flow will be observed in case of the evaporation of a multicomponent condensed phase or molecules, as well as in an intensive evaporation of the monoatomic condensed phase when multiatomic collisions are becoming important.

When a substance is evaporated from the surface of the condensed phase the temperature lowers. To maintain the constant temperature of the condensed phase surface it is necessary to bring to it the amount of the heat, which depends on the temperature of the condensed phase and the value of the potential barrier. The form of the potential barrier curve, in this case, does not play any role, since only those molecules are considered which have overcome the potential barrier.

However, we consider the evaporation process as not enough justified. Some experimental data cannot be explained within the framework of the existing theory of the heat and mass transfer during the evaporation. So, for example, in Stern's experiments on the evaporation of metal from the furnace into the vacuum, the deficit of the atoms with small speeds is observed, which shows a discrepancy of the distribution function of the departed particles from the Maxwell function[6]. It is interpreted by a deviation from the model of the dot particles and their collisions near a hole when they are leaving the furnace. The experimental data on the evaporation of the uranium melt given in[7], do not correspond to the theoretical ones.

The analysis of published articles has shown, that the regularities connecting the parameters of the condensed phase characteristics of the molecules flow which have overcome the potential barrier on the surface of the condensed phase have not been defined yet.

BASIC PRINCIPLES OF CALCULATIONS

The speed of the part of the condensed phase molecules can be sufficient for overcoming the potential barrier on the surface of the condensed phase as a result of fluctuations. These molecules leaving the surface of the condensed phase into the vacuum or gas phase. The molecules, leaving the surface of the condensed phase, some part of the kinetic energy spent for overcoming the potential barrier. Therefore, only those molecules which posses a larger kinetic energy than the value of the potential barrier can escape. The potential barrier on the surface of the condensed phase is a peculiar filter separating slow molecules from fast ones.

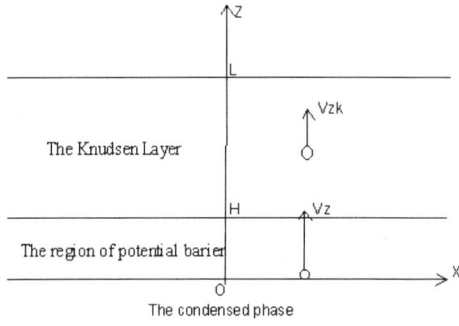

Figure 1. The general picture.

In the contributed paper the approach for finding the distribution functions of atoms on speeds and energies in the Knudsen's layer depending on the atoms mass m, the temperature T of the condensed phase and the value of the potential barrier U is offered. The paper describes the case of evaporation of the monoatomic-condensed phase into the vacuum and the Maxwell distribution of atoms in speeds in the condensed phase.

We shall assume that the condensed phase occupies half-space $z \leq 0$. The forces of interaction between the escaping atoms and the atoms on the surface of the condensed phase become negligibly small at the distance of H. Let us suppose that H is the constant value that is considerably smaller than the length of the free run of the escaped atoms (the Knudsen's layer value is L) $H \ll L$.

By computer simulations of the evaporation process, due to the principle of independent motion, we can consider only the change of z-component of the atoms speed. A part of the kinetic energy of every escaped atom is spent for overcoming the interaction forces with the atoms of the condensed phase. It follows from the energy conservation law:

$$\frac{mv_{zkn}^2}{2} = \frac{mv_z^2}{2} - U, \qquad (1)$$

where v_z and v_{zkn} are the components of atoms speed on the surface of the condensed phase and in the Knudsen's layer, accordingly.

Algorithms described in [8] were used as random number generators for the definition of atoms speeds on the surface of the condensed phase. The following technique for obtaining distribution functions of atoms energies in the Knudsen's layer was used. The components of atoms speeds on the surface of the condensed phase were defined with the help of the generator of normally distributed casual values. Then the value of atom kinetic energy in the Knudsen's layer was determined from the equation (1) and the conclusion was made about its possible appearance in the Knudsen's layer. The simulation of 1000000 atoms escape was made for definite parameters.

RESULTS AND DISCUSSION

Fig. 2 shows the schedules of dependencies of the normalized average speeds $v_n(r)$ and energies $e_n(r)$ in the Knudsen's layer on the dimensionless parameter $r = U/kT$ (where k is Boltzmann's constant).

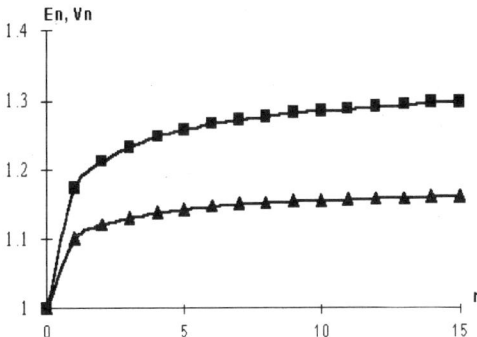
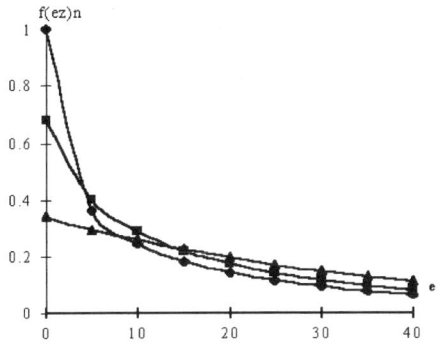

Figure 2. The dependensies of the normalized speeds $v_n(r)$ ■ and energies e_n ▲.

Figure 3. The normalized distributions z-component atoms energies: $r = 0$ ♦, $r = 0.1$ ■, $r = 5$ ▲

The normalization was made by dividing the appropriate value by the value, when $r = 0$ ($v_n(r) = v_a(r)/v_a(0)$, $e_n(r) = e_a(r)/e_a(0)$). The analysis of computer simulation showed that masses of atoms and the temperature of condensed phase do not influence on the normalized average speeds and atoms energies dependencies. Statistical deviations did not exceed 1 %. The curves growth slows down after meaning $r = 2$, which is characteristic for both regularities. The question of the existence of asymptotic meanings for the curves distributions is open. The existence of the potential barrier on the surface of the condensed phase leads not only to a simple division of slow and fast atoms, but also to paradox phenomenon such as: the average speeds and atoms energies in the Knudsen's layer exceed similar values in the condensed phase.

Fig. 3 shows normalized distributions z-component atoms energies in the Knudsen's layer for different meanings r. The distributions maxima are when $e = 0$. The maxima values decrease with the increasing of r. As r increases, part of slow atoms decreases and part of fast atoms increases.

The distribution functions of atoms in energy in the Knudsen's layer are shown in Fig. 4. The distributions curves look the same for all meanings of r. With the increasing of r the maxima decreases in height and displaces to the side of larger energies.

The obtained results enabled to understand the role of the potential barrier on the surface of the condensed phase and explain a divergence between theoretical conclusions and experimental data (in Stern's experiments, for example).

The analysis of the computer simulation showed that the probability of atoms escape from the condensed phase into the Knudsen's layer decreases with the increase in the condensed phase, i.e. with the increase of r. It happened that the probability of atoms escape from the condensed phase into the Knudsen's layer is determined by the function:

$$P(r) = 1 - F(\sqrt{r}), \qquad (2)$$

where $F(x)$ is the integrated Laplace function.

The greatest interest, from the practical point of view, is the results of power calculation for the redundant heat flow from the surface of the condensed phase:

$$G(r) = P(r)(e_a(r) - e_a(0)) \tag{3}$$

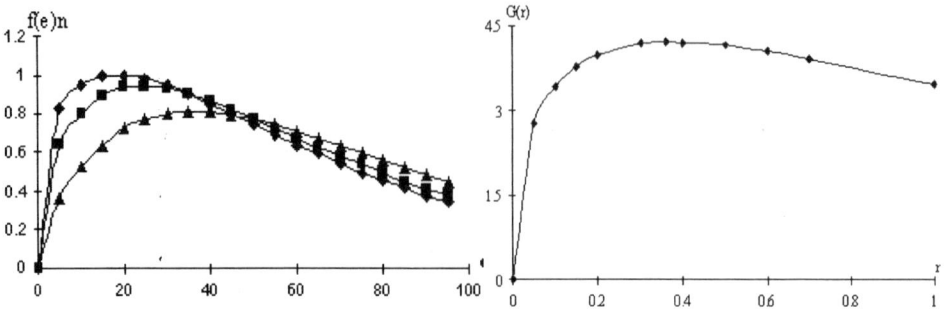

Figure 4. The energy distribution functions of atoms: $r = 0$ ♦, $r = 0.1$ ■, $r = 5$ ▲.

Figure 5. The schedule of $G(r)$ function.

The schedule of $G(r)$ function is shown in Fig. 5.

When $r \to 0$, the function $G(r) \to 0$, because $P(r) \to 1$ and $e_a(r) \to e_a(0)$. When r ($r \to \infty$) are large, the calculation results showed, that $G(r) \to 0$. Computer simulations have given an interval estimation of the maximal position for $G(r)$: $r_{max} \in (0.35; 0.37)$. The practical interest of the results obtained is that a maximum flow of heat can be retracted near maximum $G(r_{max})$ irrespective of the temperature of the condensed phase (if an appropriate substance is chosen) and minimum when $r \to 0$ or $r \to \infty$ in evaporation processes.

The importance of computer simulations made is that they enabled to look at the evaporation process in a new way from a microscopic point of view. It is shown that the escaping atoms speed energy not only for overcoming the potential barrier during the evaporation of the substance into the Knudsen's layer, but due to the escape of much faster atoms an additional kinetic energy is carried away, which can be interpreted as overcoming of the higher potential barrier.

The suggested problem solution of evaporation can be transferred, by analogy, to other physical processes connected with overcoming potential barriers, for example, ions escape from melts, electron thermoemission, particles escape from relativistic space objects, etc.

The practical significance of the obtained results lies in the possibility on their basis to create qualitatively new and to modify known methods of substance division. Selecting appropriate substances and the temperatures it is possible to create optimum modes of the heat and mass exchange in installations. The results obtained can have a great significance in the vacuum metallurgy, production of chips, determination of "true" energy values of atoms bond when escaping from the surface of the condensed phase.

REFERENCE

1. C. Cercignani. Mathematical Methods in Kinetic Theory, Plenum Press, New York (1969).
2. G.A. Bird. Molecular Gas Dynamics, Clarendon Press, Oxford (1978).
3. T. Koga. Introduction to Kinetic Theory Stochastic Processes in Gaseous Systems, Pergamon Press, Osford (1970).
4. Y. Sone, Y. Onishi, Kinetic theory of evaporation and condensation, *J. Phys. Soc. Jap.*, 44, 8: 1981 (1978).
5. I.N. Ivchenko, Generalization of the Lees method in boundary problems of transfer, *Journal of Colloid and Interface Sci.*, 135, 1: 16 (1990).
6. I. Estermann, O. Simpson, W. Stern, The free fall of atoms and the measurement of the velocity distribution in a molecular beam of cesium atoms, *Phys. Rev.*, 71: 238 (1947).
7. D. Havazelet, A. Birnboim, Evaporation of metals by a high energy dense source, *J. Phys. D.*, 16, 10: 1917 (1983).
8. G.E. Forsythe, M.A. Malcolm, C.B. Moler. Computer Methods for Mathematical Computations, Prentice-Hall, INC, New Jersy (1977).

ON OSCILLATION AND NONOSCILLATION DOMAINS FOR DIFFERENCE RICCATI EQUATION

Julia V. Elyseeva

Moscow State University of Technology "STANKIN",
101472 Moscow, Russia

INTRODUCTION

Consider the discrete Euler equations system [1]

$$\mathbf{Y}_{i+1} = \mathbf{W}_i \mathbf{Y}_i, \quad i = 0, \ldots, N \tag{1}$$

here $\mathbf{Y}_i$ is real $2n \times n$ matrix; $\mathbf{W}_i$ is real $2n \times 2n$ symplectic matrix:

$$\mathbf{W}_i' \mathbf{J}_{2n} \mathbf{W}_i = \mathbf{J}_{2n},$$

where $\mathbf{J}_{2n} = \begin{bmatrix} \mathbf{0}_n & \mathbf{I}_n \\ -\mathbf{I}_n & \mathbf{0}_n \end{bmatrix}$,

and $\mathbf{I}_n$ is the $n \times n$ identity matrix. If $\mathbf{W}_i$ is partitioned matrix with $n \times n$ blocks $\mathbf{W}^i_{kl}$ $k,l = 1, 2$ and $\mathbf{Y}_i = [\mathbf{Y}^i_1, \mathbf{Y}^i_2]'$, $\mathbf{Y}^i_2 = \mathbf{Q}_i \mathbf{Y}^i_1$, then matrix $\mathbf{Q}_i$ is a formal solution of the symplectic matrix Riccati equation

$$\mathbf{W}^i_{21} - \mathbf{Q}_{i+1} \mathbf{W}^i_{11} + \mathbf{W}^i_{22} \mathbf{Q}_i - \mathbf{Q}_{i+1} \mathbf{W}^i_{12} \mathbf{Q}_i = \mathbf{0}_n. \tag{2}$$

We assume, that the following conditions hold for the system (1).

Assumption 1. $|\mathbf{W}^i_{12}| \neq 0$, $i = 0, \ldots, N$.

Assumption 2. The system (1) is disconjugate on $[0, N+1]$, i.e. there exists at most one integer $i_0 \in [0, N]$, such that $(\mathbf{y}^{i_0}_1)'(\mathbf{W}^{i_0}_{12})^{-1} \mathbf{y}^{i_0+1}_1 \leq 0$ for any nontrivial column $[\mathbf{y}^i_1, \mathbf{y}^i_2]$ of matrix $\mathbf{Y}_i$.

According to the results of[1,2,3,4,5] we have that **Assumptions 1, 2** are equivalent to the existence of a symmetric solution of (2) such that

$$Q_i + \left(W_{12}^i\right)^{-1} W_{11}^i > 0, i \in [0, N] \tag{3}$$

It is evident, that condition (3) for the three-term recurrence relation form[2]

$$A_{i+2} Z_{i+2} - C_{i+1} Z_{i+1} + A_{i+1} Z_i = 0_n, \, A_i' = A_i, C_i' = C_i, |A_i| \neq 0, \tag{4}$$

is equivalent to $\qquad Q_i > 0.$ \hfill (5)

Where $A_{i+1} Z_{i+1} = Q_i Z_i$, because the equation (4) may be written as (1) with

$$W_i = \begin{bmatrix} 0_n & I_n \\ -I_n & C_{i+1} \end{bmatrix} \begin{bmatrix} A_{i+1} & 0_n \\ 0_n & A_{i+1}^{-1} \end{bmatrix}, Y_i = [Z_i, A_{i+1} Z_{i+1}]' \tag{6}$$

It is easy to establish, that equation (1) under **Assumption 1** may also be written by way of (6) after the following transformation of solution Y_i:

$$\begin{bmatrix} I_n & 0_n \\ \left(W_{12}^i\right)^{-1} W_{11}^i & I_n \end{bmatrix} Y_i = \tilde{Y}_i, i = 0, \ldots, N, Y_{N+1} = \tilde{Y}_{N+1}.$$

For $\tilde{Y}_i$ we obtain the system

$$\tilde{Y}_{i+1} = \begin{bmatrix} 0_n & I_n \\ -I_n & D_i \end{bmatrix} \begin{bmatrix} \left(W_{12}^i\right)^{-1'} & 0_n \\ 0_n & W_{12}^i \end{bmatrix} \tilde{Y}_i, i = 0, \ldots, N, \tag{7}$$

$$D_i = \left(W_{12}^{i+1}\right)^{-1} W_{11}^{i+1} + W_{22}^i \left(W_{12}^i\right)^{-1}, i = 0, \ldots, N-1, \, D_N = W_{22}^N \left(W_{12}^N\right)^{-1},$$

which is similar to (6). Moreover, for matrix

$$\tilde{Q}_i = Q_i + \left(W_{12}^i\right)^{-1} W_{11}^i, i \in [0, N],$$

where $\tilde{Y}_i = [\tilde{Y}_1^i, \tilde{Y}_2^i], \tilde{Y}_2^i = \tilde{Q}_i \tilde{Y}_1^i$, we have condition (5) as disconjugacy criterion for (1). To avoid complications, let the second factor in the right part of system (7) be equal to identity matrix. In general case we can introduce the sequence of transformations

$$\tilde{Y}_i = \begin{bmatrix} R_i & 0_n \\ 0_n & \left(R_i^{-1}\right)' \end{bmatrix} \hat{Y}_i, \left(R_{i+1}\right)^{-1} W_{12}^i \left(R_i^{-1}\right)' = I_n, R_0 = I_n,$$

hence, for $\hat{Y}_i$ we obtain the system

$$\hat{Y}_{i+1} = \begin{bmatrix} 0_n & I_n \\ -I_n & \hat{D}_i \end{bmatrix} \hat{Y}_i, \hat{D}_i = R'_{i+1} D_i R_{i+1}, i = 0, \ldots, N. \tag{8}$$

For (8) associated Riccati equations may be written as

$$\hat{Q}_{i+1} = \hat{D}_i - \hat{Q}_i^{-1}, \hat{Q}_i = \left(R_i^{-1}\right)' \hat{Q}_i R_i^{-1}, i = 0,\ldots, N. \tag{9}$$

Since matrices R_i are nonsingular, the existence of positively definite solutions of (9) is disconjugacy criterion for (1). We formulate in our notations some results of 1, 2, 3, 4, 5.

Theorem 1. If (1) is disconjugate $[0, N+1]$ then the following inequalities hold

$$\hat{D}_i > 0, \quad i = 0,\ldots, N-1, \qquad \hat{Q}_i < \hat{D}_{i-1}, \quad i = 1,\ldots, N+1,$$

where $\hat{Q}_i$ is any positively definite solution of (9) on $[0, N]$.

Theorem 2. The system (1) is disconjugate on $[0, N+1]$ if and only if there exists solution $P^{i,0}$ of (9) on $[0, N+1]$ such that $P^{1,0} = \hat{D}_0$ and hence, there exists solution $S^{i,N+1}$ of (9) on $[0, N]$ such that $S^{N,N+1} = 0_n$, so the following inequalities hold

$$S^{i,N+1} > 0, \quad i = 0,\ldots, N-1; P^{i,0} > 0, \qquad i = 1,\ldots, N,$$

$$S^{i,N+1} < \hat{Q}_i, \quad i = 0,\ldots, N; \hat{Q}_i < P^{i,0}, \qquad i = 1,\ldots, N+1,$$

where $\hat{Q}_i$ is any positively definite solution of (9) on $[0, N]$.

According to **Theorem 2** we can consider the domain of the initial conditions for the solutions of (9):

$$\hat{Q}_0 > S^{0,N+1}. \tag{10}$$

Any symmetric solution of (9) with initial condition from (10) is nonoscillatory on $[0, N]$. The nonoscillatory solutions domain for (9) is shown in Fig. 1 by shading. It follows from **Theorem 2** that there exist $S^{i,p}$, $i = 0,\ldots, p-1$ which are "minimal" solutions of (9) such that

$$S^{p-1,p} = 0_n, p = 1,\ldots, N+1, \tag{11}$$

and there exist $P^{i,q}$, $i = q+1,\ldots, N+1$ – "maximal" solutions of (9) such that

$$P^{q+1,q} = \hat{D}_q, \quad q = 0,\ldots, N-1..$$

These sets of solutions are shown in Fig.1 by dotted lines. The domains of the initial conditions

$$S^{0,p} < \hat{Q}_0 < S^{0,p+1}, p = 1,\ldots, N \text{ or } \hat{Q}_0 < S^{0,p+1} = 0_n, p = 0, \tag{12}$$

are oscillatory solutions domains: $\hat{Q}_i > 0$ if $i \neq p$ and $\hat{Q}_p < 0$. Some solution of (9) from (12) is shown in Fig. 1 by wavy lines.

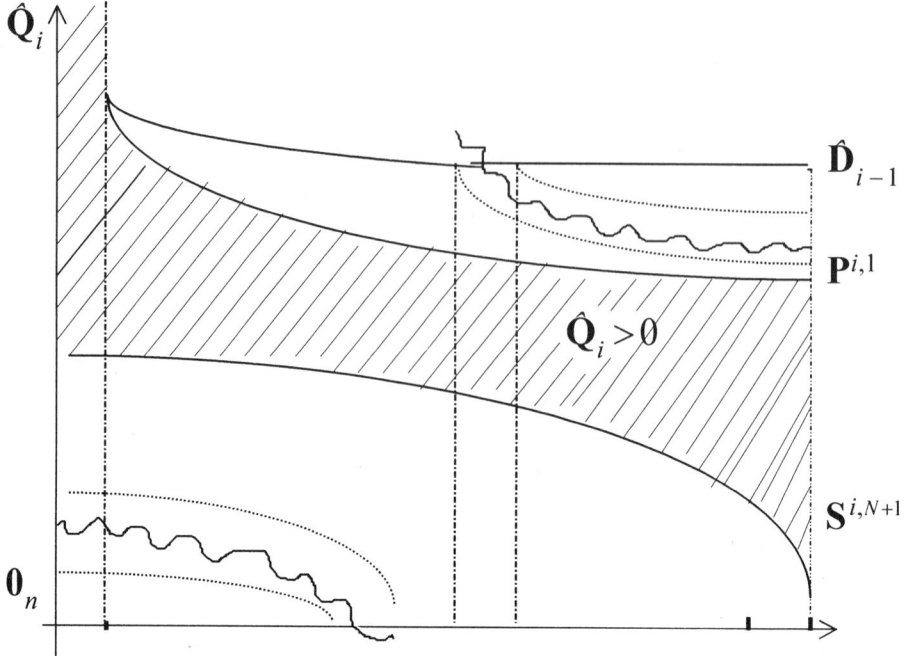

Figure 1. Oscillation and nonoscillation domains for the difference Riccati equation.

MAIN RESULTS

Further results are connected with initial conditions domains such that there exist non-oscillatory (oscillatory) submatrices of solution (9) on $[0, N]$.

We introduce a set of transformations for solutions of (9):

$$\hat{\mathbf{Q}}_j^i = \left(\mathbf{G}_j^i \hat{\mathbf{Y}}_1^i + \mathbf{F}_j^i \hat{\mathbf{Y}}_2^i\right)\left(\mathbf{F}_j^i \hat{\mathbf{Y}}_1^i - \mathbf{G}_j^i \hat{\mathbf{Y}}_2^i\right)^{-1}, \; i \in [0, N+1] \tag{13}$$

where $\hat{\mathbf{Y}}_i = \left[\hat{\mathbf{Y}}_1^i, \hat{\mathbf{Y}}_2^i\right]$ is a solution of (8) and $\mathbf{G}_j^i, \mathbf{F}_j^i$ are $n \times n$ blocks of orthogonal symplectic matrix

$$\eta^i = \eta_{j(i)}^i = \begin{bmatrix} \mathbf{F}_j^i & -\mathbf{G}_j^i \\ \mathbf{G}_j^i & \mathbf{F}_j^i \end{bmatrix}. \tag{14}$$

Here $j = j(i)$ is a function of discrete argument such that $j \in \{0, 1, ..., 2^n - 1\}$, $\mathbf{G}_j^i, \mathbf{F}_j^i$ are diagonal matrices which consist of zeros and ones. Thus, the diagonal of $\mathbf{G}_j^i$ is a binary representation for $j = j(i)$ and $\mathbf{G}_j^i + \mathbf{F}_j^i = \mathbf{I}_n$. Under condition $\left|\hat{\mathbf{Y}}_1^i\right| \neq 0$ transformation (13) can be written in form

$$\hat{\mathbf{Q}}_j^i = \left(\mathbf{G}_j^i + \mathbf{F}_j^i \hat{\mathbf{Q}}_i\right)\left(\mathbf{F}_j^i - \mathbf{G}_j^i \hat{\mathbf{Q}}_i\right)^{-1} \tag{15}$$

It is obviously that $\hat{Q}_0^i \equiv Q_i$, $\hat{Q}_{2^n-1}^i = -Q_i^{-1}$. If there exists a nonsingular $k \times k$ block Q_{11} of matrix Q_i which is located in the first k lines and columns, then there exists $\hat{Q}_j^i$ where j has binary code $(\underbrace{1...1}_{k}\underbrace{0...0}_{n-k})$ and

$$\hat{Q}_i = \begin{bmatrix} I_n & 0_n \\ Q'_{12}Q_{11}^{-1} & I_n \end{bmatrix} \begin{bmatrix} Q_{11} & 0_n \\ 0_n & \Delta \end{bmatrix} \begin{bmatrix} I_n & Q_{11}^{-1}Q_{12} \\ 0_n & I_n \end{bmatrix}, \Delta = Q_{22} - Q'_{12}Q_{11}^{-1}Q_{12}, \qquad (16)$$

$$\hat{Q}_j^i = \begin{bmatrix} -Q_{11}^{-1} & -Q_{11}^{-1}Q_{12} \\ -Q'_{12}Q_{11}^{-1} & \Delta \end{bmatrix}. \qquad (17)$$

For arbitrary j matrices $\hat{Q}_i$ and $\hat{Q}_j^i$ are connected with (16), (17) by a similarity transformation with some permutation matrix. If $j = j(i) = const$, then $\hat{Q}_j^i$ is a formal solution of the following Riccati equation:

$$\hat{Q}_j^{i+1} = F_j D_i F_j - (I_n + G_j D_i F_j)'(Q_j^i + G_j D_i G_j)^{-1}(I_n + G_j D_i F_j), \qquad (18)$$

where $G_j^i = G_j$, $F_j^i = F_j$. The system

$$\hat{Y}_j^{i+1} = \begin{bmatrix} G_j \hat{D}_i G_j & I_n - G_j \hat{D}_i F_j \\ -I_n - F_j \hat{D}_i G_j & F_j \hat{D}_i F_j \end{bmatrix} \hat{Y}_j^i, \hat{Y}_j^i = \eta_j \hat{Y}_i \qquad (19)$$

is associated with (18) (here $\eta_j = \eta_j^i$ is defined by (14)).

Now we formulate some property of operator (23).

Lemma 1. Consider system (1) with some initial condition

$$Y_0 = [Y_1^0, Y_2^0], \qquad (Y_1^0)'Y_2^0 - (Y_2^0)'Y_1^0 = 0_n, \qquad (20)$$

where Y_0 has full rank. Let (13) (where $\hat{Y}_i \equiv Y_i, \hat{Q}_j^i \equiv Q_j^i$) be transformation of the solution of (1), (20). Then there exists a set of values of $j = j(i)$, $i \in [0, N+1]$, (path of integration) such that corresponding symmetric matrix Q_j^i is uniformly bounded:

$$\|Q_j^i\| < 1 + (n-1)\sqrt{2}, i \in [0, N+1].$$

Proof. See [6,7], where continuous and discrete systems (1) are considered. Lemma 1 substantiates a numerical algorithm for solution of difference or differential equation (2) when condition $|Y_1^i| \neq 0$ is disturbed.

Lemma 2. Let A and B be two symmetric matrices satisfying

$$A < B, \; G_j A G_j + F_j > 0 \qquad (21)$$

for some $j \in \{0, 1, ..., 2^n - 1\}$. Let $\mathbf{A}_j$ and $\mathbf{B}_j$ are defined by (15) where $\hat{\mathbf{Q}}_i = \mathbf{A}$ or $\mathbf{B}$ and

$$\hat{\mathbf{Q}}^i_j = \mathbf{A}_j \text{ or } \mathbf{B}_j; \; \mathbf{G}^i_j = \mathbf{G}_j, \mathbf{F}^i_j = \mathbf{F}_j.$$

Then
$$\mathbf{A}_j \leq \mathbf{B}_j, \; \mathbf{G}_j \mathbf{B}_j \mathbf{G}_j - \mathbf{F}_j < 0. \tag{22}$$

Conversely, if (22) hold for some j then inequalities (21) hold.

Proof. Suppose, without loss of generality, that j has binary code $(\underbrace{1...10}_{k}\underbrace{...0}_{n-k})$. Then from the second inequality (21) we have $\mathbf{A}_{11} > 0$ where $\mathbf{A}_{11}$ is a submatrix of $\mathbf{A}$, which is located in the first k lines and columns and hence $0 < \mathbf{A}_{11} \leq \mathbf{B}_{11}$ from the first inequality (21). From formulae (17) rewriting for $\mathbf{A}, \mathbf{B}$ and $\mathbf{A}_j, \mathbf{B}_j$ we have

$$\mathbf{A}^j_{11} = -\mathbf{A}^{-1}_{11} \leq -\mathbf{B}^{-1}_{11} = \mathbf{B}^j_{11} < 0. \tag{23}$$

Then the second inequality (22) holds. To prove the first inequality (22) we consider the following factorization

$$\mathbf{B}_j - \mathbf{A}_j = (\mathbf{BG}_j - \mathbf{F}_j)^{-1}(\mathbf{B} - \mathbf{A})(\mathbf{G}_j \mathbf{A} - \mathbf{F}_j)^{-1}. \tag{24}$$

Hence by ***Theorem 35.2*** of [8] there exists nonsingular matrix $\mathbf{R}$ such that $\mathbf{B}_j - \mathbf{A}_j = \mathbf{R}(\mathbf{B}-\mathbf{A})\mathbf{R}'$. Matrix $\mathbf{R}$ is connected with the square root of $(\mathbf{AG}_j - \mathbf{F}_j)(\mathbf{BG}_j - \mathbf{F}_j)^{-1}$.

This lower block-triangular matrix has positive eigenvalues if (21) is hold. Hence in our case $\mathbf{R}$ is real and the first inequality (22) is true. Conversely from (22),(23),(24) we have that inequalities (21) are valid. This completes the proof of Lemma 2.

Note that more general conclusion follows from the proof of lemma 2. Thus the first inequality (22) follows from the first inequality (21) if nonsingular matrix $(\mathbf{AG}_j - \mathbf{F}_j)(\mathbf{BG}_j - \mathbf{F}_j)^{-1}$ has real positive eigenvalues corresponding mutually simple elementary divizors and the first inequality (22) follows with opposite sign if these eigenvalues are negative.

Now for the equation (18) we consider some analogs of theorems 1 and 2. Consider the conditions for $\hat{\mathbf{Q}}^i_j$ that are equivalent to condition $\hat{\mathbf{Q}}_i > 0$.

Theorem 3. The system (1) is disconjugate on $[0, N+1]$ if and only if there exists some $j \in \{0, 1, ..., 2^n - 1\}$ such that there is a solution of (18) on $[0, N]$ satisfying the following conditions for $i = 0, ..., N$:

$$\mathbf{G}_j \hat{\mathbf{Q}}^i_j \mathbf{G}_j - \mathbf{F}_j < 0, \tag{25}$$

$$\mathbf{F}_j \hat{\mathbf{Q}}^i_j \mathbf{F}_j + \mathbf{G}_j > 0, \tag{25*}$$

Proof. It follows from (16) that $\hat{\mathbf{Q}}_i > 0$ if and only if $\mathbf{Q}_{11} > 0$ and $\Delta > 0$. Then for $j = (\underbrace{1...10}_{k}\underbrace{...0}_{n-k})$ we have inequalities $(\hat{\mathbf{Q}}^i_j)_{11} < 0$ and $(\hat{\mathbf{Q}}^i_j)_{22} > 0$ which are equivalent to (25) and (25*). The proof is completed.

We introduce the following set of solutions of (18)

$$\Omega_j^p = \begin{cases} \hat{\mathbf{Q}}_j^i \text{ is the solution of (18) on } [0, p] \text{ such that} \\ \text{if } i = 1, \ldots, p \text{ then (25) holds;} \\ \text{if } i = 0, \ldots, p - 1 \text{ then (25)}^* \text{ holds.} \end{cases}, j=0,\ldots 2^n-1, p=1,\ldots N+1. \quad (26)$$

Now we consider the necessary conditions for the existence of $\hat{\mathbf{Q}}_j^i \in \Omega_j^p$.

Theorem 4. If matrix $\hat{\mathbf{Q}}_j^i \in \Omega_j^p$ exists then $\hat{\mathbf{Q}}_j^i$ satisfies the following inequalities

$$\hat{\mathbf{Q}}_j^i < \mathbf{F}_j \hat{\mathbf{D}}_{i-1} \mathbf{F}_j, i = 1, \ldots, p, \quad (27)$$

$$\hat{\mathbf{Q}}_j^i > -\mathbf{G}_j \hat{\mathbf{D}}_i \mathbf{G}_j, i = 0, \ldots, p-1. \quad (27^*)$$

Corollary 1. The system (19) is disconjugate on $[0, p]$ if and only if $\Omega_j^p \neq \emptyset$.

Corollary 2. If $\hat{\mathbf{Q}}_j^i \in \Omega_j^p$ then matrix $\mathbf{H}_i^p$ which connects $\hat{\mathbf{Q}}_j^i$ with $\mathbf{P}_j^i$ – other solution of (18):

$$\Lambda_j^i = \mathbf{L}_i \Lambda_j^k \left(\mathbf{I}_n - \mathbf{H}_i^p \Lambda_j^k \right)^{-1} \mathbf{L}_i'; \Lambda_j^i = \hat{\mathbf{Q}}_j^i - \mathbf{P}_j^i$$

$$\Delta \mathbf{H}_i^p = \mathbf{L}_i' \left(\hat{\mathbf{Q}}_j^i + \mathbf{G}_j \hat{\mathbf{D}}_i \mathbf{G}_j \right)^{-1} \mathbf{L}_i, \mathbf{H}_k^p = \mathbf{0}_n, 0 \leq k < i \leq p-1 \quad (28)$$

$$\mathbf{L}_{i+1} = \left(\left(\hat{\mathbf{Q}}_j^i + \mathbf{G}_j \hat{\mathbf{D}}_i \mathbf{G}_j \right)^{-1} \left(\mathbf{I}_n - \mathbf{G}_j \hat{\mathbf{D}}_i \mathbf{F}_j \right) \right)' \mathbf{L}_i, \mathbf{L}_k = \mathbf{I}_n$$

is positively definite: $\mathbf{H}_i^p > 0$, $i = k+1, \ldots, p$.

Note 1. Utilizing the condition $\mathbf{H}_i^p > 0$ we can formulate the comparison theorem for the solutions of (18). Thus, if $\Lambda_j^k = \hat{\mathbf{Q}}_j^k - \mathbf{P}_j^k \geq 0$ for some $k \geq 0$ and $\hat{\mathbf{Q}}_j^i \in \Omega_j^p$ then Λ_j^i does exist and $\Lambda_j^i \geq 0$ for $p \geq i > k$. Combining **Corollary 2** and **Lemma 2** we can transform inequality $\hat{\mathbf{Q}}_j^i \leq \mathbf{P}_j^i$ which holds for $j \neq 0$ in the solution space for $j = 0$ and vice versa.

Note 2. From (27) and (27*) we have the necessary condition for matrix $\hat{\mathbf{D}}_i$:

$$\mathbf{G}_j \hat{\mathbf{D}}_i \mathbf{G}_j + \mathbf{F}_j > 0, \quad \mathbf{F}_j \hat{\mathbf{D}}_{i-1} \mathbf{F}_j + \mathbf{G}_j > 0, i = 1, \ldots, p-1.$$

Proof of theorem 4. We rewrite equation (18) in form

$$\mathbf{F}_j \mathbf{D}_i \mathbf{F}_j - \mathbf{Q}_j^{i+1} = \left(\mathbf{I}_n + \mathbf{G}_j \mathbf{D}_i \mathbf{F}_j \right)' \left(\hat{\mathbf{Q}}_j^i + \mathbf{G}_j \mathbf{D}_i \mathbf{G}_j \right)^{-1} \left(\mathbf{I}_n + \mathbf{G}_j \mathbf{D}_i \mathbf{F}_j \right) \quad (29)$$

and assume that $j = (\underbrace{1\ldots1}_{k}\underbrace{0\ldots0}_{n-k})$. Utilizing Frobenius' inversion of partitioned matrices we can show that matrices $\mathbf{F}_j \hat{\mathbf{D}}_i \mathbf{F}_j - \hat{\mathbf{Q}}_j^{i+1}$ and $\left(\hat{\mathbf{Q}}_j^i + \mathbf{G}_j \mathbf{D}_i \mathbf{G}_j \right)^{-1}$ are congruent with matrix

$$diag\left[-\left(Q_j^{i+1}\right)_{11}, \left(Q_j^i\right)_{22}^{-1}\right], \qquad i=0,\ldots,p-1 \tag{30}$$

Since $\hat{Q}_j^i \in \Omega_j^p$ then matrix (30) does exist and it is positively definite for $i=0,\ldots,p-1$. This completes the proof of theorem 4.

Proof of corollary 1. It is obvious that necessary and sufficient condition for disconjugacy on $[0, p]$ for system (19) is

$$Q_j^i + (I_n - G_j \hat{D}_i F_j)^{-1} G_j \hat{D}_i G_j = Q_j^i + G_j \hat{D}_i G_j > 0, \, i \in [0, p-1],$$

which is equivalent to (27^*). Moreover from (27^*) we have (27) for any solution of (29). Thus disconjugasy of system (19) follows from condition $\Omega_j^p \neq \emptyset$. Conversely from both inequalities (27), (27^*) we have that $\hat{Q}_j^i \in \Omega_j^p$. The proof is completed.

The following theorem is an analogue of theorem 2.

Theorem 5. The system (19) is disconjugate on $[0, N]$ if and only if there exists $P_j^{i,0}$ is solution of (18) on $[1, N]$ satisfying (25) for $i=2,\ldots,N$ and (25^*) for $i=1,\ldots,N-1$ such that $P_j^{1,0} = F_j \hat{D}_0 F_j$, and hence, there is solution $S_j^{i,N}$ of (18) on $[0, N-1]$ satisfying (25) for $i=1,\ldots,N-1$ and (25^*) for $i=0,\ldots,N-2$, such that $S_j^{N-1,N} = -G_j \hat{D}_{N-1} G_j$, so the following inequalities hold

$$\hat{Q}_j^i < P_j^{i,0}, i=1,\ldots,N, \quad \hat{Q}_j^i > S_j^{i,N}, i=0,\ldots,N-1 \tag{31}$$

for any $\hat{Q}_j^i \in \Omega_j^N$.

Note 3. It follows that under conditions of theorem 5 there exist $S_j^{i,p}, p=1,\ldots,N$, $i=0,\ldots,p-1$ is "minimal" solutions of (18) satisfying $S_j^{p-1,p} = -G_j \hat{D}_{p-1} G_j$ and "maximal" solutions $P_j^{i,q}$, $q=0,\ldots,N-2$, $i=q+1,\ldots,N$ satisfying $P_j^{q+1,l} = F_j \hat{D}_q F_j$. If the system (1) is disconjugate on $[0, N+1]$ then matrix $P_j^{i,q}$ satisfies additional condition (25^*) for $i=N$ and matrix $S_j^{i,p}$ satisfies (25) for $i=0$.

Proof of **Theorem 5** is similar to the proof of **Theorem 2**. Note that $P_j^{i,q}$ corresponds to the solution of (19) such that

$$\left(Y_j^i\right)_1 = 0_n, \left(Y_j^i\right)_2 = I_n, \hat{Y}_j^i = \left[\left(Y_j^i\right)_1, \left(Y_j^i\right)_2\right], i=q. \tag{32}$$

For the solution of (19) corresponding to $S_j^{i,p}$ we have (32) for $i=p$.

Proof of note 3. In this case it follows from theorem 3 that there exists $\hat{Q}_j^i \in \Omega_j^N$ for which (25), (25^*) hold for $i=0,\ldots,N$. Then from (31) we have that matrix $P_j^{i,0}$ satisfies additional condition $(25)^*$ for $i = N$. Since $\mathbf{P}_j^{i,q} > \mathbf{P}_j^{i,0}, i = q+1, N$ then for any $\mathbf{P}_j^{N,q}$ holds (25^*). For $\mathbf{S}_j^{i,p}$ the proof is analogous.

Theorem 6. If the system (1) is disconjugate on $[0, N+1]$ then (19) is disconjugate on $[0, p]$, $p = 1, ..., N$. For any ordered sequence $0 = j_0 \prec j_1 \prec ... \prec j_n = 2^n - 1$ (the sequence is ordered if it is reduced to sequence $\tilde{j}_k = (\underbrace{1...1}_{k}\underbrace{0...0}_{n-k})$ $k = 0, 1, ..., n$ by the same permutations of components for any j_k in its binary code) corresponding sequence of "minimal" solutions satisfies the following inequalities

$$0 \leq \mathbf{\hat{S}}_{j_0}^{i,p} \leq \mathbf{\hat{S}}_{j_1}^{i,p} \leq ... \leq \mathbf{\hat{S}}_{j_n}^{i,p} \equiv \mathbf{S}_{j_0}^{i,p+1}, \quad p = 1,...,N, \quad i = 0,1,...,p-1, \quad (33)$$

$$rank\left(\mathbf{\hat{S}}_{j_q}^{i,p} - \mathbf{\hat{S}}_{j_k}^{i,p}\right) = |q - k|, \quad q,k = 0,...,n. \quad (34)$$

Here $\mathbf{S}_j^{i,p}$ is reduced to $\mathbf{\hat{S}}_j^{i,p}, j \neq j_0$ by inverse transformation (15), $\mathbf{S}_{j_0}^{i,p} \equiv \mathbf{S}^{i,p}$ is defined by (11).

Proof. From disconjugacy of (1) and **Theorem 3** we have that $\Omega_j^p \neq \varnothing, p = 1, ..., N$. Then from **Corollary 1** system (19) is disconjugate on $[0, p]$ $p = 1, ..., N$. Without loss of generality we consider sequence $j_k = (\underbrace{1...1}_{k}\underbrace{0...0}_{n-k})$ $k = 0, 1, ..., n$. Corresponding sequence of "minimal" solutions satisfies the following initial conditions

$$\mathbf{\hat{S}}_{j_k}^{p-1,p} = diag\left\{\left(\mathbf{\hat{D}}_{p-1}^{11}\right)^{-1}, \mathbf{0}_{n-k}\right\}, \quad p = 1,...,N, \quad (35)$$

where $\mathbf{\hat{D}}_{p-1}^{11}$ is the $k \times k$ block of $\mathbf{\hat{D}}_{p-1} > 0$. It follows from equation (9) that

$$\mathbf{S}_{j_0}^{p-1,p+1} = \left(\mathbf{\hat{D}}_{p-1}\right)^{-1}, p = 1,...,N. \quad (36)$$

Utilizing Frobenius' inversion of partitioned matrix it is easy to prove that initial conditions (35), (36) satisfy (33), (34). By **Note 3** matrix $\mathbf{S}_{j_k}^{i,p}$ satisfies (25), (25^*) for $i \in [0, p-2]$. Hence from **Theorem 3** we have $\mathbf{\hat{S}}_{j_k}^{i,p} > 0, \ i \in [0, p-2]$. From comparison theorem for any pair of solution $\mathbf{S}_{j_k}^{i,p}, \mathbf{S}_{j_q}^{i,p}$ we will have (33), (34).

Note 4. For "maximal" solutions $\mathbf{P}_j^{i,q}$ under the conditions of **Theorem 6** we have

$$\mathbf{\hat{P}}_{j_0}^{i,q} \leq \mathbf{\hat{P}}_{j_1}^{i,q} \leq ... \leq \mathbf{\hat{P}}_{j_n}^{i,q} \equiv \mathbf{\hat{P}}_{j_0}^{i,q+1} \leq 0, \quad q = 0,...,N-2, \quad i = q+1,...,N, \quad (37)$$

where $\mathbf{\hat{P}}_{j_k}^{i,q}$ is the image of $\mathbf{P}_{j_k}^{i,q}, j_k \neq j_n$ in the solution space with number $j = j_n$. Also we can transform (33) and (37) in any solution space with number $j \neq 0, 2^n - 1$ for index $j_k \succ j$ and $j_k \prec j$, correspondingly.

Under **Assumption 2** let us consider for (18) the domain of initial conditions

$$\hat{\mathbf{Q}}_j^0 > \mathbf{S}_j^{0,N}, \ \mathbf{G}_j\hat{\mathbf{Q}}_j^0\mathbf{G}_j - \mathbf{F}_j < 0. \tag{38}$$

For any solution of (18) from domain (38) the condition (25) holds for $i = 0, ..., N$. Hence this solution has nonoscillatory submatrix on $[0, N]$. The domain of these solutions for arbitrary $j > 0$ is shown in Fig. 2 by shading. The domain of nonoscillatory solutions from Fig. 1 is shown in Fig. 2 by double shading. For $j = 2^n - 1$ these domains are the same (see Fig. 3). The domains of initial conditions

$$\mathbf{S}_j^{0,p-1} < \hat{\mathbf{Q}}_j^0 < \mathbf{S}_j^{0,p}, \ p = 2,...,N, \tag{39}$$

$$\hat{\mathbf{Q}}_j^0 < \mathbf{S}_j^{0,p}, \ p = 1 \tag{40}$$

$$\hat{\mathbf{Q}}_j^0 > \mathbf{S}_j^{0,N}, \ \mathbf{G}_j\hat{\mathbf{Q}}_j^0\mathbf{G}_j + \mathbf{F}_j > 0, \ j > 0 \tag{41}$$

define the solutions of (18) with oscillatory submatrices. For solutions of (18) from (39), (40) condition (25) holds for $i \neq p$ and it holds with the opposite sign if $i = p$. For the solution satisfying (41) we have (25) for $i > 0$. Some "minimal" and "maximal" solutions are shown in Figs. 2, 3 by dotted lines. A solution of (18) from (39) is shown in Figs. 2, 3 by wavy lines.

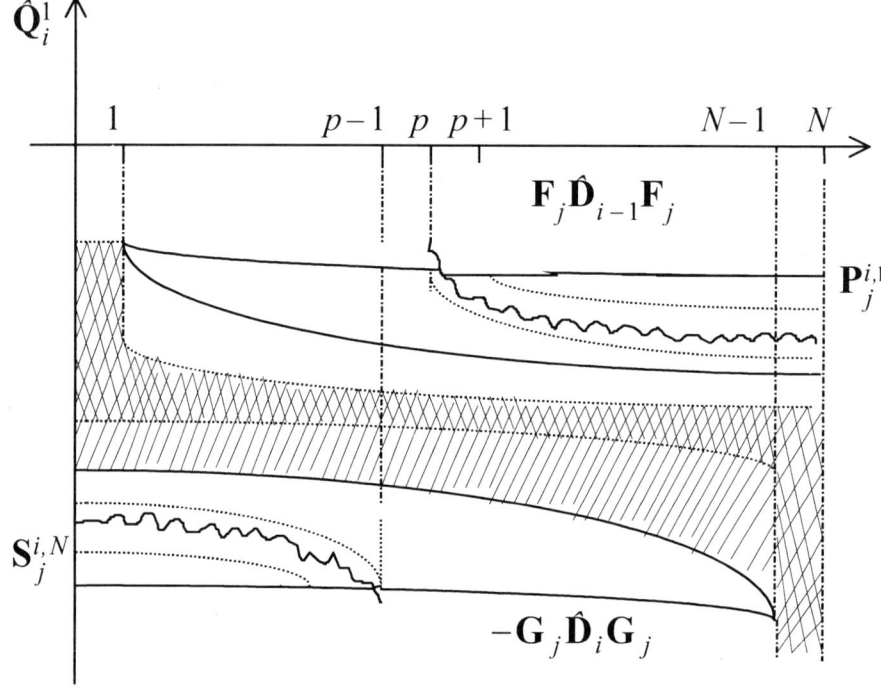

Figure 2. Domains of existence of nonoscilatory and oscillatory submatrix for arbitrary $j > 0$.

CONCLUSIONS

The necessary conditions for disconjugacy of (1) on $[0, N+1]$ are disconjugacy of (19) for any $j = 0,...,2^n - 1$ on $[0, N]$ and the existence of "minimal" solutions of (18) on

$[0, N-1]$ with the supplementary condition (25) for $i = 0$. In the solution space for $j = 0$ it permits to investigate the initial conditions which are not comparable with matrices (11).

Consider any ordered sequence $0 = j_0 \prec j_1 \prec \ldots \prec j_n = 2^n - 1$ and sequence $\hat{\mathbf{S}}^{i,p}_{j_k}$ corresponding to the "minimal" solutions of (18). By theorem 6 we have the following "net" of matrices in the initial conditions domain

$$\mathbf{0}_n = \hat{\mathbf{S}}^{0,0}_{2^n-1} = \mathbf{S}^{0,1}_0 \leq \hat{\mathbf{S}}^{0,1}_{j_1} \leq \hat{\mathbf{S}}^{0,1}_{j_2} \leq \cdots \leq \hat{\mathbf{S}}^{0,1}_{2^n-1} = \mathbf{S}^{0,2}_0 \leq \hat{\mathbf{S}}^{0,2}_{j_1} \leq \hat{\mathbf{S}}^{0,2}_{j_2} \leq \cdots$$

$$\cdots \leq \hat{\mathbf{S}}^{0,2}_{2^n-1} = \mathbf{S}^{0,3}_0 \leq \cdots \leq \hat{\mathbf{S}}^{0,N-1}_{2^n-1} = \mathbf{S}^{0,N}_0 \leq \hat{\mathbf{S}}^{0,N}_{j_1} \leq \cdots \leq \hat{\mathbf{S}}^{0,N}_{2^n-1} = \mathbf{S}^{0,N+1}_0.$$

Here $\hat{\mathbf{S}}^{i,p}_0 \equiv \mathbf{S}^{i,p}$ are defined by (11) and shown in Fig.4 by solid lines. Matrices $\hat{\mathbf{S}}^{i,p}_{j_k}$, $j_k \neq 0, \ldots, 2^n - 1$ are shown in Fig. 4 by dotted lines.

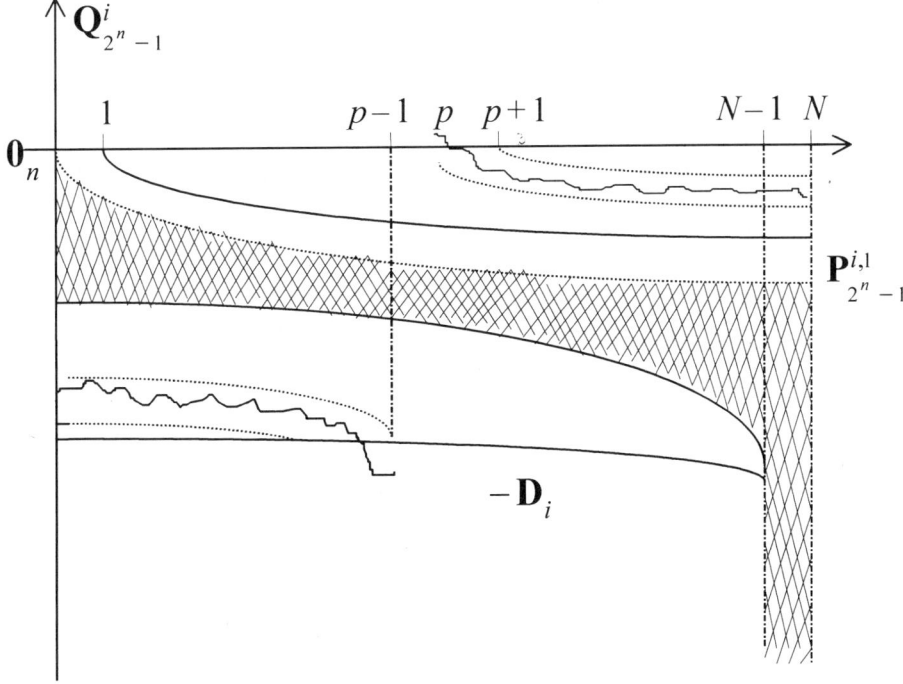

Figure 3. Oscillation and nonoscillation domains for $j = 2^n - 1$.

1. If there exists $\hat{\mathbf{S}}^{i,p}_{j_k}$ such that $\hat{\mathbf{Q}}_0 > \hat{\mathbf{S}}^{i,p}_{j_k}$ then $\hat{\mathbf{Q}}_i$ exists on $[0, p]$. Solution $\hat{\mathbf{Q}}_i$ has the nonoscillatory $k \times k$ block on $[0, p]$ and $\hat{\mathbf{Q}}_i > 0$ on $[0, p-1]$. Particularly for $p = N$ we have that $\hat{\mathbf{Q}}_i$ has the nonoscillatory $k \times k$ block on $[0, N]$.

2. If for $\hat{\mathbf{Q}}_0 > 0$ inequalities $\hat{\mathbf{S}}^{0,p-1}_{j_k} < \hat{\mathbf{Q}}_0 < \hat{\mathbf{S}}^{0,p}_{j_k}$, $p = 2, \ldots, N$ hold then $\hat{\mathbf{Q}}_i$ exists on $[0, N]$ and has the oscillatory $k \times k$ block such that the condition

$$\mathbf{G}_{i_k} \mathbf{Q}_i \mathbf{G}_{i_k} + \mathbf{F}_{i_k} > 0 \qquad (42)$$

holds for $i \neq p$ and it is valid with opposite sign for $i = p$.

3. If initial condition $\mathbf{Q}_0$, $rank\hat{\mathbf{Q}}_0 > 0$ isn't positively definite we can investigate the question of the localization of $\hat{\mathbf{Q}}_j^0$ in domains defined by (40), (41). If (40), (41) hold for some $j = j_k$ then $\hat{\mathbf{Q}}_i$ exists on $[0, N]$ and (42) holds for $i \neq 1$ or $i \neq 0$ correspondingly. Moreover for indefinite initial condition we can consider the question of localization of $\hat{\mathbf{Q}}_j^i$ in Ω_j^p, $p = 1, \ldots N$. Note that there may exist the cases when the inequality $\mathbf{Q}_j^0 > \mathbf{S}_j^{0,p}$ holds but $\hat{\mathbf{Q}}_0$ isn't comparable with $\hat{\mathbf{S}}_j^{0,p}$.

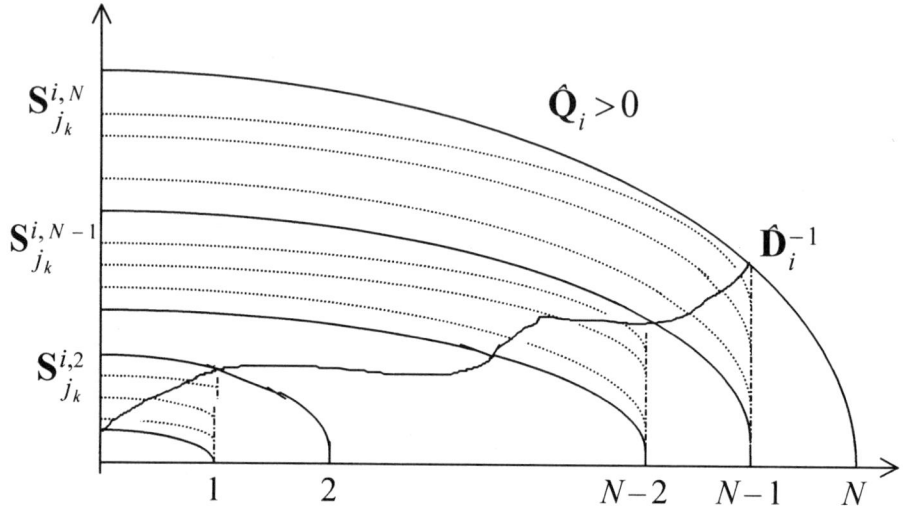

Figure 4. Images of "minimal" solutions in solution space for $j = 0$

This work is supported by Federal Program "Integration", project №43, 2.1 .

REFERENCES

1. C.D. Ahlbrand, Asymptotics of discrete time Riccati equations, robust control and discrete linear Hamiltonian systems, *Panamer. Math. J.*, 5: 1 (1995).
2. C.D. Ahlbrand, Dominant and recessive solutions of symmetric three term recurrences, *J.Differential Equations*, 107: 238 (1994).
3. M. Bohner, Linear Hamiltonian difference systems: disconjugacy and Jacobi-type conditions, *J. Math. Anal. and Appl.*, 199: 804 (1996).
4. L. Erbe, P. Yan, Disconjugacy for linear Hamiltonian difference systems, *J. Math. Anal. and Appl.*, 167: 35 (1992).
5. L. Erbe, P. Yan, On the discrete Riccati equation and its application to discrete Hamiltonian systems, *Rocky Mount. J. Math.*, 25: 167 (1995).
6. J.V. Elyseeva, On algorithm for solution of the symplectic matrix Riccati equation, *Vestnik Moskovskogo Universiteta. Seriya XV: Vychislitel'naya Matematika i Kibernetika*, 2: 14 (1990).
7. J.V. Elyseeva, On algorithm for solution of the difference matrix Riccati equation, *J.Computational mathematics and Mathematical physics*, to appear.
8. C.C. Mac Duffee. *The Theory of Matrices*, Verlag Vonjulius Springer, Berlin (1933).

Nonlinear Phenomena in Physics

DUSTY PARTICLE INTERACTION IN PLASMA, PLACED IN MAGNETIC FIELD

Andrej P. Nefedov[1], Viktor D. Lakhno[2]

[1]High Energy Density Research Center RAS
127412 Moscow, Russia

[2]Institute of Mathematical Problems of Biology RAS
142292 Pushchino, Moscow Region, Russia

The lecture is devoted to an entirely new field "Physics of dusty plasma"[1-3] which is just beginning to develop. Dusty plasma is a low-temperature plasma containing macroparticles or, to put it simply, dusty particles. Clouds of dusty plasma gas such as dusty interstellar clouds and nebulae, comet's tails, dusty coats of stars, etc. occure widely in space plasma. Dusty plasma is of great cosmogonic significance. Thus, according to current concepts, collapse of gaseous dusty clouds gives rise to aggregations and associations of young stars.

Another example of dusty plasma is rarefield low-temperature plasma consisting of neutral gas, micron-sized dusty particles, ions and electrons. In what follows we will deal just with this type of plasma. However, many regularities revealed are true for space dusty plasma as well.

Typical parameters of dusty plasma studied under earthly conditions fall in the range: pressure at ~1 $Torr$, room temperature for a low-pressure discharge and pressure ~ 10^3 $Torr$, 1700–2600 K for thermal plasma. The extent of plasma ionization is rather small and approximates 10^{-7}. Dusty particles brought into such plasma add on ions and electrons and thus acquire a charge which can be extremely large. It depends on the parameters of a particle and, for example, for a particle of a micron size it can be as much as hundreds or thousands of electron charges. An analog of dusty particles in atmospheric phenomena is drops of water in thunderclouds. Large charge of dusty particles is responsible for their strong interaction with one another and with plasma particles. According to recent investigations this interaction can modify significantly the properties of dusty plasma. A spectacular proof of the vital role of this interaction was provided by the discovery of a plasma-dusty crystallization. The conditions for crystallization of dusty plasma were found theoretically by Ikezi (USA). However, an actual "plasma crystal" was observed in high-frequency discharge plasma only 10 years later. This ordered structure is formed in a radio-frequency discharge near the bottom electrode at the boundary of the volume charge layer. The lattice constant of the plasma crystal is in the range of fractions of a millimeter which makes it visible to the naked eye. Fig. 1 shows a scheme of an experimental set up to study ordered structures in gas discharge plasma.

Plasma crystals show a diversity of unique making them an indispensable tool to study strongly imperfect plasma as well as fundamental of crystals. We can state that observations of self-organized dusty structures open up a new avenue of investigations which could naturally be called "superchemistry" or "supercondensed medium" and where dusty particles play the role of atoms and bound states of dusty particles act as simplest molecules.

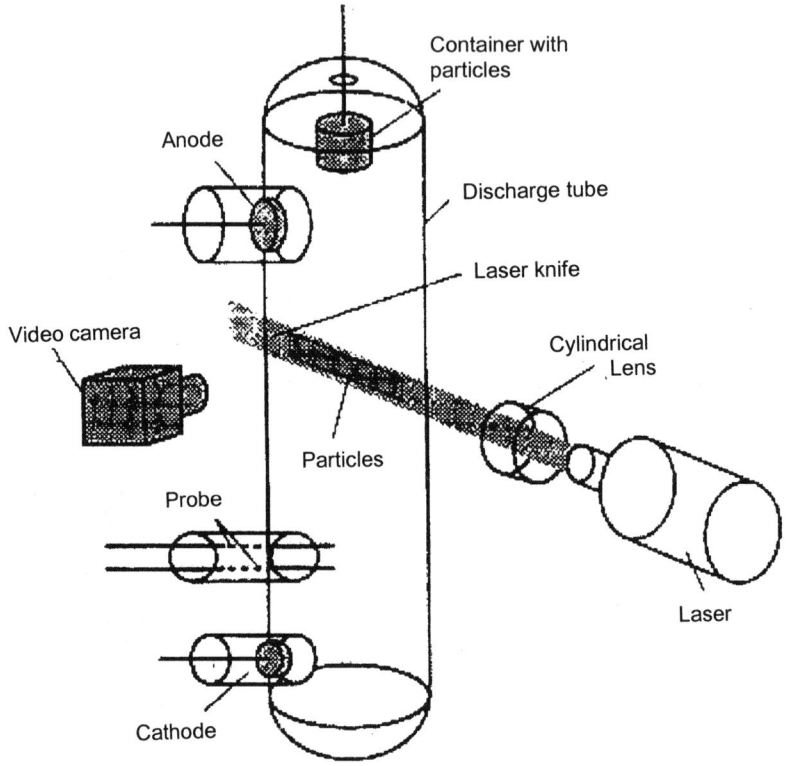

Figure 1. The setup for experimental investigation of ordered structures in gas discharge plasma.

In this connection of particular interest is to investigate the forces responsible for the interaction between dusty particles and, primarily, to understand what forces make dusty particles of the same charge to attract one another yielding a bound system, such as dusty molecule or a dusty crystal.

To unravel the situation it is very important to realize that the answer depends on the experimental conditions. Until very recently all the experiments with dusty plasma were performed under earthly conditions. There the interaction is mainly contributed by three types of forces, i.e. electrostatic, gravity and ionic ones. In the simplis case when the discharge is formed between two horizontally mounted electrodes, the cloud of dusty particles is confined in a volume charge layer near the electrode with negative potential where an equilibrium between gravity and electrostatic forces is established (Fig. 2).

In this case only two forces, i.e. the electrostatic repulsion and gravity contribute into the interaction of dusty particles. If so, the particles can be considered to occure in an electrostatic trap. From below they are confined by the electrode's repulsive field. Neither can they spread horizontally due to the repulsive field produced by the negative potential of the walls of the gas discharge tube. The answer to the question as to whether these particles are in a gaseous (disordered) state or in a crystallized (ordered) one depends on the parameter $\gamma_p = (ze)^2/\bar{r}T$, where T is the plasma temperature, ze is the particles charge, $r = (4\pi n_p/3)^{-1/3}$, n_p is the particle concentration. An ordered state arises only for $\gamma_p > 171$. Under conditions of experiments with dusty plasma γ_p lies in the range $10^4 \div 10^6$, where crystallization of dusty particles a priori takes place.

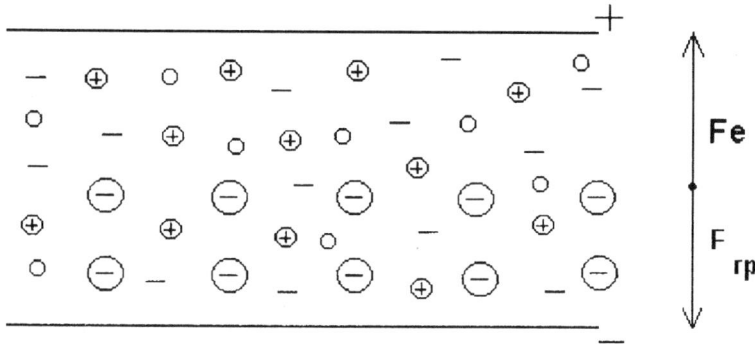

Figure 2. $\ominus$ – dusty particle, $\oplus$ - positive ions, – – electrons, $\bigcirc$ -- neutral gas molecules.

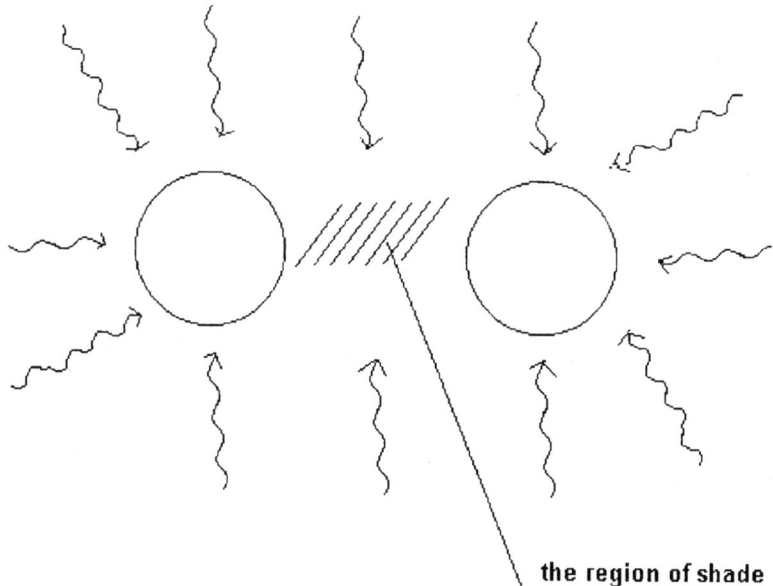

Figure 3. Illustration of Lessage forces.

To understand whether dusty particles can crystallize with no electrostatic trap the experiment should be transferred into space. In the absence of gravity the electrostatic repulsion of dusty particles is balanced by the "ionic" force alone. Let us consider more carefully this type of attractive forces acting through distances more then Debye radius, when the electric interaction may be neglected. The prototype of such forces was considered with in an old and practically forgotten theory which, in essence, was in follows. In 1784 Swiss mathematician George Louis Lessage suggested a simple and elegant mechanical substantiation for Newton's law of universal gravitation. According to views of those times ether consisted of small elementary atoms moving every which way. Lessage's main idea was that when penetrating a massive body, ether atoms lost a small portion of their kinetic energy. If the body is single it is hit by the same number of particles from every side and the total force acting on the body is zero. If there is another body nearly, it reduces the impulse of the ether to the former body, resulting in the effective attraction of the bodies (Fig. 3).

At large distances the decrease of the flow, and hence, the attractive force is proportional to the solid angle from which the second body is seen as viewed from the first one, i.e. it is inversely proportional to the squared distance. Of course, this simple model cannot account for real gravitation. The suggested mechanism would have led to friction of planets against ether and, finally, to their failing on the Sun. We have dwelt on this story at such length because Lessage's model almost exactly simulates dusty plasma, where electrons, ions and the neutral component play the role of ether atoms. In the case of single dusty particle the flow of plasma onto the surface is spherically-symmetrical and the average bombarding force is zero. Attraction between two dusty particles arises due to the fact that the particles cover each other from the plasma flow. As a result the effective pressure of plasma on the side of each particle presented to the neighbour decreases (Fig. 3).

Fig. 4 illustrates the experimental evidence for the existence of Lessage's forces in plasma. Two squared plates of size 25×25 mm were suspended by thin copper wires parallel to each other at some distance. If plasma in the space around the plates is absent the plates are separated from each other (Fig. 4a). If plasma is generated in the region of the plates they attach each other (Fig. 4b). This effect is independent of the type of material from which the plates are fabricated. The experiments where performed with both metal and dielectric plates and illustrate the existence of Lessage forces in plasma[4].

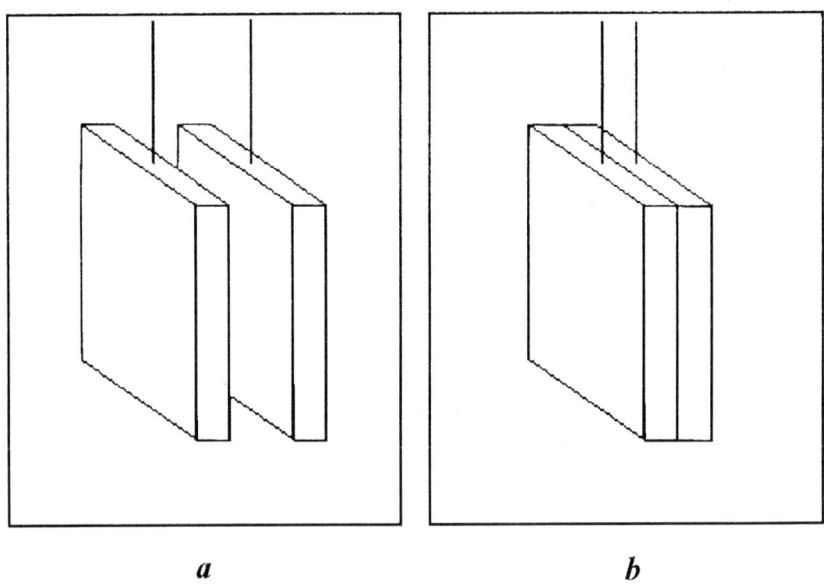

a   *b*

Figure 4. Experimental evidence for the existence of Lessage's forces in plasma.

To calculate this attractive force let us consider the case when the distance r between dusty particles greatly exceeds the particle's radius a. To simplify the calculations we invert the problem and assume that an ion colliding with a particle is not absorbed but emitted by the particle. Ions emitted by the particle bombard another particle, resulting in the repulsion of the particles. The repulsive force taken with the opposite sign is just the sought-for force of attraction between dusty particles. If the distance between the particles is much larger than their radius, the first particle can be considered as a point. Let the ions emitted by the point particle move at velocity V. Denote the number of ions emitted by the point particle in a unit time by I.

Then the number of ions passing through another dusty particle of radius a (Fig. 5) is equal to dI:

$$dI = I \frac{\pi a^2}{4\pi r^2} \tag{1}$$

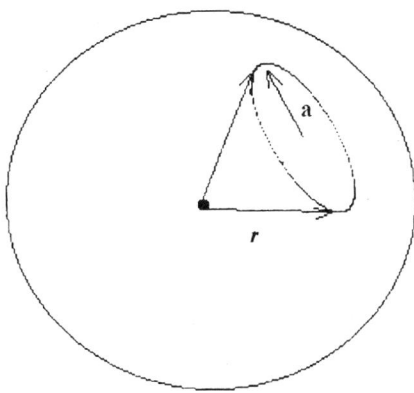

Figure 5.

To calculate the bombarding force F_i acting on the second particle we should multiply the number of ions falling on this particle dI (given by (1)) by the impulse of each ion mV:

$$F_i = mV dI \tag{2}$$

Bearing in mind, that under equilibrium conditions the number of ions absorbed by a dusty particle in a unit time is $I = \pi n_i V a^2$ (where n_i is the concentration of ions in plasma) and in view of (1), (2) we express the repulsive force of dusty particles as:

$$F_i = \frac{\pi}{4} n_i m V^2 \frac{a^4}{r^2} \tag{3}$$

Electrons and the neutral component can be considered in a similar way. Note, that under equilibrium conditions when the temperature of the neutral component and a dusty particle is equal, neutral molecules do not contribute into the total force of bombardment of dusty particles. The reason is that at equilibrium the number of neutral molecules falling on a dusty particle is equal to the number of emitted molecules. According to (3) the contribution of electrons is also much less than that of ions, since the electron mass is much less than the ion mass. Accordingly, the impulse transmitted by electrons appears to be much less than that transmitted by ions.

The force of ion bombardment F_i is much less than electrostatic repulsion of dusty particles if the particles are separated by a distance of the order of Debye radius or less. Only in the case when the distance between the dusty particles significantly larger than the Debye radius, F_i exceeds the electrostatic repulsion. Since at such a large distance the bombarding forces are small, the attractive force of dusty particles is also small. Nowadays the question as to whether a dusty crystal with an open boundary can exist remain open.

The situation can change radically if we place dusty plasma into a magnetic field. To understand what changes it will cause let us neglect collisions of ions with one another, with electrons and with the neutral component. As before, we assume that upon colliding with a dusty particle ions are absorbed by the particle. We will also believe that the distance between dusty particles greatly exceeds the Debye radius, i.e. ions move mainly in the region where an electrostatic field is lacking.

Thus, the only force acting on the ions is the Lorentz force:

$$F = \frac{q}{c} V_\perp H \tag{4}$$

where q is the ion charge, c is the speed of light, H is the intensity of the magnetic field, $V_\perp$ is the ion velocity component perpendicular to the magnetic field. The motion of an ion in the magnetic field presents a helical line with the radius $mV_\perp c/qH$, and the spacing $2\pi mcV \cos\alpha/qH$, where α is the angle between the direction of the initial velocity and the direction of the magnetic field.

A homogenous magnetic field has a focusing effect on the beams of charged particles lying in the plane perpendicular to the field and on the beams making a small angle with the direction of the line of force.

As in the case of the absence of a magnetic field let us consider an invert problem. Suppose, two dusty particles are located along the line of force. Then the ions emitted from the first particle and capable of falling on the second one form a beam slightly divergent wiyh respect to the line of force which serves as the axis of the beam (Fig. 6).

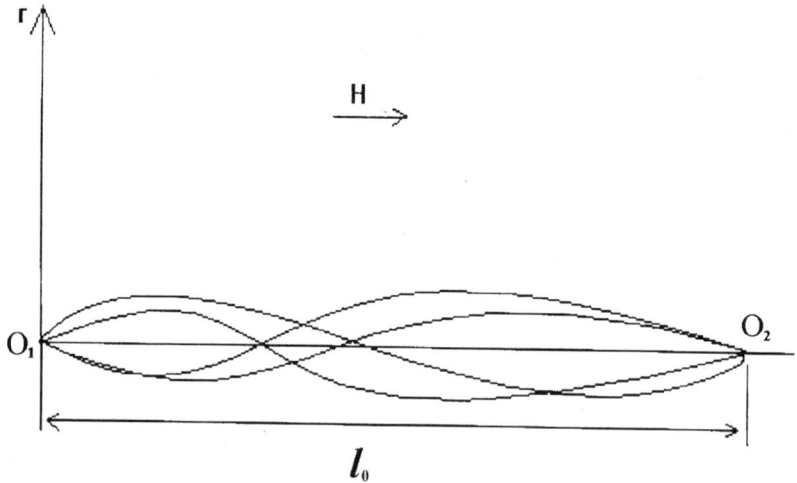

Figure 6. Focusing action of the magnetic field.

Any ion issuing out of the point O_1 where the first dusty particle resides will intersect again the line of force passing through this point after the time $T = 2\pi mc/qH$. The distance which it will travel along the axis, i.e. the period of the helical line is determined by the formula (at small angle α):

$$l = 2\pi \frac{mc}{qH} V \left(1 - \frac{\alpha^2}{2}\right) \qquad (5)$$

Thus, all the ions issuing out of the point O_1 at small angles to the line of force will fall on this line within a narrow interval Δl:

$$\Delta l = \pi \frac{mc}{qH} V \alpha^2 \qquad (6)$$

near the point O_2 separated from O_1 by the distance:

$$l_0 = 2\pi mcV/qH \qquad (7)$$

Accordingly, the radius of the "spot" Δp where the ions emitted from O_1 are focused in the range of angles $(0, \alpha)$ on the plane perpendicular to the axis and passing through O_2 is equal to:

$$\Delta p = a\Delta l \qquad (8)$$

The focusing effect of the magnetic field can enhance many-fold the attraction between the dusty particles if the second point is placed at the point O_2.

To access how many times the force of ion bombardment of a dusty particle F_H at the focal point O_2 exceeds the corresponding force F_0 without a magnetic field let us consider the case of a weak magnetic field $R \gg a$, where $R = mcV/qH$ is the Larmor radius. Note, that relation (8) enables us to access the maximum value of the angle a_{max}, at which the ions emitted from the first dusty particle fall on the second one. To do this the size of the spot Δp where the ions focus should not exceed the radius of the dusty particle a. It follows:

$$a_{max} = \sqrt[3]{\frac{a}{\pi R}} \qquad (9)$$

The number of ions emitted by the first dusty particle and falling on the second one is proportional to the ratio of the area cut out by a cone with the aperture a_{max} on the surface of sphere to the area of this sphere and equal to:

$$dI = \frac{a_{max}^2}{4} I \qquad (10)$$

Substituting (9), (10) into the expression for the force (2) we have:

$$F_H = \frac{mV}{4} \left(\frac{a}{\pi R}\right)^{2/3} I \qquad (11)$$

Comparing (11) and (3) we obtain the following ratio for the forces of ion bombardment of dusty particles separated from each other by the distance $l_0 = 2\pi R$:

$$\frac{F_H}{F_0} = 4\pi^{4/3}\left(\frac{R}{a}\right)^{4/3} \tag{12}$$

In view of (12) and the condition $R \gg a$ the attractive force in the magnetic field F_H greatly exceeds that in the absence of the field.

To summarize, a dusty molecule in a magnetic field is stable if the distance between dusty particles is proportional to l_0, i.e. equal to $l = n l_0$, $n = 1, 2, \ldots$ Since under conditions of a glow discharge at pressure ~ 1 mm Hg and normal temperature ($T = 300\ K$) the ion free path length is $\lambda \sim 10^{-2}\ cm$, the considered motion of ions without collisions: $\lambda \geq l_0$ will take place for the magnetic field values $H \geq 10^3\ o.e.$

It follows from (12) that the ratio between the force of ion bombardment in a magnetic field and that in the absence of a field increases as the field grows: $F_H/F_0 \sim H^{-4/3}$. Therefore, at first glance, the effect should be especially pronounced just in weak fields. In actual particle, however, this is not the case. Relation (12) was desired for the case when dusty particles are separated by the distance $R \sim H^{-1}$. According to (3): $F_0 = R^{-2}$. Substituting this result into the relation F_H/F_0, we find that the force of ion bombardment is $F_H \sim H^{2/3}$, i.e. it increases as the magnetic field grows. The effect is maximum in the limit of strong magnetic fields when the Larmor radius R becomes much less than the radius of the dusty particle a. This situation is shown in Fig. 7. In the absence of dusty particles, ions (given a strong external magnetic field) are not free to move in the direction perpendicular to the lines of force. The trajectory of each ion is would around the line of force, therefore the motion assumes markedly anisotropic, directed character. In the case under analysis, as is seen from Fig. 7, all the ions falling on dusty particles push them towards one another.

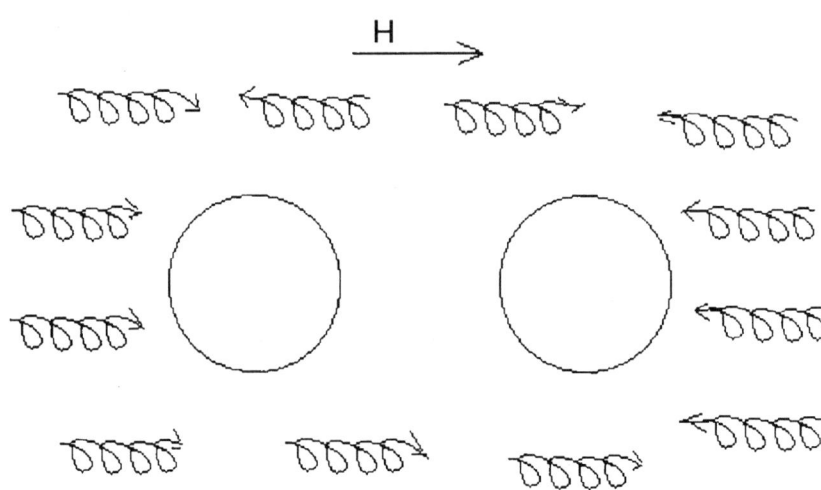

Figure 7. Illustration of dusty particle attraction in strong magnetic field.

In the above considered case of a weak magnetic field this mechanism of attraction involves only a fraction of emitted ions falling on a dusty particle, these are the ions which fall within the small angle α in Fig.5. It follows from Fig.6 that in the case of a strong magnetic field the force between dusty particles is:

$$F_i = \frac{1}{4} mVI = \frac{1}{4} \pi n_i mV^2 a^2 \qquad (13)$$

Thus, in this limiting case the force of attraction between dusty particles is maximum and independent of the distance between the particles. According to (13), this force is also independent of the magnetic field value, which is fully consistent with the scheme in Fig.6. For dusty particles of radius $a \cong 10 mcm$, the condition of a strong field when expression (13) holds, is fulfilled for $H \geq 10^4 \sqrt{A}$ (o.e.), where A is the atomic weight of an ion.

It is appropriate to compare the force F_i with the force of Coulomb repulsion between two dusty particles: $F_k = z^2 e^2 / l^2$, where l is the distance between the particles. It follows from (13) that the attractive force F_i caused by ion bombardment becomes equal to the force of Coulomb repulsion F_k at the distance l:

$$l = \frac{2ze}{\sqrt{\pi n_i mV^2 a^2}} \qquad (14)$$

For a particle with $a = 10 mcm$ and the dusty plasma parameters $n_i = 10^{10}$, $V = 2,5 \cdot 10^5 \, cm/s$ (hydrogen plasma), $m = 1,8 \cdot 10^{-24} \, gr$, $e = 4,8 \cdot 10^{-10} \, SGSE$, $z = 10^4$ in the equilibrium state this equation yilds $l = 0,16 cm$. Actually, the electrostatic repulsion of dusty particles is decreased due to Debye screening. For the above plasma parameters the Debye radius is $r_d = \sqrt{T/4\pi n_i e^2} \approx 10^{-3} cm$, i.e. much less than the calculated distance l. Thus, in real dusty plasma equilibrium distances between dusty particles in a strong magnetic field will be approximately equal to the Debye screening radius.

Properly speaking, it is this fact that makes the study of the effect of an external magnetic field on dusty molecules and crystals so difficult. As we pointed out at the beginning of the lecture, under earthly conditions a dusty crystal occurs in an electrostatic trap and the distance between dusty particles is of the order of the Debye screening radius. In this case application of a magnetic field practically does not change the crystal's structure. In space the situation is quite different. There the equilibrium distance between dusty particles can exceed considerably the Debye radius and the role of an external magnetic field may be decisive.

To turn back to dusty-plasma clouds in interstellar space mentioned at the beginning of the lecture, the considered mechanism of interaction in earthly dusty plasma can be of cosmogonic significance as well. For example, the early stages of the formation of stellar systems from dusty-plasma clouds are still obscure nowadays. It is not improbable that gravity forces alone are not sufficient to start gravity compression in so rarefield system as interstellar gas clouds. Conceivably, the interstellar magnetic field may also play a significant role in these processes.

REFERENCES

1. A.P. Nefedov, O.F. Petrov, V.E. Fortov, *Uspekhi Fizicheskih Nauk*, 167, 11: 1251 (1997).

2. *Advances in dusty plasmas*, P.K. Shukla, D.A. Mendis, T. Desai, ed., World Scientific, Singapore (1997).
3. V.E. Fortov, V.I. Molotkov, A.P. Nefedov, in: *Progress in Physics of Clusters*, V.D. Lakhno, A.P. Nefedov, G.N. Chuev, ed., World Scientific, Singapore (1998).
4. A.E. Dubinov, V.S. Zhdanov, A.M. Ignatov, S.Y. Kornilov, S.A. Sadovoy, V.D. Selemir, in: *Proceedings of Conference on Low Temperature Plasma*, Petrozavodsk State Univ. Press. Ed. A.D. Khakhaev (1998).
5. Ya.K. Hodataev, R. Binghem, V.P. Tarakanov, V.N. Tsytovich, *Physics of Plasma*, 22, 11: 1028 (1996).

TUNNEL TRANSPORT OF ELECTRONS AT ANHARMONIC ACCEPTING MODE

Dmitrii S. Chernavskii[1], Nina M, Chernavskaya[2], Ludmila A. Uvarova[3]

[1]P.N. Lebedev Physical Institute Russian Academie of Scienties,
117934 Moscow, Russia

[2]Moscow State University,
Moscow, Russia

[3]Moscow State University of Technology "STANKIN",
101472 Moscow, Russia

INTRODUCTION

Tunnel carrying of electron in biological systems possesses some particularities. Firstly, electron does from one localized state to another localized state. Secondly, energy levels of these states are different and excess of the energy is transferred to so named Accepting Mode (hereinafter AM). Part of the surplus energy dissipates, but part of it moves over to other forms and is saved. Dissipation is required for ensuring the direction of the process. Conservation of the energy is required for ensuring energy need of alive system.

These particularities are connected with the fact that biological macromolecules (protein-enzyme, in particular - electron couriers) are represented as constructions of nanometric scales. The theory developed for nonliving objects, where important role play such notions, as "zone" etc. is not applicable for molecular constructions. Exception is artificial constructions used now in nanometric electronics.

The following problems appear under the theoretical approach:
i) Problem irreversibility (dissipation of the energy) in the quantum mechanics. It pertains to fundamental problem and hitherto not solved. There are only phenomenological approaches, allowing to get plausible results. The most popular of them is method of complex hamiltonian which we shall use.
ii) In this instance theory must describe both dissipation of the energy, and conservation and transformation of part of the energy. These tasks are complementary and, however, both must be dared, in united models.

iii) Popular in the quantum mechanics perturbation method in this instance is not applicable (however it is used), so more correct approach is needed.

In proposed message we will consider a variant of the model, allowing to describe both dissipation of energy, and its conservation, not resorting to perturbation theories.

PHYSICAL MODEL OF THE PROCESS

Theories of the tunnel transport of electrons is devoted extensive literature [1-7], in which important place occupies a question about transformations of the energy of the excited electron in other forms: energy an metastable (tense) conformation, "energized" proton and others [8-12].

Physical model, provided a basis for the estimations of the velocity of the process, is kept two subsystems: electronic and AM. On the Fig. 1 potential relief of the electronic subsystem is represented by two square-wave pits. At an initial moment the electron inheres in the left pit I. At the end of the process this electron is localized in the pit II.

Profile of the free energy AM is submitted on the Fig. 1 on the right. In harmonic approach this profile is represented by a parabola, which position of the minimum depends on localization of the electron. Interaction of the electron with AM brings about the shift of the minimum of the parabola on the value Δq. On the Fig. 1 parabola "in" corresponds to a profile AM, when electron is localized in the pit I and parabola "fin" – localization of the electron in the pit II.

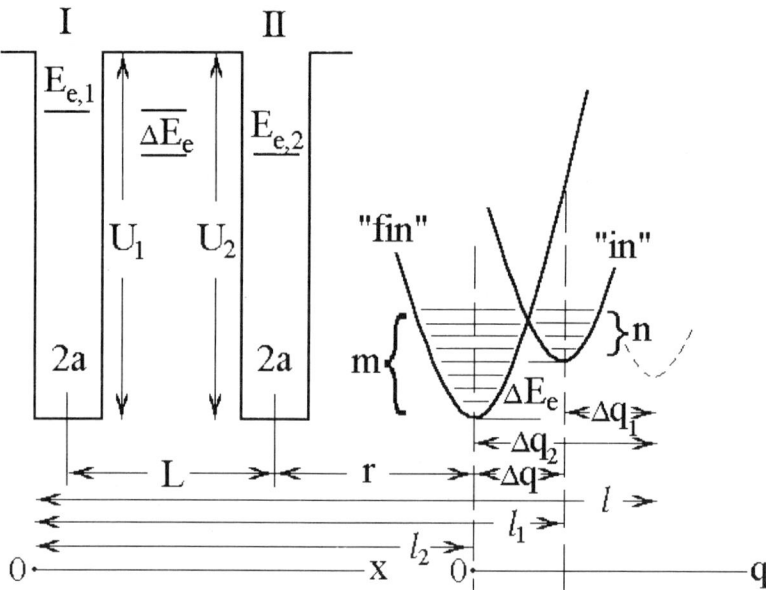

Figure 1. Physical model of electron tunnel transition under harmonic AM. Square pits 1 and 11 – electronophilic groups, 2a – width of pits, U_1 and U_2 – dehths of pits, L – distance between the pit 1 and 11. l is position of centre of gravity of AM in the absence of electron, l_1 is position of centre of gravity of AM when electron is localized in the pit I. l_2 is position of centre of gravity AM when electron is localized in the pit II.

Usually take [1-7] that probability of transition from the initial state, in which energy of the electron is $E_{e,1}$ and energy AM is $E_{AM,1} = n\hbar\omega$, in to final one, with the energy: $E_{e,2}$ and $E_{AM,2} = m\hbar\omega$ (n and m are numbers of quantum's and ω is a frequency of oscillation of AM), has a following structure:

$$W_{n,m} = \frac{2\pi}{\hbar}|E|J_e^2 \cdot F_{n,m}(\Delta q) \cdot \rho(\Gamma, \Delta E) \qquad (1)$$

where J_e^2 is a square of the overlap integral of electronic functions, (so-called barrier factor), which is equal to: $J_e^2 = exp\left[-2L\sqrt{2m_e(U-E)}/\hbar\right]$; $F_{n,m}(\Delta q)$ is a square of the overlap- integral of the overlapping of functions A.M. (factor of the Franc-Condone); $\rho(\Gamma, \Delta E)$ is density of electron levels in final state, $\rho(\Gamma, \Delta E) = \frac{\Gamma}{\pi(\Delta E^2 + \Gamma^2)}$, which depends on damping factor Γ ($\Gamma = h/t$, t is a time of damping) and differences of the energy ΔE in initial and final states $\Delta E = E_{e,1} + n\hbar\omega - (E_{e,2} + m\hbar\omega)$, E is a certain distinctive energy, which value usually is not rendered concrete.

On the form an expression (1) corresponds to the first order of the perturbation theory that is to say Fermi gold rule. Strictly speaking, perturbation theory here not applicable and so value E still in (1) is as an free parameter.

Kinetically coefficient of transition in accordance with (1) is:

$$k_{i \to f} = \sum_{n,m} \frac{exp(-n\hbar\omega/kT)}{N} W_{n,m} = \frac{2\pi}{\hbar}|E|^2 J_e^2 \cdot \sum \frac{exp(-n\hbar\omega/kT)}{N} F_{n,m}(\Delta q)\rho(\Delta E, \Gamma), \quad (2)$$

here $exp(n\hbar\omega/kT)/N$ is probability that at the temperature T in the initial state n oscillatory quantum's AM are present and $N = (1 + exp(-n\hbar\omega/kT))$ is normalizing factor.

Expression (2) well describes experimental dates and is often used, however, question on its justification and on the value E remains presently opened sense perturbation theory is not applicable. We use other approach, which is used at the test of two-level systems. Under $\hbar\omega \ll G$ such approach is justified.

TRANSITION COEFFICIENT IN THE TWO-LEVEL SYSTEM, HARMONIC MODE

In the absence of dissipation an irreversible transitions are forbidden, but reversible transition are permitted only between states with the equal energies. In this case a model of the two-level system, containing electron and AM (refer to Fig. 1) is used.

The Hamiltonian of the systems will represent in the form:

$$\hat{H} = \frac{p^2}{2m} + U_1(x) + U_2(x) + \frac{P^2}{2M} + \frac{1}{2}M\omega^2 q^2 - \frac{e^2}{\kappa(L+q-x)}, \qquad (3)$$

here: q is a coordinate AM, counted out from the point $L+2a+r$, M is a mass AM, x is a coordinate of the electron, counted out from the point 0, m is a mass of the electron, e is an elementary charge, $k = 2$ is microdielectrical constant. Accepted that electron interacts with AM to the account of electrostatic power, AM charged positively and its center of gravity in the absence of interactions situated is in $x=L+2a+r$.

Potentials $U_1(x)$ and $U_2(x)$ are:

$$U_1(x) = \begin{cases} 0, & \text{under } x < -a, \\ -U_1 & \text{under } -a < x < a, \\ 0, & \text{under } a < x, \end{cases} \quad U_2(x) = \begin{cases} 0, & \text{under } x < L-a, \\ -U_2 & \text{under } L-a < x < a, \\ 0, & \text{under } L+a < x, \end{cases}$$

where L is a distance between potential pits.

Let us introduce two auxiliary functions $F_1(x, q)$ and $F_2(x, q)$, which are own functions of Hamiltonians H_1 and H_2 with the energies E_1 and E_2:

$$H_1\,\Phi_1(x, q) = E_1\,\Phi_1(x, q);\ H_2\,\Phi_2(x, q) = E_2\,\Phi_2(x, q);\ E_1 = E_2 = E \qquad (4)$$

Function $\Phi_1(x, q)$ describes an initial state, when electron inheres in the pit I, AM is in state "in" and keeps n quantum. Function $\Phi_2(x, q)$ describes an final state, when electron inheres in the pit II, AM is in state "fin" and is keeps m quantum. So energies and functions of the states must contain indexes n and m. However, in order to avoid crockhood we in this section shall not draw these indexes.

Full energies of the states are equal: $E_1 = E_2 = E$, as far as otherwise transition between them is forbidden. Herewith electronic levels $E_{e,1}$ and $E_{e,2}$ and energy AM in states "in" and "fin", generally speaking, different. Full Hamiltonian is possible to represent in two variants: as: $H = H_1 + \Delta H_1$, or as: $H = H_2 + \Delta H_2$ (concrete form of Hamiltonians H_1 and H_2, as well as ΔH_1 and ΔH_2 we will bring later). The self functions of full Hamiltonian (3) $\Psi_+(x, q)$ and $\Psi_-(x, q)$ will represent as superposition of functions $\Phi_1(x, q)$ and $\Phi_2(x, q)$:

$$\Psi_+(x,q) = \frac{1}{\sqrt{2}}(\Phi_1(x,q) + \Phi_2(x,q));\ \Psi_-(x,q) = \frac{1}{\sqrt{2}}(\Phi_1(x,q) - \Phi_2(x,q)) \qquad (5)$$

Self energies are E_+ and E_-.

$$H\Psi_+(x, q) = E_+\,\Psi_+(x, q);\ H\Psi_-(x, q) = E_-\,\Psi_-(x, q).$$

Energies E_+ and E_- differ from the energy E on the value E_{sp}, so called splitting energy: $E_+ = E + E_{sp}$; $E_- = E - E_{sp}$. Splitting energy is:

$$E_{sp} = E_+ - E = \int \Psi_+ \hat{H} \Psi_+ dx \cdot dq - E = EJ +$$
$$+ \frac{1}{2}\left[\int \Phi_1 \Delta H_1 \Phi dx \cdot dq + \int \Phi_2 \Delta H_2 \Phi_2 dx \cdot dq + \int \Phi_1 \Delta H_2 \Phi_2 dx \cdot dq + \int \Phi_2 \Delta H_1 \Phi_1 dx \cdot dq\right] \qquad (6)$$

where: $J = \int \Phi_1 \Phi_2 dx \cdot dq$ is an integral of the overlapping of wave functions of the electron and AM. This integral is a small: $J \ll 1$, rest integrals in (6) are of the same order or less. So splitting energy is $E_{sp} \ll E$.

Let us consider a time evolution of functions $\Phi(x, q, t)$, which at an initial moment is $\Phi(x, q, t=0) = \Phi_1(x, q)$. For this we represent $\Phi(x, q, t)$ as superposition on own functions jf full Hamiltonian

$$\Phi(x,q,t) = \frac{1}{\sqrt{2}}\left[\Psi_+(x,q) \cdot \exp\left(-\frac{i}{\hbar}E_+ t\right) + \Psi_-(x,q) \cdot \exp\left(-\frac{i}{\hbar}E_- t\right)\right]; \qquad (7)$$

Other functions full Hamiltonian do not give the contribution to the decomposition (7), as far as their energy differ from the energy E on $\hbar\omega \gg E_{sp}$. From (7) and (6) follows that under $t=0$ initial conditions are kept.

Using (5) and (6) we represent (7) in the form of:

$$\Phi(x,q,t) = \frac{1}{2} \exp\left(-\frac{i}{\hbar} E_+ t\right) \left[\Phi_1\left(1 + \exp\left(-\frac{2i}{\hbar} E_{sp} t\right)\right) + \Phi_2\left(1 - \exp\left(\frac{2i}{\hbar} E_{sp} t\right)\right)\right]. \quad (8)$$

From (8) it follows, that function $F(x, q, t)$ in the course of time feels oscillations: at a moment of the time $t = \pi\hbar/2E_{sp}$ it moves over to $\Phi_2(x, q)$ and then through ditto time returns back. Period of oscillations is: $T = \pi\hbar/E_{sp}$. This effect is known under the name "quantum-mechanical oscillations"[13] and with reference to biochemistry is discussed in[14]. Thence clear what is understood under reversible transition in the quantum mechanics.

Transition can become irreversible, if time of daping of the oscillations is much period T less. For the study this will consider a behavior of the function $\Phi_1(x, q, t)$ in the time interval $\hbar/E \ll t \ll T$. We represent function $\Phi_1(x, q, t)$ in this interval in a form:

$$\Phi(x,q,t) = \exp\left(-\frac{i}{\hbar} E_+ t\right)\left(\Phi_1(x,q) + \frac{i}{\hbar} E_{sp} t \cdot \Phi_2(x,q)\right). \quad (9)$$

Amplitude of probability to observe a system in the state $\Phi_2(x, q)$ at moment t, if at a moment $t=0$ it inhere in state $\Phi_1(x, q)$ is:

$$p_{2,1}(t) = \int \Phi_2(x,q)\Phi(x,q,t)dx \cdot dq = \exp\left(-\frac{i}{\hbar} E_+ t\right)\left(J + \frac{i}{\hbar} E_{sp} t\right) \quad (10)$$

Corresponding to this amplitude probability, is:

$$\rho_{2,2}(t) = p_{2,1}^2 \rho_{1,1} = \left(J^2 + \frac{1}{\hbar^2} E_{sp}^2 \cdot t^2\right)\rho_{1,1}, \quad (11)$$

here $\rho_{1,1}$ is probability to catch a system in state $\Phi_1(x, q)$ at an initial moment of the time.

The First member in (11) does not depend on a time and means, that even there is small probability at an initial moment to catch a system in state $\Phi_2(x, q)$. The Second member means that this probability grows with a time. Gain of probability of per unit time is:

$$\frac{d}{dt}\rho_{2,2,m}(t) = k_{i,n \to f,m}\rho_{1,1,n} = \frac{2}{\hbar^2} E_{sp} \cdot t \cdot \rho_{1,1,n} \quad (12)$$

Thence, transition coefficient is:

$$k_{i,n \to f,m} = \frac{2}{\hbar^2} E_{sp}^2 \cdot t \quad (13)$$

Time t in (13) came up for the result of absences of damping, it reflects that circumstance that process is reversible one and for an limited time $t = T/2$ system will move over to the state Φ_2, but then return back.

Take into account now damping, we shall take that for a time $t \ll T$ normal fluctuations in the final state will be damped and their energy will dissipate. For this we shall take, that frequency of fluctuations has a small imaginary part: $\omega \to \omega + i\gamma$ where damping $\gamma = \tau^{-1}$. Time t in (13) will represent in the manner of:

$$t = \frac{1}{2}\lim_{E_i-E_f \to 0}\left\{\int_0^t \exp(i(E_i - E_f)\cdot t/\hbar)dt + c.c.\right\} = \frac{1}{2}\lim_{E_i-E_f \to 0}\frac{-i\hbar}{E_i - E_f} + c.c.,$$

here $E_i = E_{e,1} + \left(n + \frac{1}{2}\right)\hbar\omega$ and $E_f = E_{e,2} + \left(m + \frac{1}{2}\right)\hbar(\omega - i\gamma)$ are energies of initial and final states.

Under real values E_i and E_f (γ=0): $t = \frac{1}{2}\lim_{E_i-E_f \to 0}\left(\frac{i\hbar}{E_i - E_f} + c.c.\right) = \pi\hbar\delta(E_i - E_f)$,

that corresponds to an exact conservation of the energy. At the complex values E_f

$$t = \frac{1}{2}\lim_{E_i-E_f \to 0}\left(\frac{i\hbar}{E_i - E_f} + c.c.\right) = \frac{\hbar\Gamma}{(E_i - E_f)^2 + \Gamma^2} = \pi\hbar\rho(\Delta E, \Gamma), \quad (15)$$

here $\Delta E = E_i - E_f = E_{e,1} + n\hbar\omega - (E_{e,2} + m\hbar\omega)$ and Γ are an uncertainty of the energy of the final state, stipulated by damping. Under $(E_i - E_f) < \Gamma$ time $t \cong \tau$. This means that time t in (13) should be understood as a time of damping of the oscillations τ.

At the thermodynamic equilibrium conditions of on oscillatory degrees of freedom probability of the finding in the initial state n quantum is: $\exp[-n\hbar\omega/kT]N^{-1}$. So:

$$\rho_{1,1,n} = \rho_i \exp[-n\hbar\omega/kT]N^{-1} \quad (16)$$

Here: $\rho_{1,1,n}$ is full probability to catch an electron in the initial state or, other words, probability that that electron inheres on the left, and in the AM is kept any amount of quantum.

Transition coefficient from any initial state, containing n quantum, in the respective final state, containing m quantum, is:

$$k_{i \to f} = \sum_{m,n} k_{i,n \to f,m} \exp\left(-\frac{n\hbar\omega}{kT}\right)N^{-1}\rho(\Gamma, \Delta E) = \frac{2\pi}{\hbar}\sum_{m,n} E_{sp,n,m}^2 \exp\left(-\frac{n\hbar\omega}{kT}\right)N^{-1}\rho(\Gamma, \Delta E) \quad (17)$$

In the expression (17) stays undetermined value $(E_{sp})^2$.

For the calculation E_{sp} use a scheme, submitted for the Fig 1. In its indications are used, sense which clear from the figure.

Convert an energy of the interaction in (3) will take into account that amplitude of oscillation q much other distances less and will hold decomposition on it.

In the initial state Φ_1 energy of interaction suitable to represent in the manner of:

$$-\frac{e^2}{\kappa(l + q - x)} = -\frac{e^2}{\kappa(l_1 - x)} + \frac{e^2}{\kappa l_1^2}q' + \frac{e^2(2l_1 - x)}{\kappa(l_1 - x)^2 l_1^2}q'x, \quad (18)$$

where used indications: $q' = q + \Delta q_1$; $l_1 = l - \Delta q_1$; $\Delta q_1 = e^2/\kappa(l_1)^2 M\omega^2$.

Full Hamiltonian will represent in the manner of:

$$\hat{H} = \frac{p^2}{2m} + U_1(x) + U_2(x) + \frac{P^2}{2M} + \frac{1}{2}M\omega^2 q'^2 - \frac{e^2}{\kappa(l_1 - x)} + \frac{1}{2}M\omega^2\Delta q_1^2 + \frac{e^2(2l_1 - x)}{\kappa(l_1 - x)^2 l_1^2}q'x.$$

Last member in the initial state is small, as far as wave function of electron is concentrated in the region of $x = 0$. However, it can give a contribution to cross integrals. So we shall not include this member in Hamiltonian H_1, but will take into account it as an additive. Potential $U_2(x)$ in Hamiltonian H_1 is absent according to definition. So hamiltonian H_1 is:

$$\hat{H}_1 = \frac{p^2}{2m} + U_1(x) + \frac{P^2}{2M} + \frac{1}{2}M\omega^2 q'^2 - \frac{e^2}{\kappa(l_1-x)} + \frac{1}{2}M\omega^2 \Delta q_1^2 \qquad (19)$$

and full hamiltonian is: $H = H_1 + \Delta H_1$ where:

$$\Delta H_1 = U_2(x) + F_1(x)q' \; ; \; F_1(x) = \frac{e^2(2l_1-x)}{\kappa(l_1-x)^2 l_1^2} \cdot x . \qquad (20)$$

From (19) follows that Hamiltonian H_1 contains two operators: first depends on the coordinate of the electron x only and second on the coordinate AM – q only:

$$\hat{H}_1 = H_{1,e} + H_{1,AM}$$

$$\hat{H}_{1,e} = \frac{p^2}{2m} + U_1(x) - \frac{e^2}{\kappa(l_1-x)} + \frac{1}{2}M\omega^2 \Delta q_1^2 \qquad (21)$$

$$\hat{H}_{1,AM} = \frac{P^2}{2M} + \frac{1}{2}M\omega^2 q'^2$$

So function Φ_1 possible represent in the manner of product: $\Phi_1(x, q) = \psi_1(x)\varphi_{1,n}(q`)$, where $\psi(x)$ is a wave function of the electron and $\varphi_{1,n}(q)$ is- a function of the oscillator, containing n quantum. Thence is follows also that energy of the initial condition E_1 it is possible to represent as an amount of the energy of the electron and AM: $E_1 = E_e + E_{AM}$.

Value $-\frac{e^2}{\kappa(l_1-x)} \approx \frac{e^2}{\kappa l_1}$ in (19) corresponds to the electrostatic energy of the interaction of the electron with AM. Value $\frac{1}{2}M\omega^2 \Delta q_1^2$ is a work, made at the shift AM on the distance Δq_1; this is constant value (not dependent nor on x nor on q). It, generally speaking, can be with the equal sense referred both to the energy of the electron, and energy AM. Comfort of referring depends on following circumstances:

If shift AM occurs within elasticity limit and energy of the shift can not be removed from the system, both energy should be prefixed to the electron. Really, in this case both energies give a contribution to measured value red/ox potential. Remind, at the measurement of the red/ox potential a medium is selected, in which electron has same energy, as in the medium. The electron moves over to this medium and is deleted from the system. Energy of the shift of AM returns to the electron and raises its energy (lowers red/ox potential). Electrostatic interaction, in opposite to if, prevents a removing an electron that lowers its energy (raises red/ox potential). In this case energy E_e, including both mentioned member, corresponds to a measure value red/ox potential, energy E_{AM} only energy of quantum.

$$E_1 = E_{e,1} + (n + 1/2)h\omega \qquad (22)$$

If the center of gravity of AM is shifted sufficiently, so that value Δq is beyond the scope of elastic deformation, the energy of the shift already does not return to electron under its removing, but stays in AM. This is possible in the event of unharmonical AM, which we will consider later.

In the final state suitable to enter variables: $q`` = q + \Delta q_2$; $l_2 = 1 - \Delta q_2$; $\Delta q_2 = e^2/\kappa r^2 M\omega^2$ and energy of the interaction represent in the manner of:

$$-\frac{e^2}{\kappa(l+q-x)} = -\frac{e^2}{\kappa(r-(x-L_1))} + \frac{e^2}{\kappa r^2}q`` + \frac{e^2(2r-(x-L_1))}{\kappa r^2[r-(x-L_1)]^2}q``(x-L_1) \qquad (23)$$

Then full Hamiltonian will take a form:

$$\hat{H} = \frac{p^2}{2m} + U_1(x) + U_2(x) + \frac{P^2}{2M} + \frac{1}{2}M\omega^2 q``^2 - \frac{e^2}{\kappa(r-(x-L_1))} + \frac{1}{2}M\omega^2\Delta q_2^2 + \frac{e^2(2r-(x-L_1))}{\kappa r^2[r-(x-L_1)]^2}q``(x-L_1) \qquad (24)$$

On same considerations, as above, Hamiltonian H_2 can be represented in the manner of:

$$\hat{H} = \frac{p^2}{2m} + U_1(x) + U_2(x) + \frac{P^2}{2M} + \frac{1}{2}M\omega^2 q``^2 - \frac{e^2}{\kappa(r-(x-L_1))} + \frac{1}{2}M\omega^2\Delta q_2^2 + \frac{e^2(2r-(x-L_1))}{\kappa r^2[r-(x-L_1)]^2}q``(x-L_1) \qquad (25)$$

Then full Hamiltonian is: $H = H_2 + \Delta H$, where:

$$\Delta H_2 = U_1(x) + F_2(x)q`` \; ; \; F_2(x) = \frac{e^2[2r-(x-L_1)]}{\kappa r^2[r-(x-L_1)]^2}(x-L_1). \qquad (26)$$

Hamiltonian (25) (either as (19)) is possible to represent in the manner of sum; $\hat{H}_2 = \hat{H}_{2,e} + \hat{H}_{2,AM}$, where:

$$\hat{H}_{2,e} = \frac{p^2}{2m} + U_2(x) - \frac{e^2}{\kappa(r-(x-L_1))} + \frac{1}{2}M\omega^2\Delta q_2^2 \; ; \; \hat{H}_{2,AM} = \frac{P^2}{2M} + \frac{1}{2}M\omega^2 q``^2. \qquad (27)$$

So variable are separated and function is $\Phi_2(x, q) = \psi_2(x)\varphi_{2,n}(q``)$; where $\psi_2(x)$ is a wave function of the electron in the final state and $\varphi_{2,n}(q``)$ is a function of the oscillator depended on the shift argument, containing m quantum.

Either as in previous case constant member is a work of shift. It is suitable to refer to the energy of the electron. Then energy of the final state E_2 possible represent as an sum of the energy of the electron and energy of oscillation of AM:

$$E_2 = E_{e,2} + (m + 1/2)\hbar\omega. \qquad (28)$$

Variable $q`` = q` + \Delta q$, where difference $\Delta q = \Delta q_2 - \Delta q_1$ corresponds to a shift AM through carrying an electron from the pit I in the pit II; it is:

$$\Delta q = \frac{e^2}{\kappa l_1^2 r^2 M\omega^2}\left(l_1^2 - r^2\right) \tag{29}$$

Relative shift $\Delta = \Delta q / R_0$ (where $R_0 = \sqrt{h/M\omega}$) is an amplitude of zero oscillations) is:

$$\Delta = \frac{e^2\left(l_1^2 - r^2\right)}{\kappa l_1^2 r^2 \sqrt{\hbar \omega M\omega^2}}. \tag{30}$$

At the calculation E_{sp} we use the expression (6) and take in to account (5), (20) and (26), we get:

$$E_{sp,n,m} = J_e \cdot F_{n,m}^{1/2}\left\{E - \frac{1}{2}(U_1 + U_2)\right\} +$$

$$+ \int \tilde{\psi}_2(x) F_1(x) \psi_1(x) dx \int \tilde{\varphi}_{2,m}(q - \Delta q) q \varphi_{1,n}(q) dq +$$

$$+ \int \tilde{\psi}_1(x) F_2(x) \psi_2(x) dx \int \tilde{\varphi}_{1,n}(q) q \varphi_{2,m}(q - \Delta q) dq + \tag{31}$$

$$+ \int \tilde{\psi}_1(x) F_1(x) \psi_1(x) dx \int \tilde{\varphi}_{1,n}(q) q \varphi_{1,n}(q) dq,$$

here: $J_e = \int \tilde{\psi}_1(x) \psi_2(x) dx \ll 1$ is an integral of the overlapping of functions of the electron.

$$U_1 = \frac{1}{J_e} \int \tilde{\psi}_1(x) U_1(x) \psi_2(x) dx \; ; \qquad U_1 = \frac{1}{J_e} \int \tilde{\psi}_2(x) U_2(x) \psi_1(x) dx$$

where $F_{n,m} = \left|\int \varphi_{2,m}(q - \Delta q) \varphi_{1,n}(q) dq\right|^2$ is a square of the integral of the overlapping of functions of the oscillator, (the Franc-Condon factor).

Integrals $\int |\Psi_1(x)|^2 U_2(x) dx + \int |\Psi_2(x)|^2 U_1(x) dx \approx (J_e)^2$ have the values of the following order of magnitude, we shall not take them into account.

Integrals $\int |\varphi_1(q)|^2 q dq = \int |\varphi_2(q'')|^2 q'' dq'' = 0$ due to the properties of functions of the oscillator.

Integrals, containing functions $F_1(x)$ and $F_2(x)$ on the order of the value are:

$$\int \tilde{\psi}_2(x) F_1(x) \psi_1(x) dx \int \tilde{\varphi}_{2,m}(q - \Delta q) q \varphi_{1,n}(q) dq \approx J_e \cdot F_{n,m}^{1/2} \cdot \frac{e^2}{\kappa r} \frac{\Delta q}{r} \frac{L_1^2}{l_1^2},$$

$$\tag{31}$$

$$\int \tilde{\psi}_1(x) F_2(x) \psi_2(x) dx \int \tilde{\varphi}_{1,n}(q) q \varphi_{2,m}(q - \Delta q) dq \approx J_e \cdot F_{n,m}^{1/2} \cdot \frac{e^2}{\kappa r} \frac{\Delta q}{r} \cdot \frac{l_1 + r}{l_1}.$$

In general case they can be significant. However, in our models, they much electronic energy less. So, saving in (П.31) main members we obtain:

$$E_{sp,n,m} = J_e \cdot F_{n,m}^{\frac{1}{2}} \left\{ E - \frac{1}{2}(U_{0,1} + U_{0,2}) \right\}. \tag{32}$$

Substituting (32) in (17) we found a final expression for the transition coefficient:

$$k_{i \to f} = \frac{2\pi}{\hbar} \sum_{m,n} J_e^2 \cdot F_{n,m} \left[E - \frac{1}{2}(U_1 + U_2) \right]^2 \cdot exp\left(-\frac{n\hbar\omega}{kT}\right) \cdot N^{-1} \cdot \rho(\Gamma, E_i, E_f) \tag{33}$$

Expression (32) practically complies with (2) with that only difference that instead of the vague energy $\bar{E}$ there is wholly determined value:

$$E = \left\{ E - \frac{1}{2}(U_{0,1} + U_{0,2}) \right\}. \tag{34}$$

Here is the time to make a remark. In expressions (22) and (28) the energy of the interaction and shift are enclosed in the energy of the electron. However, there is another case, which is realized, which a proton plays role of AM, which is shifted sufficiently powerfully, so that value Δq is beyond the limit of elastic deformation. This means, that potential energy of the proton, both before tunneling an electron, and after, are not described by parabolas, but have two minimum. Shift Δq corresponds to a turning a proton on the nearby group with the close value pK. Decomposition of the energy of the interaction on q herewith is already not justified. However, and in this case in the end condition it is possible to select an energy of oscillation of the proton and energy of the interaction and shift. Last should be referred to the energy of the proton, as far as they define a value pK groups, where it inheres. Remind: measurement of the value pK, as well as red/ox potential, based on removing a proton in the medium with corresponding to acidity at the condition that electron stays put. In this case law of the conservation of the energy is possible to write:

$$E_1 = E_{e,1}\ E_1 + (n + 1/2)h\omega = E_2 = E_{e,2} + (m + 1/2)h\omega + \Delta E_p,\ \Delta E_p = -2{,}3\kappa T\Delta pK$$

UNHARMONICAL AM AND PROCESS OF THE ENERGY TRANSFORMATION

On the Fig. 2 it is represented physical model of the process of the transformation of the energy of exited electron. Electronic subsystem is represented by two pits: I and II. In the pit I there is two levels: main and excited, it corresponds to an chromophore, capable to absorb a quantum the light of the energy $h\nu$. There is one level in the pit II, which position depends on the state of AM. Potential AM, as in initial state $V_{in}(q)$, so and in final one $V_{fin}(q)$ has two minimums, which divided by the barrier. Coordinates of the minimum are marked by q_1 and q_2. Distance between them $\Delta q_{1,2} = q_1 - q_2$ is much more then shift of each of them when carrying an electron from the pit I in the pit II: $\Delta q_{1,2} \gg \Delta q_1, \Delta q_2$.

Potential $V_{in}(q)$ in the point q_2 is higher then that in the point q_1 on the value $V_{in}(q_2) - V_{in}(q_1) = \Delta \tilde{E}_e$, equal to reduction of electron level in the pit II. Exactly this part of the energy of the excited electron stores in AM. Shall hereinafter let us consider, that this energy is not small, it is of order characteristic for biologics of the energy $\Delta \tilde{E}_e = 0{,}5$ эВ. Thereby, AM can exist in two states: stable (under $q = q_1$) and metastable (under $q = q_2$). Cycle of the transformation of the energy is reduced to that excited electron returns on the main level, but AM moves over to the metastable state. Presence of the barrier $E^{\#}$ in potential $V_{in}(q)$ is a necessary condition of the transformation of the energy, as far as it ensures stability an metastable state. Potential $V_{fin}(q)$ equals:

$$V_{fin}(q) = V_{in}(q) - \Delta E_{int}(q), \tag{35}$$

where: $\Delta E_{int}(q)$ is a difference of the interaction energy of the AM with the electron in the cases, when electron inheres in pits I and II. In general case this function falls with the growing of variable q. In the most simplest case of the electrostatic interaction AM with the electron (disregarding other charges) it is:

$$\Delta E_{int}(q) = \frac{e^2}{\kappa} \cdot \frac{x_{II} - x_I}{(q - x_I)(q - x_{II})} = \frac{e^2}{\kappa} \cdot \frac{x_{II}}{q(q - x_{II})}, \tag{36}$$

here $x_I = 0$ and x_{II} are coordinates electronic pits; q is a coordinate AM, counted out from the point $x_I = 0$.

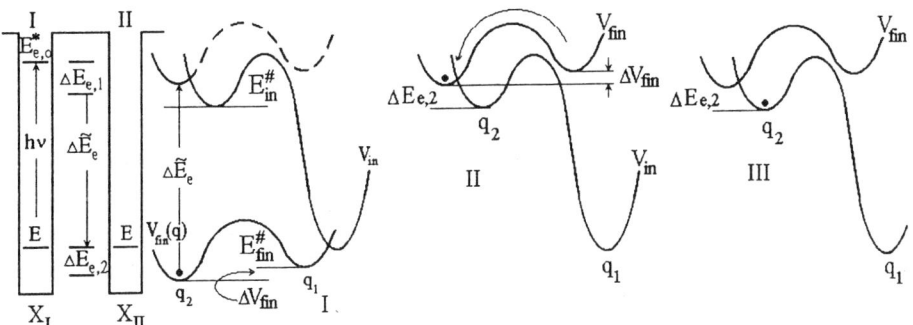

Figure 2. Physical model of energy transformation in the case of unharmonical AM. The intermolecular cycle with participation of chromophore

From (35) and (36) it follows, that curve $V_{fin}(q)$ is situated below $V_{in}(q)$, moreover left pit (under $q=q_2$) is lowered more strong, than right one (under $q=q_1$) and barrier $E^{\#}_{fin}$ is below barrier $E^{\#}_{in}$. Under sufficiently strong interaction a potential $V_{fin}(q_2)$ appears to be below potential $V_{fin}(q_1)$, i.e. in presence of the electron in the pit II metastable state becomes stable, as it is shown on the Fig. 2. This condition also is needed for transformations of the energy, in opposite case transition AM in the state with $q=q_2$ is impossible.

Let us consider a process step by step.

First stage is a tunneling an excited electron from the pit I in the pit II. The center of gravity of AM is displaced on small value Δq_1. Energy level of the electron is lowered on $\Delta E_{e,1}$ (small in comparison with $\Delta \tilde{E}_e$), which moves over to the energy of oscillation of AM and dissipates. Kinetics of this stage can be described by the expression (32), as far as shifts of AM are small, processes in AM are played in the field closed to $q = q_1$, where potentials $V_{in}(q)$ and $V_{fin}(q)$ can be represented in harmonic approach

Second stage is conformational transition, that is to say changing a coordinate of AM on the value $\Delta q_{1,2} = q_1 - q_2$ without changing the localization of the electron. Value $V_{fin}(q)$ changes weakly (it is shown on the Fig. 2). Difference $\Delta V_{fin} = V_{fin}(q_1) - V_{fin}(q_2)$ moves over to the energy of oscillation and dissipates, as ensures a direction of transition. However, values ΔV_{in} and ΔE_{int} in (35) change powerfully:

$$\Delta V_m = V_m(q_2) - V_m(q_1) \tag{37}$$

$$\Delta(\Delta E_{int}) = \Delta E_{int}(q_2) - \Delta E_{int}(q_1) = \frac{e^2}{\kappa} \frac{q_1 - q_2}{(x_{II} - q_1)(x_{II} - q_2)} = \Delta \tilde{E}_e$$

Value ΔV_{in} is an energy of metastable state, that energy, which at the end of the process stores in AM.

Value $\Delta(\Delta E_{int})$ corresponds to a work on displacement of AM, which makes an electrical field of the electron. From the balance of the energy follows that level of the electron is lowered on same value.

In previous section a work on shift AM is already discussed. In this instance it is possible to note the following:
- firstly, this energy is not small;
- secondly, expression $\frac{1}{2} M\omega^2 \Delta q_1^2$ for work of shift in unharmonical potential is already not fair. Moreover, analytical expression for this work in general event case is impossible to show (as far as it depends on concrete parameters unharmonical potential). However, in this instance analytical expression is not needed, as far as this functioning does not depend on x and q and is equal to $\Delta \tilde{E}_e$;
- thirdly, earlier we refer this work to the energy of the electron on the ground that when removing an electron (measurement red/ox./ox potential) this energy returns to the electron. In this case it does not return to the electron due to presence of the barrier. So it should refer to the energy of AM.

Formally this means that Hamiltonian H_2 (in (25)) in this instance must be represented in the manner of:

$$\hat{H}_2 = \hat{H}_{2,e} + \hat{H}_{2,AM}$$

$$\hat{H}_{2,e} = \frac{p^2}{2m} + U_2(x) - \frac{e^2}{\kappa(q_2 - x)}; \hat{H}_{2,AM} = \frac{P^2}{2M} + \frac{1}{2} M\omega^2 q''^2 + \Delta \tilde{E}_e, \tag{38}$$

here $q``$ is a coordinate of AM, counted out from the point of the minimum of potential V_{fin} (q_2). In the hamiltonian AM is present an energy of oscillations, which appears when transition and then dissipates; their energy is ΔV_{fin}.

From (38) it follows, that level of the electron in the final state is lowered to the account of the energy of the interaction (as far as work of shift is not enclosed in it), but energy AM increases approximately on the same value. On the fig. 2 this circumstance is taken into account that that potential V_{fin} (q) shifted vertically on the constant value upwards.

The kinetic coefficient of the conformational transition can be computed, using expression (33) and considering following particularities:
- -firstly, integral of the overlapping of electronic functions is an unit, as far as
- localization of the electron is not changed.
- - secondly, main contribution to the sum gives a term, in which. factor of the Franc-Condon is an unit. Thence it follows that process can run only under different from the zero temperature T;
- -thirdly, the parameter $\overline{E}$ in (33) does not contain parameters of the electronic subsystem. Main contribution to this value gives an energy of fluctuations in the initial state. At the temperature T this energy is of the order $\overline{E} = kT$;

– -fourthly, at the temperature T time of damping of the fluctuations is of the order of the time heat fluctuations; so $\Gamma \cong kT$, and density of levels $\rho(\Gamma) \cong 1/\Gamma = 1/kT$.

Substituting these valuations in (33), we come to well known in the chemical kinetics expression for velocities of transition:

$$k_{i \to f} = \frac{kT}{\hbar} exp\left(-\frac{E_{fin}^{\#}}{kT}\right) \qquad (39)$$

Third stage is a tunneling of an electron from the lower level of the pit II on the main level of the pit I. Herewith AM goes from the left minimum of V_{fin} potential to left (metastable) potential level V_{in}. Thereby it appears energized state of AM, what is the result of transformations of the energy.

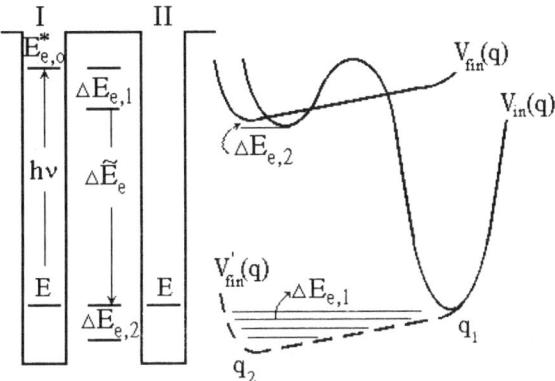

Figure 3. Physical model of one-step temperature independent process of energy transformation. Explanation is in the text.

Center of gravity of AM is herewith shifted little, process runs in the field of variable q, close to q_2, where both potentials can be approximated by parabolas. So factor of transition can be computed in harmonic approach, by the formula (33).

Thereby, under installment consideration of the process with the participation an unharmonical AM each stage is possible to describe, using expression (33), taking into account the particularities of each stage. However, division on stages is possible only in the case of presence of the barrier in potential $V_{fin}(q)$.

Let us consider particular, but important case, when barrier is absent and both first stages, including transformation of the energy, are occurred simultaneously. Corresponding to this potential $V_{fin}(q)$ is shown on Fig. 3. Curve $V_{fin}(q)$ is shifted on the value $\Delta \tilde{E}_e$ upwards, so that its left minimum is situated above metastable minimum of $V_{in}(q)$ (that it is required for the following stage). Direction of the process on first stages is ensured by excitation (and following dissipation) of unharmonical oscillations. For illustrations this on the Fig. 3 dotted lines brought by $V_{fin}(q)$ not shifted upwards, which allows graphically to represent an overlapping of wave functions AM when turning from the state "in" to "fin".

In this case it is possible to compute transition coefficient using method stated above. Herewith it is possible to get an expression, similar to (33), but with following differences:
– firstly, functions $\varphi_{fin,m}(q)$ in this case are not own functions harmonic oscillator with the shift argument. So values $F_{n,m} = \int \tilde{\varphi}_{fin,m}(q) \varphi_{in}(q) dq$ are not the Franc-Condon factors

and must be computed taking into account characteristics of potential V_{fin} in each concrete event;

– secondly energy of oscillations of AM in the state "fin" is not mhω, as far as levels are not equidistant and frequencies of oscillations are not alike. So conditions of the resonance transition takes a type: $E_{e,0} + nh\omega = E_{e,1} + \varepsilon_{AM} + \Delta \overline{E}_e$ where: E_{AM} - an energy of oscillations of AM in state "fin".

Really in one-step processes energy barrier on the first stage is small or completely is absent (curves $V_{fin}(q)$ crosses $V_{in}(q)$ in the point of its minimum, as it is shown on the Fig. 3). Then only one term with $n=0$ and $n=0$ и $\varepsilon_{AM} = V_{in}(q_1) - V_{fin}(q_2)$ is sufficient in the sum (33) and the factor $F_{n,m} = 1$. Expression (33) in this case is simplified:

$$k_{i \to f} = \frac{2\pi}{\hbar} |E|^2 J_e^2 \frac{1}{\Gamma} = \frac{2\pi}{\hbar^2} |E|^2 J_e^2 \tau, \qquad (40)$$

where τ is a time of damping of the unharmonical oscillations.

Thus, the proposed method allows to qualitative describe velocities of processes of the transformation and accumulation of energy at the participation of unharmonical AM. For quantitative evaluations it is necessary to know concrete terms.

REFERENCES

1. L.N. Grigorov., D.S. Chernavskii, *Biophysika*, 17, 2: 195 (1972) (in Russian).
2. J.J Jortner, *J. Chem. Phys.*, 64: 4860 (1976).
3. J.J Jortner, *Biochimica et biophysica acta*, 594: 193 (1980).
4. J.J Hpfield. Electrical phenomena of the biological membrane. *Proceed. 29th uecting of the Society de Chem. Physique*.472, Elsevier, Amsterdam (1976).
5. J.J. Hopfield, *Proc. Natl. Acad. Sci. USA*, 9, 71: 3640 (1974).
6. R.R. Dogonadze, M.G. Zakharaia, *Biophysika*, 29, 4: 549 (1984) (in Russian).
7. "19 Jerusalem Symp. on Quantum Chemistry Tunneling", ed. B. Pullman, J.Jortner, Reidel, Dordrecht, (1986).
8. L.A. Blumenfeld, D.S. Chernavskii, *J. Theor. Biol.*, 38: 1, (1973).
9. N.M. Chernavskaya, D.S Chernavskii, Tunnel transport of electron in Photosynthesis. MGU, Moscow (1977).
10. N.M. Chernavskaya, D.S. Chernavskii, L.C Yagujinsky, *Biophysika*, 27, 1: 114 (1982) (in Russian).
11. K.V. Shaitan, A.B. Rubi, *Biophysika*, 16, 6: 1004 (1982).
11. A.B. Rubin, K.V. Shaitan, A.A. Kononenko, Electron-conformational interactions in the primary processes of phonosynthesis, *VINITI, Itogi nauki i techniki*, 21: 161 (1987).
13. L.D. Landau, E.M. Lifshitc, Quantum Mechanics. OGIS, Moscow (1948) (in Russian).
14. L.A. Blumenfeld, V.I. Goldanskiy, M.I. Podgoretskiy, D.S Chernavskii, *Strukturnaia Khmia*, 8: 854 (1967) (in Russian).

ATOMIC OPERATOR FORMALISM OF ELEMENTARY GATES FOR QUANTUM COMPUTATION AND IMPURITY-INDUCED EXCITON QUANTUM GATES

Shozo Takeno[1] and Hideaki Matsueda[2]

[1]Department of Information Systems
Osaka Institute of Technology
Hirakata, Osaka, Japan

[2]Department of Information Science
Kochi University
Kochi, Japan

INTRODUCTION

The information-processing capability of quantum systems has long been of theoretical interest in physics and computer science. Feynman considered gate arrays with computing operations based on principles of quantum mechanics[1]. Deutch and his co-workers sought formulating quantum Turing machines[2,3] and quantum complexity theory[4]. Recently, considerable progress has been achieved in the latter line of approach where quantum computers were shown to be qualitatively much stronger than classical ones, culminating in Shor's discovery of quantum polynomial-time algorithms for factoring and discrete logarithm[5]. This has led several-physicist groups to make attempts to. implement quantum computers by using trapped ions[6,7], quantum dots, nuclear spins using multiple pulse resonance techniques[8] and so on. Appreciation of the power of quantum computing was quickly tempered by the realization that preserving quantum coherence made the implementation of practical quantum computers unlikely within decades.

Two very recent developments have changed situations. The first is the recognition that quantum error correction can be used to compute with imperfect computers[9,10]. The second is that it is possible to decrease the influence of decoherence by computing with mixed-state ensembles rather than isolated systems in a pure state. This can be done by using quantum spins[9], and space[11]. Thus, the change of the situations revives hopes of physicists in their attempts at implementing physically realizable quantum computers.

The purpose of this paper is two-fold: (1) To develop an atomic operator formalism for basic operations in elementary quantum gates and (2) to propose impurity-induced exciton quantum gates to make a step toward possible physical realization of solid state or molecular quantum gates. Our approach using atomic operators as a basis of formulation is different from the previous ones in the quantum gate theory[1-4, 12] in the following two important respects: Firstly, the atomic operators describing elementary excitations in atoms, molecules and solids[13] are shown to have direct correspondence to elementary operations in a generalized version of the Deutch gates. When studying quantum computing, this makes the atomic operators better qualified quantities than boson-like operators and matrix representations employed by Feynman[1] and Barenco et al.[12], respectively. Secondly, introduction into exciton systems of suitably chosen impurities as an active agent of quantum computation provides us with many choices in seeking physical arenas for performing quantum computation and at the same time more freedom in controlling the system than treating identical atoms in pure solids or molecules as a hole. Situations here are somewhat similar to the role of impurities (donors and acceptors) in semi-conductors physics and technology.

TWO-LEVEL ATOMS, ATOMIC OPERATORS AND FRENKEL EXCITONS

We consider a set of two-level atoms with multipole interactions among themselves. Let $H_0(n)$ be the Hamiltonian of an n th free two-level atom, and also let its ground state and its excited state be denoted by the symbols $\mathbf{0}_n$ and $\mathbf{1}_n$, respectively. Then, the eigenvalue $\varepsilon_n(\lambda_n)$ and the corresponding eigenfunction $|\lambda_n\rangle$ ($\lambda_n = \mathbf{0}_n, \mathbf{1}_n$) of $H_0(n)$ satisfy the eigenvalue equation,

$$H_0(n)|\lambda_n\rangle = \varepsilon_n |\lambda_n\rangle, \text{ with } \langle\lambda_n|\mu_n\rangle = \Delta(\lambda_n, \mu_n) \tag{1}$$

where the symbol Δ is the Kronecker delta. We introuce atomic operators $\sigma_{n\lambda\mu}$ by the equation[13]

$$\sigma_{n\lambda\mu} = |\lambda_n\rangle\langle\mu_n|, \text{ with } \sigma_{n\lambda\mu}\sigma_{n\gamma\delta} = \sigma_{n\lambda\delta}\Delta(\mu,\gamma) \tag{2}$$

Then, σ_{n01} and σ_{n10} are identified as annihilation and creation operators, respectively, of the excited state $|\mathbf{1}_n\rangle$ of the n-th atom. The vector $\sigma_n = (\sigma_n^x, \sigma_n^y, \sigma_n^z)$ formed by the $\sigma_{n\lambda\mu}$'s,

$$\sigma_n^x = \frac{1}{2}(\sigma_{n10} + \sigma_{n10}), \sigma_n^y = \frac{1}{2i}(\sigma_{n10} - \sigma_{n10}), \sigma_n^z = \frac{1}{2}(\sigma_{n11} - \sigma_{n00}), \tag{3}$$

are identified as the Pauli operator.

Let $Q(m, n)$ be the multipole interaction energy between the atom at the site n and that at site m in the system. Then, the Hamiltonian H can be written in the form[14]

$$H = \sum_n \{\varepsilon_0(n)\sigma_{n00} + \varepsilon_1(n)\sigma_{n11}\} + \sum_n \sum_m \theta_{\lambda\mu,\gamma\delta}(n,m)\sigma_{n\lambda\lambda}\sigma_{m\gamma\gamma}, \tag{4}$$

where the energy of the ground state, $\varepsilon_n(\mathbf{0}_n)$, and that of the excited state, $\varepsilon_n(\mathbf{1}_n)$, (cf, Eq,(1)) have been rewritten as $\sigma_0(n)$ and $\sigma_1(n)$, respectively. Let us consider the multipole interactions up to quadrupole-quadrupole interactions to rewrite (34) as

$$H = \sum_n \{\varepsilon_0(n)\sigma_{n00} + \varepsilon_1(n)\sigma_{n11}\} - \sum_{n,m} J(n,m)\sigma_n^x\sigma_m^x -$$

$$-\sum_{n,m} K(n,m)[\sigma_{n11}\sigma_m^x + \sigma_n^x\sigma_{m11}] + \sum_{n,m} I(n,m)\sigma_{n11}\sigma_{m11}. \tag{5}$$

Here $J(n, m)$, $K(n, m)$ and $I(n, m)$ represent dipole-dipole, dipole-quadrupole and quadrupole-quadrupole interaction energies, respectively. The symbol Σ_{nm} stands for sums over n and m with $n \neq m$. It is seen that the model Hamiltonian for Frenkel excitons are written in terms of the Pauli spin operators.

QUBITS

Let us consider a possible application of Frenkel excitons to quantum computing. As is well known, the conventional bit in the classical computation has two values, 0 and 1. Corresponding to this, two states of an atom having two accessible quantum states (two-level atom) can be taken as a quantum bit or a qubit. Thus, a quantum network consisting qubits can be considered as having close correspondence to a system of Frenkel excitons. We note, first of all, that a NOT operator N_n and an identity operator I_{n1} associated with the n-th qubit are written in terms of the $\sigma_{n\lambda\mu}$'s, as

$$N_n = \sigma_{n01} + \sigma_{n10} = 2\sigma_n^x \quad \text{and} \quad I_{n1} = \sigma_{n00} + \sigma_{n11} \tag{6}$$

It is seen that the not operator N_n in quantum computing corresponds the transition dipole moment of the nth atom. A simple calculation using Eqs. (2) yields

$$I_{n1}|\lambda_n\rangle = |\lambda_n\rangle, \qquad N_n|0_n\rangle = |1_n\rangle, \qquad N_n|1_n\rangle = |0_n\rangle. \tag{7}$$

The operator N_n so introduced satisfies the relations

$$N_n^p = i_{n1} \text{ for } p = \text{even}, \quad N_n^p = N_n \text{ for } p = \text{odd}, \tag{8}$$

$$N_n \sigma_{n00} N_n = \sigma_{n11}; \qquad N_n \sigma_{n11} N_n = \sigma_{n00}$$

$$\sigma_{n00} N_n \sigma_{n00} = \sigma_{n11} N_n \sigma_{n11} = 0$$

$$\sigma_{n00} N_n = N_n \sigma_{n11} = \sigma_{n00} N_n \sigma_{n11} = \sigma_{n01}$$

$$\sigma_{n11} N_n = N_n \sigma_{n00} = \sigma_{n11} N_n \sigma_{n00} = \sigma_{n10}$$

$$N_n \sigma_{n10} = \sigma_{n01} N_n = \sigma_{n00}; \quad \sigma_{n10} N_n = N_n \sigma_{n01} = \sigma_{n11} \tag{9}$$

QUANTUM COMPUTATIONAL NETWORK AND OPERATOR ALGEBRA

We consider a network composed of m control qubits and a single target qubit at p defined by the evolution operator $S_m(U_p)$ operating on the composite quantum states $|\lambda_1, \lambda_2, ... \lambda_m, \lambda_p\rangle$ as

$$S_m(U_p)|\lambda_1, \lambda_2, ... \lambda_m, 0_p\rangle =$$

$$= \begin{cases} u_{00}|\lambda_1, \lambda_2, ..., \lambda_m, 0_p\rangle + u_{01}|\lambda_1, \lambda_2, ..., \lambda_m, 1_p\rangle, & \text{if } \wedge_{n=1}\lambda_n = 1 \quad (10) \\ |\lambda_1, \lambda_2, ..., \lambda_m, 0_p\rangle, & \text{if } \wedge_{n=1}\lambda_n = 0 \quad (11) \end{cases}$$

$$S_m(U_p)|\lambda_1,\lambda_2,...\lambda_m,\mathbf{1}_p\rangle =$$

$$= \begin{cases} v_{10}|\lambda_1,\lambda_2,..., \lambda_m,\mathbf{0}_p\rangle + v_{11}|\lambda_1,\lambda_2,..., \lambda_m,\mathbf{1}_p\rangle, & \text{if } \wedge_{n=1}\lambda_n = 1 \quad (12) \\ |\lambda_1,\lambda_2,..., \lambda_m,\mathbf{1}_p\rangle, & \text{if } \wedge_{n=1}\lambda_n = 0 \quad (13) \end{cases}$$

for all $\lambda_1, \lambda_2,...\lambda_m, \lambda_p \in \{0, 1\}$, where U_p is a unitary matrix to be introduced soon. Here, the symbol $\wedge_{n=1}\lambda_n$ denotes the AND of the Boolean variables $\{\lambda_n\}$, and $U_p = (U_{ij})$ $(i,j = 0, 1)$ is a 2×2 unitary matrix. In terms of the $\sigma_{n\mu\mu}$'s and the u_{ij}'s, the operator $S_m(U_p)$ can be written in the form

$$S_m(U_p) = I_m + \prod_{n=1}^{m} \sigma_{n11} U_p \qquad (14)$$

where

$$I_m = \prod_{n=1}^{m}(\sigma_{n00} + \sigma_{n11}) - \prod_{n=1}^{m}\sigma_{n11} \equiv I - \prod_{n=1}^{m}\sigma_{n11} \qquad (15)$$

$$U_p = u_{00}\sigma_{p00} + u_{01}\sigma_{p01} + u_{10}\sigma_{p10} + u_{11}\sigma_{p11} = Tr(u_p \sigma_p) \qquad (16)$$

The characteristics of the formulation in terms of the atomic operators here is that the atomic operators describing elementary excitations in atoms, molecules and solids[13] have direct correspondence to elementary operations in the generalized version of the Deutch gates[2-4, 12]. In particular, Eqs. (14) have been written entirely in terms of the Pauli operator σ_{n11} for the population of the excited states except for the U_p's. Thus, our formulation can be made more transparent by introducing simplified symbols

$$\sigma_{n11} \equiv x_n; \quad \sigma_{n00} = 1 - \sigma_{n11} \equiv x'_n = 1 - x_n \qquad (17)$$

$$X_m = \prod_{n=1}^{m}\sigma_{n11} = \prod_{n=1}^{m} x_n; \quad X'_m \equiv I'_m = I - \prod_{n=1}^{m}\sigma_{n11} = 1 - X_m \qquad (18)$$

with

$$x_n^2 = x_n; \quad x_n'^2 = x'_n; \quad x_n x'_n = x'_n x_n = 0 \qquad (19)$$

$$X_n^2 = X_n; \quad X_n'^2 = X'_n; \quad X_n X'_n = X'_n X_n = 0 \qquad (20)$$

Namely, Eq. (14) is rewritten as

$$S_m(U_p) = X'_m + X_m U_p \qquad (21)$$

A full explicit expression for $S_m(U_p)$ in terms of the x_n''s and the x_n's is written as

$$S_m(U_p) = x'_1 x'_2 x'_3 ... x'_m + x_1 x'_2 x'_3 ... x'_m + x'_1 x_2 x'_3 ... x'_m + ... x'_1 x'_2 x'_3 ... x'_{m-1} x_m +$$

$$+ \ldots + x_1' x_2 x_3 \ldots x_m + x_1 x_2' x_3 \ldots x_m + \ldots + x_1 x_2 x_3 \ldots x_m' + x_1 x_2 x_3 \ldots x_m U_p \quad (22)$$

In what follows, we will use such a compact expression for $S_m(U_p)$ as that given by Eq. (21).

The result obtained above can be generalized to a more general situation, where equations corresponding to Eqs. (21), (18) and (20) take the form

$$S_{im}(u_p) = X'_{xm} + X_{im} U_p \quad (23)$$

$$X_{im} = \prod_{n=1}^{m} \sigma_{n11} = \prod_{n=1}^{m} x_n; \quad X'_{im} \equiv I_{im} = 1 - \prod_{n=1}^{m} \sigma_{n11} = 1 - X_{im} \quad (24)$$

and

$$X_{in}^2 = X_{in}; \quad X'^2_{in} = X'_{in}; \quad X_{in} X'_{in} = X'_{in} X_{in} = 0, \quad i < n, \quad n = 1, 2, \ldots, m, \quad (25)$$

respectively. Equation (23) also represents a generalized Deutch gate with $(m-i+1)$ control bits and a target bit at p.

For a specific case in which the operator U_p takes the form

$$U_p = N_p, \quad \text{or} \quad U_{00} = U_{11} = 0 \quad \text{and} \quad U_{01} = U_{10} = 1 \quad (26)$$

the operators $S_m(Xp)$ and $S_{im}(Xp)$ are written in terms of the NOT operator N_p as

$$S_m(N_p) = X'_m + X_m N_p \quad (27)$$

$$S_{im}(N_p) = X'_{im} + X_{im} N_p \quad (28)$$

Equation (27) is a Toffoli gate with m controle bits and the target bit at the p-th ($p > m$) site. For $m = 1$ and 2,

$$S_1(N_2) = x_1' + x_1 N_2 = 1 - x_1 + x_1 N_2 \quad (29)$$

and

$$S_2(N_3) = X_2' + X_2 N_3 \quad (30)$$

are identified as a XOR gate and a three-bit Toffoli gate, respectively.

CONSTRUCTION OF QUANTUM NETWORK

The characteristics of the atomic-operator formalism for the quantum gate can be appreciated by considering a construction of a $(m + 1)$-qubit generalized Deutch gate $S_m(U_{m+1})$,

$$S_m(U_{m+1}) = X'_m + X_m U_{m+1} \quad (31)$$

in terms of a set of those of lower orders. For this purpose, let us consider, as an example, a decomposition of 4-qubit Deutch gates in terms of a set of 2-qubits gates. This amounts to setting

$$S_3(U_4) = X_3' + X_3 U_4 = (x_1' + x_1 A_1)(x_2' + x_2 A_2)(x_3' + x_3 A_3)(x_1' + x_1 B_1) \times$$

$$\times (x_2' + x_2 B_2)(x_3' + x_3 B_3)(x_1' + x_1 C_1)(x_2' + x_2 C_2)(x_3' + x_3 C_3) \quad (32)$$

with

$$x_3' = x_1' x_2' x_3' + x_1 x_2' x_3' + x_1' x_2 x_3' + x_1' x_2' x_3 + x_1 x_2 x_3' + x_1' x_2 x_3 + x_1 x_2' x_3 \quad (33)$$

Then, a straightforward calculation using Eq. (23) leads to the result that a non-trivial decomposition can be achieved if the U_i's and U_i'''s are chosen as

$$A_1 B_1 C_1 = A_2 B_2 C_2 = A_3 B_3 C_3 = A_1 A_2 B_1 B_2 C_1 C_2 = A_2 A_3 B_2 B_3 C_2 C_3 = A_1 A_3 B_1 B_3 C_1 C_3 = 1, \quad (34)$$

$$U_4 = A_1 A_2 A_3 B_1 B_2 B_3 C_1 C_2 C_3 \quad (35)$$

It is seen from the result obtained above that the construction of a generalized Deutch gates from simpler ones or decomposing it into its lower orders can in principle be performed in a straightforward manner by using the atomic operator algebra. For example, all the results obtained by Barenco et al.[12] can be obtained in a more straightforward manner by using the present atomic operator algebra formalism.

IMPURITY-INDUCED EXCITON QUANTUM GATES

One of physically acceptable systems which is adapted to the present atomic operator formalism of quantum gates is Frenkel excitons with exciton transfer by multipole interactions described by Eq. (4). With their possible physical realization in mind, we consider a molecular crystal or a macromolecule as a candidate for implementing solid-state or molecular quantum gates. We rewrite Eq, (4) as

$$H = -\left\{ \sum_n \varepsilon(n) \sigma_{n00} + \sum_{n,m} k(n,m)[N_n \sigma_{m11} + \sigma_{n11} N_m] \right\} + \left\{ \sum_{n,m} J(n,m) \sigma_n^x \sigma_m^x + \sum_{n,m} I(n,m) \sigma_{n11} \sigma_{m11} \right\}, \quad (36)$$

where

$$\varepsilon(n) = \varepsilon_1(n) - \varepsilon_0(n) \quad (37)$$

is the excitation energy of the two-level atom at the site n. It is seen that a sub-system governed by the effective Hamiltonian

$$H^*_{XOR} = \sum_n \sigma_{n00} + \sum_n \frac{K(n,m)}{\varepsilon(n)} [\sigma_{n11} N_m + N_n \sigma_{m11}] \quad (38)$$

is of the form of XOR gates by itself, while the other one defined by the Hamiltonian

$$H_{trans} = -\sum_{n,m} J(n,m)\sigma_n^x \sigma_m^x + \sum_{n,m} I(n,m)\sigma_{n11}\sigma_{m11}, \tag{39}$$

describes the excitation transfer from one site to another by dipole-dipole interactions and by quadrupole-quadrupole interactions.

In spite of its XOR-like form, the system described by H^*_{XOR} does not appear by itself suitable for physical implementation of quantum computing. This is due to the fact that computing operations here amounts to applying actions to the whole atoms in the system. In view of this, let us consider an exciton system in which only impurity atoms play a relevant role in quantum computation. We assume that a solid or molecular system under consideration is composed of two kinds of atoms, a host atom with the excitation energy ε and an impurity atom with the excitation energy ε', the latter being considered to be a principal agent of computing operations. To explore this, we use the symbols $i, j, ..$ and $m, n, ..$ for the sites of the impurity atoms and those of the host atoms, respectively, separating Eq.(38) into two parts, an impurity-part $H^*_{XOR}(i)$ and a host-atom-part $H^*_{XOR}(h)$, i.e.

$$H^*_{XOR} = H^*_{XOR}(i) + H^*_{XOR}(h) \tag{40}$$

with

$$H^*_{XOR}(i) = \sum_i \left\{ \sigma_{i00} + \sum_j \frac{K(i,j)}{\varepsilon'} [\sigma_{i11} N_j + N_i \sigma_{j11}] + \sum_n \frac{K(i,n)}{\varepsilon'} [\sigma_{i11} N_n + N_i \sigma_{n11}] \right\}, \tag{41}$$

$$H^*_{XOR}(h) = \sum_n \left\{ \sigma_{n00} + \sum_m \frac{K(i,m)}{\varepsilon'} [\sigma_{n11} N_m + N_n \sigma_{m11}] + \sum_i \frac{K(n,i)}{\varepsilon'} [\sigma_{n11} N_i + N_n \sigma_{i11}] \right\}, \tag{42}$$

Similarly, the transfer-Hamiltonian H_{trans} is separated into three parts,

$$H_{trans} = H_{trans}(ii) + H_{trans}(hh) + H_{trans}(ih) \tag{43}$$

with

$$H_{trans}(ii) = -\sum_{i,j} J(i,j)\sigma_i^x \sigma_j^x + \sum_{i,j} I(i,j)\sigma_{i11}\sigma_{j11}, \tag{44}$$

$$H_{trans}(hh) = -\sum_{n,m} J(n,m)\sigma_{ni}^x \sigma_{mj}^x + \sum_{n,m} I(n,m)\sigma_{n11}\sigma_{m11}, \tag{45}$$

$$H_{trans}(ih) = -\sum_{n,i} J(n,i)\sigma_n^x \sigma_i^x + \sum_{n,i} I(n,i)\sigma_{n11}\sigma_{i11}. \tag{46}$$

We pay particular attention to a subsystem governed by an effective Hamiltonian

$$H^*(XOR) = \sum_i \left\{ \sigma_{i00} + \sum_n \frac{K(i,n)}{\varepsilon'} \sigma_{in11} N_n \right\} \equiv \sum_i h_i^*(XOR). \tag{47}$$

Physically, the quantity $h^*_i(XOR)$ represents the effective Hamiltonian for a XOR gate operating between an impurity at the site i and its neighboring host atoms. Let us assume that the impurity atoms in the systems are solely responsible for operations of quantum computing. Conditions that such an impurity-induced quantum gates (IIQG) is considered to be a working approximation to physically realizable quantum gates are listed below:

(a) An impurity atom is chosen such that its excitation energy ε' is much smaller than the excitation energy ε of the host atom, yet its excited state lies far above the ground state of the host atom.

(b) The frequency ω of an external radiation field to be applied for operations of computing is tuned to the excitation energy ε' of the impurity atom, i.e.

$$\hbar\omega = \varepsilon'. \tag{48}$$

(c) Both of the dipole moment of the impurity atom and the quadrupole moment of the host atom are much larger than the dipole moment of the host atom.

(d) The concentration $n^\wedge$ of the impurity atom is much smaller than the concentration nh of the host atom, i.e.

$$n_i \ll n_h. \tag{49}$$

Under these conditions, an impurity-induced quantum computing proceeds as follows:

(1) Ail the atoms are first set in their ground states.

(2) An external, monochromatic radiation field satisfying Eq. (48) is applied to pump solely into the excited state of the impurity atoms from their ground state to attain states in which the excited state of the impurity atom is occupied while ail its surrounding host atoms remain in their ground state. Obviously, the items (a) and (b) in the conditions correspond to picking up the factor $H^*_{XOR}(i)$ from H^*_{XOR}. Then, (49) ensures picking up H^*_{XOR} from H^*_{XOR}.

Let us examine the energy transfer due to impurity excitations. Then, this and inequality (49) ensure neglecting $H_{trans}(ii)$ and $H_{trans}(hh)$ in Eq. (43), i.e.

$$H_{trans} \approx H_{trans}(ih) = -\sum_{n,i} J(n,i)\sigma_n^x \sigma_i^x + \sum_{n,i} I(n,i)\sigma_{n11}\sigma_{i11}. \tag{50}$$

This is to be compared with the Hamiltonian $H(XOR)$ corresponding to $H^*(XOR)$ defined by

$$H^*(XOR) = \sum_i \left\{ \varepsilon' \sigma_{i00} + \sum_n K(i,n)\sigma_{in11} N_n \right\}. \tag{51}$$

The condition that the system itself can be regarded as a set of impurity-induced XOR gates is written as

$$H(XOR) \gg H_{trans}. \tag{52}$$

Let d_i (q_i) and d_h (q_h) be the transition-dipole (quadrupole) moment of the impurity atom and the host atom, respectively. Then, we obtain

$$J(n,i) \approx d_h d_i / R_{ih}^3, \quad k(n,i) \approx d_i q_h / R_{ih}^4, \quad I(n,i) \approx q_i q_h / R_{ih}^5, \tag{53}$$

where R_{ih} is an average distance between the impurity atom and the host atom. Therefore the condition (52) is satisfied provided

$$K(n, i) \gg J(n, i), \qquad K(n, i) \gg I(n, i). \tag{54}$$

Combining (54) with (53) yields

$$d_h \gg q_h/R_{ih} \gg d_h. \tag{55}$$

Inequality (55) is equivalent to the condition (c). Namely, in order for the condition (52) to be satisfied, we choose such an impurity atom whose transition dipole moment and the corresponding quadrupole moment of the host atom are much larger than the transition dipole moment of the host atom.

CONCLUDING REMARKS

A theory formulated in this paper is mainly divided into two parts. The first is an atomic operator formalism for a network of two-level atoms in which atomic operators originally introduced in quantum mechanics to describe electronic states of atoms, molecules and solids are shown to have direct correspondence to basic elements of operations in a generalized version of the Deutch gates. The second is to consider Frenkel exciton systems as a candidate for arenas where physically such a correspondence can, in principle, be made possible by paying attention to roles played by impurity excitons rather than considering identical atoms as a whole in pure systems. The characteristic point of the former is that we can formulate a theory of quantum gates with reference to real physical systems, atoms, molecules or solids, in contrast to the previous approaches[1-4, 14], where formulations have been made from the viewpoint of mathematics based on principles of quantum mechanics. The latter, which is our preliminary step toward possible physical realization of solid state quantum gates, is an introduction of a specific kind of impurity atoms to apply radiation fields selectively to them in order to perform and control gate operations. In this paper, this. was discussed very briefly in the, previous section. In order for our idea turns out to be physically reasonable and acceptable, more detailed examination of subtle "host-guest combination" is required. This will be done in the future.

Another very important issue to be examined is how to maintain coherence during computing operations against quantum-mechanical and thermal decoherence effects. Two things naturally come to our mind in connection with this fact. One is to choose an impurity, atom whose relevant excited state has long life time (for example having channel to triplet states) and the other is to apply intense radiation fields to the exciton systems so as to attain steady-state coherent states where balance can exists between in-coming energy from the radiation fields and out-going one to surroundings[15]. This is worth separate studying and will also be examined in the future.

REFERENCES

1. R.P. Feynman, Opt. News 11: 11 (1985).
2. D. Deutch, *Proc. Roy. Soc.*, A400: 96 London (1985).
3. D. Deutch and R. Rozsa: *Proc. Roy. Soc.*, A439: 553 London (1992).
4. D. Deutch, *Proc. Roy. Soc.*, A425: 73 London (1989).
5. P.W. Shor, in *Proc. of the 5th Annual Symposium on the Foundation of Computer Science,*: 116 (IEEE Computer Society, Los Alamos (1994).
6. L. Cirac and P. Zoiler, *Phys. Rev. Lett.*, 74: 4091 (1995).

7. J. Monroe et al., *Phys. Rev. Lett.*, 75: 4717(1995).
8. N.A. Gershenfeld and I.L. Chuang, *Science*, 275: 350 (1997).
9. A. Steane, *Proc. Roy. Soc.*, A452: 2551 London (1996).
10. A.R. Calderbank and P.W. Shor, *Phys.Rev.*, A54: 1098 (1996).
11. D. Cory, A. Fahmy and T. Havel, *Proc. Nat. Acad. Sci. U.S.A*, 94: 1634 (1997).
12. A. Barenco, C.H. Bennett, R. Cleve, D.P. DiVincenzo, N. Margolus, P. Shor, T. Sleator, J.A. Smolin, and H. Weinfurter, *Phys. Rev.*, 52A: 3457 (1995).
13. R. London, *The Quantum Theory of Light*: 174 Clarendon Press (1973).
14. S. Takeno, *J. Phys. Soc. Jpn.*, 62: 289 (1992).

NONLINEAR SPIN WAVES AND MAGNET-ACOUSTIC RESONANCE IN THE MODEL OF HEISENBERG MAGNET

Kh. Kh. Muminov[1] and Valirii K. Fedyanin

Laboratory of Theoretical Physics,
Joint Institute for Nuclear Research,
141980 Dubna, Russia

[1]Permanent address: Tajik State National University,
734025 Dushanbe, Tajikistn

INTRODUCTION

Investigation of nonlinear properties of magnet crystals attracts a great attention in the past decades [1–10]. Mainly, this interest has been initiated both by rapid development of the theory of nonlinear differential equations (such branches as inverse scattering method, algebraic and geometrical methods, of integrating, numerical experiments and so on), new experimental data, and the possibility of their wide application in the different branches of applied science and technology.

It should be mentioned that the most popular model used in the investigations of nonlinear particle-like excitations in magnets is the model introduced by Landau and Lifshitz [9]. The Landau – Lifshitz equation (LLE) could be written, in particular isotropic case, in the following form

$$i\hbar S_t = \frac{1}{2}[S, S_{xx}], \qquad (1)$$

where we introduce

$$S = \begin{pmatrix} S^z & S^- \\ S^+ & -S^z \end{pmatrix} = \vec{S} \cdot \vec{\sigma}, \qquad (2)$$

here $\vec{S} = (S^x, S^y, S^z)$ is the vector of magnetization $S^{\pm} = S^x \pm iS^y$ are the corresponding value of the projection of the classical spin vector, or exactly, of the magnetization vector (i.e. we allow that $\vec{S} \equiv \vec{M}$),

$$\hat{\sigma}_x = \begin{pmatrix} 0 & 1 \\ 1 & 0 \end{pmatrix}, \hat{\sigma}_y = \begin{pmatrix} 0 & -i \\ i & 0 \end{pmatrix}, \hat{\sigma}_z = \begin{pmatrix} 1 & 0 \\ 0 & -1 \end{pmatrix}$$

are Pauli operators, and the brackets [...] ({...}) correspond to (anti)commutator.

This equation (1) give us the macroscopic description of magnetization dynamics in ferromagnets and represents the equation of motion of the magnetization vector in non — dissipative media.

On the other hand, it is well known, that in the microscopic level the most popular models for description of magnetic properties of a crystal are the spin models of Heisenberg — Frenkel magnet

$$\hat{H} = -\sum_{i,k} J_{ik} \hat{\vec{S}}_i \hat{\vec{S}}_k, \qquad (3)$$

where J_{ik} are exchange integrals, $\hat{\vec{S}}_j$ are the spin operators of the atom in the j-th site.

The problem of the correspondence between classical and quantum concepts and the transition procedure from quantum — mechanical description of ferromagnet (3) to the description on the classical, macroscopic level (1) is not trivial. This problems has been discussed by many authors (see for example, [1] and the papers cited there). At the same time, the presence in the spin Hamiltonian (3) of such physical parameters as exchange integrals, constants of anisotropy, values of atom spin and so on, makes more favorable to investigate magnets starting from quantum — mechanical Hamiltonian (3). The present study is devoted to investigation of the particle — like (localized) excitation in magnets describing by the soliton solutions of nonlinear differential equations. Namely the presence of particle — like excitations in spin systems can explain a number of peculiarities of the slow neutron scattering on magnets, dynamic structure form factors and so on [10 — 12] in the presence of interaction of spin subsystem with phonon one.

The derivation of the equation of motion of the magnetization vector (or, "the vector of classical spin"), starting from microscopical description is not trivial. Concerning the formal procedure of transition from spin Hamiltonian (3) to the quasiclassical description it should be noted, that this procedure could be based, for example, on the bozonization of spin Hamiltonian (by use of Holstein — Primakoff transformation [13, 14] or other [13]) and after that the obtained bozonized Hamiltonian should be averaged using the Glauber coherent states [4, 6]. As the result of carrying out of the above mentioned procedure we obtain a classical Hamiltonian of the model. As it was shown in the paper [6] this approach is permissible (and correct) for the magnets with sufficiently large values of spins ($S \gg 1$, or for $\delta S \gg 1$ [14], δ is the constant of anisotropy). It should be noted that in this approach the appearance of nonphysical degrees of freedom due to the truncation of the infinite expanding series could lead to uncontrollable mistake [6].

In the case of magnets with the spin value $S = 1/2$, and also for magnets with the spin value $S > 1/2$ in the presence of exchange anisotropy only (i.e. when the multipole spin dynamics is frozen), direct use of the generalized coherent states which is constructed on the operators of $SU(2)$ group becomes possible. In this case it is not necessary to carry out the bozonization procedure of the spin Hamiltonian, because both the Hamiltonian and the coherent state are constructed on the operators of the same group. Also it should be mentioned, that in the case of magnets with the spin value $S = 1/2$, the correct use of exact Wigner — Seitz transformation allow us to write the Hamiltonian (3) in terms of Fermi operators and then using the transition procedure we obtain generalized nonlinear Schroedinger equation for the probability amplitudes of the spin excitations [11].

Investigation of the magnets with the spin values $S \geq 1$ taking into consideration the single—ion and some other types of anisotropy in the spin Hamiltonian (3) is more complicated due to the exciting of the multipole spin dynamics. In this case number of quasiclassical parameters required for the full macroscopic description of magnet grows up to $4S$, and the procedure of derivation of the equation of classical spin and multipole dynamics should be based on the generalized coherent states constructed on operators of $SU(2S+1)$ group (see details in papers [6, 7, 15]).

In the present paper the spin — phonon interaction in quasi — one — dimensional magnet crystals with the spin value $S = 1/2$ is investigated in the scope of the SU(2) generalized coherent states technique. As we have mentioned above this investigation is correct not only for the magnets with the spin value $S = 1/2$, but also for the magnets with the spin value $S > 1/2$ if we take into consideration exchange anisotropy only, neglecting single — ion — anisotropy, and for the case of magnets with spin value $S \gg 1$ [6 - 8].

QUANTUM AND CLASSICAL MODELS OF MAGNETS WITH MAGNETOELASTIC INTERACTION

Let us consider the model of the Heisenberg ferromagnet with single — axis anisotropy in the presence of oscillation of sites of the crystal lattice

$$\widehat{H} = \widehat{H}_s + H_p, \qquad (4)$$

where

$$\widehat{H}_s = -\sum_{j=1}^{N} \left\{ \frac{J^0}{2} \left(\widehat{S}_j^+ \widehat{S}_{j+1}^- + \widehat{S}_j^- \widehat{S}_{j+1}^+ \right) + J^z \widehat{S}_j^z \widehat{S}_{j+1}^z \right\} \equiv$$

$$\equiv -J^0 \sum_{j=1}^{N} \left\{ \overrightarrow{\widehat{S}}_j \overrightarrow{\widehat{S}}_{j+1} + \Delta \widehat{S}_j^z \widehat{S}_{j+1}^z \right\}, \qquad (5)$$

is the spin part of the Hamiltonian,

$$H_p = \sum_{j=1}^{N} \left\{ \frac{p_j^2}{2m} + \frac{k}{2} (y_{j+1} - y_j)^2 \right\} \qquad (6)$$

is the phonon part of the Hamiltonian. Here $\Delta = (J^z - J^0)/J^0$ is the constant of exchange anisotropy, m and p are the momentum and mass of the atom, correspondingly, $|y_{j+1} - y_j|$ is the displacement of the j-th atom from the equilibrium position, k is the elastic constant, j is the summation index. In the expression (6) we take into consideration the harmonic oscillations of the crystal lattice only.

Let us pass over to the classical description. In order to make this we average $\widehat{H}_s$ by use of the $SU(2)$ generalized coherent states (GCS) [4, 5]. Let us remind that $SU(2)$ GCS in complex parameterization has the form

$$|\xi\rangle = \prod_j |\xi_j\rangle = \prod_j \left(1 + |\xi_j|^2\right)^{-k} \exp\left\{\xi_j \widehat{S}_j^+\right\} |k, -k\rangle, \qquad (7)$$

here k is the number of representation, ξ_j is the parameter of quasiclassical description. Spin operators averaged by use of the $SU(2)$ GCS get the following form

$$S_j^+ = \overline{S_j^-} = \langle \widehat{S}_j^+ \rangle = \frac{2\xi_j}{1 + |\xi_j|^2}, \, S_j^z = \langle \widehat{S}_j^z \rangle = \frac{1 - |\xi_j|^2}{1 + |\xi_j|^2}. \qquad (8)$$

Note, that the parameterization via more habitual angle variables is possible. In this case the values of the averaged spin operators have the form

$$\vec{S} = s(\sin\theta\cos\varphi, \sin\theta\sin\varphi, \cos\theta). \qquad (9)$$

The relationship between the complex parameters ξ and the angle one θ, φ is given by stereographycal projection

$$\xi_j = \tan\left(\frac{\theta}{2}\right)\exp\{i\varphi_j\}, \qquad (10)$$

where the values of angle parameters are restricted as $0 \leq \theta \leq \pi$ and $0 \leq \varphi < 2\pi$. By use of the vector (7) we carry out the averaging procedure of the spin Hamiltonian $\widehat{H}_s$. We have

$$H_s = \left\langle \xi \left| \widehat{H}_s \right| \xi \right\rangle =$$

$$= -s^2 \sum_{j=1}^{N} \left\{ \frac{2J^0\left(\xi_j\overline{\xi_{j+1}} + \overline{\xi_j}\xi_{j+1}\right) + J^z\left(1 - |\xi_j|^2\right)\left(1 - |\xi_{j+1}|^2\right)}{\left(1 + |\xi_j|^2\right)\left(1 + |\xi_{j+1}|^2\right)} \right\},$$

and the Hamiltonian of the system takes the form

$$H = \left\langle \xi \left| \widehat{H} \right| \xi \right\rangle = H_s + H_p. \qquad (11)$$

In order to obtain continual limit of the Hamiltonian (11) we assume

a). In the expansion of exchange integral J^z we take into account the linear terms only (we assume, that isotropic exchange integral do not depend on the lattice deformation, i.e. $J^0 = const$)

$$J^z = J^3 + \overline{J^3}|y_{j+1} - y_j|, \qquad (12)$$

b). In the expansion of ξ_{j+1} we take into account terms no more than quadratic in a_0

$$\xi_{j+1} = \xi_j + a_0\xi_{jx} + \frac{a_0^2}{2}\xi_{jxx} + ...$$

$$y_{j+1} = y_j + a_0 y_{jx} + ...$$

Then rewriting Hamiltonian (11) in spherical variables we obtain

$$H = -s^2 \sum_{j=1}^{N} \left\{ \frac{J^0}{2}\left(S_j^+ S_{j+1}^- + S_j^- S_{j+1}^+\right) + J^z S_j^z S_{j+1}^z \right\} + H_p. \qquad (13)$$

This Hamiltonian in the continual limit takes the form

$$H = H_s + H_p + H_{sp}, \tag{14}$$

where

$$H_s = s^2 J^0 \frac{a_0^2}{2} \int \frac{dx}{a_0} \left[\frac{1}{2} \left((S_x^x)^2 + (S_x^y)^2 + (S_x^z)^2 \right) - \delta (S^z)^2 \right], \tag{14.a}$$

$$H_{sp} = -s^2 J^0 \chi \int \frac{dx}{a_0} y_x (S^z)^2, \tag{14.b}$$

$$H_p = \int \frac{dx}{a_0} \left(\frac{p^2}{2m} + \frac{ka_0^2}{2} y_x^2 \right), \tag{14.c}$$

here $\delta = 2\Delta/a_0^2$ is the constant of single-axis anisotropy,

$$J^z = J^0 + J^{z'} + \overline{J_3} |y_{j+1} - y_j| = J^3 + \overline{J^3} |y_{j+1} - y_j| = J^3 + \chi J^0 \frac{|y_{j+1} - y_j|}{a_0},$$

and $\chi = \overline{J^3} a_0 / J_0$ ($\gamma = 2\chi/a_0^2$) is the dimensionless constant of spin-phonon interaction. Note that in the expression (14) we neglect the constant terms

$$-J^0 s^2 N + \int \frac{dx}{a_0} \left[-a y_x \overline{J^3} \left(a_0 S^z S_x^z + \frac{a_0^2}{2} S^z S_{xx}^z \right) \right] = const.$$

Introducing Poison brackets in the following form (see [5])

$$\{A, B\} = \int \left\{ -\varepsilon_{ijk} \frac{\delta A}{\delta S^i} \frac{\delta B}{\delta S^j} S^k - \frac{\delta A}{\delta p} \frac{\delta B}{\delta y} + \frac{\delta A}{\delta y} \frac{\delta B}{\delta p} \right\} dx, \tag{15}$$

we derive the equation of dynamics of magnetization vector coupled with the lattice (chain) oscillations

$$\hbar S_t^l = \{H, S^l\} = -\varepsilon_{ilk} \frac{\delta H}{\delta S^i} S_k,$$

$$y_t = -\frac{\delta H}{\delta p} = -\frac{p}{m},$$

$$p_t = \frac{\delta H}{\delta y} = ka_0^2 y_{xx} - \chi s^2 J^0 \left[(S^z)^2 \right]_{xx}$$

or, rewriting this system of equations in the matrix form we obtain

$$i\hbar S_t + s^2 J^0 \frac{a_0^2}{2} [S, S_{xx}] + s^2 J^0 \frac{(\Delta + \chi u)}{2} [S, \hat{\sigma}^z] \{S, \hat{\sigma}_z\} = 0, \tag{16.a}$$

$$u_{tt} - \frac{ka_0^2}{m} u_{xx} + \chi \frac{s^2 J^0}{m} \left[(S^z)^2 \right]_{xx} = 0, \tag{16.b}$$

here we put $u = y_x$,

$$S = \begin{pmatrix} S^z & S^- \\ S^+ & -S^z \end{pmatrix},$$

and brackets $[..., ...] (\{..., ...\})$ correspond to (anti)commutator.

After simple scale transformation

$$x' = bx, t' = at$$

where

$$b^2 = \frac{k\hbar^2}{ma_0^2 (s^2 J^0)^2}, a = \frac{k\hbar}{ms^2 J^0},$$

we obtain the following system of equations

$$iS_t + \frac{1}{2}[S, S_{xx}] + G\frac{(\Delta + \chi u)}{2}[S, \hat{\sigma}_z]\{S, \hat{\sigma}_z\} = 0, \qquad (17.a)$$

$$u_{tt} - u_{xx} + \lambda \left[(S^z)^2\right]_{xx} = 0, \qquad (17.b)$$

where

$$G = \left(\frac{s^2 J^0}{\omega_0 \hbar}\right)^2,$$

$$\omega_0 = \sqrt{\frac{k}{m}},$$

$$\lambda = \frac{s^2 J^0}{k a_0^2}$$

MODELS WITH PHONON UNHARMONISM

Let us now take into consideration terms of higher order in the expansion of potential energy of interaction between the atoms of the crystal lattice (chain). The energy of lattice oscillations takes the form

$$H_p = \sum_{j=1}^{N} \frac{p_j^2}{2m} + U, \qquad (18.a)$$

where

$$U = \sum_{j=1}^{N} \varphi(y_{j+1} - y_j). \qquad (18.a)$$

After making the corresponding transition procedure as above, we obtain the system of coupled Landau — Lifshitz and Boussinesq equations describing nonlinear spin excitations accompanied with the nonlinear sound mode propagating in ferromagnet

$$i\hbar S_t + s^2 J^0 \frac{a_0^2}{2}[S, S_{xx}] + s^2 J^0 \frac{(\Delta + \chi u)}{2}[S, \hat{\sigma}^z]\{S, \hat{\sigma}_z\} = 0,$$

$$u_{tt} - v_s^2 u_{xx} - \Lambda \left(u^2\right)_{xx} - a_0^2 B u_{xxxx} + \chi \frac{s^2 J^0}{m}\left[(S^z)^2\right]_{xx} = 0,$$

where

$$\Lambda = \frac{1}{2}\varphi'''(0) a_0^3,$$

$$B = \frac{v_s^2}{2},$$

$$v_s^2 = \frac{\varphi''(0) a_0^2}{m} = \frac{k a_0^2}{m}$$

is the sound velocity, k is the constant of elasticity. This system of equations after scale transformation can be written as

$$iS_t + \frac{1}{2}[S, S_{xx}] + G\frac{(\Delta + \chi u)}{2}[S, \hat{\sigma}_z]\{S, \hat{\sigma}_z\} = 0, \qquad (19.a)$$

$$u_{tt} - u_{xx} - \alpha \left(u^2\right)_{xx} - \beta u_{xxxx} + \chi\lambda\left[(S^z)^2\right]_{xx} \qquad (19.b)$$

where

$$\alpha = \frac{m}{k a_0^2}\Lambda, \beta = \frac{\hbar^2 B}{a_0^3 (s^2 J^0)^2}$$

Let us consider now multisublattice model consisting of p sublattices of ferromagnet type and q sublattices of antiferromagnet type ($M = p + q$). Interaction between the sublattices is defined by the phonon subsystem. This model can be written in the form

$$\widehat{H} = \sum_{j=1}^{M} \widehat{H}_{sj} + H_p, \qquad (20)$$

where

$$\widehat{H}_{sj} = -\sum_{i=1}^{N} \left\{ J_i^0 \left(\hat{S}_{ji}^x \hat{S}_{ji+1}^x + \hat{S}_j^y \hat{S}_{ji+1}^y \right) + J_i^z \hat{S}_{ji}^z \hat{S}_{ji+1}^z \right\}$$

and H_p is defined by the formulae (18). Carrying out the procedure as above, we get in long wave limit approximation the following system of equations of spin — phonon interaction

$$iS_{jt} + \frac{1}{2}[S_j, S_{jxx}] + G_j \frac{(\Delta_j + \chi_j u)}{2}[S_j, \hat{\sigma}_z]\{S_j, \hat{\sigma}_z\} = 0, \qquad (21.a)$$

$$u_{tt} - u_{xx} - \alpha\left(u^2\right)_{xx} - \beta u_{xxxx} + \left(\sum_{k=1}^{p} \chi_k \lambda_k (S^z)^2 + \sum_{k=p+1}^{M} \chi'_k \lambda_k (S^z)^2\right)_{xx} = 0, \qquad (21.b)$$

here

$$J_k^z = J_k^3 + \chi'_k J^0 \frac{|y_{j+1} - y_j|}{a_0}$$

and χ'_k is the dimensionless constant of interaction of the antiferromagnet subsystem with the phonon one.

Now let us take into consideration exchange interaction between the ferromagnet and the antiferromagnet sublattices

$$\widehat{H}_{int} = -\sum_{k,j=1}^{M}\sum_{i=1}^{N} I_{kj} \hat{S}_{ji}^z \hat{S}_{ki}^z, \qquad (22)$$

where I_{kj} is the integral of exchange between k-th and j-th sublattices and the Hamiltonian of the model is

$$\widehat{H} = \sum_{j=1}^{M} \widehat{H}_{sj} + \widehat{H}_{int} + H_p$$

where $\widehat{H}_{sj}$ is defined by eq.(20) and H_p is determined by eq.(18). In this case the system of coupled Landau — Lifshitz and Boussinesq equations takes the form

$$iS_{jt} + \frac{1}{2}[S_j, S_{jxx}] +$$

$$+G_j[S_j, \hat{\sigma}_z] \left(\chi_j u \{S_j, \hat{\sigma}_z\} + \sum_{k=1}^{p} \Delta_{kj} \{S_k, \hat{\sigma}_z\} + \sum_{k=p+1}^{M} \Delta'_{kj} \{S_k, \hat{\sigma}_z\} \right) = 0, \quad (23.a)$$

$$u_{tt} - u_{xx} - \alpha \left(u^2\right)_{xx} - \beta u_{xxxx} + \left(\sum_{k=1}^{p} \chi_k \lambda_k (S_k^z)^2 + \sum_{k=p+1}^{M} \chi'_k \lambda_k (S_k^z)^2 \right)_{xx} = 0, \quad (23.b)$$

here

$$\Delta_{kj} = \frac{[I_{ik} + \delta_{ik}(I_{ii} + J^3 - J_i^0)]}{J^0},$$

and for the antiferromagnet sublattice the same expression for Δ'_{kj} has been introduced, and $S_j = \begin{pmatrix} S_j^z & S_j^- \\ S_j^+ & -S_j^z \end{pmatrix}$ is magnetization vector.

Below we discuss some limit cases of the system of equations (23).
1). Quasistationary limit $v \ll 1, u_{tt} \ll u_{xx}$.

In order to simplify our consideration we put $\alpha = \beta = 0$, then the eq.(23.b) give us

$$u(x,t) = \sum_{k=1}^{p} \lambda_k \left[(S_k^z)^2 - 1\right] + \sum_{k=p+1}^{M} \chi'_k \lambda_k \left[(S_k^z)^2 - 1\right]. \quad (24)$$

Here the constants of integration defined by the boundary conditions

$$(S_k^z(\pm\infty, t))^2 = 1, u(\pm\infty, t) = 0$$

Substituting eq.(24) to eq.(23.a) we get

$$S_{jt} + \frac{1}{2}[S_j, S_{jxx}] +$$

$$+\chi_j G_j[S_j, \hat{\sigma}_z] \left(\sum_{k=1}^{p} \chi_k \lambda_k \left[(S_k^z)^2 - 1\right] + \sum_{k=p+1}^{M} \chi'_k \lambda_k \left[(S_k^z)^2 - 1\right] \right) \{S_j, \hat{\sigma}_z\} +$$

$$+G_j[S_j, \hat{\sigma}_z] \left(\sum_{k=1}^{p} \Delta_{kj} \{S_k, \hat{\sigma}_z\} + \sum_{k=p+1}^{M} \Delta'_{kj} \{S_k, \hat{\sigma}_z\} \right) = 0. \quad (25)$$

In the simplest case $M = p = 1, q = 0$ from eq.(25) we have

$$iS_t + \frac{1}{2}[S, S_{xx}] + \frac{G}{2}\left[\chi^2\lambda\left[(S_k^z)^2 - 1\right] + \Delta\right][S, \hat{\sigma}_z]\{S, \hat{\sigma}_z\} = 0, \quad (26)$$

here $S \equiv S_1$.

Note, that the Hamiltonian of this equation is defined by the expression

$$H = s^2 J^0 \int \frac{dx}{a_0} \left[\frac{a_0^2}{2} \left(\vec{S}_x\right)^2 + \left[x^2 \frac{s^2 J^0}{ka_0^2} - \Delta \right] (S^z)^2 - \chi^2 \frac{s^2 J^0}{ka_0^2} (S^z)^4 \right]. \tag{27}$$

2). Nearsonic limit $v \sim 1$.

In this case using the standard procedure

$$\partial_t^2 - \partial_x^2 \simeq -2 \left(\partial_t + \partial_x\right) \partial_x$$

we can replace this operator in the equation (23.b). Integrating this equation with vanishing boundary condition we obtain

$$u_t + u_x + \frac{\alpha}{2} \left(u^2\right)_x + \frac{\beta}{2} u_{xxxx} - \frac{1}{2} \left(\sum_{k=1}^{p} \chi_k \lambda_k \left(S_k^z\right)^2 + \sum_{k=p+1}^{M} \chi_k' \lambda_k \left(S_k^z\right)^2 \right)_x = 0. \tag{28}$$

In the harmonic phonon approximation the constants $\alpha = \beta = 0$ and this equation can be written in the form

$$u_t + u_x - \frac{1}{2} \left(\sum_{k=1}^{p} \chi_k \lambda_k \left(S_k^z\right)^2 + \sum_{k=p+1}^{M} \chi_k' \lambda_k \left(S_k^z\right)^2 \right)_x = 0. \tag{29}$$

3). Small amplitude approximation.

In this case we assume, that deviations of classical spin vector from the equilibrium position (classical vacua) is sufficiently small $|S^+|^2 \ll 1$. We introduce the function $\varphi_j(x,t) = S_j^+(x,t)$, so $|\varphi_j|^2 \ll 1$, then $S_j^z = 1 - \frac{1}{2}|\varphi_j|^2$. Substituting this relation to eq.(23) we derive

$$i\varphi_t - \varphi_{xx} + 2\chi G u \varphi + 2\Delta G \varphi - \Delta G (\overline{\varphi}\varphi) \varphi = 0 \tag{30.a}$$

$$u_{tt} - u_{xx} - \alpha \left(u^2\right)_{xx} - \beta u_{xxxx} + 2\chi \lambda (\overline{\varphi}\varphi)_{xx} = 0, \tag{30.b}$$

where

$$\varphi = (\varphi_1, ..., \varphi_M)^t, \Delta = \left(\Delta_1, ..., \Delta_p, \Delta'_{p+1}...\Delta'_M\right),$$

$$\chi = \left(\chi_1, ..., \chi_p, \chi'_{p+1}...\chi'_M\right),$$

$$\overline{\varphi}\varphi = \sum_{k=1}^{p} |\varphi_k|^2 + \sum_{k=p+1}^{M} |\varphi_k|^2,$$

$$\Delta_j = \sum_{k=1}^{p} \Delta_{kj} + \sum_{k=p+1}^{M} \Delta'_{kj},$$

We have obtained the system (30) neglecting the following terms

$$\varphi_j |\varphi_j|^2_{xx} + \frac{1}{2}\varphi_{jxx} |\varphi_j|^2 + \chi u |\varphi_j|^2 \varphi_j$$

as a terms of higher order of nonlinearity.

In harmonic approximation $\alpha = \beta = 0$, and from eq.(30) we derive

$$i\varphi_t - \varphi_{xx} + 2\chi Gu\varphi + 2\Delta G\varphi - \Delta G\left(\overline{\varphi}\varphi\right)\varphi = 0 \tag{31.a}$$

$$u_{tt} - u_{xx} + 2\chi\lambda\left(\overline{\varphi}\varphi\right)_{xx} = 0, \tag{31.b}$$

Note the most remarkable case, which is the nearsonic limit $v \sim 1$ that reduces eq.(31) to nonlinear Schroedinger equation with Yajima — Oikawa potential (see [16]) ($M = 1$)

$$i\varphi_t - \varphi_{xx} + 2\chi Gu\varphi + 2\Delta G\varphi - \Delta G|\varphi|^2\varphi = 0, \tag{32.a}$$

$$u_t + u_x + \chi\lambda\left(|\varphi|^2\right)_{xx} = 0. \tag{32.b}$$

Here we take into account relation

$$\partial_t^2 - \partial_x^2 \simeq -2\left(\partial_t + \partial_x\right)\partial_x$$

It should be mentioned, that the regular method of constructing of multisoliton solutions of eq.(23) is proposed in ref. [17].

DOMAIN-WALL AND SOLITON-LIKE SOLUTIONS

In order to derive and investigate soliton solutions of the obtained nonlinear quasi-classical models it is convenient to pass over the angle variables for the classical spin (magnetization) vector. Then the Hamiltonian averaged by use of $SU(2)$ GCS in the continual limit takes the following form

$$H = H_s + H_{sp} + H_p, \tag{33}$$

where

$$H_s = s^2 J_0 \int \frac{dx}{a_0}\left[\frac{a_0^2}{2}\left(\vec{S_x}\right)^2 - \Delta\left(S^z\right)^2\right] \tag{33.a}$$

is the spin part of Hamiltonian, (δ is the constant of exchange anisotropy),

$$H_p = \int \frac{dx}{a_0}\left(\frac{p^2}{2m} + \frac{a_0^2 k}{2}y_x^2\right) \tag{33.b}$$

is the phonon part of Hamiltonian, and

$$H_{sp} = -s^2 J_0\chi \int \frac{dx}{a_0} y_x \left(S^z\right)^2 \tag{33.c}$$

is the Hamiltonian of spin — phonon interaction.

In terms of angle variables the magnetization vector (classical spin) is

$$\vec{S} = s(\sin\theta\cos\varphi, \sin\theta\sin\varphi, \cos\theta). \tag{34}$$

Let us remind that the Hamiltonian equation of motion obtained for $SU(2)$ GCS in ref. [5] by path integral method, have the following form

$$\hbar\varphi_t = -\frac{1}{\sin\theta}\frac{\delta H}{\delta\theta}, \tag{35.a}$$

$$\hbar\theta_t = \frac{1}{\sin\theta}\frac{\delta H}{\delta\varphi}, \tag{35.b}$$

and supplementing this equations by obvious equations

$$y_t = -\frac{\delta H}{\delta p} = -\frac{p}{m}, \tag{35.c}$$

$$p_t = \frac{\delta H}{\delta y}. \tag{35.d}$$

we obtain the full set of classical equations of motion. Using the equations (35) and the Hamiltonian density

$$h = s^2 J_0 \left\{ \frac{a_0^2}{2}\left(\theta_x^2 + \sin\theta\varphi_x^2\right) - [\Delta + \chi y_x]\cos^2\theta \right\} + \frac{p^2}{2m} + \frac{a_0^2 k}{2}y_x^2, \tag{36}$$

derived from eq.(33) we obtain the following system of equations of spin — phonon dynamics

$$a_0^2 \theta_{xx} - \left[a_0^2\varphi_x^2 + 2[\Delta + \chi u]\right]\sin\theta\cos\theta + \frac{\hbar}{J_0 s^2}\sin\theta\varphi_t = 0, \tag{37.a}$$

$$a_0^2\left(\sin^2\theta\varphi_x\right)_x - \frac{\hbar}{J_0 s^2}\sin\theta\theta_t = 0, \tag{37.b}$$

$$u_{tt} - \frac{k a_0^2}{m}u_{xx} - \chi\frac{s^2 J_0}{m}\left(\cos^2\theta\right)_{xx} = 0, \tag{37.c}$$

In (37.c) we introduce the designation $u = y_x$. It should be mentioned that in order to obtain solution with physical sense of the system (37) we assume that the boundary conditions for $\theta(x,t)$ and $\varphi(x,t)$ are defined by the minima of the classical Hamiltonian (33) of easy — axis magnet ($\delta > 0$) (i.e. by the classical vacua of the system)

$$\theta = 0, \pi, x \to \pm\infty. \tag{38}$$

In order to simplify the system (37) it is convenient to use dimensionless variables

$$z = bx, \tau = at, \tag{39}$$

where

$$b^2 = \frac{k\hbar^2}{ma_0^2(s^2 J_0)^2}, a = \frac{k\hbar}{ms^2 J_0}. \tag{40}$$

Then we can rewrite eq. (37) in the form

$$\theta_{zz} - \left[\varphi_z^2 + 2\left(\frac{s^2 J_0}{\omega_0 \hbar}\right)^2\{\Delta + \chi u\}\right]\sin\theta\cos\theta + \sin\theta\varphi_\tau = 0, \tag{41.a}$$

$$\left(\sin^2\theta\varphi_z\right)_z - \sin\theta\theta_\tau = 0, \tag{41.b}$$

$$u_{\tau\tau} - u_{zz} - \chi\frac{s^2 J_0}{ka_0^2}\left(\cos^2\theta\right)_{zz} = 0, \tag{41.c}$$

215

here
$$\omega_0 = \sqrt{\frac{k}{m}}.$$

We shall obtain the soliton solution of system (41) in the following form
$$\theta(z - v\tau), u = u(z - v\tau), \varphi = \psi(z - v\tau) + \omega\tau, \qquad (42)$$

and with the boundary condition (easy — axis magnet)
$$\theta = \theta(z, \tau)|_{z(x) \to \pm\infty} = 0, \pi, \qquad (43)$$

we derive
$$\theta_{\xi\xi} - \left[\varphi_\xi^2 + 2\left(\frac{s^2 J_0}{\omega_0 \hbar}\right)^2 \{\Delta + \chi u\}\right] \sin\theta \cos\theta + \{-v\psi_\xi + \Omega\} \sin\theta = 0, \qquad (44.a)$$

$$\left(\sin^2\theta \varphi_\xi\right)_\xi + v \sin\theta \theta_\xi = 0, \qquad (44.b)$$

$$(v^2 - 1) u_{\xi\xi} + \chi \frac{s^2 J_0}{k a_0^2} \left(\cos^2\theta\right)_{\xi\xi} = 0, \qquad (44.c)$$

Note that in the case $\chi = 0$ the system (44) reduces to the system of equation (8.10), (8.16) (see [1]). Integrating the equation (44.c) taking into account boundary condition (43) and
$$u = 0, \xi \to \pm\infty, \qquad (45)$$

we obtain
$$u = \chi \frac{s^2 J_0}{k a_0^2} \frac{\sin^2\theta}{v^2 - 1}, \qquad (46)$$

or
$$u = \chi \frac{s^2 J_0}{k a_0^2} \frac{\sin^2\theta}{v^2 - v_s^2}. \qquad (47)$$

Here v is the velocity of the magnetic soliton in the system with spin — phonon interaction. Sound velocity in our designation (40) is
$$v_s = a_0 \sqrt{\frac{k}{m}} \equiv 1. \qquad (48)$$

and
$$\omega_0 = \sqrt{\frac{k}{m}} \qquad (49)$$

Thus in the case $v = v_s$ solution (46) becomes singular. This singularity we define as magnet- acoustic resonance, which means that in the case of motion of magnetic soliton with nearsonic velocities the pumping of energy of magnetic soliton to the phonon subsystem takes place in the system.

We can integrate the second equation of the system (44)
$$\left(\sin^2\theta \psi_\xi - v\cos\theta\right)_\xi = 0 \qquad (50)$$

if we take into account boundary conditions

$$\xi \to \pm\infty, \frac{d\psi}{d\xi} < 0, \theta = 0. \tag{51}$$

Then we derive the following relation

$$\psi_\xi = -\frac{v}{2}\frac{1}{\cos^2\theta/2}. \tag{52}$$

The boundary conditions (51) correspond to nonlinear excitation of bell — soliton types.

Substituting relationships (52) and (46) into (44.a) we obtain

$$\left(\frac{\theta}{2}\right)_{\xi\xi} + \sin\frac{\theta}{2}\left\{\frac{v^2}{4}\cos^{-3}\left(\frac{\theta}{2}\right) + \omega\cos\left(\frac{\theta}{2}\right)\right\} -$$

$$-\left[a_1\Delta - 4\chi^2\frac{a_1 a_2}{1-v^2}\sin^2\left(\frac{\theta}{2}\right)\cos^2\left(\frac{\theta}{2}\right)\right]\sin\left(\frac{\theta}{2}\right)\cos\left(\frac{\theta}{2}\right)\cos\theta = 0, \tag{53}$$

here

$$a_1 = 2\left(\frac{s^2 J_0}{\omega_0 \hbar}\right)^2 = 2G, \tag{54}$$

$$a_2 = \frac{s^2 J_0}{k a_0^2} = \lambda. \tag{55}$$

First integrating of eq.(53) with boundary conditions (51) gives

$$\left(\frac{\theta}{2}\right)_\xi^2 = -\frac{v^2}{4}\tan^2\frac{\theta}{2} + a_1\Delta\cos^2\frac{\theta}{2}\sin^2\frac{\theta}{2} - \omega\sin^2\frac{\theta}{2} + 2\chi^2 A\cos^4\frac{\theta}{2}\sin^4\frac{\theta}{2}, \tag{56}$$

where

$$A = \frac{a_1 a_2}{1 - v^2}. \tag{57}$$

Integrating equation (56) we obtain the elliptic integral of the following form

$$\xi - \xi_0 = \frac{1}{2}\int\frac{ydy}{(y-1)\sqrt{-\frac{v^2}{4}y^4 - \omega y^3 + a_1\Delta y^2 + 2\chi^2 A(y-1)}}, \tag{58}$$

where

$$y = \tan^2\frac{\theta}{2} + 1. \tag{59}$$

Integral in the eq.(58) can be expressed through the elliptic integral of Weierstrasse of I and III types and then the numerical calculations must be done. However we can consider some limit cases of solution of the system (41), which can be expressed by the elementary function.

a). Let us consider magnetic solitons moving with velocity v, $v^2 \ll v_s^2, (v_s = 1)$. Then $\chi^2 A \to 0$ and assuming $\chi^2 A = 0$ (i.e. magnetic solitons do not feel the deformation of lattice), we have the solution

$$\tan^2\frac{\theta}{2} = \frac{\mu^2}{\Omega\cosh^2\mu\xi - [\Omega - \Omega_1]/2}, \tag{60.a}$$

where $\xi = z - v\tau$, $\mu = \sqrt{a_1\Delta - \omega - \left(\frac{v}{2}\right)^2}$, and $\Omega = \sqrt{\Omega_1^2 + 4\gamma^2(v/2)^2}$ is the parameter, that in laboratory frame of reference performs the dimensionless frequency of precession of magnetic moment in nonlinear spin wave with the parameters v and ω, and $\Omega_1 = \omega + (v/2)^2$ defines the dimensionless frequency of the precession of magnetic moment in the soliton with the same parameters v and ω in the laboratory frame of reference. Integrating eq.(52) and taking into account (60.a) and (42) we obtain

$$\varphi = \omega\tau - \frac{v}{2}\xi + \arctan\left[\sqrt{\frac{\Omega - \Omega_1}{\Omega + \Omega_1}}\tanh\gamma\xi\right]. \qquad (60.b)$$

The solution (60.a,b) taking into account our designation (39) completely coincides with the solution obtained in ref.[1]. Solution for the deformation wave accompanying magnetic soliton we get from eq.(46) using the solution (60.a- b)

$$u = 4\chi\frac{s^2 J_0}{ka_0^2}\frac{4\mu^2}{1-v^2}\frac{\Omega\cosh^2\mu\xi - [\Omega - \Omega_1]/2}{\{\Omega\cosh^2\mu\xi - [\Omega - \Omega_1]/2 + \mu\}}. \qquad (60.c)$$

b). In the case of the magnet solitons moving with the velocity $v^2 \gg v_s^2$ as in (57) we have $A \to 0$ and assuming $A = 0$ we have solution (60).

Thus both the supersonic magnetic solitons and the solitons moving with the velocities less than velocity of sound do not feel the lattice deformation

c). Let us consider solutions of the system of equation (44) in the small amplitude spin deviations approximation. We rewrite the eq.(58) as

$$\xi - \xi_0 = \frac{1}{2}\int\frac{(x+1)\,dx}{x\sqrt{-\frac{v^2}{4}(x+1)^4 - \omega(x+1)^3 + a_1\Delta(x+1)^2 + 2\chi^2 Ax}}, \qquad (61)$$

where

$$x = \tan^2\frac{\theta}{2}$$

and assuming that the deviation of the classical spin from the equilibrium position (i. e. from the ground states of the classical model) is sufficiently small $\theta \ll 1$, we take into account in eq.(61) terms $O(x^2)$ and neglect the terms of higher order. Then the eq.(61) can be reduced to the following sum of integrals, which can be easily integrated

$$\xi - \xi_0 = \frac{1}{2}\left[\int\frac{dx}{\sqrt{R}} + \int\frac{dx}{x\sqrt{R}}\right], \qquad (62)$$

where

$$R = a + bx + cx^2,$$

$$a = \mu^2 = a_1\Delta - \omega - \frac{v^2}{4},$$

$$b = 2\chi^2 A + 2a_1\Delta - 3\omega - v^2,$$

$$c = a_1\Delta - 3\omega - \frac{3v^2}{2}.$$

Integrating the eq. (62) we obtain the solution in the following form

$$\xi - \xi_0 = \frac{2}{\sqrt{c}} \operatorname{arcsinh} \frac{2cx+b}{\sqrt{D}} - \frac{2}{\mu} \operatorname{arcsinh} \frac{2\mu^2 + bx}{x\sqrt{D}}, \qquad (63)$$

where $D = \sqrt{4\mu^2 c - b^2}$.

Finally let us find the domain — wall type solutions of the system (44). We put

$$\varphi = \varphi_0 = const,$$

and find the solution of eq.(44.a) in the following form

$$\theta = \theta(\xi), \xi = z - v\tau \qquad (64)$$

Substituting eq.(64) to the equation (44.a) we derive

$$\theta_{\xi\xi} - a_1 [\Delta + \chi u] \sin\theta \cos\theta = 0. \qquad (65)$$

Integrating eq.(65) with the vanishing boundary conditions (we consider the easy — axis model) leads us to the following expression

$$\theta_\xi^2 - a_1 \Delta \sin^2\theta - \chi^2 \frac{\Gamma}{2} \sin^4\theta = 0. \qquad (66)$$

Solution of this equation can be expressed trough the integrals

$$2(\xi - \xi_0) = \int \frac{dx}{x\sqrt{R}} + \int \frac{dx}{\sqrt{R}}, \qquad (67)$$

where

$$\Gamma = \frac{s^2 J_0}{k a_0^2} \frac{1}{v^2 - 1},$$

$$R = a + bx + cx^2,$$

$$a = c = a_1 \Delta,$$

$$b = 2\left(a_1 \Delta + \chi^2 \Gamma\right),$$

$$x = \tan^2 \frac{\theta}{2}.$$

One can easily integrate eq. (67) and get the solution in the following form

$$2\sqrt{a}(\xi - \xi_0) = \ln\left|x \frac{2\sqrt{aR} + 2a(x+1) + 2\chi^2 \Gamma}{2\sqrt{aR} + 2a(x+1) + 2\chi^2 \Gamma x}\right|. \qquad (68)$$

It is obvious, that in the case $\chi^2 = 0$ the solution (68) takes the form of well — known domain wall

$$\theta = 2 \arctan \exp\left\{\sqrt{a}(\xi - \xi_0)\right\}. \qquad (69)$$

If we put $\chi^2 \ll 1$ expanding the solution (68) we obtain the domain wall type solution in the following form

$$\tan^2 \frac{\theta}{2} \simeq \frac{1}{1 - \chi^2 \Gamma} \exp\left\{2\sqrt{a}(\xi - \xi_0)\right\}, \qquad (70)$$

CONCLUSION

Thus, in this paper we show, that magnetic solitons in deformable crystal lattice is accompanied by the deformation wave. In the case of motion of the magnetic soliton with the near – sonic velocities the effect of resonance due to the interaction of spin subsystem with the phonon one takes place in the system. We define this phenomenon as magnet – acoustic resonance, and here the energy pumping from the magnetic soliton to the deformation wave takes place. At the same time the linear approximation for the sound equation used to derive solutions (60) and (61) is doubtfully based. Neglected nonlinear terms could play a leading role in the case of near – sonic velocities. This fact can be amplified also by the requirement to take into consideration effects of dissipation of the energy of magnetic soliton. More realistic model should take into account also the presence of two additional transversal sound modes, and. consequently, the possibility of resonance by interaction of magnetic soliton with transversal waves.

The authors are grateful to N.M. Plakida, V.K. Melelnikov, V.N. Robuk, V.A. Osipov and R. Nazmitdinov for helpful discussions and remarks.

REFERENCES

1. A.M.Kosevich, B.A.Ivanov, A.S.Kovalyov, Magnetic Solitons. *Phys. Rep.*, 194, 3–4: 1 (1990).
2. V.G. Bar'yakhtar, B.A. Ivanov, A.L. Sukstansky, Phonon slowing down of the domain wall in rare earth orthoferrite, *JETP*, 75, 6(12): 2183 (1978).
3. A.V.Mitin, V.A.Tarasov, Sound generation in a multidomain ferromagnet, *JFTP*, 72: 793 (1977).
4. A.M. Perelomov, *Generalized Coherent States and Their Application*, Nauka, Moscow (1987) (in Russian).
5. H. Kuratsuji, T. Suzuki, Path integral representation of SU(2) coherent state and classical dynamics in a generalized space, *J. Math. Phys.*, 21: 472 (1990).
6. Kh.O. Abdulloev, M. Aguero, A.V. Makhankov, V.G. Makhankov, Kh.Kh. Muminov, Generalized spine coherent states as a tool to study quasiclassical behavior of the Heisenberg ferromagnet, in: *Proceedings of the IY International workshop "Solitons and Applications"*, W.S.Singapore (1990).
7. Kh.O. Abdulloev, A.T. Maksudov, Kh.Kh. Muminov, General dynamical equations in the $SU(2S+1)/SU(2S)*U(1)$ and easy-axis magnet with the spin $S=3/2$, *Solid State Physics*, 34, 2: 544 (1992).
8. Kh.O.Abdulloev, Kh.Kh.Muminov, Semiclassical description of an isotropic magnets under influence of constant external magnetic fields. *Solid State Physics*, 36, 1: 170 (1994).
9. L.D. Landau, E.M. Lifshitz, To the theory of magnetic susceptibility of ferromagnet bodies. In: Landau L.D. *Collection of Manuscripts*, Nauka, Moscow (1969).
10. V.K. Fedyanin, B.Yu. Yushankhay, Soliton mode contribution to dynamical structure factor of easy–axis ferromagnet, *Phiysics of Low Ternperature*, 7, 2: 176 (1981).
11. A.V. Makhankov, V.K. Fedyanin, Ideal gas of particle-like excitations at low temperatures. *Physica Scripta*, 28: 221 (1983).
12. V.K. Fedyanin, Dynamical formfactor of neutron scattering on solitons in quasi-one-dimensional magnets, JMM,:.31–34: 1237 (1983)
13. I.A. Akhiezer, V.G. Bar'yakhtar, S.V. Peletminsky. *Spin Waws*, Nauka, Moscow, (1967) (in Russian).
14. V.G.Makhankov, R.Myrzakulov, A.V.Makhankov, Generalized coherent states and the continuous Heisenberg *XYZ* model with one-ion anisotropy. *Physica Scripta*, 34: 163 (1986).
15. V.S. Ostrovsky, On teh nonlinear dynamics of strongly anisotropic magnets with spin $S = 1$, *JETP*, 91, 5: 1690 (1986).
16. N. Yajima,. M. Oikawa. *Progr. Theor. Phys.*, 56: 1719 (1976).
17. B.A. Dubrovin, T.M. Malanyuk, I.M. Krichever, V.G. Makhankov, Exact solutions of nonstationary Schrodinger equation with self consistent potentials. *Phys. Elemen. Particles and Atomic Nuclei*, 19, 3: 579 (1998) (in Russian).

MODELLING OF 2D ELECTRON FIELD EMISSION FROM SILICON MICROCATHODE

Valery A. Fedirko[1], and Sergey V. Polyakov[2]

[1]Moscow State University of Technology "STANKIN"
101472 Moscow, Russia

[2]Institute for Mathematical Modelling Russian Academy of Sciences
125047 Moscow, Russia

INTRODUCTION

Our paper is devoted to the mathematical modelling of non-linear non-equilibrium processes in vacuum microelectronic semiconductor devices (VMSD). VMSD have proved perspective for various applications, especially for the design of field emission flat displays. Therefore, the fabrication of field emitter arrays with sufficiently high, controlled and stable emission ability is very important task in microvacuum device technology. Vacuum microelectronics batch technology is now in progress, and silicon is considered as one of the most suitable material for microcathode arrays (see[1,2]). For a successful development of VMSD the accurate calculations of the emission characteristics of silicon cathodes are badly needed. Thus, the mathematical modelling becomes one of the most important component of VMSD technology.

In [3] field emission from silicon was investigated using $1D$ quasihydrodynamic model. It was shown that the electron heating drastically affects the field emission from semiconductor and is extremely inhomogeneous. Therefore, it is reasonable to realize two-dimensional modelling taking into account the inhomogeneous distribution of electric field at the emitting surface of a real microcathode due to its geometry. In this paper we elaborate the two-dimensional mathematical model of hot electron field emission from silicon including the effect of inhomogeneous charge carrier heating near the emitting surface.

QUASIHYDRODYNAMIC MODEL

Our mathematical model relies on the following conjectures. An intense electric field penetrates rather deep into semiconductor and induces an intense heating of charge carriers. Due to the fact that the electron density and electron temperature are expected to be sufficiently high, at least, near the emitting surface, one should use a quasi-hydrodynamic approach, and hence, add the charge carrier energy equation to the model. The electron tunnelling through the emitting surface should be described by a relevant boundary condition which results in a strong non-linearity of the problem.

Let us consider an emission microcathode whose cross-section is given in Fig. 1. We pose the model in the plane (x, y). To simplify the problem, we consider a rectangular cathode with a model dependence $E_y(x)$ of electric field at outer area of plane emitting surface. We restrict the computation domain investigating the electron transfer processes only in one half of a cathode.

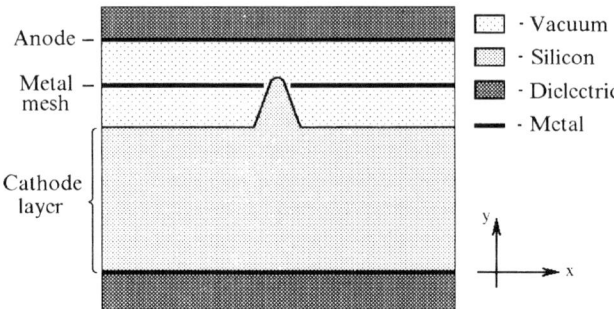

Figure 1. The cross-section of a cathode.

Thus, the initial dimensionless system of equations consists of the continuity equation for electron concentration n, energy balance equation for an electron density of energy $\frac{3}{2}nT$, and the Poisson equation for self-consistent electric field **E**. To complete the system, we add the equations for electron current **j**, flux of electron energy **Q**, and electric potential φ.

$$\frac{\partial n}{\partial t} = \text{div}\,\mathbf{j}, \quad \mathbf{j} = n\,\mu(T)\,\mathbf{E} + \nabla(D(T)n), \tag{1}$$

$$\frac{3}{2}\frac{\partial(nT)}{\partial t} = -\text{div}\,\mathbf{Q} + \mathbf{j}\cdot\mathbf{E} - \frac{3}{2}n\frac{T-1}{\tau_\varepsilon}, \quad \mathbf{Q} = -n\,\tilde{\mu}(T)\,\mathbf{E} - \nabla(\tilde{D}(T)n), \tag{2}$$

$$-\text{div}\,\mathbf{E} = \gamma(n-1), \quad \mathbf{E} = -\nabla\varphi, \tag{3}$$

$$0 < x < x_m, \quad 0 < y < 1, \quad t > 0,$$

The initial and boundary conditions have the form

$$n\big|_{t=0} = n_0(x, y), \quad T\big|_{t=0} = 1, \quad \mathbf{E}\big|_{t=0} = 0, \tag{4}$$

$$(\mathbf{j}\cdot\mathbf{v})\big|_\Sigma = j_s, \quad (\mathbf{Q}\cdot\mathbf{v})\big|_\Sigma = -Q_s, \quad (\mathbf{E}\cdot\mathbf{v})\big|_\Sigma = E_s, \tag{5}$$

In (1)–(5) T is the electron gas temperature normalised to the medium temperature T_0, $\mu(T) = [1 + \beta(T-1)]^{-1}$ and $D(T) = \mu(T)\,T$ are the coefficients of electron mobility and diffusion (β is a coefficient of heat mobility variation), $\tilde{\mu}(T) = \mu(T)\chi(T)$, $\tilde{D}(T) = D(T)\chi(T)$ and are the analogous coefficients in the energy balance equation,

$$\chi(T) = T\alpha(T) = T\left(\frac{5}{2} + \frac{\partial \ln \mu}{\partial \ln T}\right),$$ τ_ε is the energy relaxation time, N_D is the mean concentration of positively charged donors, x_m denotes the half-width of a cathode in the direction perpendicular to the external field, n_0 is the equilibrium distribution of electron concentration, the surface Σ bounds a semiconductor component of a cathode, v is the outer normal to Σ.

We put the surface electron current j_s, the energy flux Q_s, and the electric field E_s zero on the lateral area of a cathode whereas on the back and emitting surfaces (for $y = 0$ and $y = 1$ respectively) we use the following approximate expressions.

$$j_s = \begin{cases} n\mu(T) E_y^0, & y = 0, \\ A_0 n T^{1/2} \int_0^\infty D_0(E_s, T\xi) \exp(-\xi) d\xi & y = 1, \end{cases} \qquad (6)$$

$$Q_s = \begin{cases} -n\bar{\mu}(T) E_y^0, & y = 0, \\ A_0 n T^{3/2} \int_0^\infty D_0(E_s, T\xi)(\xi+1)\exp(-\xi)d\xi, & y = 1, \end{cases} \qquad (7)$$

$$E_s = \begin{cases} E_y^0, & y = 0, \\ E_0 + (E_m - E_0) f(x)\left[1 - \exp(-t/\tau_0)\right], & y = 1. \end{cases} \qquad (8)$$

Here E_y^0 is an unknown field value determined from a self-consistent solution of the Poisson equation when $\varphi(x,0) = 0$. A coefficient $D_0(E, \xi)$ describes a dependence of a tunnelling transparency on external electric field and electron energy, A_0 is a normalization coefficient, τ_0 is the time of external field establishing, E_m is the peak field value at the cathode centre. The function $f(x)$ describes the distribution of E_s in the cross coordinate direction.

$$f(x) = \begin{cases} 1, & x < \rho, \\ \sqrt{\dfrac{\rho}{x}}, & x > \rho. \end{cases}$$

A parameter ρ corresponds to the cathode tip radius. In [4,5] the coefficient $D_0(E, \xi)$ was found numerically using the Shrodinger equation with potential. To simplify our calculations, in this work we take the well known formula

$$D_0(E,\xi) = \begin{cases} \exp\left[-\alpha_0(1-\beta_0\xi)^{3/2}\theta(\eta)/E\right], & \xi < \beta_0^{-1}, \\ 1, & \xi \geq \beta_0^{-1}, \end{cases}$$

$$\theta(\eta) = max\left(0, 1 - 0.07\eta - 0.739\eta^2 - 0.191\eta^{15}\right), \quad \eta = \gamma_0 \sqrt{E}/(1-\beta_0\xi),$$

where α_0, β_0 and γ_0, are relevant constants.

NUMERICAL METHOD AND PARALLEL IMPLEMENTATION

First of all let us transform the equation (2) for energy density to the divergent form. Introducing the new flux $\mathbf{Q}_T = -\mathbf{Q} + D(T)n\mathbf{E}$, and taking into account the formula for current we get

$$\frac{3}{2}\frac{\partial(nT)}{\partial t} = \text{div}\mathbf{Q}_T - \text{div}(D(T)n\mathbf{E}) + \mu(T)n\,\mathbf{E}^2 + \nabla(D(T)n)\cdot\mathbf{E} - \frac{3}{2}n\frac{T-1}{\tau_\varepsilon},$$

Then we recall the form of $D(T)$ and $\mu(T)$ and apply the Poisson equation for field. As a result, we obtain

$$\frac{3}{2}\frac{\partial(nT)}{\partial t} = \text{div}\mathbf{Q}_T + \mu(T)n\,\mathbf{E}^2 + \mu(T)\,nT\,(n - N_D) - \frac{3}{2}n\frac{T-1}{\tau_\varepsilon}.$$

The corresponding boundary condition from (5) has the form

$$(\mathbf{Q}_T \cdot \mathbf{v})\big|_\Sigma = Q_s + \mu(T)\,nT\,E_s.$$

Taking $\frac{3}{2}nT$ for w and noting that

$$\mathbf{Q}_T = \mu(T)(1+\alpha(T))\,nT\,\mathbf{E} + \nabla(D(T)\alpha(T)nT)$$

and Q_s is proportional to nT we can rewrite the initial equation (2) once more

$$\frac{\partial w}{\partial t} = \text{div}\mathbf{Q}_T + f, \quad f = \mu(T)\,n\,|\mathbf{E}|^2 + \frac{2}{3}\mu(T)\,w(n-N_D) - \frac{w}{\tau_\varepsilon} - \frac{3n}{2\tau_\varepsilon}. \tag{9}$$

The operators in (1) and (9) have uniformly divergent structure which makes it possible to construct a conservative monotone scheme using the approach proposed in [6].

To solve the considered differential problem numerically, we choose a splitting technique. The regular grid ω_t with the step τ in t direction is introduced. The proposed numerical algorithm starts with the initial values (4) and passes on to the next $t + 2\tau$ layer in three stages. At the first stage a grid analogue of φ denoted by φ_h is computed at the $t + 2\tau$ layer by means of a grid analogue of the Poisson equation (3) with the field values at the emitting surface $E_{sh}(t+2\tau)$ and concentration values $n_h(t)$. At the second stage the functions n_h and w_h at the intermediate $t + \tau$ layer are determined as the solution of a grid analogous of (1) and (9) in y direction. At the third stage n_h and w_h are computed at the $t + 2\tau$ layer as the solutions of the same grid equations but in x direction. At the stages 2 and 3 we use a smoothing procedure allowed to get solution even for large gradients.

Now we describe the proposed numerical algorithm in terms of the theory of difference schemes. Our problem has a number of the scales of electric field variation. Therefore, it is reasonable to construct an irregular grid ω_y in y direction and a regular grid ω_x in x direction. These grids consist of integer nodes. The grids ω'_x, ω'_y with half-integer nodes are intended to provide a sufficient accuracy of a smoothing procedure. To smooth a grid solution we introduce the integral grid functions N_h, W_h, F_h on $\omega'_x \times \omega_y$ ($\omega_x \times \omega'_y$) at the second (third) stage. Derivatives of these functions with respect to y (x) coincide with n_h, w_h, f_h at the second (third) stage of our algorithm. We use an over-line marking for the functions at intermediate t-layer $t + \tau$ and a hat marking for that at $t + 2\tau$ layer. Finally we write the following chain of difference equations.

$$\Lambda\varphi_h = \gamma(n_h - 1), \tag{10}$$

$$\frac{N_h - \overset{\smile}{N}_h}{2\tau} = \Lambda_y N_h, \quad n_h = N_{hy}, \quad n_h = N_{hy}, \tag{11}$$

$$\frac{W_h - \overset{\smile}{W}_h}{2\tau} = \Lambda_y W_h + \frac{F_h}{2}, \quad w_h = W_{hy}, \quad w_h = W_{hy}, \quad f_h = F_{hy}, \tag{12}$$

$$\frac{\hat{N}_h - \overset{\smile}{N}_h}{2\tau} = \Lambda_x \hat{N}_h, \quad \hat{n}_h = \hat{N}_{hx}, \quad n_h = N_{hx}, \tag{13}$$

$$\frac{\hat{W}_h - \overset{\smile}{W}_h}{2\tau} = \Lambda_x \hat{W}_h + \frac{\hat{F}_h}{2}, \quad \hat{w}_h = \hat{W}_{hx}, \quad w_h = W_{hx}, \quad \hat{f}_h = \hat{F}_{hx}. \tag{14}$$

In (10)–(14) Λ denotes a common approximation of the $2D$ Laplace operator. The $1D$ approximations Λ_x, and Λ_y are defined as follows

$$\Lambda_y N_h = (D(T_h) N_{hy})_y + r_h^+ N_{hy} + r_h^- N_{hy}, \quad r_h^\pm = |r_h| \pm r_h, \quad r_h = -\mu(T_h) \phi_{hy},$$

$$\Lambda_x \hat{N}_h = (D(T_h) \hat{N}_{hy})_y + r_h^+ \hat{N}_{hy} + r_h^- \hat{N}_{hy}, \quad r_h^\pm = |r_h| \pm r_h, \quad r_h = -\mu(T_h) \phi_{hx}.$$

The Poisson equation is solved by means of FFT in x direction, whereas we use the sweep method in y direction. Solutions (11)–(14) are searched for by means of the sweep method. As the system (10)–(14) is non-linear, we realize it by iterations. Let us observe the basic characteristics of the proposed scheme. It is conservative in the sense of the validity of the grid analogue of the charge conservation law. Moreover, the described scheme holds the property of n_h and w_h non-negativity regardless of grid steps.

For the parallel implementation of this numerical method we propose the "block-conveyor" sweep algorithm (BCS-algorithm). This algorithm being applied to an equation computes sequentially α, β coefficients as well as a solution at each processor node. It is not effective for a single finite-difference equation. However, it may be useful for a solving of systems consisting of a large number N_{eq} of finite-difference equations. To apply the BCS-algorithm, one should divide an equation system into blocks. A number L_b of these blocks must be sufficiently large. Using this parallelization procedure we can reach the 90–95 % efficiency when a processor number n_p is fixed. The BCS-algorithm works efficiently when the following conditions are valid.

$$n_p \ll L_b \ll N_{eq}, \quad M_b = O\left(\frac{N_d L_b}{n_p}\right) \to min,$$

where N_d is a size of each equation, M_b is a necessary memory size for a storing of each block. For our problem $N_{eq} = N_x N_y$, $N_d \equiv N_y$, $M_b = \dfrac{6 N_y L_b}{n_p}$.

NUMERICAL RESULTS

We used the Parsytec CC workstation with 12 processors to conduct a series of the numerical experiments. We computed the stationary distributions of electric field potential, electron concentration, and electron temperature, as well as the corresponding distributions of electron current density for different values of external field. Two models are compared.

The first one corresponds to the equations (1)–(3) and describes the inhomogeneous charge carrier heating effects. The second model does not include the energy balance equation. In this case a homogeneous electron heating is described by some fitted temperature. It was chosen sufficiently high to provide the emission ability compared with that of the first model for the same value of the surface electric field E_s.

In the case of the first model we examined the sample with sizes $L_x = 1$ μm ($x_m = 0.5$ μm), $L_y = 1$ μm. The donor concentration 10^{18} cm^{-3} and the electron mobility 280 cm^2/(Vs) for the initial temperature $T_0 = 300$ K correspond to typical parameters of real microcathodes. Under these conditions the energy relaxation time τ_ε is $\approx 4\times 10^{-13}$ s. In the case of the second model we take the same sample with electron mobility and tunnelling coefficient corresponding to the electron temperature $T = 10500$ K. In both cases the steady-state value of an external field E_s is of the order of 10^6 V/cm. We take the curvature radius $\rho = 10^{-2}$ μm. The time of an external field establishing τ_0 is equal to 10^{-14} s.

In our calculations the time step τ is equal to 1.38×10^{-14} s, the x step $h_x = 3.9\times 10^{-3} L_y$, and y minimum and maximum steps are equal to $4\times 10^{-4} L_y$, and $1.88\times 10^{-2} L_y$,. We assumed that the system is in a steady-state condition when the time derivatives of concentration and temperature do not exceed some given small value. The typical relaxation time was approximately 2000 time steps.

The Fig. 2–5 illustrate the steady-state distributions for the values of an external field $E_m = 5\times 10^6$ V/cm and $E_m = 5\times 10^5$ V/cm. These distributions are typical for the sufficiently high values of an external field ($>10^6$ V/cm) corresponding to the values of an emission current in a microcathode.

In Fig. 2 one can see the domain of a strongly inhomogeneous electric field at the 0.2 μm distance from the emitting surface ($y = 1$). The concentration of free charge carriers in that domain essentially exceeds an equilibrium value as seen in Fig. 3, and the distribution of these carriers is extremely inhomogeneous. The marked inhomogeneous electron heating takes place near the emitting surface which results in a high emission current density (see Fig. 4). Note that the emission current would be negligibly small if there is no electron heating. The electric field is much lower and almost homogeneous far from the emitting surface. Moreover, the electron concentration and temperature are close to the equilibrium ones. Near the surface all the distributions are also inhomogeneous along the x coordinate. It is caused by the conditions (6)–(8).

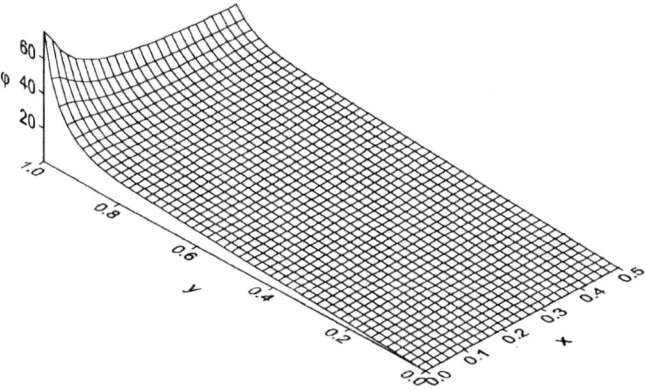

Figure 2. The electric potential distribution in the calculation domain.

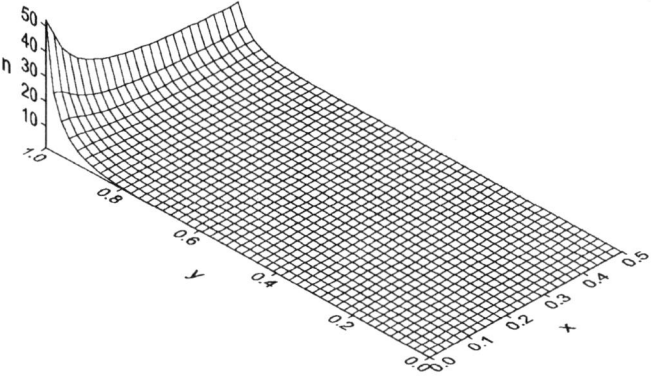

Figure 3. The profile of a free electron concentration.

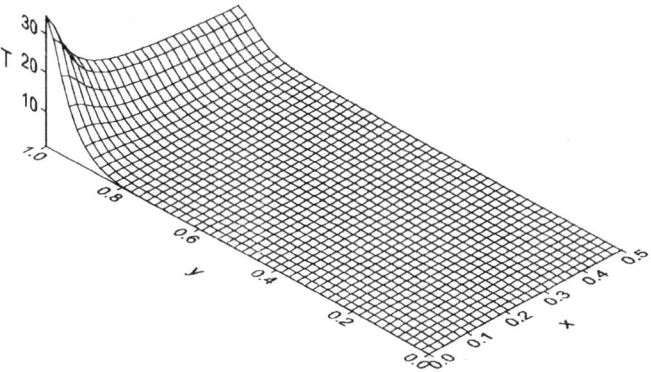

Figure 4. Distribution of electron temperature.

At the back surface ($y = 0$) the electron concentration is close to its equilibrium value. The electric field **E** is directed along the y coordinate ($E_x \approx 0$) and is equal to $\approx 3 \times 10^3$ V/cm. Here an electron overheat is less than 10% and is almost homogeneous. Note that the distributions of an electric potential as well as electron temperature and concentration along y are essentially different from the 1D distributions obtained in [3] for the same parameters.

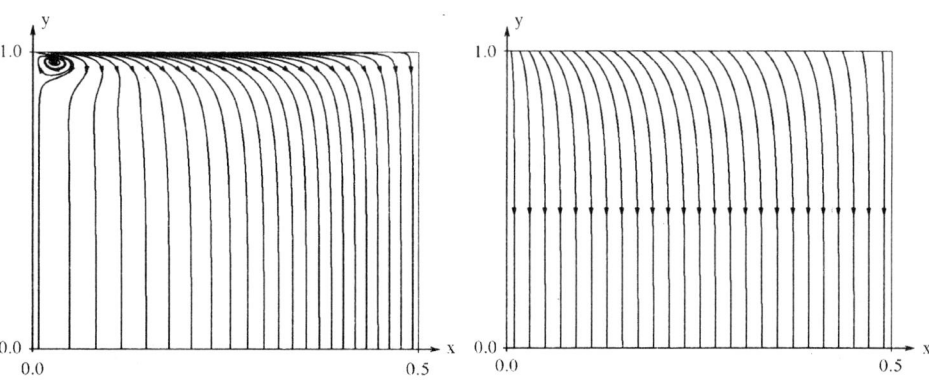

Figure 5. Current density distribution in microemitter volume.

Figure 6. Current density distribution obtained in model with homogeneous carrier heating for electron temperature $T = 35T_0$.

As it is seen from Fig. 5, the distribution of a current density is notably inhomogeneous, especially at a neighbourhood of the point $x = 0$, $y = 1$ where the most strong field is observed. The maximum value of an emission current density defined by the formula (6) is equal to 3.7×10^6 A/cm^2 at the point $x = 0$, $y = 1$. The current density decreases with the distance x due to the decrease of electric field (8), local electron heating and concentration (as seen in Fig. 3, 4). $j = 12$ A/cm^2 at the point $x = 0.5$, $y = 1$. At the back surface the current density is almost homogeneous and its mean value is 1.65×10^5 A/cm^2. This is in accordance with that estimated in the framework of a weak overheat of electron gas for the given electric field value.

The complicated distribution of current density is apparently connected with heat transfer and thermo-diffusion processes in electron subsystem. Fig. 6 illustrates the distribution of current density found from the equations (1) and (3) if a heat transfer is not taken into account. The fixed temperature $T = 35T_0$ coincides with the maximum value of the steady-state distribution given in Fig. 4. One can see that the current distribution is almost homogeneous even at the emitting surface. The potential distribution also is quite different from that given in Fig. 2, the voltage decrease on the cathode (along y) is being ten times as much as compared with that of Fig. 2. Note that these distributions are similar to the $1D$ distributions. Thus, to describe field emission from a real wedge microcathode accurately, the $2D$ modelling is needed. The computational results demonstrate a high effectiveness of the elaborated numerical scheme. It should be useful for technologic computations especially when parallel algorithms are applied.

ACKNOWLEDGEMENTS

The Parsytec CC workstation used to provide all computations was delivered to the Institute for Mathematical Modelling in the framework of the equipment grant of European Economic Community (project No. ESPRIT 21042). The investigation is supported by Russian Foundation for Basic Research (grants No. 96-0100979, 97-0100070, 96-15-97226) and by Federal Program "Integration" (project No 43–2.1).

REFERENCES

1. H.S. Uh, S.J. Kwon, J.D. Lee, Process design end emission properties of gates n^+ polycristalline silicon field emission arrays for flat-panel display application, *J. Vacuum Sci. & Techn. B*, 15 (2): 472 (1997).
2. H. Takemura, N. Furutake, M. Nisimura et al, Fully large-scale integration process compatible *Si* field emitter technology with high controllability of emitter neigh and sharpness, *J. Vacuum Sci. & Techn. B*, 15 (2): 488 (1997).
3. V.A. Fedirko, V.A. Nikolaeva, Field emission from silicon, *Matematicheskoe modelirovanie*, 9 (9): 75 (1997), (in Russian).
4. V.I. Makhov, V.A. Fedirko, N.A. Duzhev, S.A. Kazmina, Field emission through thin oxide layer, in *Techn. Digest of IVMC'91*, Nagahama, Japan, (1991).
5. V.A. Fedirko, V.D. Shadrin, Peculiarities of the silicon electron tunneling through the oxid layer by action of strong field, *Phys. Solid State*, 39(2):337 (1997).
6. Yu.N. Karamzin, I.G. Zakharova, New additive difference method for solving semiconductor problems, *Russ. J. Numer. Anal. Math. Modelling*, 11 (6): 477 (1996).

ECHO–PHENOMENON IN FERROELECTRIC SOLID AND LIQUID CRYSTALS

Sergei S.Lapushkin[1], Alexander R. Kessel[2]

[1]Moscow State Academy of Instrument-Engineering and Informatics,
Moscow, Russian

[2]Physical-Technical Institute of Russian Academy of Sciences,
Kazan, Russian

INTRODUCTION

The electroacoustical echo (EAE) is named the totality of phenomena, connected with the revival of the coherency of preliminary exited electroacoustical oscillations of substances under the influence of the posterior irradiation by electromagnetic pulses. EAE was observed by two research groups in St. Peterburg[1] and Kazan[2], which have investigated the general features of phenomenon end suggested its elementary model[2]. The following investigations of echo-phenomenon form is a new chapter in solid state physics with grate amount of publications. EAE description in a form of big specialized article entered in the Physical Acoustic Encyclopaedia[3].

On the basis of EAE-effect the multifunctional devices of information storage and processing are developed: the volume[4] and surface[5] waves parametric amplifiers, memory cells[6–8], switched over delay lines on HF and UHF diapasons, cells of electromagnetic pulses compression[9], convolution[10] and correlation[11].

PHYSICAL NATURE OF EAE

It is known well that the deformations are arose in piezoelectrics, when a constant electric field is affected them. Under the influence of an alternating electric field of frequency ω_0 the electroacoustical oscillation is produced which frequency ω_{EA} is near ω_0. The electroacoustical oscillation is accompanied by the periodical electrical polarization $P_{EA}(t)$. The amplitudes of all periodic values are the resonance function of ω_0: they sharply increase when ω_0 coincides with the mechanical resonance frequency Ω of piezoelectric sample.

What is happened, when the electromagnetic field will be switched off at some moment $t = t_1$? It is obvious that from the moment $t = t_1$ the equilibrium state begins to establish in the substance by means of a transient process, which has a form of monotonous damping oscillations. Time scale of this damping (it is named) depends on sample form and the properties of its surface, on samples internal properties and its defect structure, and also on the surrounding environment.

The damping electroacoustical oscillations give rise of the periodic electrical polarization of the piezoelectric. The last can be detected through its electrical dipole irradiation. Such signals, being named "electroacoustical ring", are well known for a long time.

Most simply observation of the "electroacoustical ring" can be done by irradiation the piezoelectric with pulsed electromagnetic field. The pulse duration have to be choose much longer than oscillation period to imitate the stationary irradiation, and the intervals between the pulses have to be longer than the damping time T_2, to provide the independence of each pulse effect. At these conditions each pulse of external electromagnetic field would be accompanied by easy detectable "electroacoustical ring".

Nevertheless the nonstationary, transient processes in piezoelectric materials are considerably richer then it can be seen from the outlined picture. It is turned out that there will arise additional splashes of dipole irradiation at moments $t = 2\tau, 3\tau,...$ besides the "electroacoustical ring", if two pulses influenced the piezoelectric material at moment $t = 0$ and $t = \tau$. These additional slashes were named electroacoustical (polarization) echo (EAE) signals.

Effect EAE has a couple of valuable properties, making it perspective phenomenon for scientific investigations and technical applications[1, 2]. (A) Physical nature: echo signals are exited by electrical component of electromagnetic pulses, the coherency revival mechanism is connected with the nonlinear electroacoustic interactions, observed electromagnetic signal is produced by electric dipole moment oscillation of the sample. (B) Bright spreading: effect takes place, practically, in all piezoelectrics, including such important classes as ferroelectrics, piezosemiconductors, liquid crystals and etc., both in powder and crystalline samples. (C) Universality: effect was observed in wide frequency range from 10^5 to $10^{10}\ s^{-1}$ and in temperature range from 4.2 to 720 K; there are no principle prohibitions for its existence at all frequencies of solid state acoustical vibrations and at all temperatures of solid sate existence. (D) Intensity: the amplitudes of EAE-signals considerably exceed the amplitudes of other nature echo and exceed noise amplitudes. (E) This phenomenon demonstrates existence of the long-time (with the characteristic times $T_1 \gg T_2$) phase memory: in UHF-diapason T_1 has the order of milliseconds[12] and in RF-diapason T_1 has the order of hours at room temperatures[13].

An outline of EAE phenomena models in ferroelectrics will be done down for displacement – type powders, order-disorder type crystals and liquid crystals. The initial motion equations will be listen and the way of obtaining the most simple theirs solutions, being necessary for EAE description, will be given.

MATHEMATICAL MODELLING OF EAE IN DISPLACEMENT TYPE FERROELECTRICS

For mathematical modelling of EAE-phenomenon in solid ferroelectrics one have to know, that they exist in two forms being rather different in physical nature: displacement - type and order-disorder type. Correspondingly the mathematical models have to differ strongly. There electrical dipole moments in elementary crystalline cell in displacement – type ferroelectrics are absent. When the temperature fall lower than sartein value T_2 the ions displacement in the new equilibrium cell position take place. As a result the cell dipole electrical moments and electrical polarization $P(r,t)$ of hole sample arise.

In the crystalline cells of order-disorder ferroelectrics electrical dipole moments exist even at the high temperatures. However they are arbitrary orientated, so the polarization of sample as a whole turns to equal zero. When the temperature falls lower then the critical one, the dipole-dipole interaction between cells orient dipoles in one direction and nonzero sample polarization arises. The same interactions determine the specific material electrical properties at all temperatures.

Piezoelectric powder is irregular form crystalline ensemble. The linear sizes of powder particles are distributed usually with approximately equal intensity round the mean value l_0. Let ξ is characteristic width of this distribution. The frequencies of acoustical resonance Ω_0, which are in a complicated manner depended on the powder particles size and form, also have the similar distribution, corresponding parameters are Ω_0 and θ. The powder samples in EAE experiments are placed between the plates of oscillation contour condenser, which linear size L is much grater than powder size ($L \gg l_0$), so the latter are in a space uniform alternating electrical field.

The ferroelectric materials are the piezoelectric ones with strong nonlinear electroacoustic interactions. This means that some coefficients in free energy expansion on the thermodynamic coordinates are rather big, so that terms, connected with them, are not small. Choosing the electric field $E(r,t)$ and deformation tensor $S(r,t)$ as the thermodynamic coordinates, one can present the free energy expansion in the form

$$F(\mathbf{S},\mathbf{E}) = F_L(\mathbf{S},\mathbf{E}) + F_{NL}(\mathbf{S},\mathbf{E})$$

$$F_L(\mathbf{S},\mathbf{E}) = (1/2)c\,\mathbf{S}^2(r,t) - (1/2)\,\mathbf{E}^2(r,t) + g\mathbf{S}(r,t)\,\mathbf{E}(r,t) \quad (1)$$

$$F_{NL}(\mathbf{S},\mathbf{E}) = \sum \sum B_n \mathbf{S}^m(r,t)\,\mathbf{E}^{n-m}(r,t),$$

where for the qualitative picture illustration only the absolute values of vector $\mathbf{E}$ and tensor $\mathbf{S}$ were used in expansion (1). In expression (1) value c is the elastic modulus, g is piezoelectric constant, ε is dielectric permeability and the sense of the linear (L) and nonlinear (NL) expansion parts will be evident later. Now one can determine strain tensor $\sigma(r,t)$ and dielectric polarization $P(r,t)$ of material:

$$\sigma(r,t) = \sigma_L(r,t) + \sigma_{NL}(r,t),$$

$$\sigma_L(r,t) = c\mathbf{S}(r,t) + g\,\mathbf{E}(r,t), \quad (2)$$

$$\sigma_{NL}(r,t) = \delta\,F_{NL}(\mathbf{S},\mathbf{E})/\delta\,s(r,t)$$

$$P(r,t) = P_L(r,t) + P_{NL}(r,t),$$

$$P_L(r,t) = \varepsilon\,\mathbf{E}(r,t) - g\,\mathbf{S}(r,t), \quad P_{NL}(r,t) = -\delta\,F_{NL}(\mathbf{S},\mathbf{E})/\delta\mathbf{E}(r,t) \quad (3)$$

The usage strain tensor in form (2) in elasticity theory equations and polarization vector (3) in the electrodynamics equations leads to the nonlinear coupled equation system in the partial derivatives. From the physical point of view initial conditions have to be chosen as free acoustic vibration and periodic stationary electromagnetic field on the powder particle surface (during the electromagnetic pulse action); the RF-field on the surface have to be put equal zero in the free oscillation interval. One should assumed the zero initial condition for $S(r, t)$ and $E(r, t)$ before the first pulse and theirs value at the pulse end ($t = t_1$) serve as a initial conditions for free oscillation interval and so on. Obvious that to get an analytic solution of problem in such general setting is very difficult.

For this risen one have to use different type of the problem simplifications: (A) The powder is considered as an ensemble of the same material crystalline rods, performing longitudinal vibrations. (B) The internal electric field is neglected in comparison with external alternating electrical field of exiting pulses. (C) The perturbation theory in $S(r, t)$ and $E(r, t)$ is used. (D) The RF-frequency dispersion $\theta_{puls} \gg \theta$ is powder ensemble eigenfrequences dispersion, so that one may consider, that resonance excitation of electroacoustical vibrations takes place in the powder.

In the framework of these assumptions after the separation the variables $S(r,t) = S(t)U(r)$ one can get in the first approximation from the full system of equations of piezoelectric rod electroacoustical vibrations the following equation[14]

$$d^2 S(t)/dt^2 + \gamma dS(t)/dt + \Omega^2 S(t) = (g\Omega^2 \chi / \rho v^2) E_n \sin \omega_0 (t - \tau_n) \quad (4)$$

where $\gamma = 1/T_1$ is the vibration dumping parameter, E_n is the amplitude of n-the pulse of RF-field, ω_0 and τ_n is its carrying frequency and switching on moment, V is sound velocity, $\chi \sim 1$ is the numerical value, which can be specified in more detail description. The suitable solution of equation (4) at the zero initial condition is

$$S(t) = (g\chi / 2\rho v^2)\omega_0 (t - \tau_n)\{[1 + \gamma(t - \tau_n)]^2 + (\gamma/\omega_0)^2\}^{-1} \times$$
$$\times \{[1 + \gamma(t - \tau_n)]\cos \omega_0 (t - \tau_n) - (\gamma/\omega_0)\sin \omega_0 (t - \tau_n)\} \quad (5)$$

The solution of equation for free damping vibrations is

$$S(t) = \exp\{-\gamma(t - t_0)\}\{S(t_0)\cos \omega_0^*(t - t_0) + (1/\omega_0^*)[\dot{S}(t_0) + \gamma S(t_0)]\sin \omega_0^*(t - t_0)\} \quad (6)$$

where t_0 is the beginning of system free evolution, $\omega_0^* = \omega_0 [1 - (\gamma/\omega_0)^2]^{1/2}$.

The transient processes pulsed excitation demands the fulfillment of following conditions:

$$\omega_0 T_N \gg 1, \qquad \gamma t_n \ll 1, \qquad \omega_0 \gg \gamma \quad (7)$$

where t_n is the n-the pulse length. This means that pulse can not be monochromatic, and represents wave packet of the length $\omega^* \approx 1/t_n$. Under the influence of a few pulses the following vibration regime is established in the powder sample, if the conditions (7) are fulfilled

$$S(t) \propto \sum_n \theta(t - \tau_n) E_n t_n \exp\{-\gamma(t - t_n - \tau_n)\} \cos \omega_0 (t - \tau_n) + 0(\gamma / \omega_0) \qquad (8)$$

where $\theta(t - \tau_n)$ is Heviside function. Let us assume for simplicity: (E) RF-pulse envelope has the form of normal distribution with the second moment ω^*.

So the expression for sample polarization in the free evolution regime can be obtained by integrating the expression (8) on this distribution

$$D(t) = VP(t) = (g^2 \chi V / 2\rho v^2) \sum_n \theta(t - \tau_n) E_n t_n \cos \omega_0 (t - \tau_n) \times$$
$$\times \exp\{-\omega^{*2}(t - \tau_n)^2 / 2\} \exp\{-\gamma(t - t_n - \tau_n)\}, \qquad (9)$$

where V is the sample volume. The damping of the oscillating dipole moment (9) is depended mostly of the exiting pulse shape, because of $\omega^* \gg \gamma$ one of inequalities (7). Therefore expression (9) has maximums at $t = \tau_n$, that is immediately after the n-th pulse action. These oscillations give the rise the ring signal.

One have to go out of the perturbation theory for obtaining the expressions for echo-responses. It can be done most simply, taking in account one of nonlinear terms in polarization expansion (3): $P = b_4 S^3(r,t) E(r,t)$. Than a few additional terms appear in expression (9), one of which has the form

$$\sim b_4 g^2 E_1 E_2^2 t_1 t_2^2 \cos \omega_0 (t - \tau_2) \exp\{-\omega^{*2}(t - 2\tau_2)^2 / 2\} \exp\{-\gamma(t - 3t_1 - 2\tau_2)\}$$

This expression describes in general EAE, excited by two RF-pulses. Really, it is connected with the nonlinear electroacoustical vibrations, give rise the electrical dipole radiation, depends as $E_1 E_2^2$ on the amplitudes and as $t_1 t_2^2$ on the duration of the pulses. These dependencies are well observed in experiment.

Above outlined variant of EAE theory is probably the simplest. It can be shown that one can get the similar results for two pulses echo, if the nonlinearity will be introduced in theory though the motion equations, for example by means of nonlinear term $S^3(r,t)$ in the polarization (or by term $S^3(r,t)E(r,t)$ in the free energy)[14]. In this way the expression for polarization for calculation the output signal can contain the linear term $P_L(r,t)$ only; so the signal output from the sample form the linear channel whereas in previous consideration the output was connected with nonlinear channel. The deep mathematical analysis is given by Lapushkin[15], who undertakes the calculation and comparison echo responses for nonlinear oscillator with quadratic and cubic nonlinearities, which correspond to $S^3(r,t)$ and $S^4(r,t)$ terms in the free energy expansion.

MATHEMATICAL MODELLING OF EAE IN ORDER–DISORDER TYPE FERROELECTRICS

The ferroelectric of order-disorder type is the piezoelectric material, having in the crystalline cells nonzero dipole moments, coupled strongly each other by exchange interaction. Partially the crystals with the hydrogen connections belong to this class. The protons, positive ions of such substances have the possibility to be placed in one of pair potential minimums on the connection α of a cell j. The proton jump into another minimum leads to changing the cell electrical moment sign. The electrical properties of such ferroelectrics are determined by the ordering the cells dipoles.

The experimental scheme in this case differ from described in previous paragraph9. The monocrystalline samples are used, cutting of as rod with perfect polishing plane-parallel edges. The RF field of the first pulse produce the electroacoustic oscillations spreading into the rod from one of the rods edge. After the first pulse action and the free evolution regime is finished, the second RF-pulse, switched on at moment $t = \tau_2$, is effected over the all length of the rod. Homogeneous in space electrical field of second pulse interacts with the electroacoustic oscillations, produced by first pulse, and creates the electroacoustic wave spreading in the opposite direction. This wave will come back to the initial edge after time τ_2 from second pulse beginning and so will produce at time $t = 2\tau_2$ RF-field by means of inverse transformation. This field is electroacoustic echo response.

Thermodynamic and dynamic properties of order-disorder type ferroelectrics are described by the Hamiltonian of Ising model

$$H = \sum (g_1 S_j^x + g_3 S_j^z) + \sum J(j-k) S_j^z S_k^z \tag{10}$$

where g_1 is tunneling integral, g_3 is energy interaction with the external electric field, $J(j-k)$ is integral of exchange interaction between psevdospin j and k. The psevdospin operators S_j^x, S_j^y, S_j^z, containing in expression (10), can expressed through the second quantization operators of proton, being on the connection α be in the cell j. It is also known that essential role in the ferroelectric properties plays the interaction of electric subsystem with lattice vibrations, describing by the Hamiltonian

$$H = (1/2m)\sum P_i^2 + (1/2m)\sum \Omega_{ij}^2 [Q_i - Q_j]^2 + \sum F_{ij}[Q_i - Q_j] S_i^z \tag{11}$$

where P_i and Q_i are the mechanical impulse and displacement of i-th cell, which mass is m, F_{ij} are interaction constant. The acoustical vibration subsystem was taken into account for the explanation of the EAE existence: in the absence of external influence electroacoustical system oscillate on the eigenfrequensies, which are much greater than the exiting frequencies. That is way the repeated pulse, interacting with the previously produced excitations can not form the echo response. The strong electroacoustic interaction leads to the existence in substance not only quasielectric vibration brunch, but also quasiacoustic one. The last can interaction with RF-pulses effectively.

Hamiltonian (10) was used by many authors for describing the elementary excitations in the exchange systems in the frame of long-wave approximation ($<S_j> \to <S(x)>$) and random phase approximation ($<S(x)S(x)> \to <S(x)><S(x)>$). With these approximations one can obtain the following system of equations for the mean values $<S(x)>$ of psevdospins operators, which motion is managed by full Hamiltonian of the problem[16]:

$$(\partial Z_1 / \partial t) = (g_3 + J[Z_3 + (a^2/2)(\partial^2 Z_3 / \partial x^2)]) Z_2 + hZ_2(\partial Z_4 / \partial x),$$

$$(\partial Z_2 / \partial t) = -(g_3 + J[Z_3 + (a^2/2)(\partial^2 Z_3 / \partial x^2)]) Z_1 - HZ_1(\partial Z_4 / \partial x) + g_1 Z_3,$$

$$(\partial Z_3 / \partial t) = -g_1 Z_2, \tag{12}$$

$$(\partial^2 Z_4 / \partial t^2) - v^2(\partial^2 Z_4 / \partial x^2) = -(h/m)(\partial Z_3 / \partial x),$$

$$Z_1 = <S^x(x)>, \quad Z_2 = <S^y(x)>, \quad Z_3 = <S^z(x)>, \quad Z_4 = <Q(x)>,$$

where $h = a_1 F_{ij}$ and a is the lattice constant.

This system of equations was solved by method similar to Bogolubov-Mitropolskiy one for pulse regimes RF excitation and free evolution, and then the expressions for EAE were obtained. In the case, when the second RF pulse has the same frequency as the first one, the qualitatively similar dependencies EAE on the excitation pulse parameters were obtained. The only exclusion gives the temperatures dependence of the EAE amplitude, which in this type of ferroelectric has the specific character and dramatically changes after the phase transition. In the case, when the second pulse frequency is twice the first pulse one, the EAE amplitude turns out be proportional to first power of field E_2. Note, that in the paraphrase of ferroelectric effect EAE is absent at the both excitation types, if the constant electric field is not applied to the sample.

MATHEMATICAL MODELING OF EAE IN LIQUID CRYSTAL FERROELECTRICS

The possibility of EAE observation in liquid crystals (LC) was shown only theoretically for smectic LC (SLC) in distinction with solid ones. From the physical point of view and from nonequalities (7) one can conclude that at least three conditions have to be fulfilled for EAE observation. There have to be strong nonlinear electroacoustic interactions, the exciting pulse frequencies have to correspond to frequency branches of the material and relaxation times of these frequencies have to be rather long.

The nonlinear interactions are traditionally strong in LC where phase transitions are sharp pronounced. It is not difficult to fulfil the second condition in LC possessing wide spectrum of the physical properties. So long relaxation time is the most complicated requirement. For example relaxation time of order 10^{-12} s is typical for molecular vibration in nematic LC.

Nevertheless one can find suitable oscillations which have relaxation parameters that satisfy the necessary conditions. For example the damping of the molecular long axis (director) rotation around the smectic cone in SLC occurs on the times of order 10^{-2} s. This is the type of motion which can be used to induce echo.

Molecules in SLC are placed in smectic layers and its directors are oriented along z-axis perpendicular to these planes. SLC have the liquid properties inside the smectic planes but they possess nearly solid properties in z direction. Molecule dipole moments are arranged in the smectic planes.

The most suitable experimental scheme for EAE observation in SLC is probably the experiment being described in previous section. Let us consider that smectic layers lie in the xy-plane and the first pulse acts on the boundary one $z = 0$ so that the strength vector is in the layer plane and induces only small variations of the director from equilibrium. These oscillations run inward the sample. The second pulse inverts their wave vector. The inverted oscillations reaching the plane $z = 0$ give rise to coherent oscillations of surface electric dipoles. This should be recorded as echo.

The free energy density at the phenomenological approach can be written as[17]:

$$F = \frac{a}{2}\theta^2 + \frac{b}{4}\theta^4 + \frac{g}{2}\left(\frac{\partial \theta}{\partial z}\right)^2 + \frac{K\theta^2}{2}\left(\frac{\partial \phi}{\partial z}\right)^2 + \lambda \theta^2 \frac{\partial \phi}{\partial z} +$$

$$+\frac{P^2}{2\chi} - \theta\left(\varsigma - \mu\frac{\partial\phi}{\partial z}\right)(P_x\sin\phi - P_y\cos\phi) - \mathbf{PE} \tag{13}$$

where $a = \alpha(T - T_c)$, $b > 0$, E is external electric field, K and g are elastic constants, ς and μ are piezoelectric constants, λ is helicity, χ is dielectric susceptibilit.

By minimization of the free energy on the polarization vector one can obtain expressions for its components:

$$P_x = \chi\left(E_x + \theta\sin\phi\left(\varsigma - \mu\frac{\partial\phi}{\partial z}\right)\right),$$
$$P_y = \chi\left(E_y - \theta\cos\phi\left(\varsigma - \mu\frac{\partial\phi}{\partial z}\right)\right) \tag{14}$$

The angles θ and ϕ can be chosen as thermodynamic variables after the substitution of P_x and P_y in (13). These angles are the transformation parameters at the transitions of SLC into different phases and paraphase. The dipole ordering manifistates itself as secondary phenomenon being due to connection (14) of polarization with θ and ϕ.

At the fixed temperature free energy density $F(\theta,\phi)$ plays the role of the potential energy. The kinetic energy per unit volume is

$$T = \frac{I}{2}\left(\theta^2\left(\frac{\partial\phi}{\partial t}\right)^2 + \left(\frac{\partial\theta}{\partial t}\right)^2\right) \tag{15}$$

where I is inertia moment density. Using Lagrange function $L = T - F$ one can obtain the equation of motion in usual manner.

$$I\frac{\partial^2\theta}{\partial t^2} = g\frac{\partial^2\theta}{\partial z^2} + I\theta\left(\frac{\partial\phi}{\partial t}\right)^2 - a\theta - b\theta^3 - K\theta\left(\frac{\partial\phi}{\partial z}\right)^2 - \lambda\theta\frac{\partial\phi}{\partial z} +$$
$$+\chi\left(\varsigma - \mu\frac{\partial\phi}{\partial z}\right)(E_x\sin\phi - E_y\cos\phi),$$
$$I\frac{\partial}{\partial t}\left(\theta^2\frac{\partial\phi}{\partial t}\right) = K\frac{\partial}{\partial z}\left(\theta^2\frac{\partial\phi}{\partial z}\right) + \lambda\frac{\partial}{\partial z}(\theta^2) + \chi\varsigma\theta(E_x\cos\phi + E_y\sin\phi) +$$
$$+\chi\mu\frac{\partial\theta}{\partial z}(E_x\sin\phi - E_y\cos\phi) \tag{16}$$

To determine the boundary conditions let us assume that first pulse acts only on the boundary layer at $z = 0$ and influence of internal layers can be neglected. Then the motion inside boundary layer will be described by the solution of equations (16) in which the coordinate dependence is omitted:

$$\phi(0,t) = -\frac{\chi\varsigma}{I\theta_0\omega^2} E_{1x} e^{i\omega t} + \text{c.c.},$$

$$\theta(0,t) = \theta_0 + \frac{\chi\varsigma}{I\omega^2 + 2a} E_{1y} e^{i\omega t} + \text{c.c.}$$

(17)

here E_{1x} denote the x component of first pulse.

We will find the solution of the equation of motion (16) in the simplest case when all molecules have the same direction and the interval τ between the pulses satisfies the non equalities $T_\phi > \tau > T_\theta$, where T_ϕ and T_θ are damping times for azimuthal and polar oscillations. Due to these inequalities the polar oscillations are damped before the beginning of the second pulse in contrast to azimuthal ones.

In the case of small deviations from the equilibrium the solution of equation (16) with the boundary condition (17) is

$$\phi(z,t) = -\frac{\chi\varsigma}{I\theta_0\omega^2} E_{1x} \Phi(z - Vt) e^{i(\omega t - kz)} + \text{c.c.},$$

(18)

where ω and k are coupled by dispersion relation:

$$I\omega^2 = Kk^2$$

(19)

and $V = \omega/k$.

The form of running wave envelope Φ is the same as the first pulse time envelope. It spread along z-axis without change according the equations (18, 19):

$$\left(\frac{\partial}{\partial t} + V\frac{\partial}{\partial z}\right)\Phi = 0$$

(20)

The solution in the period of the second pulse action contains the oscillations $exp\{i(\omega t + kz)\}$ besides the main oscillations $exp\{i(\omega t - kz)\}$ because of nonlinear terms in equations of motion. Equations for envelopes Φ and Ψ of these oscillations are:

$$i\left(\frac{\partial}{\partial t} + V\frac{\partial}{\partial z}\right)\Phi = \Lambda E_{2y}\Psi$$

$$i\left(\frac{\partial}{\partial t} - V\frac{\partial}{\partial z}\right)\Psi = -\Lambda E_{2y}\Phi$$

(21)

$$\Lambda = \frac{\chi\varsigma}{2I\theta_0\omega^2} \frac{a}{a + 2I\omega^2}$$

By the end of the second pulse the amplitude of inverted wave turns out to be equal:

$$\phi_{inv}(z,t) = -i\left(\frac{\chi\varsigma}{I\theta_0\omega^2}\right)^2 \frac{a}{a+2I\omega^2} \frac{\omega t_2}{2} E_{1x} E_{2y} \Psi(z-V(t-2\tau))e^{i(\omega[t-2\tau]+kz)} + \text{c.c.},$$

(22)

$$\Psi(z) = \frac{1}{Vt_1} \int_{z-Vt_2}^{z+Vt_2} J_0\left(\frac{\Lambda E_{2y}}{V}\sqrt{(z-\xi)^2 - V^2 t_2^2}\right) \Phi(\xi) d\xi$$

here $J_0(x)$ is the Bessel function of order zero.

Later, in the free evolution regime it reaches the boundary $z = 0$ and manifestate itself as echo on the frequency ω with the amplitude

$$E_x^{echo} = \frac{\varsigma}{\mu k} E_y^{echo} = \left(\frac{\chi\varsigma}{I\theta_0\omega^2}\right)^2 \frac{a}{a+2I\omega^2} \frac{\omega t_2}{2} E_{1x} E_{2y}$$

(23)

Similar results for the case when the both kind of oscillations do not damp ($T_\phi, T_\theta > \tau$) up to the moment of second pulse switching on was obtained in[20].

The estimation shows that at the standard values of the parameters $\chi = 0.2$, $\varsigma = 80\,\text{CGS}, I \approx 10^{-14}\,g/cm$, $\omega = 2\pi \cdot 10^7\,s^{-1}$, $t_1, t_2 \approx 10^{-5}\,s, E_1, E_2 \approx 100v/cm$, the magnitude of echo is of order $10^{-2} v/cm$. It is quite sufficiently to record the echo experimentally.

PRINCIPLES OF SOME PRACTICAL APPLICATIONS

One of the most promising applications of EAE effect finds in the radio signals delay lines due to its switching over lightness. The bright application of delay lines is based on the fact that the difference of sound and electromagnetic wave velocities is rather big. The delay lines usually were fulfilled as a piezoelectric rod. The signal to be delayed enter one of the rod edge, were it is transformed by means of inverse piezoeffect into the acoustic brunch of electroacoustical oscillations of substance, propagating with velocity, which is near sounds one. At other rods adge the opposite transformation of oscillation into electromagnetic signal takes place. Since both velocities are fixed in a given substance the signal delay depends only on the rods length, and practically can not be changed during device action. Delayed signal in the delay line, operating on the EAE principle, serves as first exiting pulse, and the second, reading out pulse can be applied at the arbitrary moment τ to get the delay time 2τ. The advantage of such delay line is provided by the facts that second pulse switching on time can be changed easily, life time of EAE signals can be long and theirs amplitudes can bebig.

It was experimentally shown[9], that one can reach the compression of echo-response in 500 times by effecting the lithium niobate ferroelectric rod with two pulses with linear switch over the carrying frequency. The idea of device of such type is based on the fact mentioned above: electrical field of ω frequency exits only such powder particles, which eigen-frequencies satisfy the resonance equality $\Omega_0 = \omega$. Let the k-th pulse has the frequency, changing linearly $\omega_K = \omega_0 + \upsilon_K \xi_K$, where v_K is velocity of frequency changing, ξ_K is the time, counted from the pulse beginning, $K = 1,2$. Therefore equal monochromatic oscillation are effected the sample at the times $t = \xi_1$ and $t = \tau + \xi_2, \xi_2 = (\upsilon_1/\upsilon_2)\xi_1$, so the interval between the pulses is $\delta t = \tau + \xi_2 - \xi_1$. Now one can see, that the "monochromatic echo" has to take place at $t_{echo} = 2(\tau + \xi_2) - \xi_1$ since it usually takes place at the distance δt after second pulse beginning. And finally all monochromatic oscillation will excite echo in the same time, if the condition $2\xi_2 = \xi_1$ (or $2v_1 = v_2$) is fulfilled. This circumstance gives the signal echo compression.

The memory cell construction, based on the EAE effect, uses the possibility of recording by RF-pulse the long-time phase information in the piezoelectric powder. The second, read-out pulse will excite echo-response (signal "yes"), if the cell was previously radiated at the same frequency, and echo signal will not arise (signal "not"), in the opposite case. As the piezoelectric powder particles can possess an considerable eigenfrequency dispersion, due to particles linear size dispersion, many information bits can be recorded on the different frequencies in the same powder volume. The information recording density is estimated as $10^6 \, bit/cm^3$ at the transient damping in 25 Db and delay time possible interval $10^{-6} \div 10^7 \, sec$.

REFERENCES

1. S.N. Popov, N.N. Krainik, Detecting of anomalous echo in ferroelectric *SbSJ*, *Solid State Physics*, 10: 3022 (1970).
2. A.R. Kessel, I.A. Safin, A.M. Gol'dman, Macroscopic analogue of spin echo-effect in polycrystalline ferroelectrics, *Solid State Physics*, 10; 3070 (1970).
3. "Electroacoustic echo" in "Ultrasound", *Little Encyclopaedia*, 383, ed. "Soviet Encyclopaedia", Moscow (1979).
4. V.S. Bondarenko and others, "Parametric amplifier", *Author's Certificate* No 468351, 12.28.73.
5. V.S. Bondarenko and others, "Parametric amplifier on the surface waves", Author's Certificate No 490396, 10. 26.73.
6. V.S. Bondarenko and others, "Piezoelectric memory element", Author's Certificate No 413528, 07.12.72.
7. L.V. Petrosjan and others, "Phonon echo information canceling method", Author's Certificate No 2815324, 05. 29.75.
8. *Electronics* 1: 75 (1976) (in Russian),.
9. A.I. Validov, R.G. Deminov, B.P. Smoljakov, S.L. Carevskiy, Research of elastic nonlinearity of ferroelectrics for hypersound polarizable echo signals in ultrasound field, *ZETF*, 106, 1 (7): 217 (1994).
10. B.M. Berezov and others, "Device for determining the convolution of the envelopes of modulated signals", Author's Certificate No 2666104 /18–24.
11. V.S. Bondarenko and others, "Method of recording and processing the RF-pulsed signals and the devices for its realization", Author's Certificate No 2609677/40–23/063480, 06.13.75.
12. U.H. Kopvillem, B.P. Smoljakov, R.Z. Sharipov, Polarizable echo in ferroelectric monocrystal KH_2PO_4, *ZETF Letters*, 13: 558 (1971).
13. Ja.Ja. Asadullin, V.İ. Berezov and others, Anomalous relaxation of stimulated echo in piezoelectric crystals. *ZETF Letters*, 22: 285 (1975).
14. A.R. Kessel, À.V. Lisner, V.İ. İusin The form of doublepulse echo in piezoelectric powder, *Solid State Physics*, 31: 161 (1989).
15. S.S. Lapushkin, About one nonlinear equation with fast oscillating coefficients. *Mathematics, High School News*, 11: 17 (1993); To the question of nonlinear system reaction for impulsive action. *Mathematics, High School News*, 10: 26 (1996).

16. V.A. Popov, A.R. Kessel, S.S. Lapushkin, Theory of electroacoustic echo in monocrystals of ferroelectrics of order-disorder types. *Solid State Physics*, 39, 4: 697 (1997).
17. S.A. Pikin, Structural Transformation in Liquid Crystals. Nauka, Moskow (1981) (in Russian).
18. M.B. Vinogradova, O.V. Rudenko, A.P. Sukhorukov, Wave Theory. Nauka, Moskow (1979) (in Russian).
19. R.K. Dodd, J.C. Eilbeck, H.C. Morris, Solitons and Nonlinear Wave Equations. Acad. Press, Inc. London, (1984).
20. V.A. Popov, A.R. Kessel, Echo-response in ferroelectric liquid crystals. *Solid State Physics*, 40: 1370 (1998).

INVESTIGATIONS OF BLISTERING IN SOLIDS USING STOCHASTIC MODEL

Anna L. Bondareva and Galina I. Zmievskaya

Keldysh Institute of Applied Mathematics Russian Academy of Sciences
125047 Moscow, Russia

INTRODUCTION

The problem of the kinetic description[1,11,12] of the phase transition[3,6,7] at their fluctuation stage[9] is introduced by the stochastic simulation method (SSM) in works [8-10]. The modification of SSM for computer simulation a phenomena into solids put forward in [2] and was developed by the authors [4, 5]. The insight of computer experiments in plasma physics[18] is successful, as well SSM is promising as kinetic approach in collision media modeling. The same technique is used as the basis for the phase transition investigations.

We dwell on problem of the gas penetration in a metal lattice, where helium atoms and small clusters appear here, after them due to nucleation the bubbles can grow, finally it lead to macroscopically observable material deterioration.

Previous models of this phenomena[3,6,7,17] can be extended by means of determination of the distribution function (DF) of defects inside solids (pores, gaseous atoms and its clusters and others) taking into account that both its sizes and coordinates on lattice varies in the time. DFs being nonequilibrium in the time scale are of great interest. The studies of such kind can be performed by means of solving of equations set of mathematical physics on space-time scales specific to solids. The process of the clusterization of defects according to our assumptions is similar to the Brownian motion model with alternating masses[4,5,14].

Here the purpose is the development of kinetic code for numerical investigation of the defects clusterisation into solids. General assumptions as well equations of this problem which laid foundation for SSM kinetic code was published in report[14]. List of applications related with computer simulation code is really wide, for example: a reason of the degradation of the dielectric surfaces optical properties under action of the space factors[15] or the radiation-induced blistering of metal surfaces in discharge plasmas [17].

This article concerns the process of blistering. We simulate initial stage of the process of clustering, when initial bubble sizes are near critical size and the process of the bubble formation is of fluctuational nature. A surface of solid is irradiated by ions of *He*. High temperature blistering case is examined in this work. We are demonstrating preliminary results of the computer simulations for *He-Ni* system, He bubble forms inside *Ni*–lattice. The work represents researches which precede numerical code realization. This code can describe kinetic phenomena of bubble origin in solids.

BLISTER FORMATION MODEL.

Blister is large gas bubble (its size is of the order of 1 micron) which forms under action of bombardment of a surface by gas ions in surface layer of a solid.

Blistering is observed when we irradiate metals and dielectric, crystal and amorphous materials by ions of poorly soluble gases, in particular, helium. Its formation begins with certain dozes of irradiation (critical doze Φ_{cr}). Φ_{cr} is approximately 10^{16} $ions/cm^2$ for He. If during the irradiation the layer, greater average depth of penetration of ions R_p is scattered, the concentration of gas atoms will be less critical in a superficial layer of a target and the bubbles will not be formed. A condition on a doze for occurrence of a blistering is:

$$S\,\Phi_{cr} < R_p N \tag{1}$$

where S is coefficient of surface sputtering; N is number of atoms He in cm^3. R_p may be as much as 70 nm where as a parameter of a lattice is equal $3.52\ 10^{-8}$ cm.

The destruction of gas cavities starts when doze of an irradiation Φ increases more then approximately 10^{19}.

Let us put forward the mathematical model of the high-temperature blistering, when $0{,}4T_{melting} < T < 0.6T_{melting}$, T is a temperature of a surface of an irradiated material and $T_{melting}$ is a material melting temperature.

A flow of He ions falls on a surface Ni normally to a surface with energy $E = 10\ keV$ and doze Φ is approximately equal to 10^{16} $ions/cm^2$, the sample temperature is $T = 1000\ K = = 0{,}58\ T_{melting}$.

We consider the nonequilibrum (fluctuation) stage of a bubble formation only.

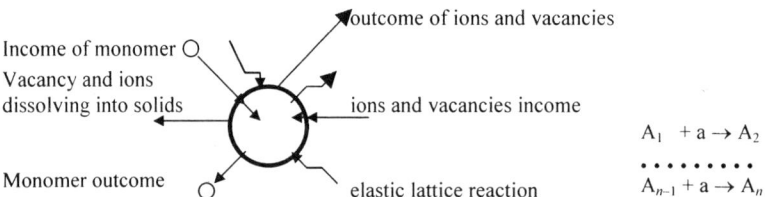

Figure 1. Scheme of model processes on a bubble (A_n – meric cluster) surface, a denoted the monomer.

The critical bubble size is determined by the competition of two factors:
1) pressure on bubble on the side of a material, which to aspire to destroy it.
2) diffusion introduced He from a material (Ni) in a bubble.

Bubble is unstable and depends on fluctuation of any of the competing factors at this stage and it can begin to grow or degrade. Thermodynamic potential, free energy Gibbs $\Delta\Phi$ depends on the bubble size, which we express in terms of "g" – number of He atoms in the bubble. There is an unstable area near to a maximum of Gibbs energy versus the size of bubbles (approximately equal to kT).

Unstable parts of Gibbs energy curve is determined by expression $\Delta\Phi(g_{cr}) = = max\Delta\Phi(g)$, where g_{cr} is critical size of bubble and left and right part of the curve $g_{left} < g < g_{rigt}$ – forms unstable region of the thermodynamic potential (Fig.2):

$$\Delta\Phi(g_{left}) = \Delta\Phi(g_{cr}) - kT \tag{2}$$

$$\Delta\Phi(g_{right}) = \Delta\Phi(g_{cr}) - kT$$

If the initial size of bubble gets in this area, then any competing factor fluctuation can be cause of bubble increase or degradation. And it does not depend on where initial bubble size localizes on the right or left curve branch. If the initial bubble size gets on the right curve branch outside of the unstable area, then we will observe bubble increase. There is its destroy if it is on left curve branch.

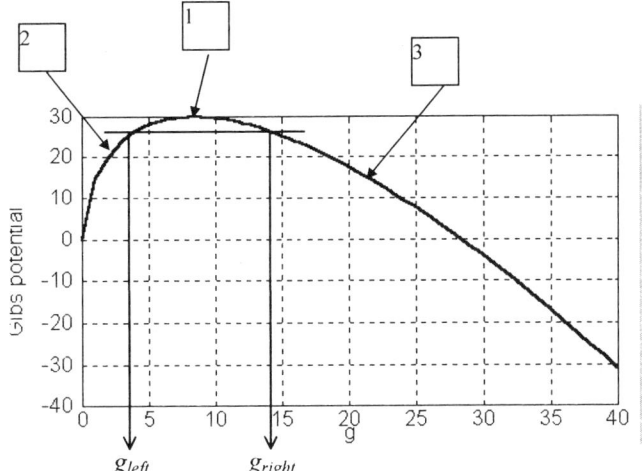

Figure 2. Gibbs energy as a function of bubble size g. 1 is unstable region; 2 is left branch of Gibbs potential; 3 is right branch of Gibbs potential; g is bubble size in number of atoms in cluster of defects.

So, in this work we consider a bubble with the size near to critical size, its growth or degradation and motion in coordinate space realizing fluctuation nature stochastic model.

BROWNIAN PARTICLES IN BLISTERING MODEL.

The following model is offered for phase transition. Bubble nucleus we consider as a Brownian particle (BP)[3] of the spherical form and alternating masses [4,5]. Thus, it is considered that bubble centers of masses can change the points in space with current of time (that is in process of growth or degradation). The method of BP is convenient because the influence can be presented by the account of stochastic force in some cases. This force operates on BP. This approach reduces system dimension and simplifies calculation. Stochastic character of process affects in evolution of probabilistic descriptions, for example, DF which depends on bubble size and displacement in space.

KINETIC DESCRIPTION OF THE BLISTERING

Let us describe BP behavior by the partial differential equations. Generalized kinetic equation (backward Kolmogorov equation) is follow:

$$\frac{\partial f}{\partial t} = L(f),\qquad(3)$$

$$f|_{t=0} = f_0,$$

where $L(f)$ is nonlinear or quasilinear integro-differential operator or superposition L_g and L_r, here r is a coordinate and g is a number of particles into bubble. f is DF of phase coordinates is formulated for the description of evolution of the alternate mass BP. Role of BP plays the bubble. The change of the bubble size is described by diffusion Markov process as a first approximation of problem. These kinetic equations are integro-differential equations of mathematical physics and it is difficult to solve. So, we use the stochastic approach. Its basic idea are replacement these kinetic equations by their stochastic analogues [8-10].

Thus kinetic equation on a discrete grid of time are solved by a method of splitting on physical processes and representation of each of stages by its stochastic analogue [4, 5].

THE FOL'MER-ZEL'DOVICH PROBLEM FOR BLISTERING.

We consider diffusion in space of the bubble sizes or the Fol'mer- Zel'dovich problem in absence of external forces and correlation on a fist time step Δt_1 (equation of the Fokker- Planck-Kolmogorov):

$$\frac{\partial f_r(g,t)}{\partial t} = \frac{\partial \left[D(g,t) \frac{\partial f_r(g,t)}{\partial g} \right]}{\partial g} + \frac{1}{kT} \frac{\partial \left[D(g,t) f_r(g,t) \frac{\partial \{\Delta \Phi(g,t)\}}{\partial g} \right]}{\partial g} \quad (4)$$

where g is number of atoms He in a bubble, i.e. bubble size in He atoms; $D(g, t)$ is diffusion factor in space of bubble sizes; $f_r(g, t)$ is distribution function of bubble sizes; $\Delta\Phi(g,t)$ is functional coefficient of free Gibbs energy (see Fig.2 and Fig.3).

It is necessary to note, that in Gibbs energy we take into account not only superficial tension on metal – bubble border and difference of chemical potentials of old and new of phases, but also elastic energy of lattice reaction, which influence has appeared essentially [4, 6, 7]. The Fig.3 illustrates elastic force influence of lattice reaction on a dependence form of Gibbs energy. Energy depends on number of atoms in a bubble.

STOCHASTIC ANALOG OF THE FOKKER-PLANCK-KOLMOGOROV EQUATION.

Stochastic analogue in differential form of this kinetic equation is:

$$\frac{\partial g}{\partial t} = -\frac{1}{kT} \frac{\partial \Delta\Phi(g,t)}{\partial g} - \frac{1}{2} \frac{\partial D(g,t)}{\partial g} + \sqrt{D(g,t)} \xi(t) \quad (5)$$

$t_0 \leq t \leq T_k, \quad g(t_0)=g_0, \quad g(t)>2,$

where ξ is stochastic function corresponding to the random Wiener process. The solution of (5) using SSM [8-10, 4, 5, 14] in case of the Stratonovich form of the Ito SDE means that the Wiener process can be replaced by standard «white noise». This assumption of SSM, introduced in deciding on a computational scheme of the solution approximation, for example[19,13]. This point of view is very important in technical problem of the realization of computer experiments of SSM.

It is important to keep in mind that in SSM we are used the functional coefficients of the kinetic equations (depend on DF) and the diffusion coefficient $D(g, t)$ and the free Gibbs energy $\Delta\Phi(g,t)$ are degree approximated function of the DF of the bubble sizes.

$$\Delta\Phi(g,t) = -(a + c) g + b g^{2/3},$$

where $-cg$ is the contribution of elastic force from [6,7] and the expression of the diffusion functional-coefficient is: $D = D g^{2/3}$.

These functions depend not only on thermodynamic parameters, which describe introduced phase (temperature and pressure inside bubble), but also on mathematical expectation of random cluster size $g = E(g)$ or otherwise speaking, on distribution function of bubbles $f_r(g, t)$. It allows to speak about solve of a nonlinear problem.

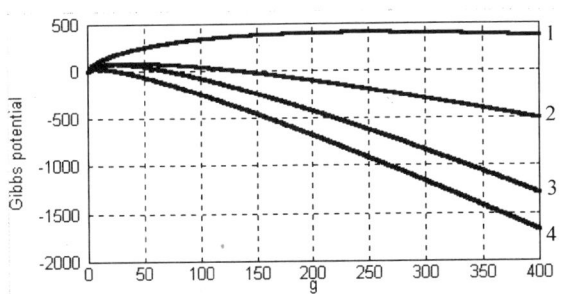

Figure 3. Dependence of the Gibbs energy $\Delta\Phi$ on a bubble size g of different case taking into account. $\Delta\Phi$ is measured in *Joule/atom*: 1 conforms to $T=700\ K$ without including elastic force of lattice response; 2 corresponds to $T=1000\ K$ without including elastic force of lattice response; 3 conforms to $T=700\ K$ with including elastic force of lattice response; 4 correspond to $T=1000\ K$ with including elastic force of lattice response.

THE KRAMERS EQUATION FOR CLUSTER DIFFUSION INTO LATTICE.

Spatial diffusion of BP is simulated on a step Δt_2. I.e. the problem the Einstein- Fokker (the Kramers–Smoluhovsky equation) in force fields of a collective nature with account of a long- influence part of bubbles interaction potential through acoustic phonon of a lattice and oscillation of electronic density $U(r)$ is solved.
The kinetics equation has form:

$$\frac{\partial f_g(\vec{r},t)}{\partial t} = \frac{\partial \left[D_r(\vec{r}) \frac{\partial f_g(\vec{r},t)}{\partial r} \right]}{\partial \vec{r}} - \frac{\partial \left[\frac{\vec{F}(\vec{r})}{M_g \gamma} f_g(\vec{r},t) \right]}{\partial \vec{r}}, \qquad (6)$$

where: $M_g = M(g,t)$ is alternating cluster mass; $D_r(r)$ is diffusion coefficient in space R; γ is the dissipation factor of friction.

$$\vec{F}(\vec{r}) = -\frac{\partial U(\vec{r})}{\partial \vec{r}}, \qquad (7)$$

here $U(\vec{r})$ is model potential (similar to the functions presented in[11,17]) but modified for bubbles interaction problem (see Fig. 4)[5]. It has strictly auxiliary nature, the attenuation of $U(\vec{r})$ is of interest to own researches.

We take diffusion coefficient by analogy with kinetic theory (see Fig.5) in form: $D_r = D_l(T)\ C(r)$, the temperature dependence of D_r for Ni we can see on Fig. 5, the function $C(r)$ is degree approximated function with r (the cluster position into lattice)[5].

STOCHASTIC ANALOG OF THE KRAMERS EQUATION.

Stochastic analogue (of integral form) of kinetic equation (6) is:

$$r(t) = r(t_0) + \int_{t_0}^{t} H(\tau,r(\tau))d\tau + \int_{t_0}^{t} \sigma(\tau,r(\tau))dW(\tau), \qquad (8)$$

here

$$H = -\frac{1}{\gamma M_g}\frac{\partial U(r)}{\partial r} - \frac{1}{2}\frac{\partial D_r}{\partial r} \qquad (9)$$

and

$$\sigma_r(r,t) = \sqrt{D_r(r,t)} \qquad (10)$$

with initial and boundary conditions : $t_0 \leq t \leq T_k$, $r(t_0)= 0$, $r(T_k) < 300a$, a is lattice parameter and dW is a increment of the Wiener process during Δt, when we solved this equation we replaced it by standard white noise $\xi(t)$; r is site of mass center of a bubbles in the following coordinates system (Fig. 6).

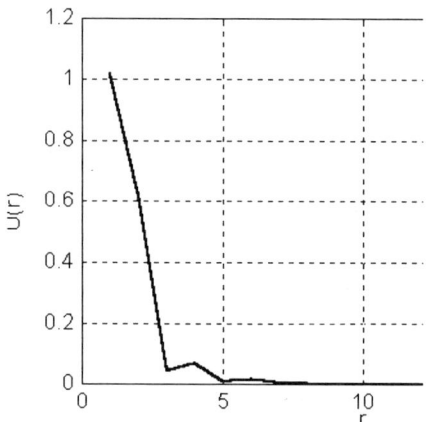

Figure 4. This figure schematically shows imentionless potential $U_l(r)^{11,16}$. r is distance between cluster in a lattice with parameter of lattice a $U(r)$ is potential interaction between bubbles. $U(r) \sim U_l(r)$. If $r=a$ then $U_l(r)=1$.

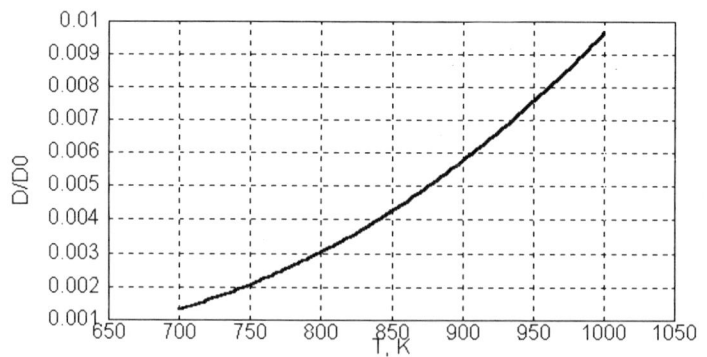

Figure 5. Diffusion coefficient D as a function of the temperature T. D is coefficient of diffusion[17] $D=D_lC(r)$, $C(r)$ is the exponential function and $D_1=D_0 e^{-a/(kT)}$, where a=const is fited to the data of Ni metal.

246

We choose a point in unit of crystal lattice on depth about projective *He* run in *Ni* (~ R_p) and put center of coordinates system in it. The bubbles mass centers are distributed randomly on volume units of a crystal lattice in the first instant of time. It is necessary to note, that the displacement of bubble mass center from its initial site enters into diffusion factor and distance between bubbles enters into potential of interaction.

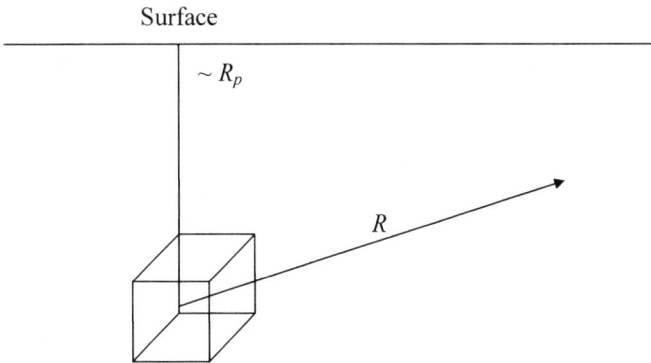

Figure 6. System of coordinates.

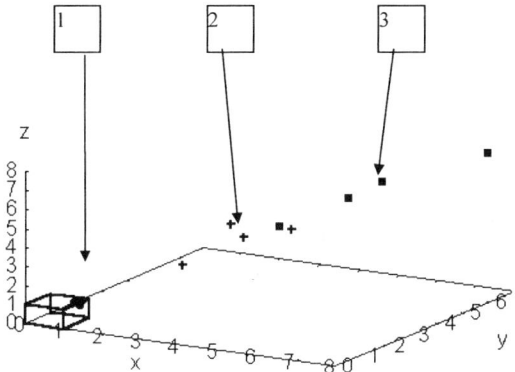

Figure 7. This figure schematically shows movement center mass of cluster *r* in phase space $\{R\} = \{x, y, z\}$. 1 means initial state where centers mass of bubbles were randomly distributed as a BP; "2" means center mass of cluster for time equal to $0.5T$ where $T=10\tau$, $\tau = 3 \cdot 10^{-11}$ sec; 3 means the final center mass of cluster positions (time is equal to *T*).

THE NUMERICAL SCHEME OF SDE SOLUTION.

For solution of the set of SDE we used[9,4,5,14] stable computational scheme of the Artem'ev [19], which is alternative to the similar known computer simulation schemes [18].
We are using the stochastic differential equations (5) and (8) in the form of the Stratonovich. General view of the one-dimentional SDE is:

$$\frac{\partial x}{\partial t} = H(x,t) + \sigma(x,t)\xi(t). \tag{11}$$

Its solution is follows:

$$x_{n+1} = x_n + \left[\hat{I} - \frac{h}{2}\frac{\partial H_n(x,t)}{\partial x}\right]^{-1} * \left[hH_n(x,t) + \sqrt{h}\sigma(x,t)\xi_n + \frac{h}{2}\frac{\partial \sigma}{\partial x}\xi_n^2\right], \quad (12)$$

here h is a step on time, $\hat{I}$ is identity matrix, ξ_n is a sequence of the normally distributed independent random numbers with zero mathematical expectation and individual dispersion, it is designed under the formula:

$$\xi_n = \sqrt{-2\log\alpha_1}\cos(2\pi\alpha_2) \quad (13)$$

α_1 and α_2 are random numbers in regular intervals distributed in an interval [0,1].

DISCUSSION & CONCLUSION REMARKS

As you can see from this work elastic part of potential plays important role in evolution and degradation of bubble.

If initial cluster size belongs to unstable part of Gibbs potential (see equations (2) and Fig. 2, Fig. 3), bubble can grow its initial size $g_0 < g_{cr}$ (see Fig. 8) and it can degrade if $g_0 > g_{cr}$.

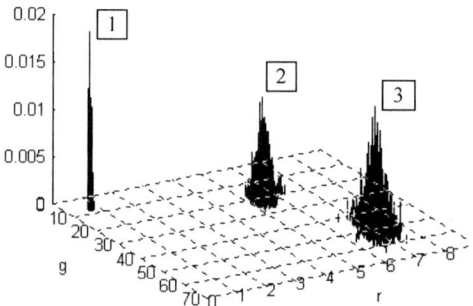

Figure 8. Two-dimensional distribution function $f(g,r)$ as function of size of bubble g and position of bubble center of mass r. Initial bubble size $g_0=6 < g_{cr}=8$, $T=1000$ K. The number of trajectories are 1000. 1 denotes the point in time approximately equal to 0; 2 shows $t = 0.5T$ where $T = 10\tau$, $\tau = 3 \cdot 10^{-11}$ sec; 3 is $f(g, r)$ for $t = T$.

As you can see if $g_{01} < g_{cr}$, $g_{02} > g_{cr}$ but g_{01} and g_{02} are near each other then $f(g, t)$ in both cases are similar but they are not equal (Fig. 8 and Fig. 9). From given figures we can see that bubbles sizes evolution more quick than theirs motion in coordinate space.

Let us demonstrate some possibilities of the successive approximations in solution of two-dimensional kinetic problems of (4) and (6) taking the sets of SDE (5) and (7).

Traditionally, the results concerning to cluster formation calculation[3,7,16,17] are responsible to knowledge of two first moments of stochastic process, namely, the mathematical expectation of the mean size of cluster as well as the its dispersion. Only SSM, presented above, related with the system of kinetic equations, gives us possibility to analyze DF of blisters into solids and to investigate the dependence of the defects clusterization processes on many model factors (thermodynamic parameters, some hypothetical mechanisms of cluster formation, initial conditions). Recent results concerning kinetic model of vapor coagulation[9] was modified to the model of cluster formation into solids. Now we can ameliorate kinetic code for future comparing with experimental data.

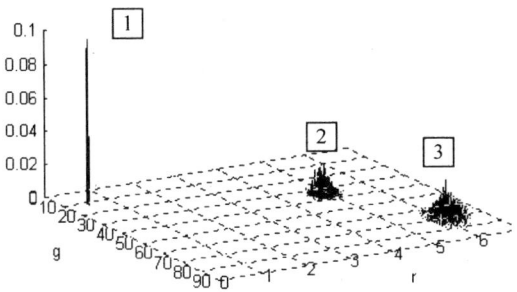

Figure 9. Two-dimensional distribution function $f(g,r)$ as a function of size of bubble g and position of the center of mass r. Initial bubble size $g_0 = 11 > g_{cr} = 8$, the parameters of model are matched by $T = 1000$ K. 1 see Fig. 8; 2 denotes the point in time $t = 0.5T$, T is the final instant of time, where $T = 10\tau$, $\tau = 3 \cdot 10^{-11}$ sec; 3 is $f(g, r)$ for $t = T$. The number of SDE (for trajectories of MP) is equal to 1000.

ACKNOWLEDGMENTS

This work is partially supported by the Russian Foundation for Basic Researches, grants N 96-02-17640 and No 97-02-17627.

REFERENCES

1. E.M. Lifshits and L.P. Pitaevskii, *Physical Kinetics, in Series.: Theoretical Physics, Volum 10*, Nauka, Moscow (1979).
2. F. Premuda, G.I. Zmievskaya and T.V. Zin'kovskaya. *Defect Clusterization Model as a Computer Simulation Method of Fluctuation Phenomena*, Preprint No 134, Keldysh Inst. of Applied Mathem., Russ. Acad. of Sci., Moscow (1995).
3. Yu.N. Devjatko and V.N. Tronin, Nonequilibrium phase transition into systems of interacting Brownian particles, *Doklady Akademii Nauk SSSR (Reports of Academy of Sci. of SU)*, 309: 85 (1989).
4. G.I. Zmievskaya and A.L. Bondareva. *Stochastic Models of Defects' Clusterization in Solids*, Preprint No 102, Keldysh Inst. of Applied Mathem., Russ. Acad. of Sci., Moscow (1997).
5. A.L. Bondareva and G.I. Zmievskaya. *Stochastic Simulation Method for the Fokker-Einstein Kinetic Equation*, Preprint № 101, Keldysh Inst. of Applied Mathem., Russ. Acad. of Sci., Moscow (1997).
6. L.B. Begrambekov, Yu.B. Gorbatov and V.N. Tronin. Research blistering on optical materials, in: *Ionized Radiation & Laser Materials*, Yu.N. Devjatko, ed., Energoatomizdat, Moscow (1982).
7. J.W. Christian, Transformation in Metals and Alloys, part 1, Pergamon Press, Oxford (1975).
8. G.I. Zmievskaya, Stochastic analogs of noneqilibrium collisional processes, *Plasma Physics Reports*, 23: 368 (1997).
9. G.I. Zmievskaya and T.V. Zin'kovskaya, A numerical stochastic model of the nucleation, *Doclady Academii Nauk SSSR (Reports of Academy of Sci. of SU)*, 309: 301 (1989).
10. G.I. Zmievskaya, Stochastic computer simulation of nonequilibrium media, in: *Dynamics of Transport in Plasmas and Charged Beam*, G. Maino and M. Ottaviani, eds., World Scientific Publishing Co, Singapore–London (1996).
11. H. Risken, *The Fokker-Planck Equation: Methods of Solution and Applications*, Springer series in Sinergetics, Vol. 18, Springer Verlag, Berlin (1983).

12. Yu.L. Klimontovich, *Kinetic Theory of Electromagnetic Processes,* Springer series in Sinergetics, Vol. 19, Springer Verlag, Berlin (1983).
13. P.E. Kloeden, E. Platen and H. Schurz, *Numerical Solution of SDE through Computer Experiment*, Springer Verlag, Berlin (1991).
14. G.I. Zmievskaya, Stochastic simulation of fluctuation stages in cluster formation, in: *Book of Abstracts of 21st Intern. Symp. on Rarefied Gas Dynamics, Marseille (France), 26-31 July, 1998*, 2: 152 (1998).
15. E.I. Vainberg, V.P. Efanov, G.I. Zmievskaya and V.P. Prianichnikov, Automated diagnostic complex based on mathematical models of the material degradation and industrial X-ray tomography, in *Proc. of Conf. «Adaptive Robots and General System Logical Theory of Intelligent Robots.» 7- 10 July 1998, Saint-Petersbourg.*, Russia (1998).
16. A.A. Berzin, A.I. Morozov and A.S. Sigov, Ligt atom diffusion and clustering of crystal surfaces, *J. Phys.: Condens. Matter,* 9: 33 (1997).
17. H. Ullmaier, The influence of helium on bulk properties of fusion reactor structural materials, *Nuclear Fusion*, 24: 1029 (1984).
18. Yu.S. Sigov and V.D. Levchenko, Coherent phenomena in the relaxation of a diffuse electron beam in open plasma systems, *Plasma Physics Reports*, 23: 325 (1997).
19. T.A. Averina and S.S. Artem'ev, Numerical solution of the SDE, *Soviet Numerical Annal. Math. Modelling,* 3: 267 (1988).

ELECTRIC POTENTIALS DISTRIBUTION FOR PARTICLES LOCATED IN SOLUTION

N.I. Gamayunov

Tver State Technical University,
170002 Tver, Russia

This work proposes a theory of electric potential distribution in hydrophilic swollen organic gels. The gel suspensions consist of the set of macromolecules associates (matrices) located in a liquid medium. The associates permeability depends on the number of "links" between the macromolecules.

Due to their thermal motion, counter-ions move from the thin, internally porous boundary layer of the associate (particle) to the external pure solvent. A diffusion flow of the counter-ions from particles to solvent, and an inverse flow caused by the electric potential gradient occur is appeared. These two flows are equal when

$$\frac{dc}{dx} + \frac{Fzc}{RT}\frac{d\varphi}{dx} = 0. \tag{1}$$

Let us consider a semi-limited medium ($-\infty < x < 0$), that consists of immobile potential-forming co-ions (e.g., anions) $z^- c_0^-$ and mobile counter-ions $z^+ c_0^+$, located both in the particle and out of it, in the solvent ($0 < x < \infty$) (Fig. 1). Assume the origin of coordinates is located on the particle-solvent interface. In equilibrium distribution, a constant concentration of co-ions c_{01}^+ and counter-ions $c_{02}^- < c_0^- = c_{01}^- + c_{02}^-$ sets on the boundary. The electric neutrality condition suggests that equalities $|z^- c_0^-| = |z^+ c_0^+|$; $|z^- c_{01}^-| = |z^+ c_{01}^+|$; $|z^- c_{02}^-| = |z^+ c_{02}^+|$ hold at an infinitely long distance from the phase boundary, inside the particle.

The transfer of some counter-ions into the solvent generates a negative charge in the near-to-the-boundary layer in the particle. The negative charge is equal to the positive charge of the counter-ions forming a diffusive envelope near the interface.

$$\rho_1(x) = \left|-Fz^- c_1^-(x)\right| = Fz^+ c_1^+(x), \quad (x<0), \tag{2}$$

$$\rho_2(x) = Fz^+ c_2^+(x), \quad (x>0). \tag{3}$$

Omitting further the plus, we can rewrite equation (1) as follows

$$\frac{1}{c_i}\frac{dc_i}{dx} = -\frac{Fz}{RT}\frac{d\varphi_i}{dx}, \qquad (4)$$

where, $\bar{c}_i = c_i(x)/c_{0i}$ ($i = 1, 2$).

Accounting the Poisson's equation

$$\frac{d^2\varphi_i}{dx^2} = -\frac{\rho_i(x)}{\varepsilon_0\varepsilon_i} = -\frac{Fzc_{0i}}{\varepsilon_0\varepsilon_i}\bar{c}_i, \qquad (5)$$

after equation (4) differentiation we derive

$$\frac{d}{dx}\left(\frac{1}{\bar{c}_i}\frac{d\bar{c}_i}{dx}\right) = \frac{(Fz)^2 c_{0i}}{\varepsilon_0\varepsilon_i RT}\bar{c}_i. \qquad (6)$$

Multiplying the right-hand and left-hand parts of equation (6) by dc_i/c_i, we obtain

$$d\left(\frac{1}{\bar{c}_i}\frac{d\bar{c}_i}{dx}\right)^2 = \lambda_i^2 d\bar{c}_i, \qquad (7)$$

where $\lambda_i^2 = 2(Fz)^2 c_{0i}/(\varepsilon_0\varepsilon_i RT)$. Equation (7) is solvable under the boundary conditions

$$c_i(0) = 1, \quad c_i(\pm\infty) = 0. \qquad (8)$$

Taking into account second conditions (8) and $d\bar{c}_i(|\infty|)/dx = 0$, respectively, we can write after integration the equation (7) as

$$\frac{d\bar{c}_i}{dx} = \pm\lambda_i \bar{c}_i^{3/2}. \qquad (9)$$

As $d\bar{c}_i/d|x| < 0$, the "minus" sign in equation (9) is to assume and we obtain the solution

$$\bar{c}_i = \left[1 + \lambda_i |x|/2\right]^{-2} \quad (|x| > 0). \qquad (10)$$

The values c_{0i} can be found from the equality of space charges inside and outside the particle

$$J = Fzc_{01}S\int_0^{|\infty|}\bar{c}_1(x)dx = Fzc_{02}S\int_0^{\infty}\bar{c}_2(x)dx, \qquad (11)$$

where S = the area perpendicular to the x axis. The charge density on the interface is

$$\sigma_b = J/S = 2Fzc_{0i}/\lambda_i. \qquad (12)$$

As follows from equality (12),

$$c_{01}/c_{02} = \lambda_1/\lambda_2 = \varepsilon_2/\varepsilon_1, \tag{13}$$

$$c_{01} = c_0/(1+\varepsilon_1/\varepsilon_2), \quad c_{02} = c_0/(1+\varepsilon_2/\varepsilon_1). \tag{14}$$

Substituting the value of $\bar{c}_i(x)$ into equation (5) and integrating the one twice taking into account the boundary conditions at $x = 0$

$$\varphi_1 = \varphi_2 = \varphi_h; \tag{15}$$

$$\varepsilon_1(d\varphi_1/dx) = \varepsilon_2(d\varphi_2/dx), \tag{16}$$

we obtain

$$\varphi_2(x) - \varphi_1(x) = \varphi_0 \ln \frac{1+(\lambda_2 x/2)}{1+(\lambda_1|x|/2)} = \varphi_0 \ln\left[1 - \frac{1-\varepsilon_1/\varepsilon_2}{1+2/\lambda_1|x|}\right], \tag{17}$$

where $\varphi_0 = 2RT/(Fz^*)$, φ_h is boundary potential.

Let's assume $\varphi_1(x) = \varphi_1(0) = \varphi_h$. When $x \to \infty$, $\varphi_2(\infty) = 0$, so $\varphi_h = \varphi_0 \ln(\varepsilon_2/\varepsilon_1)$. Substitution φ_h into solution (17) gives

$$\Phi_2(x) = \frac{\varphi_2(x)}{\varphi_0} = \ln\left\{\frac{\varepsilon_2}{\varepsilon_1}\left[1 - \frac{1-\varepsilon_1/\varepsilon_2}{1+2/(\lambda_1|x|)}\right]\right\}, \quad x > 0. \tag{18}$$

The assuming $\varphi_2(x) = \varphi_h$ in equation (17) produces

$$\Phi_1(x) = \frac{\varphi_1(|x|)}{\varphi_0} = -\ln\left\{\frac{\varepsilon_1}{\varepsilon_2}\left[1 - \frac{1-\varepsilon_1/\varepsilon_2}{1+2/(\lambda_1|x|)}\right]\right\}, \quad |x| > 0. \tag{19}$$

As follows from equation (19), if $|x| \to \infty$ then $\varphi_1(|\infty|)=2\varphi_h$. A single integration of equation (5) allows to find the intensities

$$E_i = -E_{ri}/(1+\lambda_i|x|/2). \tag{20}$$

The E_{bi} values are

$$E_{b1} = \{2c_0^+ \varepsilon_2 RT/((\varepsilon_1+\varepsilon_2)\varepsilon_0\varepsilon_1)\}^{1/2}, \quad E_{b2} = (\varepsilon_1/\varepsilon_2)E_{b1}. \tag{21}$$

Fig. 1 shows the concentration, potential and intensity distributions for $\varepsilon_1=56$, $\varepsilon_2=81$, $z^\pm=1$, $c_0^\pm=1,96\cdot 10^3$ mole/m^3, $\lambda_1=4,19\cdot 10^9$ m^{-1}, $\lambda_2=2,9\cdot 10^9$ m^{-1}, $\varphi_h/\varphi_0=0,396$, $E_{b1}=1,075\cdot 10^8$ V/m, and $E_{b2}=0,743\cdot 10^8$ V/m.

If the particles are situated in the electrolyte, then the counter-ions entered the particle also interact with the bound co-ions. As a result of ion exchange another dynamic equilibrium[1-4] is established different from that one in the pure solvent, and a new, excessive counter-ion concentration and electric potential distributions appear near the associate-solution boundary compared to the solution.

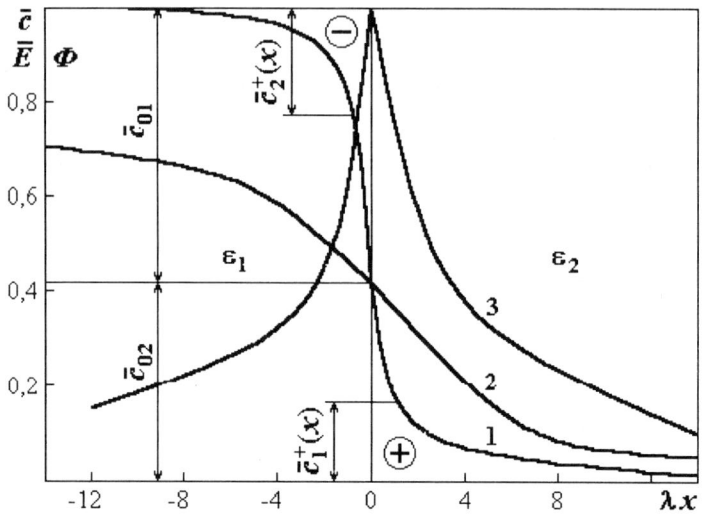

Figure 1. Distribution of relative values of concentrations $\bar{c}=c/c_0$ (1), electric potential Φ (2) and intensity $\bar{E}=E/E_b$ (3) in the vicinity of the phase boundary.

According to Fig. 1 and equation (10), no Boltzmann distribution is reached in the solvent. The approximating function being similar to the Fermi distribution

$$c = \{1 + exp[(-1)^i \lambda_i x]\}^{-1}, \qquad (22)$$

is in a good agreement with solution (10). Such a function can be obtained, if one considers the distribution of ion-releasing groups in the particle in the form of a cubic lattice. Then divide the real space inside and outside the particle into layers and cells parallel to the counter-ion flow (Fig. 2). The layers have the same number of particles $N \gg 1$ and energy characteristics. Each cell can contain only one mobile particle interacting with the co-ions located on macromolecules. A certain number of, e.g., anions n^- with their charge z^- corresponds to the mobile cations' total charge n^+z^+. The average cell volume $v = V_1/m$, where V_1 is the total volume of the swollen gel, and m is the number of dissociated ion-releasing groups in the gel. To make the distribution symmetric relative to the interface (Fig. 2, b) requires multiplying the volume v by $\varepsilon_1/\varepsilon_2$.

Near the boundary, in the solution, they are distributed quasiregularly in the form of a monoionic crystalline lattice. The cell volume in the solution is $v_2 = (cN_0)^{-1}$, where c is counter-ion concentration, and N_0 is the Avogadro constant. Thermal motion in this structure causes the counter-ions to make translational jumps into vacant "points" of the lattice with corresponding generating "holes" at their initial positions. Because of the translations in the particle and solution, each cell either contains or not one counter-ion. Taking into account this fact, as well as the facts that the counter-ions have practically the same properties, and the number of the cells is $g \gg N$, the number of ways how the counter-ions can be placed[5] is

$$W = g!/[N!(g-N)!], \qquad (23)$$

which allows deriving the Fermi distribution

$$\bar{c} = N/g = \{1 + exp[(E-\mu)/kT]\}^{-1}. \tag{24}$$

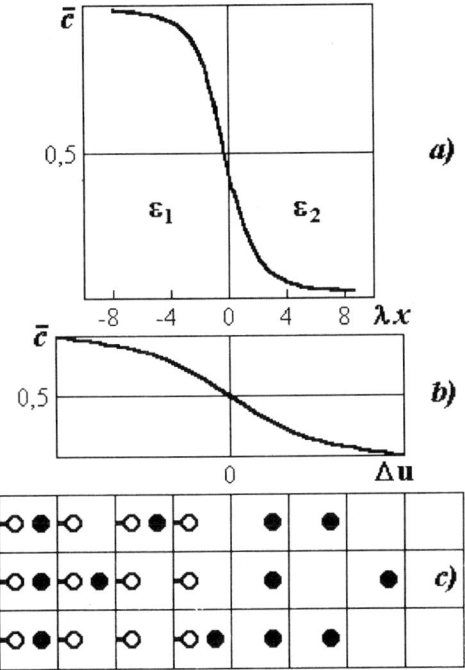

Figure 2. *a)* Distributions of counter-ion concentrations calculated from equation (10); *b)* Fermi distribution ($\Delta u = ez[\varphi_b - \varphi(x)]$; *c)* diagram of ions distribution in the cells (light and dark circles are bound co-ions and mobile counter-ions, respectively).

The energy of a mole of the counter-ions E and chemical potential μ for $T \gg 0$ depend on the potential energy of the electric field only. Let's assume $E = \mu' + ez\varphi_b$, $\mu = \mu' + ez\varphi(x)$, where μ' is the chemical potential when no electric field is available. Then equation (24) can be rewritten as follows:

$$c = \{1 + exp[Fz(\varphi_b - \varphi(x))/RT]\}^{-1}. \tag{25}$$

The potential distribution in the solvent can be approximately described as follows

$$\varphi(x) \cong \varphi_b \, exp(-\lambda_2 x) \cong \varphi_b(1 - \lambda_2 x), \quad x > 0, \tag{26}$$

and the value $\varphi_b \cong RT/(Fz^*)$.

It follows from these relationships that

$$(Fz/RT)[\varphi_b - \varphi(x)] \cong \lambda_2 x. \tag{27}$$

Write the potential's distribution in a particle so

$$\varphi(x) \cong \varphi_b(2 - exp(-\lambda_1 x)) \cong \varphi_b(1 + \lambda_1 x), |x| > 0 \tag{28}$$

and, consequently,

$$(Fz/RT)[\varphi_b - \varphi(x)] \cong -\lambda_1 x. \tag{29}$$

Substituting expressions (27) and (29) into relationship (25) we obtain equation (22).

It follows from the above calculations that the movable counter-ions distribution in the contacting environments, near the interface can be described approximately by the Fermi equation (25).

REFERENCES

1. N.I. Gamayunov, *ZhFKh, Journal of Physical Chemistry*, 61, 5: 1267 (1987) (in Russin).
2. N.I. Gamayunov, *ZhFKh, Journal of Physical Chemistry*, 64, 12: 3322 (1990) (in Russin).
3. N.I. Gamayunov, *ZhFKh, Journal of Physical Chemistry*, 69, 8: 1484 (1995) (in Russin).
4. N.I. Gamayunov, *TOKhT, Theoretical Principles of Chemical Technology*, 29, 4: 430 (1995) (in Russin).
5. Yu.B. Rumer, M.Sh. Ryvkin. Thermodynamics, Statistical Physics and Kinetics. Moscow, Nauka Publishers (1972).

STATEMENT AND SOLUTION OF A BOUNDARY VALUE PROBLEM IN A MODEL OF PLASMA GENERATOR AS CONTROLLED SYSTEM

Yurii G. Ionov[1], Alexei Yu. Ionov[2]

[1]Tver State Technical University,
Department of Automation of Technological Processes
170026 Tver, Russia

[2]Tver State University,
170002 Tver, Russia

INTRODUCTION

Different models of electric arc generators of low-temperature plasma are used for solving practical problems[1-8]. The statement of the boundary value problems in this models and solution of them are connected with generator of plasma as physical or technological object. The tasks are guided by analysis of the processes in generator and optimization of its construction.

Fig. 1 gives common illustrating of the most distributed constructions of the electric arc generators of plasma, which are being influenced by: I is electric current, G is expenditure of plasma-generating gas, Φ is electromagnetic stream. The additional factors, which determine conditions of plasma stream forming of a are: geometry of a channel, disposition of electrodes, chemical composition of the materials, sort of current and gas. Each scheme of generator of plasma is brought in correspondence to definite physical and mathematical model.

The aim of this work is to set up features of the dynamic properties of the generator of plasma and synthesize management of this object. This work is a continuation of researches[9,10].

STATEMENT OF A PROBLEM

There is a generalized base scheme of an electrical arc and flowing of plasma forming, presented on Fig. 2. The scheme corresponds to plasma generators with the beam cathode (Fig. 1 (2, 5–7, 10)). Geometry of a trunk of the arc in a cylindrical channel is conventional[5,7]. Connecting with non-stationary state of the arc, it's temporal changing of radius is also taken into account.

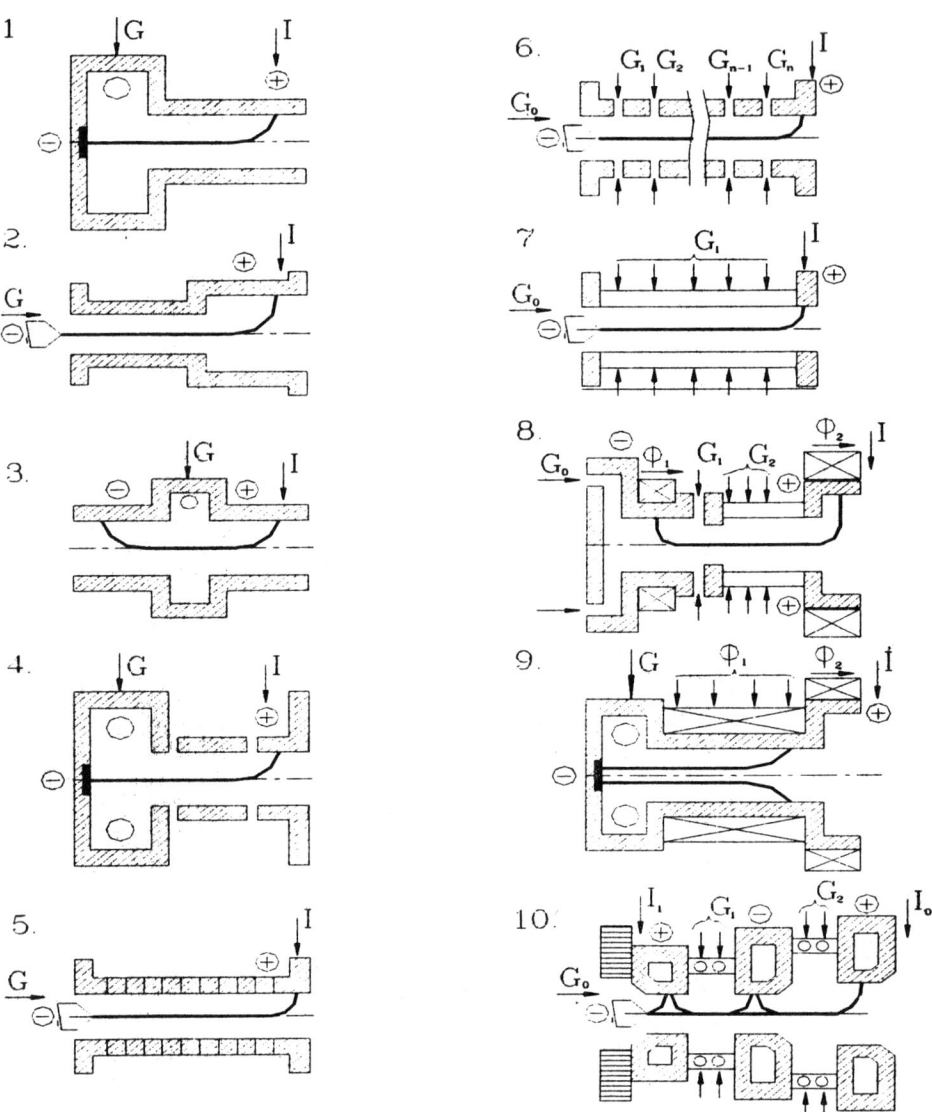

Figure 1. Schematic representation of generators.

On Fig. 2 the following labels are adopted: r_k, r_σ, r_w are a radius (r) of a cathode spot, trunk of the arc and channel of plasma generator accordingly; $z = z_0, z = \ell_\partial$ are the boundaries of cross section of the channel in a point of binding of the cathode spot and on an exit of plasma stream from the channel. Let's suppose, that the scheme of forming of the arc and flowing of plasma remains within the framework of this physical model and at distributed inflation of gas along an axes Z. Fig. 2 demonstrates in the scheme the part of the arc up to an explosive line, an entering part and a part of steady-state flowing after explosive. In the schemes of designs with distributed inflation of gas at which parameter $l_\partial / d \leq 4 \div 5$ the forming of an initial part is finished practically to an output section of the channel. That corresponds to the same scheme presented on Fig. 2.

According to the energy conservation law the temperature variation with the time is being determined by the balance between the specific power coming to a unit volume and the specific power going out due to a heat transfer, convection and radiation. According to the momentum conservation law the momentum variation of unit volume depends on the modification of viscosity of gas, pressure and also on the convective transfer of the gas mass in longitudinal and diametrical directions provoking the modification of its density.

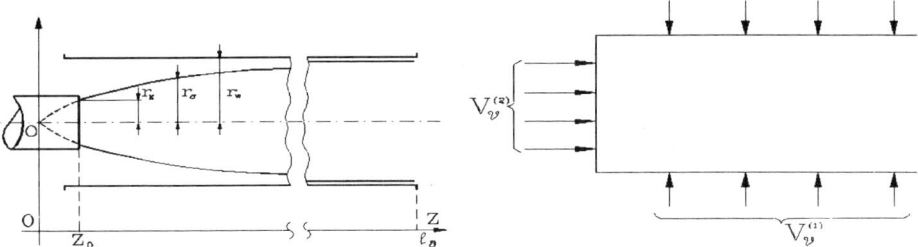

Figure 2. The scheme of flowing of plasma in a cylindrical channel

Figure 3. The scheme of boundary controls of a plasma generator.

Let's consider two limited domains in two-dimensional space $X_1 = \{0 \leq r \leq r_\sigma, z_0 \leq z \leq \ell_\partial\}$ and $X_2 = \{r_\sigma < r \leq r_w, z_0 \leq z \leq \ell_\partial\}$ which have the common boundary $r_\sigma(z,t)$.

The physical processes in plasma are determined by differential equation on domain X_1 ($t_0 \leq t \leq t_k$):

$$\rho c_p \frac{dT}{dt} + \rho u_z c_p \frac{dT}{dz} + \rho u_r c_p \frac{dT}{dr} - \frac{dP}{dt} - \sigma E^2 - \frac{1}{r}\frac{d}{dr}\left(r\lambda\frac{dT}{dr}\right) + Q = 0, \quad (1a)$$

$$\rho \frac{du_z}{dt} + \rho u_r \frac{du_z}{dr} + \rho u_z \frac{du_z}{dz} + \frac{dP}{dz} - \frac{1}{r}\frac{d}{dr}\left(r\mu\frac{du_z}{dr}\right) = 0, \quad (1b)$$

$$\frac{d\rho}{dt} + \frac{d}{dz}(\rho u_z) + \frac{1}{r}\frac{d}{dr}(r\rho u_r) = 0, \quad (1c)$$

where T is temperature, u_z and u_r are longitudinal and diametrical components of velocity.

The equation for the temperature T on domain X_2 is ($t_0 \leq t \leq t_k$):

$$\rho c_p \frac{dT}{dt} + \rho u_z c_p \frac{dT}{dz} + \rho u_r c_p \frac{dT}{dr} - \frac{dP}{dt} - \frac{1}{r}\frac{d}{dr}\left(r\lambda\frac{dT}{dr}\right) = 0, \quad (2)$$

and the equations for longitudinal and diametrical components of velocity u_z and u_r on domain X_2 will be the same ones as well as on domain X_1.

In addition, we are taking into account the equation of state:

$$P = \frac{R_0 \cdot \rho \cdot T}{M}, \quad (3)$$

which completes a equations system (1)–(2). In (1)–(3) the conventional notation[3] are used: $\rho, c_p, \lambda, \sigma, \mu, Q$ are a density, thermal capacity, thermal- and electro- conductivity, viscosity and volumetric radiation.

Functions and their derivatives included in equations should satisfy to boundaries conditions. These conditions being applied to the considered scheme of development of the arc discharge are given below (that are the conditions of the first or third kind using the heat exchange of the stream with a channel wall):

$$r = 0: \frac{\partial T(z,t)}{\partial r} = 0, \quad \frac{d u_z(z,t)}{d r} = 0, \quad \frac{\partial \sigma(z,t)}{\partial r} = 0, \quad u_r(z,t) = 0,$$

$$r = r_\sigma : T(r_\sigma, z, t) = T_\sigma = const, \quad \sigma(r_\sigma, z, t) = 0, \tag{4}$$

$$r = r_w : T(r_w, z, t) = T_w(z,t), \quad u_z(r_w, z, t) = 0, \quad u_r(r_w, z, t) = u_R(t).$$

$$\left(\lambda \frac{dT}{dr}\right)_w = \alpha\left(\overline{T} - T_w\right), \quad u_z = 0, \quad u_r = u_R,$$

$$z = z_0 : T(r,z,t) = T^0(r,t), \quad u_z(r,z,t) = u_z^0(r,t), \quad u_r(r,z,t) = u_r^0(r,t), \tag{5}$$

$$z = l_\partial : T(r, l_\partial, t) = T^0(r,t),$$

$$t = t_0 : T(r, z_0, t_0) = T_0(r), \quad u_z(r, z_0, t_0) = u_{z0}^0(r), \quad u_r(r, z_0, t_0) = u_{r0}^0(r).$$

There are two tasks in the model:
a) it is necessary to realize a numerical research of static and dynamic characteristics, and also of parameters of the object condition at given values of controlling parameters – d, l_∂, I, G (i.e., to solve a problem of the analysis and to reveal features of dynamic properties of generator);
b) to find the parameters values of controlling influences which correspond to a defined index of quality of control (i.e., to decide a problem of synthesis of controls).

The physical model of the boundary control is presented on Fig. 3, where $V_v^{(1)}$ is the vector (in common case) of controlling actions in section z_0, and $V_v^{(2)}$ is the one along the channel.

The index of quality of control by the generator is defined as functional connected with boundary (in section $z = l_\partial$) distributions of temperature and velocity:

$$\Pi_A^{Ju}(t) = C \cdot \frac{R_0}{M} \cdot T(t) \cdot G(t) = 2\pi C \cdot \frac{R_0}{M} \cdot \int_0^{r_w} T(r, l_\partial, t) \rho(r, l_\partial, t) u_z(r, l_\partial, t) r \, dr, \tag{6}$$

where C is dimensionless factor which is taking into account a type of design of generator,

$$T(t) = \frac{\int_0^{r_w} T(r, l_\partial, t) \cdot \rho(r, l_\partial, t) \cdot u_z(r, l_\partial, t) \cdot r \cdot dr}{\int_0^{r_w} \rho(r, l_\partial, t) \cdot u_z(r, l_\partial, t) \cdot r \cdot dr}, \tag{7}$$

is average mass temperature of plasma in an output section,

$$G(t) = 2\pi \cdot \int_0^{r_w} \rho(r, l_\partial, t) \cdot u_z(r, l_\partial, t) \cdot r \cdot dr, \tag{8}$$

is value of gas flow rate calculated according to the conservation law of mass.

It is important to note the following circumstance: the entered index of control quality (6) can be defined experimentally.

The synthesis of control vectors $V_v^{(1)}$, $V_v^{(2)}$ of a plasma stream is the main problem of this work. The boundary problem is formulated as follows: to find such components of vectors $V_v^{(1)}$, $V_v^{(2)}$, that the distributions $T(r, 1, t)$ and $u_z(r,l,t)$ being a solution of equations (1)–(5) are satisfying to a given value of the index of quality $\Pi_A^{Ju}(t_H)$, where t_H is a instant time of observation.

SOLUTION OF A BOUNDARY VALUE PROBLEM

Let's solve the boundary value problem (1)–(5) by preferring numerically-analytical method. Numerical method of solution (finite-difference method, for example) is inconvenient to apply. This complicated because of hard approximation of integration area near to the boundary of conducting zone $r_\sigma(z,t) = r_\sigma^*(t) \cdot f(z)$ which is moving in time. Near to this boundary thermal potential and electroconductivity of stream almost spasmodically varies ($r_\sigma^*(t)$ is radius on a steady-state part of flowing of plasma). Analytical solution gives additional possibilities. It can be checked and corrected in an interactive mode of computer handling.

The parameter $\lambda(r)$ is a nonlinear function of temperature. It is expedient to realize their linearization by entering the Kirchhoff's substitution, which apply rather often[3]:

$$S = \int_0^T \lambda \cdot dT, \qquad (9)$$

The definition of values of thermal-physical factors and factors of transposition $\rho, c_p, \lambda, \sigma, \mu, Q$ as non-linear functions of temperature is the known problem[1-7]. Usually these parameters are set as tabulated functions in the limited number of focal points of temperature[7]. It seemed naturally to determine values of factors in any points of temperature by interpolating. From the comparison of the Lagrange's method and the method of piece-linear approximation[2] in a range of temperatures up to 20000 K follows that in practice approach of such kind isn't rational (the arc high requirements to precision and limited computing resources). The proposed method of the piece-linear approximation is the development of the Maekker's method:

$$\int_{S_i}^{S_{i+1}} \sigma(S)dS = (S_{i+1} - S_i) \cdot \sigma_i + 0.5 \cdot (S_{i+1} - S_i) \cdot (\sigma_{i+1} - \sigma_i) + \varepsilon, \qquad (10)$$

where $0 \leq i \leq n$, n is the number of segments in the method of approximation, ε is precision of approximation.

Factor of approximation varying from segment to segment is obtained from (10):

$$b_\sigma \cong \frac{2}{(S_{i+1} - S_i)^2} \cdot \left(\int_{S_i}^{S_{i+1}} \sigma(S)dS - \varepsilon \right) - \frac{2}{(S_{i+1} - S_i)} \cdot \sigma_i, \qquad (11)$$

Then the dependence of electro-conductivity from a thermal potential takes the following form:

$$\sigma(S) \cong \sigma_{S_i} + b_\sigma(S - S_i), \qquad (12)$$

So it is possible to approximate other thermal-physical factors, which are included into the system (1)–(5). By introduction them in appropriate equations there are important features, which are necessary to take into account. For example, after the transformation of the first differential equation there is:

$$-B_s^p \frac{\partial S}{\partial t} + A_s + \left(\frac{\rho c_p}{\lambda} u_r + \frac{1}{r}\right)\frac{\partial S}{\partial r} + \frac{\partial^2 S}{\partial r^2} + \left(b_\sigma E^2 - b_Q\right)\cdot S = 0, \qquad (13)$$

where $A_s = \left[(a_\sigma + b_\sigma S_{a_\sigma})E^2 - (a_Q + b_Q S_{a_Q})\right]$, B_s^p are factor, defining characteristic time of the establishment of parameter S in an initial condition after removal of perturbation with taking into account pressure variation.

The normalization of an independent variable r adduce to a dimensionless radius (r is the current coordinate on a radius of the channel):

$$\gamma = k \cdot (r/r_w), \qquad \text{where} \qquad k = r_w\sqrt{b_\sigma E^2 - b_Q}, \qquad (14)$$

here are two features relating to the solution (13):
1) the radius variation scale will be a function of time, if the electric field tension **E** varies with the time:

$$E(t) = \frac{I(t)}{2\pi \int_0^{r_\sigma} \sigma(r,t) r\, dr}, \qquad (15)$$

where $I(t)$ depends on many factors and practically always is function of time.
2) if $b_Q > b_\sigma E^2$, the last term in (13) accepts a minus sign. The solution of the mentioned equation is bound up with modified cylindrical functions of Bessel.

Variant in which $b_\sigma E^2 > b_Q$ is considered below. Different equations are used on domains X_1 and X_2 at $t_0 \leq t \leq t_k$. The searching of own functions (1a) is realized with using ideas of the method of separation of variables. The searching of own functions (16) and (1c) is found by the Pikar's method and by the Galerkin's method accordingly. The solution is implemented concordantly for all equations. Thus there is a system of the nonlinear equations.

$$S(\gamma,z,t) - \left\{[S(\gamma,z_0,t) - S_w]\cdot e^{\left(\frac{\tau_s u_r}{kr_w}\right)_{z_0}\cdot \gamma}\cdot e^{\left[\int_{t_0}^{t}\frac{k}{\tau_s}dt\right]_{z_0}} + J_0(\gamma)\int_{z_0}^{z}\tilde{Q}(z)e^{-\int_{z_0}^{z}\Phi(z)dz}dz\right\} \times$$
$$\times e^{-\frac{\tau u_r}{kr_w}\cdot \gamma}\cdot e^{-\int_{t_0}^{t}\frac{k}{\tau_s}dt}\cdot e^{\int_{z_0}^{z}\Phi(z)dz} - S_w = 0 \qquad (16a)$$

$$u_z(\gamma,z,t) - \left[u_z(\gamma,z_0,t) + \int_{z_0}^{z}\Psi(\alpha,\ldots)d\alpha\right]J_0(\gamma)e^{-\frac{\tau_u u_r}{kr_w}\cdot\gamma}\cdot e^{-\int_{t_0}^{t}\frac{k}{\tau_u}dt} = 0, \qquad (16b)$$

$$u_r(\gamma,z,t) - \varphi_\gamma \gamma - (3u_r + 3k\varphi_\gamma + k\varphi_v u_r)\frac{\gamma^2}{k^2} + (2u_r + 2k\varphi_\gamma + k\varphi_v u_r)\frac{\gamma^3}{k^3} = 0, \quad (16c)$$

where $0 < \gamma < \gamma_\sigma, 0 < z < 1, t_0 < t$ are defined as well as system (16) on X_1.

By analogous method on X_2 is obtained:

$$S(\gamma,z,t) - [S(\gamma,z_0,t) - S_w] e^{\left[\left(\frac{\tau_s u_r}{\lambda_1 r_w}\right)_{z_0} - \frac{\tau_s u_r}{\lambda_1 r_w}\right]\cdot \gamma} \cdot e^{-\int_{t_0}^{t}\left[\left(\frac{\lambda_1}{\tau_s}\right)_{z_0} - \frac{\lambda_1}{\tau_s}\right]dt} \cdot e^{\int_{z_0}^{z}\Phi(z)dz} - S_w = 0, \quad (17a)$$

$$u_z(\gamma,z,t) - \left[u_z(\gamma,z_0,t) + \int_{z_0}^{z}\Psi(\alpha,...)d\alpha\right] J_0(\gamma) e^{-\frac{\tau_u U_r}{\lambda_1 r_w}\cdot \gamma} \cdot e^{-\int_{t_0}^{t}\frac{\lambda_1}{\tau_u}dt} = 0, \quad (17b)$$

$$u_r(\gamma,z,t) - \varphi_\gamma \gamma - (3u_r + 3\lambda_1\varphi_\gamma + \lambda_1\varphi_v u_r)\frac{\gamma^2}{\lambda_1^2} + (2u_r + 2\lambda_1\varphi_\gamma + \lambda_1\varphi_v u_r)\frac{\gamma^3}{\lambda_1^3} = 0, (17c)$$

where $\gamma_\sigma \leq \gamma \leq \lambda_1, z_0 < z < 1, t_0 < t$.

In systems of the nonlinear equations (16), (17) functions $\tilde{Q}(z)$, $\Phi(z)$, $\Psi(\alpha,...)$, φ_γ, φ_v are their own functions. The solution of systems is necessary to execute for each value of independent variables of the problem. If the law of modification on length of one of defining parameters is found, the problem is closed. There are the axial values of a thermal potential, longitudinal velocity or radius of conducting zone in the channel. According to the law, which established by the author[11], the supposition about modification of the radius is used. However the radius in any section varies in time.

The distributions of a thermal potential and velocity in section $z = z_0$ at $t = t_0$ are determined by the problem of synthesis solving. The boundary control in the considered section varies to find required distributions in section $z=1$. The character of distributions in the initial section $z = z_0$ is defined by the law depending on distribution of temperature in the cathode spot with diameter[4]:

$$(d_k)_{N_2} = 9.1 \cdot 10^{-3} I^{0.5}, \qquad (d_k)_{Ar} = 1.3 \cdot 10^{-2} I^{0.5}, \qquad (18)$$

These data concern to nitrogen and argon accordingly and are given for the tungsten cathode. They are used in calculated distributions, connecting a current of the arc with their parameters. The specificity of distribution of a thermal potential in section $z = z_0$ is determined not only by processes of trunk of the arc, but also by influence of the cathode. Concerning to the distribution in this section it is impossible to affirm, that its character should be determined only by initial equations. For detection of a role of the cathode processes it is required to execute additional researches. The appropriate distributions in section $z = z_0$ are given below for case $t = t_0$:

$$S(\gamma,z_0,t_0) = [S_0 - S_\sigma] \cdot J_0(\gamma) \cdot e^{\left[-\frac{\tau_s u_r}{k r_k}\right](\gamma - \gamma_\sigma)} + S_\sigma, \quad (19a)$$

$$u_r(\gamma,z_0,t_0) = (2\frac{\gamma}{k} + \frac{\gamma^2}{k^2} - 2\frac{\gamma^3}{k^3}) u_{r\sigma}, \quad (19b)$$

$$u_z(\gamma,z_0,t_0) = -\frac{G_{z_0}(t_0)}{2\pi r_w^2 \int_0^{\lambda_1} \frac{1}{k^2} \rho(S(\gamma)) \cdot J_0(\gamma) \cdot e^{\left[-\frac{\tau_s u_r}{k r_w}\right]\gamma} \gamma d\gamma} \cdot J_0(\gamma) \cdot e^{\left[-\frac{\tau_s u_r}{k r_w}\right]\gamma}, \qquad (19c)$$

where $\gamma = k \cdot (r/r_k) \leq \gamma_\sigma$.

They are recevied from the solution of problem (5), (8), (15)–(17). When $\gamma > \gamma_\sigma$ the distributions instead of k use a constant value $\lambda_1 = 2.4048$. The thermal potential in this zone has the following distribution:

$$S(\gamma,z_0,t_0) = [S_\sigma - S_w]\frac{J_0(\gamma)}{J_0(\gamma_\sigma)} e^{\left[-\frac{\tau_s u_r}{\lambda_1 r_w}\right](\gamma-\gamma_\sigma)} + S_w, \qquad (20)$$

which essentially differs from mentioned for conducting zone.

Required profiles of a thermal potential and diametrical component of velocity in the section $z = 1$ we define as interdependent by the following system:

$$S(\gamma,z=1,t) = \left\{[S(\gamma,t_0) - S_w] + J_0(\gamma) \cdot e^{-\frac{\tau_s^p u_r}{k r_w}\gamma} \cdot \int_{t_0}^{t} \tilde{Q}(t)e^{\int_{t_0}^{t}\Phi(t)dt} dt\right\} \cdot e^{-\int_{t_0}^{t}\Phi(t)dt} + S_w, \quad (21a)$$

$$u_r(\gamma,t) = u_{r_\sigma}\left[\frac{\gamma}{k} + \left(1 - \frac{\gamma}{k}\right)\frac{\gamma^2}{\gamma_\sigma^2}\right], \qquad (21b)$$

for conducting zone and

$$S(\gamma,z=1,t) = [S(\gamma,t_0) - S_w] \cdot e^{-\int_{t_0}^{t}P(t)dt} + S_w, \qquad (22a)$$

$$u_r(\gamma,t) = u_{r_\sigma}\left[\frac{\gamma}{\lambda_1} + \left(1 - \frac{\gamma}{\lambda_1}\right)\frac{\gamma^2}{\gamma_\sigma^2}\right] - \frac{\gamma - \gamma_\sigma}{\lambda_1 - \gamma_\sigma} u_{r_\sigma}, \qquad (22b)$$

for nonconducting zone.

These distributions cannot submit to any laws, and should satisfy to input equations of the solving system and correspond to the laws (8), (15). If in this case as the defined magnitude there will be a potential on an axes and wall of the channel then under the following conditions the problem is determined:

$$\begin{aligned}&\gamma = 0, u_r = 0, \\ &\gamma = \gamma_\sigma, u_r = u_{r_\sigma}, \\ &\gamma = \lambda_1, u_r = u_R, \\ &\frac{u_{r_\sigma}}{E} = \frac{r_w^2\sqrt{b_\sigma}}{\tau_s \gamma_\sigma} \cdot \ln\left[J_0(\gamma_\sigma)\frac{(S_0 - S_w)}{(S_\sigma - S_w)}\right]\end{aligned} \qquad (23)$$

where the parameter u_{r_σ} is diametrical component of velocity on the boundary of conducting zone. The last relation is found from (22a) and determines a value of the diametrical component of velocity on the boundary of conducting and nonconducting zones of current of gas in the channel. The magnitude u_R in a particular case can be equal to zero. In common case, when $V_\upsilon^{(2)} \neq 0$, the velocity should be expressed by known[7] mode through the inflated distributed gas flow rate. In this case solution of systems (16) and (17) uses corrected on (8) value of the calculated consumption.

For interface of profiles in zones in an output section, where the flowing of plasma is steady-state, it is necessary to define a radius of conducting zone. It can be found from the correlation (14). On this boundary $b_Q = 0$ and $b_\sigma = b_\sigma(S_\sigma)$.

The expression for profile of a longitudinal component of velocity in the considered section coincides expression (19c), as it also is a solution of the equation of motion. But in this case velocity of gas is determined through summarized gas flow rate in a section $z = 1$.

PARTICULAR CASES OF SOLUTION

The mathematical model of plasma generator as an object with the distributed parameters built on initial description (1)–(5) is rather complicated. The problems of synthesis of controls are connected to it. The models and algorithms of synthesis should be simple, but at the same time they should be adequate to conditions of control. It is because of aspiration for solving more simple problems which are a particular case of the considered boundary value problem[9, 10, 12, 13]. For example, at the scheme 5 (see Fig. 1) there are such values of parameters, that $l_\partial > 8 \cdot r_w$. In this case length of an entering part (see on Fig. 2 up to an explosive line) is essential less than steady-state part of flowing (after the explosive line). At constant mass gas flow rate ($\rho u_z = const$) the equation (1) is solved separately from others under boundary conditions of the third kind. The detail substantiation of this and other assumptions is given in work[14]. Taking into account simplifying assumptions about a character of plasma flowing and in case of control of it's average mass temperature we have an outcome of solution of the boundary value problem which define more exactly a model of generator as a lumped-parameter system [10, 12, 14]. The parameters of a steady-state condition are determined from the solution of the nonlinear equations, and deviation from this condition - from the linearized differential equations. The various forms of presentation of models are developed, namely: by the linear differential equations, by the transmission functions and electrical circuits of substitution. The special class of models and particular (in relation to considered) boundary value problems present the models of plasma generator as an object with parameters varying in time[13]. These models are developed on the basis of electrical circuits of substitution of the arc in plasma generator in the supposition, that the elements of the schemes change on the defined law relatively constant values in a steady-state condition. The processes of shunting called by electrodynamical forces can be the reason of the mentioned modifications (parametrical excitations). The modifications can appear as the result of influence of exterior fields of any physical nature, as the result of the modification of a dielectric permeability of plasma generating gas, volumetric charge of plasma in areas close to an electrode and so on. If to enter any law of modification of parameters of an equivalent circuit then the mathematical description can be presented by the Hill's equation, and if to enter the harmonic law, then by the Mathieu's equation[13].

The boundary value problems which appropriate considered, simplified models have rather simple solutions and give evaluations of parameters and performances of various processes useful for researches. The examples of the mentioned estimations are given in works[16, 17].

CONCLUSION

The problem, which statement and solution were considered, has essential features of numerical realization connected with analytical transformation of initial differential equations to the system of the nonlinear equations. It is expedient to solve the system numerically by the method which is not using derivatives, although the derivatives are used in the famous Netwon's method. For satisfactory stability and convergence of solution of the problem is worse using the iterative method Braune. To provide the plasma generator modelling the base variant of computing complex is created. As to the using area of basic results it is not limited only by mathematical.

REFERENCES

1. *Low-Temperature Plasma*, Multiple volume issue, ed. M.F. Zhukov, Nauka, Novosibirsk, (1990–1994) (in Russian).
2. *Moving Plasma*, The collected translations, eds. E.V. Kudryavcev and V.P. Ionov, Moscow (1961) (in Russian).
3. *Physics and Engineering of Low-Temperature Plasma*, ed. N.A. Dresvin, Atomizdat, Moscow (1972).
4. M.F. Zhukov, A.S. Koroteyev, and B.A. Uryukov, *Applied Dynamics of Thermal Plasma*, Nauka, Novosibirsk (1975) (in Russian).
5. *Mathematical Modelling of an Electrical Arc*, ed. V.S. Engelsht, Frunze (1983).
6. *The Theory of Thermal Electric Arc Plasma*, eds. M.F. Zhukov and A.S. Koroteyev, Nauka, Novosibirsk (1987) (in Russian).
7. *Low-Temperature Plasma*, 1, *The Theory of a Pole of an Electrical Arc*, eds. V.S. Engelsht and B.A. Uryukov, Nauka, Novosibirsk (1990) (in Russian).
8. *Low-Temperature Plasma*, 2, *Mathematical Methods of Researching of Dynamics and Problems of Control of Low-Temperature Plasma*, ed. M.F. Zhukov, Nauka, Novosibirsk (1991) (in Russian).
9. Yu.G. Ionov, Mathematical Models of Generators of Plasma for Synthesis of Controlled Technological Systems, *Thesises of the reports of an international scientific conference "Mathematical Models of Non-Linear Excitation, Transport, Dynamics, Control in Condensed and Other Mediums"*, Tver (1996) (in Russian).
10. Yu.G. Ionov, Mathematical description of a plasma generator as an object of automatic control, *Electromechanics*, 2 (1979).
11. A. Koller, Z. Zur, Theorie des Warmestrompotentials in angestromten Lichtbogen, *Naturforschung*, 21a, 5: 505 (1966).
12. Yu.G. Ionov, To the theory of non-stationary state of processes of an electrical arc of direct current, *Information Of Siberian Branch of Academy of Science of the USSR*, 4 (1986).
13. Yu.G. Ionov and A.S. Parhomenko, *The identification of dynamic characteristics of an electrical arc*, Cheboksari State University, Cheboksari (1988) (in Russian).
14. Yu.G. Ionov, The scheme of substitution of an electrical arc of direct current, *Electricity*, 12 (1986).
15. Yu.G. Ionov, Yu.E. Gladish and A.Yu. Ionov, Computing complex "Scalpel" – an instrumental medium for modelling of medical plasmatic tools, *The inter-university collection of the proceedings*, Tver (1998) (in Russian).
16. Yu.G. Ionov and V.D. Parhomenko, Characteristic times of some processes in arc plasma, *Generators of low-temperature plasma, Thesises of the reports of X All-Union conference*, 1, Minsk (1986) (in Russian).
17. Yu.G. Ionov, A behaviour of an electric arc dynamic system at small deviations from an equilibrium condition, *Electricity*, (1989).

Nonlinear Models in Chemical Physics and Physical Chemistry

COMPLEX REPRESENTATION OF DYNAMICS OF COUPLED NONLINEAR OSCILLATORS

Leonid I. Manevitch

N.N. Semenov Institute of Chemical Physics
of Russian Academy of Sciences,
117334 Moscow, Russia

1. INTRODUCTION

The complex representation of the classical equations of motion of a system of linear oscillators was used for the first time in a quantum mechanics and in the analysis of so-called coupled oscillations and waves in a mechanics, electronics engineering and solid state physics[1-6]. The complex conjugate linear combinations of displacements and velocities of oscillators being sought-for functions in this representation, can be visually presented as vectors of equal length rotating in opposite directions. Actually it is enough to find only one complex function for each oscillator completely defining both displacement and velocity. Such choice of variables leads in particular to very simple and natural procedure of quantization: complex conjugate functions become operators of an annihilation and birth, and their squared module –by the number of elementary excitations[1,5,8].

If there is coupling between oscillators the complex conjugate functions are included in each equation of motion, but in the theory of coupled oscillations it is supposed that the uni-directional rotations of oscillators are connected more strongly than the vectors with opposite directions[2-4]. This simplification reduces the order of the system of motion equations by the factor of two and ensures right expressions for first two terms in expansion on a coupling parameter of an exact solution of the linearized equations of motion[1,2]. In case of the coupled system of nonlinear oscillators the possibility of comparison with exact solutions is absent. In this connection the equations of motion for unidirectional rotations are usually treated phenomenologically as the elementary mathematical model of a nonlinear oscillatory system[5]. Its validity is confirmed by comparison with the averaged equations for complex amplitudes[3,5,7], qualitative reasons[4] and asymptotic estimations[7]. Naturally the application of the justified scheme of the perturbations theory has not only to reveal a domination of coupling between unidirectional rotations but also to provide a possibility of a construction of higher approximations.

In the present paper the complex representation of the motion equations of motion of the weakly coupled nonlinear oscillators is considered as the natural form for effective application of a method of two-scale expansions[10,11] in a combination with a Poincare-Lighthill type transformation[12]. Thus the coupled nonlinear equations for unidirectional rotations correspond to a principal approximation in "slow time" and the possibility of its corrections is ensured.

The paper is constructed as follows: in the second section the basic ideas of the considered approach are illustrated with example of a nonlinear oscillator with cubic anharmonicity (free oscillations with allowance for dissipation of an energy, steady-state oscillations in a periodic external field, parametrical resonance and self-excited oscillations). The third section is devoted to a system of two coupled oscillators. In the fourth and the fifth sections the nonlinear waves in finite and infinite systems of the weakly coupled nonlinear oscillators and also in an infinite chain of atoms are considered.

The most interesting physical effects in this field are connected with the space localization of vibrations and waves. This problem is the subject of the growing interest during last decades. For much extent it relates to essentially discrete vibrations and waves called as breathers (e.g. S.Flach, C.R.Willis in Phys.Rev. 295, 1998, 181; S.Aubry in PhysicaD, 103, 1997, 201). It is beyond our scope to discuss all interesting results in this field which are successfully reviewed in the above references. The main goal of this paper is to demonstrate the possibility of crucial simplification of the problems concerning the nonlinear waves and vibrations in the systems of coupled oscillators by means of complex representation. Therefore we mention only those references where this representation was used.

II. NONLINEAR OSCILLATOR WITH CUBIC ANHARMONICITY

1. Free oscillations

Let's first consider nonlinear oscillatory system with one degree of freedom described by the equation

$$m\ddot{U} + 2n\dot{U} + c_1 U + c_3 U^3 = 0 \tag{2.1}$$

with the initial conditions: $t = 0$, $U = U_0$, $\dot{U} = V_0$, where "($\cdot$)" means the derivation on time. After introduction of dimensionless variables: $\tau = \omega_0 t$, $u = U(E_0/c_1)^{-1/2}$, where $\omega_0 = \sqrt{c_1/m}$, E_0 is initial energy of an oscillator, we have

$$\frac{d^2 u}{d\tau^2} + 2\varepsilon\gamma \frac{du}{d\tau} + u + 8\varepsilon\alpha u^3 = 0 \tag{2.2}$$

$$\tau = 0: \quad u = u_0; \quad \frac{du}{d\tau} = v_0,$$

here $\varepsilon\gamma = \dfrac{n}{\sqrt{c_1 m}}$; $8\alpha\varepsilon = \dfrac{c_3}{c_1^2} \cdot E_0$.

Let's write the equation of motion (2.2) as a system of two first order equations

$$\frac{dv}{d\tau} = -2\varepsilon\gamma v - u - 8\alpha\varepsilon u^3$$

$$\frac{du}{d\tau} = v$$

Multiplying the second equation on imaginary unit and adding it with first, and then deducting from first we obtain

$$\frac{d\psi}{d\tau} - i\psi + \varepsilon\gamma(\psi + \psi^*) + i\varepsilon\alpha(\psi - \psi^*)^3 = 0$$

(2.3)

$$\frac{d\psi^*}{d\tau} + i\psi^* + \varepsilon\gamma(\psi + \psi^*) + i\varepsilon\alpha(\psi^* - \psi)^3 = 0,$$

where $\psi = v + iu$, $\psi^* = v - iu$.

Obviously, the second equation can be obtained from the first one by operation of complex conjugation. Therefore only first equation will be considered later on.

Let's change the dependent variable in (2.3): $\psi = e^{i\tau}\varphi(\tau)$. Then we obtain

$$\frac{d\varphi}{d\tau} + \varepsilon\gamma(\varphi + e^{-2i\tau}\varphi^*) + i\alpha\varepsilon(e^{2i\tau}\varphi^3 - 3|\varphi|^2\varphi + 3e^{-2i\tau}|\varphi|^2\varphi^* - e^{-4i\tau}\varphi^{*3}) = 0.$$

(2.4)

Hereinafter we assume that $\varepsilon \ll 1$. The direct expansion of the solution of the equation (2.4) by a small parameter ε should lead to the appearance of secular terms. Alternatively we introduce a " slow " time

$$\tau_1 = \varepsilon\tau[1 + \xi_1(\tau_1)\varepsilon + \xi_2(\tau_1)\varepsilon^2 + ...],$$

(2.5)

along with time τ where $\xi_i(\tau_1)$ yet not defined functions, and we will consider required complex function as function of two variables $\varphi(\tau,\tau_1)$. Let's present this function as expansion by a small parameter ε:

$$\varphi = \varphi_0 + \varepsilon\varphi_1 + \varepsilon^2\varphi_2 + ...$$

(2.6)

With allowance for (2.4–2.6) we obtain

$$\frac{\partial}{\partial\tau}(\varphi_0 + \varepsilon\varphi_1 + \varepsilon^2\varphi_2 + ...) + \varepsilon L(\tau_1,\varepsilon)\cdot\frac{\partial}{\partial\tau_1}(\varphi_0 + \varepsilon\varphi_1 + \varepsilon^2\varphi_2 + ..) + \varepsilon M(\varphi_1,\varphi_1^*,\varepsilon) = 0,$$ (2.7)

where $L(\tau_1,\varepsilon) = \dfrac{1 + \xi_1(\tau_1)\varepsilon + \xi_2(\tau_1)\varepsilon^2 + ...}{1 - \varepsilon\tau_1\left[\dfrac{d\xi_1(\tau_1)}{d\tau_1} + \dfrac{d\xi_2(\tau_1)}{d\tau_1}\varepsilon + ...\right]\cdot[1 - \xi_1(\tau_1)\varepsilon - ...]}$

$$M(\varphi_1,\varphi_1^*,\varepsilon) = \gamma\left[(\varphi_0 + \varepsilon\varphi_1 + \varepsilon^2\varphi_2 + ..) + e^{-2i\tau}(\varphi_0^* + \varepsilon\varphi_1^* + \varepsilon^2\varphi_2^* + ..)\right] +$$ (2.8)

$$+ i\alpha\Big[e^{2i\tau}(\varphi_0 + \varepsilon\varphi_1 + \varepsilon^2\varphi_2 + ..)^3 - 3|\varphi_0 + \varepsilon\varphi_1 + \varepsilon^2\varphi_2 + ..|^2(\varphi_0 + \varepsilon\varphi_1 + \varepsilon^2\varphi_2 + ...) +$$

$$+3e^{-2i\tau}\left|\varphi_0+\varepsilon\varphi_1+\varepsilon^2\varphi_2+...\right|^2(\varphi_0^*+\varepsilon\varphi_1^*+\varepsilon^2\varphi_2^*..)-e^{-4i\tau}(\varphi_0^*+\varepsilon\varphi_1^*+\varepsilon^2\varphi_2^*+...)^3\;]=0,$$

$l=0,1,...$

Now we equate to zero the coefficients at each from growing powers of the parameter ε

1) $\varepsilon^0: \dfrac{\partial \varphi_0}{\partial \tau}=0$. Therefore $\varphi_0=\varphi_0(\tau_1)$, i.e. the principal approximation is the function φ_0 depending only on "slow" time

2) $\varepsilon^1: \dfrac{\partial \varphi_0}{\partial \tau_1}+\gamma\varphi_0-3\alpha|\varphi_0|^2\varphi_0=-\dfrac{\partial \varphi_1}{\partial \tau}-\gamma e^{-2i\tau}\varphi_0^* -i\alpha\left(e^{2i\tau}\varphi_0^3-3e^{-2i\tau}|\varphi_0|^2\varphi_0^*+e^{-4i\tau}\varphi_0^{*3}\right)$ (2.9)

....

The left part of this equation depends only on slow time τ_1. The right part contains a derivative of function $\varphi_1(\tau,\tau_1)$ on fast time and function of slow time with fast varying exponential coefficients. Integrating both parts of the equation (2.9) by fast time in limits from 0 to π, we will obtain in a principal approximation with respect slow time

$$\dfrac{\partial \varphi_0}{\partial \tau_1}+\gamma\varphi_0-3i\alpha|\varphi_0|^2\varphi_0=\dfrac{-\Delta\varphi_1}{\pi}, \qquad (2.10)$$

where $\Delta\varphi_1=\varphi_1(\pi;\tau_1)-\varphi_1(0,\tau_1)$. As is shown lower the function φ_1 can be defined in such a manner that $\Delta\varphi_1=0$; then the principal approximation is described by the equation:

$$\dfrac{\partial \varphi_0}{\partial \tau_1}+\gamma\varphi_0-3i\alpha|\varphi_0|^2\varphi_0=0. \qquad (2.11).$$

We pay attention to that important circumstance, that in a main approximation for $\varphi_0(\tau)$, in difference from phenomenological model, the terms describing inertial, dissipative and nonlinear elastic forces have the same (zero) order on a parameter ε, the linearly-elastic force in the equation is absent.

To obtain in the explicit form a solution of the equation (2.11) we introduce a new change of a dependent variable

$$\varphi_0=e^{-\gamma\tau_1}\Phi_0. \qquad (2.12)$$

After a substitution (2.12) in (2.11) we obtain

$$\dfrac{d\Phi_0}{d\tau_1}-3i\alpha e^{-2\gamma\tau_1}|\Phi_0|^2\Phi_0=0. \qquad (2.13)$$

At $\gamma=0$ the equation (2.13) and equation conjugate to it, can be written in the Hamiltonian shape:

$$\dfrac{d\Phi_0}{d\tau_1}=\dfrac{dH}{d\Phi_0^*}; \quad \dfrac{d\Phi_0^*}{d\tau_1}=-\dfrac{dH}{d\Phi_0}; \qquad (2.14)$$

where $H = \frac{3}{2} i\alpha |\Phi_0|^4$, and, hence, admit the first integral $H = const$.

However, in general case there is another integral of the equation (2.13) also, which can be found, multiplying this equation and conjugate to it accordingly on Φ_0^* and Φ_0, and then summing the obtained expressions. As a result, it appears that after transition to variable Φ_0, despite of a dissipation presence, $|\Phi_0|^2 = N = const$. Hence, the equation (2.13) is the integrable system and its solution has a form:

$$\Phi_0 = \sqrt{N} e^{-\frac{3i\alpha}{2\gamma} N(e^{-2\gamma\tau_1} - 1)}.$$

The appropriate functions φ_0, ψ_0 are written as follows:

$$\varphi_0 = \sqrt{N} e^{-\gamma\tau_1} e^{\frac{3i}{2\gamma} N(e^{-2\gamma\tau_1} - 1)}; \qquad \psi_0 = \sqrt{N} e^{-\gamma\varepsilon\tau} e^{i\left[1 - \frac{3\alpha}{2\gamma} N(e^{-2\gamma\varepsilon\tau} - 1)\right]}. \qquad (2.15)$$

After selection of real and imaginary parts this outcome completely coincides with a solution obtained in the 1-st approximation on the basis of the real equations by an averaging method in[13].

The advantages of the complex equations, noticeable already at this stage (reducing of the system order by factor two), especially are exhibited at a construction of higher approximations. For the determination of function φ_1 we return to the equation (2.8). After a substitution of expressions for φ_0 in the left part (2.8) the latter has a form

$$\frac{\partial \varphi_1}{\partial \tau} = -\gamma e^{-2i\tau} \varphi_0^* - i\alpha \left(e^{2i\tau} \varphi_0^3 + 3 e^{-2i\tau} |\varphi_0|^2 \varphi_0^* + e^{-4i\tau} \varphi_0^{*3} \right).$$

Then $\int_0^\pi \frac{\partial \varphi_1}{\partial \tau} d\tau = 0$ (i.e. $\Delta\varphi_1 = 0$ really) and the function $\varphi_1(\tau, \tau_1)$ can be written as follows:

$$\varphi_1 = -\frac{1}{2} i\gamma \varphi_0^* e^{-2i\tau} - \frac{1}{2} \alpha \left(\varphi_0^3 e^{2i\tau} - 3 |\varphi_0|^2 \varphi_0^* e^{-2i\tau} - \frac{1}{2} \varphi_0^{*3} e^{-4i\tau} \right) \qquad (2.16)$$

Let's consider still equation for the function $\varphi_2(\tau,\tau_1)$, equating zero coefficients at ε^2 in (2.7)

$$\frac{\partial \varphi_2}{\partial \tau_1} = -\frac{\partial \varphi_1}{\partial \tau_1} - \left[\tau_1 \frac{d\xi_1}{d\tau_1} + \xi_1(\tau_1) \right] \frac{d\varphi_0}{d\tau_1} - \gamma\varphi_1 - \gamma e^{-2i\tau} \varphi_1^* - 3i\alpha e^{2i\tau} \varphi_0^2 \varphi_1 +$$

$$+ 3i\alpha \left(2|\varphi_0|^2 \varphi_1 + \varphi_0^2 \varphi_1^* \right) - 3i\alpha e^{-2i\tau} \left(2|\varphi_0|^2 \varphi_1^* + \varphi_0^{*2} \varphi_1 \right) + 3i\alpha e^{-4i\tau} \varphi_0^{*2} \varphi_1^* \qquad (2.17)$$

In the right part of the equation (2.17) all magnitudes except for function $\xi_1(\tau_1)$ are known. It contains the resonance terms which lead to solution for $\varphi_2(\tau,\tau_1)$ growing in fast time. However, due to presence of not yet defined function $\xi_1(\tau_1)$, it is possible to convert in zero a sum of resonance terms. In limiting cases $\alpha = 0$ and $\gamma = 0$ the solutions for function $\xi_1(\tau_1)$ are degenerated in constants, in particular $\xi_1(\tau_1) = \frac{1}{2}i\gamma$ for linear system.

Generally function $\xi_1(\tau_1)$ satisfies to the linear differential equation of the first order:

$$\frac{d\xi_1}{d\tau_1} + \frac{1}{\tau_1}\xi_1 = -\frac{Q(\tau_1)}{\tau_1 \frac{d\varphi_0}{d\tau_1}},$$

which particular solution has a form:

$$\xi(\tau_1) = \frac{1}{\tau_1}\int \frac{Q(\tau_1)}{\frac{d\varphi_0}{d\tau_1}} d\tau_1,$$

where $Q(\tau_1)$ is the sum of resonance terms (i.e. terms, not containing fast time, in the equation (2.17)), which we do not write out here because of its unwieldy form.

2. Nonlinear oscillator in a harmonic external field.

Let oscillator, considered in item 1, is now in a periodic external field $F(t) = 2\varepsilon f \cos\omega t$. By introducing of dimensionless variables $\tau = \omega_0 t$, $u = \frac{U}{f/c_1}$ we rewrite the equation of motion as

$$\frac{d^2 u}{d\tau^2} + 2\varepsilon\gamma \frac{du}{d\tau} + u + 8\alpha\varepsilon u^3 = \varepsilon\left(e^{i\frac{\omega}{\omega_0}\tau} + e^{-i\frac{\omega}{\omega_0}\tau}\right), \qquad (2.18)$$

where

$$\varepsilon\gamma = \frac{n}{\sqrt{c_1 m}}; \quad 8\alpha\varepsilon = \frac{c_3}{c_1^3}f^2.$$

Let's assume that $\varepsilon \ll 1$ and $l\omega - m\omega_0 = \beta_1\varepsilon\omega_0$, where l and m are integer mutually prime numbers $\beta_1 = O(1)$. Case 1) $l = m = 1$ corresponds to a principal resonance; cases 2) $m = 1$, $l > 1$, 3) $l = 1$, $m > 1$ and 4) $l > 1$, $m > 1$ – sub-harmonic, ultra-harmonic and sub-ultra-harmonic resonances accordingly. Then the complex representation of the equation of motion has a form:

$$\frac{d\psi}{d\tau} - i\psi + \varepsilon\gamma(\psi + \psi^*) + i\alpha\varepsilon(\psi - \psi^*)^3 = \varepsilon\left(e^{i\frac{1}{l}(m+\beta_1\varepsilon)\tau} + e^{-i\frac{1}{l}(m+\beta_1\varepsilon)\tau}\right) \qquad (2.19)$$

After the change of a variable

$$\psi = e^{i\tau}\varphi(\tau) \qquad (2.20)$$

we obtain

$$\frac{d\varphi}{d\tau} + \varepsilon\gamma\,(\varphi + e^{-2i\tau}\varphi^{*}) + i\alpha\,(e^{2i\tau}\varphi^{3} - 3|\varphi|^{2}\varphi + 3e^{-2i\tau}|\varphi|^{2}\varphi^{*} - e^{-4i\tau}\varphi^{*3}) =$$

$$= \varepsilon\left(e^{i\left(\frac{m}{l}-1\right)\tau}e^{i\beta_{1}\varepsilon\tau} + e^{-i\left(\frac{m}{l}+1\right)\tau}e^{-i\beta\varepsilon\tau}\right); \qquad \beta = \beta_{1}/l \qquad (2.21)$$

As well as in case of a free oscillator we introduce now a slow time (2.5). Also search for a solution as an expansion on a small parameter ε (2.6).
Substituting (2.5), (2.6) in the equation (2.21) we have

$$\frac{\partial}{\partial\tau}(\varphi_{0} + \varepsilon\varphi_{1} + \varepsilon^{2}\varphi_{2} + \ldots) + \varepsilon L(\tau_{1},\varepsilon)\cdot\frac{\partial}{\partial\tau_{1}}(\varphi_{0} + \varepsilon\varphi_{1} + \varepsilon^{2}\varphi_{2} + \ldots) +$$

$$+ \varepsilon M(\varphi_{1},\varphi_{1}^{*},\varepsilon) = \varepsilon r(\tau,\tau_{1},\varepsilon), \qquad (2.22)$$

where

$$r(\tau,\tau_{1},\varepsilon) = \varepsilon\left(e^{i\left(-1+\frac{m}{l}\right)\tau}e^{i\beta\,(1-\xi_{1}(\tau_{1})\varepsilon+\ldots)\tau_{1}} + e^{-i\left(1+\frac{m}{l}\right)\tau}e^{-i\beta\,(1-\xi_{1}(\tau_{1})\varepsilon+\ldots)\tau_{1}}\right) \qquad (2.23)$$

1) Main resonance.

If $\dfrac{m}{l} = 1$ we come to case of a main resonance. Then, equating zero coefficients at various powers of a small parameter ε we obtain:

ε^{0}: $\dfrac{\partial\varphi_{0}}{\partial\tau} = 0$, therefore $\varphi_{0} = \varphi_{0}(\tau_{1})$

ε^{1}: $\dfrac{d\varphi_{0}}{d\tau_{1}} + \gamma\varphi_{0} - 3i\alpha|\varphi_{0}|^{2}\varphi_{0} - e^{i\beta\tau_{1}} = -\dfrac{\partial\varphi_{1}}{\partial\tau_{1}} - \gamma e^{-2i\tau}\varphi_{0}^{*} -$

$$\qquad (2.24).$$

$$-i\alpha(e^{2i\tau}\varphi_{0}^{3} + 3e^{-2i\tau}|\varphi_{0}|^{2}\varphi_{0}^{*} - e^{-4i\tau}\varphi_{0}^{*3}) + e^{-2i\tau}e^{-i\beta\tau_{1}}$$

Integrating both parts of the equation (2.24) on a fast variable τ from 0 up to π we obtain in a principal approximation on "slow" time τ_{1} (as well as in the case of free oscillations, the function φ_{1} is determined in such a manner that $\Delta\varphi_{1} = 0$):

$$\frac{d\varphi_{0}}{d\tau_{1}} + \gamma\varphi_{0} - 3i\alpha|\varphi_{0}|^{2}\varphi_{0} = e^{i\beta\tau_{1}} \qquad (2.25)$$

Let's introduce in (2.25) the change of variable

$$\varphi_0 = i\Phi_0(\tau_1)e^{i\beta\tau_1}. \tag{2.26}$$

Then we come to the autonomic equation:

$$\frac{d\Phi_0}{d\tau_1} + (\gamma + i\beta)\Phi_0 - 3i\alpha|\Phi_0|^2\Phi_0 + i = 0,$$

which at $\gamma = 0$ together with conjugate equation form the integrable Hamiltonian system:

$$\frac{d\Phi_0}{d\tau_1} = \frac{\partial H}{\partial \Phi_0^*}; \quad \frac{d\Phi_0^*}{d\tau_1} = -\frac{\partial H}{\partial \Phi_0},$$

admitting the first integral:

$$H = -i\beta|\Phi_0|^2 + 3/2\, i\alpha|\Phi_0|^4 - i(\Phi_0 + \Phi_0^*) = const.$$

In the general ($\gamma \neq 0$) the representation of required function as

$$\Phi_0 = a(\tau_1)e^{i\delta(\tau_1)},$$

where $a(\tau_1)$ and $\delta(\tau_1)$ are real functions, leads to the system of the nonlinear equations:

$$\frac{da}{d\tau_1} + \gamma a - \sin\delta = 0$$

$$a\frac{d\delta}{d\tau_1} + \beta a - 3\alpha a^3 + \cos\delta = 0$$

The stationary points of this system corresponding to steady-state oscillations satisfy to a cubic equation for the squared amplitude:

$$b^3 - \alpha_2 \beta^2 + \alpha_1 \beta - \alpha_0 = 0,$$

where

$$b = a^2; \quad \alpha_0 = \frac{1}{9\alpha^2}; \quad \alpha_1 = \frac{1}{9\alpha^2}(\gamma^2 + \beta^2); \quad \alpha_2 = \frac{2\beta}{3\alpha},$$

and also to transcendental equation:

$$\sin\delta = \gamma a,$$

determining the phase δ through amplitude.

The simple analysis shows, that if $\alpha \leq 2/3\gamma^3$, the condition of uniqueness of the real solution of the given cubic equation is fulfilled irrespective of magnitude of frequency parameter β. If $\alpha > 2/3\gamma^3$, at each fixed pair of α and γ, there are regions of rather small and rather large values β in which the real solution of a cubic equation also is single, but between these regions the values β corresponding to three real roots, i.e. three steady state regimes (resonance, unstable and nonresonance), are located. Thus the maximal amplitude of a resonance mode is determined by a relation $a = 1/\gamma$.

If $\gamma \to 0$ the value β corresponding to upper (at $\alpha > 0$) and low (at $\alpha < 0$) boundaries of a nonuniqueness region go to $\pm \infty$, as well as amplitude of a resonance mode.

The lower (at $\alpha > 0$) and upper (at $\alpha < 0$) boundaries of this region are determined by values of frequency parameter $\beta = \mp(9/2)^{2/3} \alpha$ (if $\alpha < 0$ the boundary value β is also negative).

At $\gamma = 0$, when the equation concerning function Φ_0 represents integrable system, the phase of the stationary solution is determined by a condition $\delta = k\pi$ $(k = 0, \pm 1,...)$, so the amplitude satisfies to a cubic equation:

$$3\alpha a^3 - \beta a \mp 1 = 0.$$

Obviously, in this case the stable stationary points on a phase plane (a, δ), due to presence of first integral $H = const$, are enclosed by a set of trajectories, which equation has a form

$$\beta a^2 - \frac{3}{2}\alpha a^4 - a\cos\delta = C.$$

The regions close to various unstable stationary points, corresponding to some critical values of a constant C, are divided by separatrixes transiting through unstable stationary points.

The analysis of a main resonance in the principal approach is in a full accordance with the results for phenomenological model[5]. However, within the framework of the asymptotic approach the opportunity of construction of higher approaches is unclosed also.

For determination of the first correction to principal approximation, determined by the function φ_1, let us consider the equation (2.24). After substitution in it the expression for φ_0 one can obtain:

$$\frac{\partial \varphi_1}{\partial \tau} = -\gamma c^* e^{-2i\tau} e^{-i\beta\tau_1} - i\alpha \left(e^{2i\tau} c^3 e^{3i\beta\tau_1} + 3e^{-2i\tau} |c|^2 c e^{-i\beta\tau_1} - e^{-4i\tau} c^{*3} e^{-3i\beta\tau_1} \right) + e^{-2i\tau} e^{-i\beta\tau_1}.$$

From here the first correction to a principal approximation can be found by integration on a fast variable τ:

$$\varphi_1(\tau, \tau_1) = -\frac{1}{2} i\gamma c^* e^{-2i\tau} e^{-i\beta\tau_1} - \frac{\alpha}{2}\left(c^{*3} e^{2i\tau} e^{3i\beta\tau_1} - 3e^{-2i\tau} c^2 c e^{-i\beta\tau_1} + \frac{1}{2} e^{-4i\tau} c^{*3} e^{-3i\beta\tau_1} \right) + \frac{1}{2} i e^{-2i\tau} e^{-i\beta\tau_1}.$$

Thus the value $\Delta\varphi_1$ really is equal to zero. At a determination second and consequent corrections there are resonance terms which compensating is achieved by an appropriate choice of functions $\xi_i(\tau_1)$, similarly to case of a free oscillator.

2) Secondary resonances.

At $m/l \neq 1$ the resonant effect consists in excitation of steady-state oscillations with frequency, close to a natural frequency of linearized system, under action of external force, far on frequency. As the analysis shows it is possible only under condition that the intensity of external action has zero, and dissipative force -- at least, second order on parameter ε. Really, in such case of the equations for the first two approaches look like:

$$\frac{\partial \varphi_0}{\partial \tau} = e^{i\left(-1+\frac{m}{l}\right)\tau} e^{i\beta\tau_1} + e^{-i\left(1+\frac{m}{l}\right)\tau} e^{-i\beta\tau_1} \qquad (2.27)$$

$$\frac{d\varphi_0}{d\tau_1} - 3i\alpha\varphi_0|^2 \varphi_0 = -\frac{\partial\varphi_1}{\partial\tau} - \gamma e^{-2i\tau}\varphi_0^* - i\alpha\left(e^{2i\tau}\varphi_0^3 + 3e^{-2i\tau}|\varphi_0|^2\varphi_0^* - e^{-4i\tau}\varphi_0^{*3}\right), \qquad (2.28)$$

The general solution of the linear nonhomogeneous equation (2.27) is written as follows:

$$\varphi_0 = \tilde{\varphi}_0 + i\left(\frac{1}{1-\frac{m}{l}} e^{i\left(-1+\frac{m}{l}+\varepsilon\beta\right)\tau} + \frac{1}{1+\frac{m}{l}} e^{-i\left(1+\frac{m}{l}+\varepsilon\beta\right)\tau}\right). \qquad (2.29)$$

Substituting this expression in the equation (2.28) and fulfilling integration on fast variable τ in limits from 0 up πl we obtain in main approach on slow variable τ_1:

$$\frac{d\tilde{\varphi}_0}{d\tau_1} - \frac{24i\alpha}{\left[1-(m/l)^2\right]^2}\tilde{\varphi}_0 - 3i\alpha|\tilde{\varphi}_0|^2\tilde{\varphi}_0 = 0 \qquad (2.30)$$

(as well as in case of a main resonance it is possible to be convinced, that $\Delta\varphi_1 = 0$).

After transformation $\tilde{\varphi}_0 = e^{\frac{24i\alpha}{\left[1-(m/l)^2\right]^2}}\Phi_0$ we come to already considered in section (2.1) equation (at $\gamma = 0$)

$$\frac{d\Phi_0}{d\tau_1} - 3i\alpha|\Phi_0|^2\Phi_0 = 0$$

with the stationary solution:

$$\Phi_0 = \sqrt{N} e^{3i\alpha N \tau_1}.$$

Then $\tilde{\varphi}_0 = \sqrt{N} e^{i\left(\frac{24\alpha}{\left[1-(m/l)^2\right]^2}+3\alpha N\right)\tau_1}$, and the requirement of periodicity is satisfied, if

$$N = \frac{\beta}{3\alpha} - \frac{8}{\left[1-(m/l)^2\right]^2}.$$

At realization of this requirement in a steady-state mode the combination of oscillations with frequency, close to a natural frequency of linear system and frequency of external excitation is implemented though the period of external excitation in l/m times more (less) than a period of natural linear oscillations. Or else, as well as in case of a main resonance the speech goes about "supported" natural oscillations, but accompanied in this case by forced oscillations with frequency far from a natural frequency of linear system.

The analysis is valid so long as the magnitude m/l itself does not become a small or large parameter. In the latter case it should be taken into account from the very beginning at the asymptotic analysis.

As for existence of combination resonances it is required still, as it was underlined above, the smallness of dissipative forces on a comparison with driving forces (influence former ones should be exhibited only in the equation concerning the second order corrections φ_2), practically observable there can be only the resonances of low order. Let's note, that using the complex representations the research of resonance conditions becomes considerably simpler on a comparison with a standard averaging method[13], due to reducing of the equation order by coefficient two. Especially the treatment of combination resonances becomes noticeably simpler.

3. Parametrical excitation of a nonlinear system.

Let's consider now case of parametrical excitation of a nonlinear oscillator, when the equation of motion in dimensionless variables has a form:

$$m\ddot{U} + 2n\dot{U} + (c_1 + p\cos\omega t)U + c_3 U^3 = 0 \qquad (2.31)$$

Again we introduce the change of independent and dependent variables,

$$\tau = \omega_0 t, \quad u = \frac{U_0}{\sqrt{E_0/c_1}}, \quad E_0 = \frac{1}{2}m\dot{U}_0^2 + \frac{1}{2}c_1 U_0^2 + \frac{1}{4}c_3 U_0^4,$$

$$\frac{d^2 u}{d\tau^2} + 2\varepsilon\gamma\frac{du}{d\tau} + \left(1 + 4\varepsilon\cos\frac{\omega}{\omega_0}\tau\right)u + 8\varepsilon\alpha\, u^3 = 0, \qquad (2.32)$$

where $\varepsilon\gamma = \dfrac{nm_0}{c_1}$; $\varepsilon = \dfrac{p}{c_1}$, $8\varepsilon\alpha = \dfrac{c_3 E_0}{c_1}$,

We write now the equations for complex functions: $\psi(\tau), \psi^*(\tau)$:

$$\frac{\partial\psi}{\partial\tau} - i\psi + \varepsilon\gamma(\psi+\psi^*) + i\alpha\varepsilon(\psi-\psi^*)^3 - i\varepsilon\left(e^{i\frac{\omega\tau}{\omega_0}} + e^{-i\frac{\omega\tau}{\omega_0}}\right)\times(\psi-\psi^*) = 0 \qquad (2.33)$$

Let $l\omega - m\omega_0 = \beta_1 \varepsilon \omega_0$ where l and m – integer mutually prime numbers, $\varepsilon \ll 1$, $\beta_1 = O(1)$. As well as earlier we introduce the change of a dependent variable $\psi = e^{i\tau}\varphi(\tau)$. Then

$$\frac{d\varphi}{d\tau} + \varepsilon\gamma_1\,(\varphi + e^{-2i\tau}\varphi^*) + i\alpha\varepsilon(e^{2i\tau}\varphi^3 - 3|\varphi|^2\varphi + 3e^{-2i\tau}\varphi^2\varphi^* - e^{-4i\tau}\varphi^{*3}) -$$

$$-i\varepsilon\left(e^{i\left(-1+\frac{m}{l}\right)\tau}e^{-i\beta\varepsilon\tau} + e^{-i\left(1+\frac{m}{l}\right)\tau}e^{-i\beta\varepsilon\tau}\right)\cdot\left(\varphi e^{i\tau} - e^{-i\tau}\varphi^*\right) = 0\,; \qquad \beta = \frac{\beta_1}{l} \qquad (2.34)$$

Again we introduce a "slow" time

$$\tau_1 = \tau\varepsilon\left[1 + \xi_1(\tau_1)\varepsilon + \xi_2(\tau_1)\varepsilon^2 + \ldots\right] \qquad (2.35)$$

Also we present a solution as

$$\varphi(\tau,\tau_1) = \varphi_0 + \varepsilon\varphi_1 + \varepsilon^2\varphi_2 + \ldots \qquad (2.36)$$

With allowance for (2.34), (2.35) and (2.36) we obtain

$$\varepsilon L(\tau_1,\varepsilon)\cdot\frac{\partial}{\partial\tau_1}(\varphi_0 + \varepsilon\,\varphi_1 + \varepsilon^2\varphi_2 + ..) + \varepsilon N(\varphi_1,\varphi_1^*,\varepsilon) -$$

$$\varepsilon L(\tau_1,\varepsilon)\cdot\frac{\partial}{\partial\tau_1}(\varphi_0 + \varepsilon\,\varphi_1 + \varepsilon^2\varphi_2 + ..) + \varepsilon N(\varphi_1,\varphi_1^*,\varepsilon) - \qquad (2.37)$$

$$-\varepsilon r(\tau,\tau_1,\varepsilon)\times\left[e^{i\tau}(\varphi_0 + \varepsilon\varphi_1 + \varepsilon^2\varphi_2 + ..) - e^{-i\tau}(\varphi_0^* + \varepsilon\varphi_1^* + \varepsilon\varphi_2^* + ..)\right] = 0$$

The main parametrical resonance will be realized at $m/l=2$. As well as in the previous cases, $\varphi_0 = \varphi_0(\tau_1)$, and the equation of a principal approximation in slow time has a form:

$$\frac{d\varphi_0}{d\tau_1} + \gamma\varphi_0 - 3i\alpha|\varphi_0|^2\varphi_0 + ie^{i\beta\tau_1}\varphi_0^* = 0. \qquad (2.38)$$

By the change of variables $\varphi_0 = \Phi_0 e^{(-\gamma + i\beta)\tau_1}$ we come to the equation:

$$\frac{d\Phi_0}{d\tau_1} + i\left(\frac{\beta}{2}\Phi_0 + \Phi_0^*\right) - 3i\,\alpha e^{-2\gamma\tau_1}|\Phi_0|^2\Phi_0 = 0. \qquad (2.39)$$

Let's assume $\Phi_0 = a(\tau_1)e^{-i\delta(\tau_1)}$, where $a(\tau_1)$ and $\delta(\tau_1)$ – real functions. Substituting this expression in the equation of motion (2.39) after simple transformations we obtain the system of real equations:

$$\frac{da}{d\tau_1} = -a\sin 2\delta$$

$$\frac{d\delta}{d\tau_1} = -\beta/2 - \cos 2\delta + 3\alpha\, a^2 e^{-2\gamma\tau_1}. \qquad (2.40)$$

At $\gamma = 0$ the equation (2.39) and equation conjugated to it form the Hamiltonian system:

$$\frac{d\Phi_0}{d\tau_1} = \frac{dH}{d\Phi_0^*}; \quad \frac{d\Phi_0^*}{d\tau_1} = -\frac{dH}{d\Phi_0};$$

where

$$H = \frac{i}{2}\left[-\beta|\Phi_0|^2 + 3\alpha|\Phi_0|^4 - \left(\Phi_0^2 + \Phi_0^{*2}\right)\right]$$

and, hence, admit first integral $H = const$.

Hence, variable $a(\tau_1)$ and $\delta(\tau_1)$ are connected by a relation:

$$-\beta a^2 + 3\alpha a^4 - 2a^2 \cos 2\delta = const,$$

representing the equation of a phase trajectory on a plane (a,δ). The stationary points on a phase plane are determined as follows:

$$\delta = 0, \pm 1, \pm 2, \ldots; \qquad a^2 = \pm\frac{1}{3\alpha} + \frac{\beta}{6}.$$

It is obvious that for $\alpha \neq 0$ all trajectories on a phase plane are restricted. The conditions for determination of stationary regimes are

$$\alpha \neq 0 \quad 1) \qquad \delta = 0; \qquad a^2 = \frac{1}{3\alpha} + \frac{\beta}{6}.$$

$$2) \qquad \delta = \pi; \qquad a^2 = -\frac{1}{3\alpha} + \frac{\beta}{6}.$$

The first type regime is stable (it exists when $\dfrac{|\beta|}{2} < \dfrac{1}{\alpha}$), but the second type one is unstable (it appears when $\dfrac{|\beta|}{2} > \dfrac{1}{\alpha}$).

The close equation is postulated (without a derivation) in [5], where its full analysis is performed. Obviously, it is applicable only in a neighbourhood of a main resonance. The asymptotic approach allows to obtain the equation for secondary parametrical resonances also.

4. Self-excited oscillations.

We will show now how it is possible to obtain the complex equation in slow time for a self-oscillating system described in a dimensionless form by the Van der Pol equation:

$$\frac{d^2u}{d\tau^2} - 2\varepsilon(1-2bu^2)\frac{du}{d\tau} + u = 0, \quad b > 0 \tag{2.41}$$

Let's introduce the complex functions ψ and ψ^* again. Then

$$\frac{\partial \psi}{\partial \tau} - i\psi - \varepsilon(\psi + \psi^*)\left[1 + \frac{1}{2}b(\psi - \psi^*)^2\right] = 0 \tag{2.42}$$

After a change of dependent variable $\psi = e^{i\tau}\varphi(\tau)$ we obtain

$$\frac{\partial \varphi}{\partial \tau} - \varepsilon(\varphi + e^{-2i\tau}\varphi^*)\left[1 + \frac{1}{2}b(e^{i\tau}\varphi - e^{-i\tau}\varphi^*)^2\right] = 0 \tag{2.43}$$

Itroducing of "slow" time

$$\tau_1 = \varepsilon\tau\left[1 + \xi_1(\tau_1)\varepsilon + \xi_2(\tau_1)\varepsilon^2 + \ldots\right] \tag{2.44}$$

and considering the function φ as function of two times we search for a solution as

$$\varphi = \varphi_0 + \varepsilon\varphi_1 + \varepsilon^2\varphi_2 + \ldots \tag{2.45}$$

After a substitution (2.44) and (2.45) in the equation (2.43) we obtain:

$$\frac{\partial}{\partial \tau}(\varphi_0 + \varepsilon\varphi_1 + \varepsilon^2\varphi_2 + \ldots) + \varepsilon L(\tau_1,\varepsilon) \times \frac{\partial}{\partial \tau_1}(\varphi_0 + \varepsilon\varphi_1 + \varepsilon^2\varphi_2 + \ldots) -$$

$$- \varepsilon\left[(\varphi_0 + \varepsilon\varphi_1 + \varepsilon^2\varphi_2 + \ldots) + e^{-2i\tau}(\varphi_0^* + \varepsilon\varphi_1^* + \varepsilon^2\varphi_2^* + \ldots)\right] \times$$

$$\times \left\{1 + \frac{1}{2}b\left[e^{i\tau}(\varphi_0 + \varepsilon\varphi_1 + \varepsilon^2\varphi_2 + \ldots) - e^{-i\tau}(\varphi_0^* + \varepsilon\varphi_1^* + \varepsilon^2\varphi_2^* + \ldots)\right]^2\right\} = 0$$

As well as earlier $\frac{\partial \varphi_0}{\partial \tau} = 0$ and $\varphi_0 = \varphi_0(\tau_1)$. A condition of equality to zero of coefficient at the first power of a parameter ε leads to equation

$$\frac{d\varphi_0}{d\tau_1} - \varphi_0(1 - b|\varphi_0|^2) = -\frac{\partial \varphi_1}{\partial \tau} + e^{-2i\tau}\varphi_0^* \cdot b(e^{2i\tau}\varphi_0^3 + e^{-2i\tau}\varphi_0^2\varphi_0^* + e^{-4i\tau}\varphi_0^{*3}) \tag{2.46}$$

Integrating the equation (2.46) on fast time in limits from 0 up to π we come to the equation of a principal approximation in slow time:

$$\frac{d\varphi_0}{d\tau_1} - \varphi_0(1 - b|\varphi_0|^2) = 0 \tag{2.47}$$

with a particular solution $\varphi_0 = b^{-1} e^{-i\theta_0}$ (limit cycle), where θ_0 – is arbitrary phase. Returning to the variable ψ we have

$$\psi_0 = b^{-1} e^{i(\tau - \theta_0)}.$$

The non-stationary solution (approaching to the limit cycle) is also easily obtained from (2.47). Let's note that in[5] the equation close to (2.47) is obtained as a result of transformation of some modified system which is distinct from the equation (2.41). It appears however that (2.47) is a main approximation to the Van der Pol equation in "slow" time.

III. SYSTEM OF TWO WEAKLY COUPLED NONLINEAR OSCILLATORS

The efficiency of complex representation of the equations of motion becomes more and more obvious with complication of the system considered. It was convincingly shown in the monograph[5] that this representation allows treating with a maximum simplicity the complicated problems concerning to a bifurcation of solutions in a system of two weakly coupled anharmonic oscillators. However, in this case the complex representations are also considered phenomenologically as model systems, which at a qualitative level only are connected with known realistic models.

Let's consider a system of two identical weakly coupled nonlinear oscillators. Their dynamics is described by the equations:

$$m\ddot{U}_1 + 2n\dot{U}_1 + c_1 U_1 + c_3 U_1^3 + c_{12}(U_1 - U_2) = 0$$

$$m\ddot{U}_2 + 2n\dot{U}_2 + c_1 U_2 + c_3 U_2^3 + c_{12}(U_2 - U_1) = 0 \qquad (3.1)$$

In-phase $(U_2 = U_1)$ and out-of phase $(U_2 = -U_1)$ nonlinear normal modes can be immediately found due to the symmetry of a considered system. Because the principle of superposition in the nonlinear theory is not valid, the corresponding partial solutions can not be used for a construction of a general solution.

Using the complex representation of unknown functions, we obtain

$$\frac{d\psi_j}{d\tau} - i\psi_j + \varepsilon\gamma i\left(\psi_j + \psi_j^*\right) + i\alpha\varepsilon\left(\psi_j - \psi_j^*\right)^3 - i\beta\varepsilon\left[\left(\psi_j - \psi_j^*\right) - \left(\psi_k - \psi_k^*\right)\right] = 0, \quad (3.2)$$

$$(j, = 1,2; \qquad k = 3 - j)$$

where

$$\psi_j = v_j + iu_j; \qquad \psi_j^* = v_j - iu_j \qquad (j = 1,2); \qquad v_j = \frac{dv_j}{d\tau}; \qquad i = \sqrt{-1}$$

$$\tau = \omega_0 t; \quad u_i = \frac{U_i}{\sqrt{E_0/c_1}}; \quad E_0 \text{ is initial energy of a system,}$$

$$\varepsilon\gamma = \frac{n}{\sqrt{c_1 m}}; \quad 8\alpha\varepsilon = \frac{c_3}{c_1^2} \cdot E_0; \quad 2\varepsilon\beta = \frac{c_{12}}{c_1}$$

Introducing the new variables $\psi_j = e^{i\tau}\varphi_j$ and "slow" time

$$\tau_1 = \varepsilon\tau\left[1 + \xi_1(\tau_1)\varepsilon + \xi_2(\tau_1)\varepsilon^2 + \ldots\right], \tag{3.3}$$

we consider φ_j as a functions of two times τ and τ_1. Let's present a solution as a form of power expansion

$$\varphi_j(\tau,\tau_1) = \varphi_{j,0} + \varepsilon\,\varphi_{j,1} + \varepsilon^2\,\varphi_{j,2} + \ldots \tag{3.4}$$

The same transformations and operations as in case of an isolated oscillator lead to the following main approximation in slow time:

$$\frac{d\varphi_{j,0}}{d\tau_1} + \gamma\varphi_{j,0} - 3\alpha i |\varphi_{j,0}|^2 \varphi_{j,0} - i\beta(\varphi_{j,0} - \varphi_{k,0}) = 0; \qquad \psi_{j,0} = e^{i\tau_1}\varphi_{j,0} \tag{3.5}$$

The system which is close to (3.5) (at $\gamma = 0$) was considered in[5] as an example of phenomenological model ("a discrete model with self-localization"). It was shown that this system is integrable. Besides, the localized normal modes arise in addition to in-phase and out-of phase normal modes at some values of a coupling parameter β. Their appearance is a consequence of instability of in-phase or out-of phase normal modes dependent on the sign of parameter α)[3-5].

Let's discuss the influence of damping forces in more details. Introducing in (3.5) the new variable,

$$\varphi_{j,0} = e^{(i\beta-\gamma)\tau_1}\Phi_{j,0} \qquad (j=1,2),$$

we obtain

$$\frac{d\Phi_{1,0}}{d\tau_1} + i\beta\Phi_{2,0} - 3i\alpha e^{-2\gamma\tau_1}|\Phi_{1,0}|^2\Phi_{1,0} = 0$$

$$\frac{d\Phi_{2,0}}{d\tau_1} + i\beta\Phi_{1,0} - 3i\alpha e^{-2\gamma\tau_1}|\Phi_{2,0}|^2\Phi_{2,0} = 0. \tag{3.6}$$

Let's note, that the Lagrangian corresponding to the equations of motion (3.6) at $\gamma = 0$ has a form:

$$L = \frac{1}{2}i\sum_{j=1}^{2}\left(\Phi_{j,0}\frac{d\Phi^*_{j,0}}{d\tau_1} - \Phi^*_{j,0}\frac{d\Phi_{j,0}}{d\tau_1}\right) - \beta\left(\Phi_{1,0}\Phi^*_{2,0} + \Phi_{2,0}\Phi^*_{1,0}\right) + \frac{3}{2}\alpha\sum_{j=1}^{2}\Phi^4_{j,0}, \tag{3.7}$$

and the integral of energy is written as follows:

$$H = \beta(\Phi_1\Phi_2^* + \Phi_2\Phi_1^*) - \frac{3}{2}\alpha\sum_{j=1}^{2}|\Phi_{j,0}|^4 = const \qquad (3.8)$$

Exact integrability of a system (3.6) at $\gamma = 0$ is a consequence of existence of energy integral (3.8) and integral[5]:

$$N = \sum_{j=1}^{2}|\Phi_{j,0}|^2 = const \qquad (3.9)$$

Returning to a full system (3.6) ($\gamma \neq 0$) we come to a conclusion that the integral (3.9) is preserved in common case also. Really, multiplying 1st and 2nd equations (3.6) respectively on $\Phi_{1,0}^*$ and $\Phi_{2,0}^*$, fulfilling the operation of conjugation and combining all four equations we obtain

$$\frac{dN}{d\tau_1} = 0, \text{ or } N = const \qquad (3.10)$$

Therefore, as in the case of zero damping [3,5], the unknown functions $\Phi_{1,0}$ and $\Phi_{2,0}$ can be presented as

$$\Phi_{1,0} = \sqrt{N}\cos\frac{\theta}{2}e^{i\delta_1(\tau_1)}; \Phi_{2,0} = \sqrt{N}\sin\frac{\theta}{2}e^{i\delta_2(\tau_1)}$$

Substituting these expressions in the equation (3.6) we obtain

$$-\frac{1}{2}i\,tg\frac{\theta}{2}\frac{d\theta}{d\tau_1} - \frac{d\delta_1}{d\tau_1} - \beta\,tg\frac{\theta}{2}e^{i(\delta_2-\delta_1)} + 3\alpha\,N\,e^{-2\gamma\tau_1}\cos^2\frac{\theta}{2} = 0$$

$$\frac{1}{2}i\,ctg\frac{\theta}{2}\frac{d\theta}{d\tau_1} - \frac{d\delta_2}{d\tau_1} - \beta\,ctg\frac{\theta}{2}e^{i(\delta_1-\delta_2)} + 3\alpha\,Ne^{-2\gamma\tau_1}\sin^2\frac{\theta}{2} = 0 \qquad (3.11)$$

Equating the real and imaginary parts of these relations to zero we come to the equations for real functions $\theta(\tau)$ and $\Delta\delta(\tau) = \delta_1(\tau) - \delta_2(\tau)$:

$$\frac{d\theta}{d\tau_1} = 2\beta\sin\Delta\,;$$

$$\frac{d\Delta}{d\tau_1} = 2\beta\,ctg\,\theta\cos\Delta + 3\alpha\,Ne^{-2\gamma\tau_1}\cos\theta \qquad (3.12)$$

The stationary points of a system in which right sides of the equations (3.12) are equal to zero are:

1) $\Delta = 0, \quad \theta = \frac{\pi}{2}$ \qquad 2) $\Delta = \pi, \quad \theta = \frac{\pi}{2}$

They correspond to in-phase and out-of-phase cooperative normal modes which are preserved in the presence of energy dissipation also. In the case $\gamma = 0$, the additional out-of-phase ($\alpha > 0$) or in-phase ($\alpha < 0$) normal modes appear[3-5]. They correspond to values,

$$\Delta = \pi \ \Delta = \pi \ (\alpha > 0), \Delta = \pi \ (\alpha < 0)$$

$$\sin\theta = \frac{2\beta}{3\alpha N} = \frac{N_0}{N},$$

where $N_0 = \frac{2\beta}{3\alpha}$, and describe the nonlinear oscillations which are more localized (with increase N) on a mass. Really, the ratio of complex functions $\Phi_{1,0}$ and $\Phi_{2,0}$ corresponding to additional modes is written as follows:

$$\frac{\Phi_{1,0}}{\Phi_{2,0}} = -\frac{\sqrt{1+\sqrt{1-\rho^2}}}{\sqrt{1-\sqrt{1-\rho^2}}}, \qquad \text{where } \rho = \frac{N_0}{N}$$

Obviously such modes arise only at $\rho < 1$, and at $\rho \to 0$ (that, naturally, falls outside the limits of applicability of the principal approximation) the full localization of excitation on a mass would be reached. All these aspects for phenomenological model were considered in [4] and [5]. The consideration of a dissipation reveals a new important aspect of the problem. In this case at $N \cdot e^{-2\gamma\tau_1} \geq N_0$ the energy of the system can also be initially localized on a mass. But because of presence of an exponential coefficient in the second equation of motion, in the instant $\tau_1 = \frac{1}{\gamma}\left(\ln\frac{N}{N_0}\right)$ the localization of excitation becomes impossible and the energy should be redistributed between both masses. This important effect can be observed in computer experiment (in some instant the domination of one mass is replaced by a redistribution of the energy between masses[14]).

Let's note that the determination of higher approximations can also be made without difficulty, as well as the principal approximations for cooperative and localized normal modes are known.

The direct computation of localized normal modes becomes difficult if the number of degrees of freedom is more than two. Therefore generalization on more complicated cases requires the more common approach to revealing the existence conditions of such modes. These conditions can be connected with a bifurcation of solutions and instability of cooperative modes. For the system considered the bifurcation analysis becomes more simple the equations (3.11), i.e. in terms of real functions $\theta(\tau_1)$, $\delta(\tau_1)$[5]. However, the appropriate generalization on the systems with many degrees of freedom becomes impossible. In this connection let us perform a bifurcation analysis using the complex equations (3.6) at $\gamma = 0$.

$$\frac{d\Phi_{1,0}}{d\tau_1} + i\beta\Phi_{2,0} - 3i\alpha|\Phi_{1,0}|^2\Phi_{1,0} = 0,$$

$$\frac{d\Phi_{2,0}}{d\tau_1} + i\beta\Phi_{1,0} - 3i\alpha|\Phi_{2,0}|^2\Phi_{2,0} = 0. \tag{3.13}$$

Let's stop on the analysis of a stability of out-of phase cooperative mode

$$\Phi_{2,0} = -\Phi_{1,0}$$

We consider the finite perturbations of this mode w_1 and w_2:

$$\Phi_{1,0} + w_1 \quad \text{and} \quad -\Phi_{1,0} + w_2 \tag{3.14}$$

Substituting expressions (3.14) in the equations of motion and taking into account relations (3.13), we obtain:

$$\frac{dw_1}{d\tau_1} + i\beta w_2 - 3i\alpha\left(\Phi_{1,0}^2 w_1^* + 2|\Phi_{1,0}|^2 w_1 + 2\Phi_{1,0}|w_1|^2 + w_1^2\Phi_{1,0}^* + |w_1|^2 w_1\right) = 0,$$

$$\frac{dw_2}{d\tau_1} + i\beta w_1 - 3i\alpha\left(\Phi_{1,0}^2 w_2^* + 2|\Phi_{1,0}|^2 w_2 - 2\Phi_{1,0}|w_2|^2 - w_2^2\Phi_{1,0}^* + |w_2|^2 w_2\right) = 0, \tag{3.15}$$

where $\Phi_{1,0} = \sqrt{N}e^{i(\beta+3\alpha N)\tau_1}$

By combining these equations and by denoting through W_1 and W_2 a sum and difference of w_1 and w_2 accordingly, we get the system of two nonlinear equations:

$$\frac{dW_1}{d\tau_1} + i(\beta - 6\alpha N)W_1 - 3i\alpha\Big[Ne^{-2i(\beta-3\alpha N)\tau_1}W_1^* +$$

$$+ \sqrt{N}e^{-i(\beta-3\alpha N)}\left(W_1W_2^* + W_2W_1^* + W_1W_2\right) + \frac{1}{2}\left(|W_1|^2 W_1 + W_2^2 W_1^*\right)\Big] = 0$$

$$\frac{dW_2}{d\tau_1} - i(\beta + 6\alpha N)W_2 - 3i\alpha\Big[Ne^{-2i(\beta-3\alpha N)\tau_1}W_2^* + \sqrt{N}e^{i(\beta-3\alpha N)\tau_1} \times \tag{3.16}$$

$$\times \left(|W_1|^2 + W_2W_2^* + W_1W_2\right) + \frac{1}{2}\left(|W_2|^2 W_2 + 2|W_1|^2 W_2 + 2|W_2|^2 W_1\right)\Big] = 0$$

The linearization of this system leads to two independent equations of parametrical oscillations concerning functions W_1 and W_2, describing perturbation of an initial mode on in-phase and out-of-phase modes accordingly.

$$\frac{dW_1}{d\tau_1} i(\beta - 6\alpha N) W_1 - 3i\alpha \left[Ne^{2i(\beta+3\alpha N)\tau_1} \right] W_1^* = 0$$

(3.17)

$$\frac{dW_1}{d\tau_1} - i(\beta + 6\alpha N) W_1 - 3i\alpha \left[Ne^{2i(\beta+3\alpha N)\tau_1} \right] W_1^* = 0$$

At a value of out-of-phase mode amplitude $\sqrt{N} = \sqrt{\beta/3\alpha}$, the conditions of parametrical resonance[5] are satisfied (the boundary is reached between the stability and instability regions) for first equation (3.17), i.e. the intensive transfer of an energy to in-phase mode occurs. In the second equation (3.17), as it is easy to be convinced, the instability occurs at infinitesimal amplitudes, but it is realized as a phase shift of out-of phase oscillations.

The same value of complex amplitude has appeared to be critical at a formulation of an existence condition of localized modes with the use of exact solution of the equations (3.6). But as it was shown below, the applied approach allows finding the conditions of energy localization not only in finite-dimensional systems of the high dimensionality, but also in infinite-dimensional models of quasi–one-dimensional crystal.

IV. NONLINEAR DYNAMICS OF AN INFINITE CHAIN OF COUPLED OSCILLATORS

The equations of motion of an infinite system of weakly coupled nonlinear oscillators with damping have a form:

$$m\ddot{U}_j + 2\tilde{\gamma}\dot{U}_j + c_1 U_j + c_3 U_j^3 + \tilde{c}(2U_j - U_{j-1} - U_{j+1}) = 0, \qquad (\infty < j < \infty). \qquad (4.1)$$

Let's introduce a change of variables:

$$\tau = \sqrt{\frac{c_1}{m}}; \quad u_j = \frac{U_j}{r_0}, \qquad (4.2)$$

where r_0 is distance between particles.
Then we have

$$\frac{d^2 u_j}{d\tau^2} + 2\varepsilon\gamma \frac{du_j}{d\tau} + u_j + 8\alpha\varepsilon u_j^3 + 2\varepsilon\beta(2u_j - u_{j-1} - u_{j+1}) = 0, \qquad (4.3)$$

where,

$$8\alpha\varepsilon = \frac{c_3}{c_1} \cdot r_0^2, \gamma\varepsilon = \frac{\tilde{\gamma}}{\sqrt{c_1 m}}, \varepsilon\beta = \tilde{c}/c_1, \varepsilon \ll 1.$$

Let's use now the complex representation of the motion equations:

$$\frac{d\psi_j}{d\tau} - i\psi_j + \varepsilon\gamma(\psi + \psi^*) + i\alpha\varepsilon(\psi - \psi^*)^3 - \varepsilon i\beta \left[2(\psi_j - \psi_j^*) - (\psi_{j-1} - \psi_{j-1}^*) - (\psi_{j+1} - \psi_{j+1}^*) \right] = 0$$

where $\psi_j = v_j + iu_j$; $\psi_j^* = v_j - iu_j$

By introducing new change of variables

$$\psi_j = e^{i\tau}\varphi_j,$$

we come to a set of equations

$$\frac{d\varphi_j}{d\tau} + \varepsilon\gamma\left(\varphi_j + e^{-2i\tau}\varphi_j^*\right) + i\alpha\alpha\left(e^{2i\tau}\varphi^3 - 3|\varphi|^2\varphi + 3e^{-2i\tau}|\varphi|^2\varphi^* - e^{-4i\tau}\varphi^{*3}\right) - $$
$$- i\varepsilon\varepsilon\left[\left(2\varphi_j - \varphi_{j-1} - \varphi_{j+1}\right) - e^{-2i\tau}\left(2\varphi_j^* - \varphi_{j-1}^* - \varphi_{j+1}^*\right)\right] = 0. \qquad (4.4)$$

Let's introduce a "slow" time

$$\tau_1 = \varepsilon\tau[1 + \xi_1(\tau_1)\varepsilon + \xi_2(\tau_1)\varepsilon^2 + \ldots]. \qquad (4.5)$$

Also we will consider φ_j as function of "fast" time τ and "slow" time τ_1. By presenting the unknown functions as expansions on the parameter ε:

$$\varphi_j = \varphi_{j,0} + \varepsilon\varphi_{j,1} + \varepsilon^2\varphi_{j,2} + \ldots \qquad (4.6)$$

and substituting (4.6) in the equations of motion we obtain

$$\frac{\partial}{\partial \tau}(\varphi_0 + \varepsilon\varphi_1 + \varepsilon^2\varphi_2 + \ldots) + \varepsilon L(\tau_1,\varepsilon)\cdot\frac{\partial}{\partial \tau_1}(\varphi_0 + \varepsilon\varphi_1 + \varepsilon^2\varphi_2 + \ldots) + \varepsilon M\left(\varphi_{j,l},\varphi_{j,l}^*,\varepsilon\right) - $$
$$- i\varepsilon\beta\{[2(\varphi_{j,0} + \varepsilon\varphi_{j,1} + \ldots) - (\varphi_{j-1,0} + \varepsilon\varphi_{j-1,1} + \ldots) - (\varphi_{j+1,0} + \varepsilon\varphi_{j+1,1} + \ldots) - \qquad (4.7)$$
$$- [2(\varphi_{j,0}^* + \varepsilon\varphi_{j,1}^* + \ldots) - (\varphi_{j-1,0}^* + i\varphi_{j-1,1}^* + \ldots) - (\varphi_{j+1,0}^* + \varepsilon\varphi_{j+1,1}^* + \ldots)]e^{-2i\tau}\} = 0,$$

where L and M were presented in the section II.

Equating to zero the coefficients at various degrees of a small parameter ε we have:

$$\varepsilon^0: \quad \frac{\partial \varphi_{j,0}}{\partial \tau} = 0, \quad \varphi_{j,0} = \varphi_{j,0}(\tau_1),$$

$$\varepsilon^1: \frac{d\varphi_{j,0}}{d\tau_1} + \gamma\varphi_{j,0} - 3\alpha i|\varphi_{j,0}|^2\varphi_{j,0} - i\beta(2\varphi_{j,0} - \varphi_{j+1,0} - \varphi_{j-1,0}) = -\frac{\partial\varphi_{j,1}}{\partial\tau} - \gamma e^{-2i\tau}\varphi_{j,0}^* - \qquad (4.8)$$

$$- i\alpha\left(e^{2i\tau}\varphi_{j,0}^3 + 3e^{-2i\tau}|\varphi_{j,0}|^2\varphi_{j,0}^* + e^{-4i\tau}\varphi_{j,0}^{*3}\right) + 2\beta\left(2\varphi_{j,0}^* - \varphi_{j-1,0}^* - \varphi_{j+1,0}^*\right)e^{-2i\tau}$$

$$\varepsilon^1: \frac{d\varphi_{j,0}}{d\tau_1} + \gamma\varphi_{j,0} - 3\alpha i|\varphi_{j,0}|^2\varphi_{j,0} - i\beta(2\varphi_{j,0} - \varphi_{j+1,0} - \varphi_{j-1,0}) = \frac{\partial\varphi_{j,1}}{\partial\tau} - \gamma e^{-2i\tau}\varphi_{j,0}^* -$$

$$-i\alpha\left(e^{2i\tau}\varphi_{j,0}^3 + 3e^{-2i\tau}|\varphi_{j,0}|^2\varphi_{j,0}^* + e^{-4i\tau}\varphi_{j,0}^{*3}\right) + 2\beta(2\varphi_{j,0}^* - \varphi_{j-1,0}^* - \varphi_{j+1,0}^*)e^{-2i\tau}$$

....

Integration of the second equation on the fast time in limits from 0 to π leads to equations of the principal approximation in "slow" time:

$$\frac{d\varphi_{j,0}}{d\tau_1} + \gamma\varphi_{j,0} - 3\alpha i|\varphi_{j,0}|^2\varphi_{j,0} - i\beta(2\varphi_{j,0} - \varphi_{j+1,0} - \varphi_{j-1,0}) = 0, \tag{4.9}$$

(the value $\Delta\varphi_1 = \frac{1}{\pi}[\varphi_{j,0}(\pi) - \varphi_{j,0}(0)]$, as it was shown below, may be put equal to zero).

Let's introduce now a new change

$$\varphi_{j,0} = e^{-\gamma\tau_1}\Phi_{j,0}. \tag{4.10}$$

Then the set of equations in the principal approximation is presented as follows:

$$\frac{d\Phi_{j,0}}{d\tau_1} - 3\alpha i e^{-2\gamma\tau_1}|\Phi_{j,0}|^2\Phi_{j,0} + i\beta(2\Phi_{j,0} - \Phi_{j+1,0} - \Phi_{j-1,0}) = 0 \tag{4.11}$$

Let's search for the functions $\Phi_{j,0}$ as

$$\Phi_{j,0} = f_0(\tau_1)e^{i(kr_0 j + \theta_0)} \tag{4.12}$$

where r_0 – the distance between particles, θ_0 – the phase corresponding to $j = 0$. Substituting (4.12) in (4.11), we obtain

$$\frac{df_0}{d\tau_1} - 3i\alpha e^{-2\gamma\tau_1}|f_0|^2 f_0 - 4i\beta\sin^2\frac{kr_0}{2}\cdot f_0 = 0 \tag{4.13}$$

After the change of variable

$$f_0 = e^{4i\beta\sin^2\frac{kr_0\tau_1}{2}} w_0(\tau_1)$$

we come to the equation

$$\frac{dw_0}{d\tau_1} - 3i\alpha e^{-2\gamma\tau_1}|w_0|^2 w_0 = 0, \tag{4.14}$$

which coincides with equation (2.10) for a system with one degree of freedom. Its solution has the form

$$w_0 = \sqrt{N} e^{-\frac{3i\alpha}{2\gamma}N(e^{-2\gamma\tau_1}-1)},$$

so that

$$\Phi_{j,0} = \sqrt{N}\Phi_{j,0} = \sqrt{N} e^{i\left[4\beta\sin^2\frac{kr_0}{2}\tau_1 - 3\alpha\frac{N}{2\gamma}(e^{-2\gamma\tau_1}-1) + kr_0 j + \theta_0\right]} \qquad (4.15)$$

Returning to the initial complex variable we obtain

$$\psi_{j,0} = \sqrt{N} e^{-\gamma\varepsilon\tau} e^{i\left\{\left[\tau - 3\alpha\frac{N}{2\gamma}(e^{-2\gamma\varepsilon\tau}-1)\right] + kr_0 j + 4\beta\sin^2\frac{kr_0+\theta_0}{2}\tau_1 + \theta_0\right\}} =$$

$$= \sqrt{N} e^{-\gamma\varepsilon\tau} e^{i\left[\left(\tau + 4\varepsilon\beta\tau\sin^2\frac{kr_0}{2}\right) - 3\alpha\frac{N}{2\gamma}(e^{-2\gamma\varepsilon\tau}-1) + kr_0 j + \theta_0\right]}. \qquad (4.16)$$

Thus, the analysis of the principal approximation in slow time leads to conclusion that in an infinite chain of the weakly coupled damped nonlinear oscillators the quasi-harmonic waves with exponentially decreasing amplitude and variable frequency can spread. At $\gamma = 0$ and $\alpha = 0$ they turn to usual harmonic waves

$$\psi_{j,0} = \sqrt{N} e^{i(\omega\tau + kr_0 j + \theta_0)}, \qquad (4.17)$$

where $\omega = 1 + 4\varepsilon\beta\sin^2\frac{kr_0}{2}$, with the spectrum of wave numbers $0 \leq kr_0 \leq \pi$, spectrum of frequencies $1 \leq \omega \leq 1 + 4\varepsilon$ and by constant amplitude. Spectrum of simple harmonic waves, determined by expression (4.17), is characterized by double degeneration of all modes, except of corresponding to wave numbers $k = 0$ and $k = \pi/r_0$. It means that the modes with various phases θ_0 having identical frequencies and wave numbers can be obtained by a superposition of two such modes. The introduction of nonlinearity removes the degeneration[5], since a dependence of normal oscillations frequencies on distribution of the particles energy arises which, generally speaking, is different for various values θ_0.

Let's note that the obtained solution turns out to be asymptotically grounded if the lengths of considered waves are not too large as compared with interparticle distances. In further we accept r_0 as unit of length, then this condition will be noted as $k \approx 1$.

In case when $k \ll 1$ there is a one more small parameter in the problem that requires revision all procedure of asymptotic expansion. It can not be avoided if, counting wave numbers from $k_{max} = \pi$, to introduce the change of variables

$$\Phi_{j,0} = (-1)^j e^{4i\beta\tau_1} \tilde{\Phi}_{j,0}, \qquad (4.18)$$

where the functions $\tilde{\Phi}_{j,0}$ describe modulations of a saw-tooth mode (out-of-phase mode with minimum wavelength and wave number $k = \pi$).

Really, substituting (4.18) in (4.11) we obtain

$$\frac{d\tilde{\Phi}_{j,0}}{d\tau_1} - i\beta\left(2\tilde{\Phi}_{j,0} - \tilde{\Phi}_{j+1,0} - \tilde{\Phi}_{j-1,0}\right) - 3i\alpha e^{-2\gamma\tau_1}\left|\tilde{\Phi}_{j,0}\right|^2 \tilde{\Phi}_{j,0} = 0 \qquad (4.19)$$

The system (4.19) differs from (4.11) only by the sign before the coefficient β, therefore its solution after returning to initial variables has the form:

$$\psi_{j,0} = \sqrt{N}(-1)^j e^{-\gamma\varepsilon\tau} e^{i\left\{\left[1+4\varepsilon\beta\left(1-\sin^2\frac{k_1}{2}\right)\right]\tau - 3\alpha\frac{N}{2\gamma}(e^{-2\gamma\varepsilon\tau}-1) + k_1 r_0 j + \theta_0\right\}} \qquad (4.20)$$

where $k_1 = \pi - k$

This solution is not justified for the long waves. Thus, equations (4.11) and (4.19) in combination allow to consider the nonlinear waves in a system of the weakly coupled oscillators in short waves.

In the case of intermediate coupling $(\bar{c}/c_1 = O(1))$ the complex representation becomes justified (i.e. the rotations in opposite directions are divided in a principal approximation), if the coupling between oscillators is "effectively weak" because of a relative smallness of the second differences in the equations of motion. It means that the area of applicability of complex representation in that case coincides with the area of applicability of the continual approximation. The equations of motion (4.3) are replaced in this case by relations:

$$\frac{d^2 u_j}{d\tau^2} + 2\varepsilon\gamma\frac{d u_j}{d\tau} + u_j + 8\varepsilon u_j^3 + 2\beta(2u_j - u_{j-1} - u_{j+1}) = 0$$

where $\beta = \bar{c} c_1 = O(1)$. The distances between particles are expedient now for measuring in terms of $r_0 \varepsilon$, so that we have for the second differences in continual approximation the expansions:

$$-(2u_j - u_{j-1} - u_{j+1}) = -\varepsilon\frac{\partial^2 u}{\partial \zeta^2} - \frac{\varepsilon^2}{12}\frac{\partial^4 u}{\partial \zeta^4} - \ldots$$

Thus, in spite of the fact that $\beta \approx O(1)$, the gradient terms contain now the small parameter and the complex representation turns out to be justified at use of a continual approximation. The similar situation arises in the case of short waves. As a result the equations in the principal approximation take a form:

1) $k \ll 1$

$$i\frac{\partial \Phi_0}{\partial \tau} - \beta\frac{\partial^2 \Phi_0}{\partial \zeta^2} + 3\alpha e^{-2i\tau} |\Phi_0|^2 \Phi_0 = 0 \qquad (4.21)$$

2) $k_1 \ll 1$

$$i\frac{\partial \tilde{\Phi}_0}{\partial \tau} + \beta\frac{\partial^2 \tilde{\Phi}_0}{\partial \zeta^2} + 3\alpha e^{-2i\tau} |\Phi_0|^2 \Phi_0 = 0 \qquad (4.22)$$

At last, if the coupling between oscillators is strong, the complex representation of the equations of motion is not adequate from a point of view of asymptotic analysis and the simplification is achieved after the transition to continual approximation in the initial system (4.3). Really, measuring a distances between particles in terms of $\varepsilon^{-\frac{1}{2}} r_0$, we come furthermore to the same asymptotic representation of the second difference as in the case of intermediate coupling. But now coefficient at the second derivative in the equations of motion is not small and consequently the coupling between oscillators is not even "effectively weak".

Therefore the principal approximation corresponds (at a weak dissipation and anharmonicity) to linear Klein-Gordon equation. If the nonlinearity is not small, we come in a principal approximation to nonlinear Klein-Gordon equation, then the influence of a weak dissipation can be appreciated with use of a method of two-scale expansions in real representation. Let's note else that, as well as in case $\bar{c}/c_1 \approx O(1)$, the continual equation for modulation waves in a short-wave limit can be obtained.

Relations (4.21) and (4.22) at $\gamma = 0$ will be converted to the nonlinear Schrodinger equation (NSE), which is the integrable system[15]. Its solution in a wide class of the initial conditions can be obtained by inverse scattering method[15]. It is known[15,16] that, if $\alpha > 0$, at a relation of coefficients signs, which has a place in case of the equation (4.22), all periodic wave packets are unstable and as a result the localized soliton-like waves ("envelope solitons") are formed:

$$\Phi(\zeta,\tau_1) = \sqrt{\frac{2S}{\alpha/\beta}} e^{i(k\xi - \omega\tau_1)} \operatorname{sec} h\left[S^{1/2}(\xi - v\tau_1) \right]$$

where $k = \frac{v}{2\beta}$; $\omega = \frac{v^2}{4\beta^2} - S$,

amplitude and velocity of the soliton are the independent magnitudes here.

At $\gamma = 0$ and $\alpha < 0$ ("soft" nonlinearity) the periodic wave packets described by the equation (4.22), are stable and solitonic solutions do not exist. At signs of coefficients in NSE, corresponding to the equation (4.21), on the contrary, the wave packets are unstable that leads to the formation of solitons and are stable, if $\alpha > 0$ (solitonic solutions exist). As to case of strongly coupled oscillators when the description of the system considered is reduced to the nonlinear Klein-Gordon equation, the wave packets are unstable at $\alpha < 0$[16] (there are localized soliton-like waves) and are stable at $\alpha > 0$ (soliton-like solutions, satisfying to a conditions of decay at $\zeta \to \pm\infty$, are absent).

Thus if the coupling between oscillators is not weak, the conditions of the formation of the localized excitations are well known and are formulated in terms of a continual approximation for waves of the strain or modulation waves.

How the matter is in the case of weak coupling, when it is necessary to use a discrete approach?

As well as in the system with two degrees of freedom, in the infinite chain of nonlinear oscillators apart from cooperative modes the waves with some space localization can be realized. They result from instability of cooperative modes by the mechanism of parametrical resonance. Considering small perturbations of cooperative nonlinear normal modes, we substitute in the equations (4.11) instead of functions of $\Phi_{j,0}$ the expressions

$$\Phi_{j,0} + w_{j,0}. \tag{4.23}$$

Then, taking into account that the functions $\Phi_{j,0}$ satisfy to the equations (4.11) and $|w_{j,0}| << |\Phi_{j,0}|$, we obtain, supposing $\gamma = 0$:

$$\frac{dw_{j,0}}{d\tau_1} - i\beta(2w_{j,0} - w_{j+1,0} - w_{j-1,0}) - 3i\alpha\left(\Phi_{j,0}^2 w_{j,0}^* + 2|\Phi_{j,0}|^2 w_1\right) = 0, \tag{4.24}$$

where the cooperative mode is determined by expression (4.15)

Let's assume now, that instead of an infinite system of coupled oscillators there is a finite chain of length L with number of atoms $N = 2n$ ($n = 0,1,2...$), and the conditions of periodicity are fulfilled

$$\Phi_{0,0} = \Phi_{N,0} \tag{4.25}$$

Then the above mentioned relations remain valid, but it is necessary to take into account restrictions $k \geq \frac{2\pi}{N}$ and $k = \frac{2\pi}{N}m$, where $m = 0, 1, 2, ..., N/2$. Let's consider solutions of the system (4.24), satisfying to relations,

$$w_{j+1,0} = e^{ik_1 r_0} w_{j,0}$$

$$w_{j-1,0} = e^{-ik_1} w_{j,0}, \tag{4.26}$$

$$k_1 = \frac{2\pi}{N} m_1 \quad (m_1 = 0, 1, 2, ..., N\,2).$$

The system (4.24) after substitution (4.26) becomes uncoupled

$$\frac{dw_{j,0}}{d\tau_1} = 2i\left(2\beta\sin^2\frac{k_1}{2} + 3\alpha N\right)w_j + 3i\alpha N e^{2ik_j} e^{2i\left(3\alpha N + 4\beta\sin^2\frac{k}{2}\right)\tau_1} w_{1,0}^* \tag{4.27}$$

Thus, the study of stability of the cooperative modes characterized by wave number m, respect to mode with wave number m_1 is reduced to the equation of parametrical oscillations considered above. The criterion of instability of this cooperative mode (i.e of reaching of the boundary between regions of stability and instability) is determined by the condition

$$\left(4\beta \sin^2 \frac{k}{2} + 3\alpha N\right) = 9\alpha N + 4\beta \sin^2 \frac{k}{2} \tag{4.28}$$

From here for the magnitude of the squared complex amplitude, corresponding to instability of the cooperative mode with wave number m respect to mode with wave number m1, we obtain the expression

$$N = \frac{2}{3}\frac{\beta}{\alpha}\left(\sin^2 \frac{k}{2} - \sin^2 \frac{k_1}{2}\right), \tag{4.29}$$

where $k = \frac{2\pi}{N}m$, $k_1 = \frac{2\pi}{N}m_1$.

As follows from (4.29), at $\alpha > 0$ each mode with wave number m with increasing of the intensity of excitations becomes unstable sequentially respect to modes with wave numbers $m_1 = m - 1, m - 2,...$, smaller than the wave number of the considered nonlinear normal mode. The threshold excitation energy decreases with decreasing m_1 at fixed m. It is necessary to stop especially on the case $m_1 = m$, when the critical value of an excitation energy is equal to zero. In this case instability at no matter how small amplitudes reflects not that other, as the Lyapunov instability, and actually reduces only in a small phase shift of an analyzable mode (similar situation arose in the system with two degree of freedoms – section III). The most important is the first nontrivial instability respect to mode with the wave number $m_1 = m - 1$, corresponding to minimal (if to eliminate the case $m_1 = m$) excitation energy.

The matter is that at superposition of modes with close wave numbers m and m_1 (space beatings!) there is a tendency to localization of excitation, which becomes more and more noticeable with increase of the energy. Eventually, the highest possible localization of excitation is reached. With a decrease of wave number m_1, when the critical energy will increase, the localization becomes weaker. Let's consider, for example, stability out-of phase mode of minimum length ($m = N/2$). Then at an excitation energy corresponding to square of the module of complex amplitude

$$N = \frac{4}{9}\frac{\beta}{\alpha}\left[1 - \sin^2 \frac{2\pi\left(\frac{1}{2}N - 1\right)}{N/2}\right]$$

the instability is realized which leads, eventually, to localization of oscillations practically on one particle, just as it happens in a system to two degree of freedoms. It is clear that the formed mode is multiply degenerated, as the localization of excitation can be realized on any from masses. The comparison with models consisting of three and four particles[5], allows to conclude, that all critical values of an energy correspond to formation of unstable modes.

It is necessary to note that the solution (4.5) for cooperative modes, as it was mentioned above, becomes inconsistent in the region of long waves, when $k \ll 1$. In that case it is possible to be converted to a system (4.19), which solution is selfconsistent just at $k \ll 1$. It appears necessary at $\alpha < 0$, when, as it follows from the relation (4.29), instability of long waves becomes the most important

Now, returning to an initial model, we will consider a limiting case $N \to \infty$. With increase N the arguments of trigonometrically functions for m and m_1 (4.29) differ ever less and less, so that in the limit the energetic barrier, corresponding to the first instability, tends to zero. But it means that all cooperative modes in an infinite chain are unstable (except for a homogeneous mode for which there are no the values m_1, satisfying to condition $m < m_1$) and spatially localized oscillations and waves turn out to be the unique elementary excitations

If anharmonicity is negative $(\alpha < 0)$, the situation is quite similar. In this case each from cooperative modes, except for out-of-phase mode with the minimal wavelength, is unstable in an infinite limit respect to the modes with wave numbers $m_1 > m$. As a result of this instability the localized modes are formed again.

Let's note that numerous applications of representations about localized modes to theory of oscillatory spectra in polyatomic molecules and molecular crystals are presented in [3-5,17-28]. A common concept of localized nonlinear oscillations with applications to the problems of a mechanics is presented in [29,30].

V. NONLINEAR DYNAMICS OF AN INFINITE ATOMIC CHAIN

Let's consider now to an infinite chain of atoms, the interaction between which being described by a gradient-type symmetric potential

$$U = \frac{1}{2}c_1 \sum_{j=-\infty}^{\infty}(U_j - U_{j-1})^2 + \frac{1}{4}c_3 \sum_{j=-\infty}^{\infty}(U_j - U_{j-1})^3 \qquad (5.1)$$

We discuss briefly in the beginning the case of long waves when the complex representation is not efficient.

The system of equations of motion after transformation to dimensionless variables accepts the form:

$$\frac{d^2 u_j}{d\tau^2} + (2u_j - u_{j-1} - u_{j+1})\{1 + \alpha[(u_{j+1} - u_{j-1})^2 + (u_j - u_{j+1})(u_j - u_{j-1})]\} = 0, \qquad (5.2)$$

where

$$\tau = \sqrt{\frac{c_1}{m}} t, \quad u_j = \frac{U_j}{r_0}, \quad \alpha = \frac{c_3 r_0^2}{c_1}; \quad \varepsilon = << 1.$$

Let's measure a distance between atoms in terms $r_0/\sqrt{\varepsilon}$ and introduce the appropriate space coordinate ζ. Then, representing the difference expressions in (5.2) by their Taylor expansions, we obtain one equation with partial derivatives instead of infinite system of the ordinary differential equations:

$$\frac{\partial^2 u}{\partial \zeta^2} - \varepsilon \frac{\partial^2 u}{\partial \zeta^2}\left[1 + 3\alpha\varepsilon\left(\frac{\partial u}{\partial \zeta}\right)^2 + ...\right] - \frac{\varepsilon^3}{12}\frac{\partial^4 u}{\partial \zeta^4} + ... = 0, \qquad (5.3)$$

where $u = u(\zeta, \tau)$, and "..." corresponds to terms of the higher order of smallness on the parameter ε. Here, in difference from case of long waves in the chain of coupled oscillators there are no reasons for using of complex representation, as the equation (5.2) contains only gradient terms. Natural way of the use of the smallness of the parameter ε consists in the introduction of new space and temporal variables:

$$\eta = \zeta - \varepsilon^{1/2} \cdot \tau; \quad \tau_1 = \varepsilon^{5/2} \cdot \tau \qquad (5.4)$$

After substitution (5.4) in (5.3) we have

$$\frac{\partial^2 u}{\partial \tau_1 \partial \eta} + \frac{3}{2}\alpha \frac{\partial^2 u}{\partial \eta^2}\left(\frac{\partial u}{\partial \eta}\right)^2 + \frac{1}{24}\frac{\partial^4 u}{\partial \eta^4} + \ldots = 0$$

The new space coordinate η is counted from front of the linear wave, and the new time τ_1 is slow on comparison with τ.

Introducing a notion $\dfrac{\partial u}{\partial \eta} = w$ and limiting by a main approximation, we come to the modified *Korteveg-de Vries* (mKdV) equation

$$\frac{\partial w}{\partial \tau_1} + \frac{3}{2}\alpha w^2 \frac{\partial w}{\partial \eta} + \frac{1}{24}\frac{\partial^3 u}{\partial \eta^3} = 0 \qquad (5.5)$$

In this equation all terms have the same order on the parameter ε and it ensures the most simple exposition of nonlinear dynamics for atomic chain in the case of symmetric anharmonicity. As is known, mKdV equation has localized solutions (solitons, breathers and multisoliton waves[31]).

In the case of short waves the situation drastically changes: the complex representation becomes efficient again. The equations of motion (5.2) have an exact solution in the form of nonlinear standing wave with minimal wavelength:

$$u_j = (-1)^j w(\tau); \quad -\infty < j < \infty \qquad (5.6)$$

Meaning the analysis of short-wave modes we use, following to[7], the change of variables

$$u_j = (-1)^j w_j.$$

Then the equations of motion are written as follows:

$$\frac{d^2 w_j}{d\tau^2} + (2w_j + w_{j-1} + w_{j+1})\{1 + \varepsilon\alpha[(w_{j+1} - w_{j-1})^2 + (w_j + w_{j+1})(w_j + w_{j-1})]\} = 0$$

Let's introduce the function of two variables $w(\zeta,\tau)$ describing in a continual limit the nonlinear dynamics of crystal in the supposition that the modulations of a nonlinear mode with a minimal wavelength have a characteristic space scale essentially exceeding the distance between atoms, which in selected units is equal ε. Thus

$$w_{j\pm 1}(j,\tau) = w_j(\zeta,\tau) \pm \varepsilon \frac{\partial w}{\partial \zeta} + \frac{1}{2}\varepsilon^2 \frac{\partial^2 w}{\partial \zeta^2} + \ldots \qquad (5.7)$$

where $j\varepsilon = \zeta$.

With the following continual equation of motion is obtained

$$\frac{\partial^2 w}{\partial \tau^2} + 4(w + \alpha\varepsilon w^3) + \varepsilon \frac{\partial^2 w}{\partial \zeta^2} + \ldots = 0, \qquad (5.8)$$

where "…" corresponds to terms of higher powers of the parameter ε. Now, in difference from the case of long waves, the equation of motion contains a non-gradient term, coupling between atoms along a chain described by gradient term becomes "effectively weak" and the use of complex representation is justified. The similar situation arose at intermediate coupling between oscillators in section IV, when the complex representation was introduced in a continual model of a system of oscillators. Thus, from a mathematical point of view the problem of nonlinear short-wave dynamics for an atomic chain is equivalent to the dynamical problem for a system of coupled oscillators in the case of intermediate coupling. After a change of variables $\tau_1 = 2\tau$ we use the complex functions

$$\psi(\zeta,\tau_1) = \frac{\partial w}{\partial \tau_1} + iw; \quad \psi^*(\zeta,\tau_1) = \frac{\partial w}{\partial \tau_1} - iw, \qquad (5.9)$$

for which two conjugate equations are obtained. Let's consider one of them

$$\frac{\partial \psi}{\partial \tau_1} - i\psi - \frac{1}{8} i\varepsilon \left(\frac{\partial^2 \psi}{\partial \zeta^2} - \frac{\partial^2 \psi^*}{\partial \zeta^2} \right) + \frac{1}{8} \alpha i\varepsilon (\psi - \psi^*)^3 + \ldots = 0 \qquad (5.10)$$

By introducing of new function φ, so that

$$\psi = e^{i\tau_1} \varphi(\zeta,\tau_1)$$

and slow time

$$\tau_2 = \varepsilon \tau_1 [1 + \xi_1(\tau_2)\varepsilon + \xi_2(\tau_2)\varepsilon^2 + \ldots]$$

we are searching for a solution as

$$\varphi = \varphi_0 + \varepsilon \varphi_1 + \cdots,$$

where $\varphi = \varphi(\zeta,\tau_1,\tau_2)$

We come in a principal approximation to NSE in slow time

$$i\frac{\partial \varphi_0}{\partial \tau_1} + \frac{1}{8}\frac{\partial^2 \varphi_0}{\partial \zeta^2} + \frac{3}{8}\alpha |\varphi_0|^2 \varphi_0 = 0.$$

The signs of coefficients in this equation are same as for the system of oscillators with intermediate coupling between them in short wave approximation. Therefore there are preserved the same conditions for existence of localized excitation.

VI. CONCLUSION

The complex representation of dynamics of coupled nonlinear oscillators becomes the most efficient at the asymptotic analysis of nonlinear systems with weak and intermediate coupling. In the first case the equations of principal approximation remain discrete and in the second case the asymptotic approach reduces in continual description in terms of a nonlinear Schrodinger equation and its generalizations. If the coupling between oscillators is strong, the complex representation of the equations of motion ceases to be effective and dynamics of a nonlinear system in a principal approximation is described by the real continual Klein-Gordon equation. It is shown, that gradient systems, in which nonlinearity is exhibited only in interaction of atoms, are equivalent in the region of short waves to systems of nonlinear oscillators with average coupling. Other possible generalizations concern to chains with an asymmetric anharmonic potential[32-35]. At last, the significant interest represents the analysis with use of complex representation of nonlinear dynamics in the case of complicated chains forming, e.g. example, the polymeric crystals[36].

I am grateful to O.V. Gendelman and A.I. Manevich for useful discussions.

The work is supported by RBRF (grant N 98-03-33366)

REFERENCES

1. M.I. Rabinovich, D.I. Trubetskov, *Introduction to the theory of vibrations and waves* (in Russian), Moscow, Nauka, (1984).
2. W.H. Louisell, *Coupled mode and parametric electronics,* New York: Wiley, (1960).
3. A.A. Ovchinnikov, *Sov.Phys.JETP*, 30: 147 (1970).
4. A.C. Skott, P.S. Lomdahl, *Chemical Physics Letters,* 311: 29 (1985).
5. A.M. Kosevich, A.S. Kovalyov, *Introduction to Nonlinear Physical Mechanics* (in Russian), Kiev: Naukova dumka, (1989).
6. J.K. Pierce, *J.Appl.Phys,* 25: 179, (1954).
7. A.M. Kosevich, A.S. Kovalyov, *Sov.Phys.JETP,* 67: 1793, (1974).
8. W.H. Louisell, *J.Appl.Phys.,* 33: 2435, (1962).
9. L.I. Manevitch, in *Dynamical Systems: Theory and Applications* (Ed. J. Awrejcewicz), Poland: Lodz, (1997).
10. J.D. Cole, *Perturbation methods in applied mathematics,* Toronto-London: Blaisdell Publishing Company, (1968).
11. A.H. Nayfeh, *Introduction to Perturbation Techniques* New York: Wiley, (1981).
12. S.S. Tsyayn in *Advances in Applied Mechanics, ed. H.L. Dryden, Th.Von. Karmen,* 4, New York: Academic Press Inc, (1956).
13. N.P. Bogolyubov, Yu.A. Mitropolsky, *Asymptotic methods in the theory of nonlinear vibrations(in Russian),* M.: Nauka, (1963).
14. O.V. Gendelman, A.F. Vakakis, *Private communication.*
15. V.E. Zaharov, S.P. Novicov, B.V. Shabat, L.P. Pitaevsky, *Theory of solitons,* (in Russian) Moscow: Nauka, (1980).
16. G.B. Whitham, *Linear and nonlinear waves,* New York: Wiley, (1974).
17. B.R. Henry, W. Siebrand, *J.Chem. Phys.* 49: 5369 (1968).
18. B.R. Henry in *Vibrational Spectra and structure* (Ed. J.R. Durig), Amsterdam: Elsevier, (1986).
19. M.I. Saje, J. Jornter, *Advan.Chem.Phys.* 47: 293 (1981).
20. D.F. Heller, *Chemical Physics Letters,* 61: 583 (1979).
21. H.S. Moller, O.S. Morbensen, *Chemical Physics Letters,* 66: 5339 (1979).
22. M.S. Child, R.T. Lawbon, *Chemical Physics Letters,* 87: 221 (1982).
23. C. Jaffe, P. Brumer, *J.Chem.Phys,* 73: 5646 (1980).
24. E.L. Sibert, W.P. Reinhardt, J.T. Hynes, *J.Chem.Phys,* 77: 3583 (1982).
25. I. Benjiamin, R.D. Levine, *Chemical Physics Letters,* 101: 518, (1983).

26. O.S. van Rosmalen, F. Lachello, R.D. Levine, A.E.L. Dieperim, *J.Chem.Phys,* 79: 2515 (1983).
27. E. Thiele, D.J. Wilson, *J.Chem.Phys,* 35: 1256 (1961).
28. A. Collinns, *Advan.Chem.Phys.*, 532: 225 (1983).
29. L.I. Manevitch, Yu.V. Mikhlin, V.N. Pilipchuk, *Normal vibration technique for strongly nonlinear systems,* Moscow: Nauka, (1989).
30. A.F. Vakakis, L.I. Manevitch, Yu.V. Mikhlin, V.N. Pilipchuk, A.A. Zevin, *Normal Modes and Localzation in Nonlinear Systems,* New York: Wiley, (1996).
31. J.G. Lam, *Elements of soliton theory,* New York: Wiley, (1980).
32. L.I. Manevitch, V.V. Smirnov, *Phys. Lett. A,* 165: 365 (1992).
33. L.I. Manevitch, A.V. Savin, V.V. Smirnov, S.N. Volkov, *Phys.Uspehkhi,* 37: 859 (1994).
34. L.I. Manevitch, A.V. Savin, *Exper. Theor. Phys.,* 80(4): 706 (1995).
35. O.V. Gendelman, L.I. Manevitch, *Exper. Theor. Phys.,* 85(4): 824 (1997).
36. L.I. Manevitch, A.V. Savin, *Phys. Rev. E,* 55: 4713 (1997).

CONDUCTING CHANNELS STRUCTURE AND DIELECTRIC-METAL SWITCHING STABILITY IN THIN AMORPHOUS FILMS

Elena S. Shikhovtseva

Institute of Molecules and Crystals Physics, Ufa Research Centre of RAS, 450075 Ufa, Russia

INTRODUCTION

Since the first discovery of dielectric-metal switching effect in amorphous materials (As-Te-I system)[1,2] in 1962 there have been published numerous reports on switching phenomena in various mostly low-conducting amorphous films. In all known cases the conducting state was imagined to be related to spatially separated current paths of irregular shape (filaments). Then bistable switching effects were summarized in review articles (see, for example,[3,4]). As it turns out to be the bistable switching may occur in metal-insulator-metal systems with an insulator material of thickness 10 nm to 2 μm [3] or 5 nm to 1 μm [4]. In any case thickness of 1–3 μm is a threshold value for such effects. Some early switching models require filamentary current paths[3-6]. The background insulator-metal state switching was described as growth mechanism, involving dielectric relaxation and ion motion[3], regrowth[5], two-stage process: regrowth and switching[6]. The inverse switching to the insulating state is rupture[5,6] or thermal rupture[3]. Such "percolation"-type switching models have been developed for individual switching systems (MIM-metal/insulator/metal sandwich[3], Al-polystyrene-Al[5], Au on quartz glass[6]). As a rule amorphous film switching requires preliminary treatment (so called "forming"). It is a voltage larger than some threshold value. The earlier models of forming process suppose ion injection from anode to insulator[7], conducting channel growth through insulating film[8], local dielectric-electrode alloy and following ion injection[9]. Detailed review of forming models contains, for instance, in paper[3]. Very interesting and important result of this analysis is the forming voltage independence on insulating film thickness d and current peak $I_{max} \sim d^{-3}$ (of course, it takes place for $d < 1$ μm only)[3,10]. This facts will be discussed in the last paragraph.

Very promising class of switching systems is chalcogenide glassy semiconductors. The simplest one is a-Se. In addition, a-Se is one of the chalcogenide semiconductors which have been widely studied. It was presented the model for the structure and properties of active centres in lone-pair (LP) semiconductors[11]. On the base of the model atomic configurations and configuration energies were obtained for the a-Se molecular chains. It was

Mathematical Models of Non-Linear Excitations, Transfer, Dynamics, and Control in Condensed Systems and Other Media, edited by Uvarova *et al.*, Kluwer Academic / Plenum Publishers, New York 1999.

shown, that a-Se chain consists of C_2^0 simple bonding unit configurations (here the upper index means the charge state of unit configuration and the lower one is a number of nearest neighbors). That is the lowest-energy unit configuration C_2^0 is the ordinary, neutral, two-fold-co-ordinated Se atom, which has an energy $-2E_b$, $-E_b$ is the energy per electron of any σ orbital. It turns out to be C_3^0 configuration has the energy $-2E_b + \Delta$, $\Delta \ll E_b$. The low energy splitting Δ allows their mutual transformations. Se atoms are connected by strong covalent bonds between the nearest neighbors in the chain. Covalent bonds form the central and lower parts of valence band. The upper part of valence band is made by LP-electrons wave functions overlapping for LP-electrons of next-nearest neighbors in the chain. Valent angle fluctuations can cause local decreasing of LP-electrons wave functions overlapping. So the local large resistant regions may arise. The perfect a-Se chain has 90° – valent angles (by the way, for example, S-As-S valent angle in a-As_2S_3 system is equal to 98°[12]). In perfect chain Se atoms have low sp-hybridization and small dipole moment. Fluctuations cause large sp-hybridization and local increasing of dipole moment. External electric field can turn the dipole moments and corresponding LP-electrons wave functions. This process may create regions increasing conductivity[13]. The model [13] supposes the simultaneous turns of all dipole moments. However, it seems to be more probable the conducting state propagation along the molecular chain. Such propagation may take place due to the local structure units interaction in the chain (Se atoms in a-Se chains as well as pyramids and tetraheda in two- and three-component systems, for instance, As-S, Ge-S and Ge-S-I, Ge-S-Br, As-S-I). In fact the local structure units of amorphous materials were investigated by means of IR-spectroscopy and such units are pyramids[14], tetraheda[14] and rings [15]. When the film material has chalcogenide atoms in plenty, a number of bonds increases and separated local structure units form chains[15,16]. The medium range order configurations are a result of the local structure units connection. Such configurations are offered on the base of X-ray scattering methods[17-20]. In addition to the base structure units may rotates with regards to each other in presented configurations. Akin chain structures may be the case in oxides. Alike molecular chains can play the role of filaments in MIM diode configurations[4]. So, the switching model may be based on the chemical structure of molecular chains in amorphous materials. The switching occurs due to conducting state propagation along the molecular chains. It is proved to be such process is described by the perturbed sine-Gordon equation, hence the conducting state propagation has a soliton-like nature.

Similar mechanism was offered in[21, 22] for the dielectric-metal switching in the thread-like structure polymers with the side chain molecules (for example, polyarylenephthalides). Soliton propagation along the chain causes the conducting band structure transformations. The energy spectrum owing to single soliton as well as double soliton propagation along the molecular chain may be investigated. In the case of amorphous films time and coordinate dependence of conducting band structure are the same as in[22].

HAMILTONIAN AND CONDUCTING CHANNEL STRUCTURE

The following statements are important for the model of bistable switching.
1. The amorphous materials contain quasi-one-dimensional structures or molecular chains of local structure units (pyramids, tetraheda or single atoms).
2. The local structure units may have a few stable states in the configuration. The stable states are characterized by relative turning of local structure units. The energy and conducting state are functions of this turning.
3. The local structure units interaction may cause the switching propagation along the molecular chain.

The effective Hamiltonian of quasi-one-dimensional molecular chain may be written as:

$$H = \sum \varepsilon_n a^+_{i,n,\sigma} a_{i,n,\sigma} + \sum J\left(\cos\left(\tfrac{1}{2}\lambda\varphi_{L,n}\right)\right) a^+_{i,n,\sigma} a_{i,n+g,\sigma} +$$
$$+ \sum_{i\neq j} J_{ij} a^+_{i,n,\sigma} a_{j,n,\sigma} + \frac{1}{2}\sum V_{ijnm} a^+_{i,n,\sigma} a_{i,n,\sigma} a^+_{i,m,\sigma_1} a_{i,m,\sigma_1} +$$
$$+ \frac{1}{2}\sum \frac{P^2_{L,n}}{I} + \frac{\beta}{2}\sum\left[\left(\varphi_{L,n+1}-\varphi_{L,n}\right)^2 + \left(\varphi_{L,n-1}-\varphi_{L,n}\right)^2\right] +$$
$$+ \frac{U}{\lambda}\sum\left(1-\cos(\lambda\varphi_{L,n})\right). \tag{1}$$

Here $a^+_{i,n}$ and $a_{i,n}$ are a creation and annihilation operators for an electron of ε_n energy on the ith molecular chain and nth local structure unit, $\varphi_{L,i}$ is the valence angle deviation of the nth bridge atom in the ith molecular chain or the nth local structure unit in the ith configuration rotation. $J(\cos(\lambda \varphi_{i,L}/2))$ is an exchange integral between n-th and $(n\pm1)$-th local structure units, J_{ij} is an exchange integral between the different molecular chains or different configurations of the medium-range order in two- and tree-component systems, V_{ijnm} is Coulomb interaction of electrons,. $M^z_{i,n}$ is nth bridge atom in the ith molecular chain or the nth local structure unit in the ith configuration angular momentum operator, I is the moment of inertia of local structure unit with regards to rotation of angle $\varphi_{L,i}$, b is an indirect exchange interaction parameter for neighboring local structure units due to electron configuration transformations. The last but one term in Hamiltonian (1) describes the local structure unit rotation energy for the case of equal stable states. The last term must be introduced in the case of stable states energy splitting. λ is determined by the symmetry of local structure unit under consideration.

EQUATION FOR CONDUCTING STATE PROPAGATION

The $\varphi_{L,i}$ angle value along the molecular chain characterizes the local conductivity. The equation for $\varphi_{L,i}$ may be obtained from Hamiltonian (1) by means of standard treatment:

$$i\frac{d}{dt}\varphi_{L,n} = [\varphi_{L,n}, H] = -\frac{i}{I} M^z_{L,n} = \frac{1}{I}\frac{\partial}{\partial\varphi_{L,n}} ;$$

$$i\frac{d}{dt}M^z_{L,n} = [M^z_{L,n}, H]$$

Thus, in the continuum approximation Hamiltonian (1) yields the equation of motion for $\varphi_{L,i}$:

$$\frac{\partial^2 y}{\partial t^2_1} = -\sin(y) + \frac{\partial^2 y}{\partial x^2_1} - \sum_j \frac{W_j}{U}\cdot\sin\left(\frac{y}{k_j}\right), \tag{2}$$

303

here $\lambda\varphi(x,t) = y(x, t)$, $t_1 = t\,(U\lambda/I)^{1/2}$, $x_1 = x\,(2U\lambda/\beta L^2)^{1/2}$, L is the distance between the local structure units and $x = L\,n$ is a distance along the molecular chain.

Obviously, Eq. (2) is the perturbed sine-Gordon equation. So the switching in amorphous materials may be described as a result of a soliton-like conducting state propagation along the molecular chain. Such attempt is in a good agreement with an experimental results of switching time being strong dependent on film thickness [13, 23]. Unlike pressure and electric field induced switching in polymer films, where the switching threshold voltage is independent on external pressure[24, 25], in amorphous films the switching threshold voltage decreases essentially when external pressure increases[26]. It is conditioned by the difference of polymer conducting channels and amorphous molecular chains. In the polymer film the polymer molecule is a base of conducting channel and this structure is naturally prepared. In amorphous materials connected chains of enough length are artificial structures and external pressure makes easier the junction process.

With the dissipation equation (2) takes the form:

$$\frac{\partial^2 y}{\partial t_1^2} = -\sin(y) + \frac{\partial^2 y}{\partial x_1^2} - \sum_j \frac{W_j}{U}\sin\left(\frac{y}{k_j}\right) - \sigma\frac{\partial y}{\partial t_1} - \sum_{k=1}^{N} \mu_k\,\delta(x_1 - \xi_k)\cdot F(y). \tag{3}$$

Two last terms in equation (3) do not follow from (1) and this terms are introduced with the purpose of the dissipation calculation. The last but one term describes the dissipation because of the local structure unit interaction with the molecular chain and other neighbouring molecular chains. Obviously, the σ value does not depend on y, it is positive for any type of the local structure unit. The last term in (3) describes the solition (antisoliton) energy loses due to interchain transitions. The function $F(y)$ characterizes dependence of the energy losses on the state of the local structure unit. k is the number of the interchain transition in the channel from one electrode to another, the total number of transitions in the channel being equal to N, x_k is the coordinate of kth transition and μ_k is the energy loss factor in the kth transition.

Thus, the $\varphi_{L,i}$ angle value along the molecular chain characterizes the thin amorphous film conductivity and the solutions of equation (3) determine the dielectric-metal switching regime.

SOLITON PULSE STABILITY

The single soliton and many-soliton solutions of perturbed sine-Gordon equation were investigated in a great number of works (see, for example,[27]). In this paper solutions of perturbed sine-Gordon equation were determined as the perfect sine-Gordon equation solutions with time-dependent velocity. So, the single soliton solutions of equation (3) have the form:

$$y = 4\tan^{-1}\left[\exp\left(\pm\frac{x_1 - X(t_1)}{\sqrt{1 - u^2(t_1)}}\right)\right], \tag{4}$$

$$X(t^1) = \int^{t^1} u(t')dt' + x^0(t')$$

Here and below the sign "+" corresponds to the soliton movement in the positive direction (the molecular chain transforms from conducting state when $\varphi_{L,i} = \pi/2$ to the dielectric one when $\varphi_{L,i} = 0$) and "–" corresponds the antisoliton movement in the positive direction in x (the inverse transformation from $\varphi_{L,i} = 0$ to $\varphi_{L,i} = \pi/2$). The soliton velocity u (t) can take any value between –1 and +1, that is $-1 < u < +1$. The factor $(1 - u^2)^{1/2}$ determines the time-depending width of the soliton pulse. The chemical structure of material determines the $F(y)$ function.

$F(y) = \sin y$ and $F(y) = $ const are the simplest dependencies of energy loss.

In this case $F(y) = \sin y$, $\mu_k > 0$, $j = 1$, $k_j = 2$ using (3)–(4), we have the equation system for $u(t_1)$ and $X(t_1)$:

$$\frac{du}{dt_1} = \mp \frac{W}{2U}(1-u^2)^{3/2} - \sigma u(1-u^2) -$$

$$-\frac{1}{2}(1-u^2)\sum_i \mu_i \operatorname{sech}^2\left(\frac{\xi_i - X}{\sqrt{1-u^2}}\right) \operatorname{th}\left(\frac{\xi_i - X}{\sqrt{1-u^2}}\right), \quad (5)$$

$$\frac{dX}{dt_1} = u - \frac{u}{2}(1-u^2)\sum_i \mu_i (\xi_i - X)\operatorname{sech}^2\left(\frac{\xi_i - X}{\sqrt{1-u^2}}\right)\tanh\left(\frac{\xi_i - X}{\sqrt{1-u^2}}\right)$$

When absorption is independent on φ and $F(y) = const$, instead of (5) we have:

$$\frac{du}{dt_1} = \mp \frac{\varepsilon}{2}(1-u^2)^{3/2} \pm \sigma_{10}\frac{\pi}{4}|u|(1-u^2) - \sigma_2 u(1-u^2) -$$

$$-\frac{1}{2}(1-u^2)\sum_i \mu_i \operatorname{sech}^2\left(\frac{\xi_i - X}{\sqrt{1-u^2}}\right) \operatorname{th}\left(\frac{\xi_i - X}{\sqrt{1-u^2}}\right), \quad (6)$$

$$\frac{dX}{dt_1} = u - \frac{u}{2}(1-u^2)\sum_i \mu_i (\xi_i - X)\operatorname{sech}^2\left(\frac{\xi_i - X}{\sqrt{1-u^2}}\right)\tanh\left(\frac{\xi_i - X}{\sqrt{1-u^2}}\right)$$

Similar attempt was developed in[28, 29] for the dielectric-metal switching stability investigation in the thread-like structure polymers with the side chain molecules (for example, polyarylenephthalides).

RESULTS AND DISCUSSION

Solutions of equation system (5)–(6) are determined by constants σ, U, W, ξ_i, μ_i as well as velocities of soliton movement per interchain transition. For equation system (5) in Figure 1 and 2 the soliton phase trajectories are shown for the solitons produced in $X_0 = -2$ and initially moving in positive direction per interchain transition in $\xi = 0$. In the case of large dissipation (Fig. 1, $\sigma = 0.4$) three solitons are reflected and four solitons propagate through the transition point $\xi = 0$. When the dissipation decreases a number of reflected solitons decreases too. In the case of $\sigma = 0.1$ (Fig.2) one can see only one reflected soliton. Figures 1 and 2 show the phase trajectories for the energy degenerated conducting and insulating states of local structure units. Nonequal energy states give some new effects (Figures 3–6). $W > 0$ (Figures 3 and 4) corresponds to the case of conducting state energy being large than insulator state one. Negative parameter W ($W < 0$ in Fig. 5 and 6) determines the inverse energy level relation. The sign of W is determined by the local structure units of film material. As it does, when the ground state of local structure unit in configuration is low conducting ($\varphi_{L,i} = 0$) and the excited one is high conducting ($\varphi_{L,i} = \pi/2$) then $W > 0$. Much more likely such relation between the dielectric and metal state takes place in polymer films[21, 22]. Obviously, amorphous films can have both $W > 0$ and $W < 0$ because of the medium range configurations of amorphous materials are very complicated and various. Moreover all combinations of $j \neq 1$ and $k_j = 3, 4,...$ must be considered for different types of local structure units symmetry.

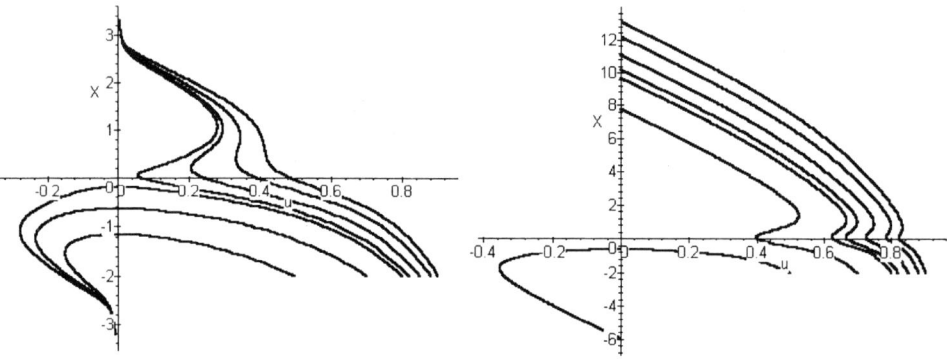

Figure 1. Phase portraits (u,X) for the solitons produced in $X_0 = -2$ with initial velocities $u_0 = 0.5$; 0.7; 0.8; 0.82; 0.85; 0.88; 0.9. $\sigma = 0.4$, $\mu_i = 0.8$; $\xi_i = 0$, $i = 1$. $W = 0$.

Figure 2. Phase portraits (u,X) for the solitons produced in $X_0 = -2$ with initial velocities $u_0 = 0.5$; 0.7; 0.8; 0.82; 0.85; 0.88; 0.9. $\sigma = 0.1$, $\mu_i = 0.8$; $\xi_i = 0$, $i = 1$. $W = 0$.

The set of figures 1–6 is in a position to give sufficiently complete notion of switching regimes. The comparison in figure pairs 1–2, 3–4 and 5–6 shows the dissipation (s) dependence. Series of figures 1, 3, 5 and 2, 4, 6 display the influence of stable states energy splitting for different local structure units.

An important characteristics of soliton propagation is equilibrium values of velocities. Such values may be determined by means of relation $du/dt=0$ from the equation system for $u(t1)$, $x(t1)$. The first equations of systems (5) and (6) give (as it had to be) $u = \pm 1$ and under some conditions $|u_{eq}| \neq 1$ can occur.

When $W \neq 0$ equation system (5) has two equilibrium values of the velocity (Figures 3–6):

$$u_{s,a} = \pm \frac{W}{U \cdot \left(1+4\sigma^2\right)^{1/2}} \qquad (7)$$

(here u_s is the soliton velocity and u_a is the antisoliton velocity). Pinning trajectory of Figure 6 (soliton capture) describes the evolution of the initial velocity value $v_0 = 0.8$ soliton.

Unlike the set of equations (5) the equation system (6) permits more analytical results. Except for equilibrium values of the velocity $u_{s,a}$ being the same as (7), it may be obtained the pinning point coordinate X_p and the oscillation frequency ω.

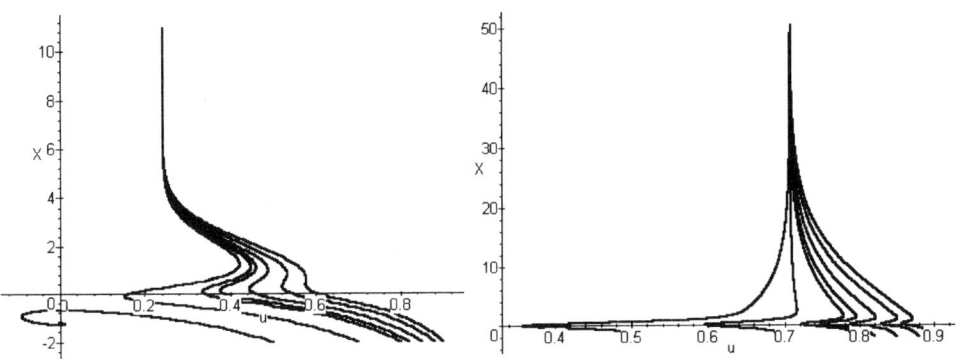

Figure 3. Same as Fig. 1, $W = 0.2\ U$.

Figure 4. Same as Fig. 2, $W = 0.2\ U$.

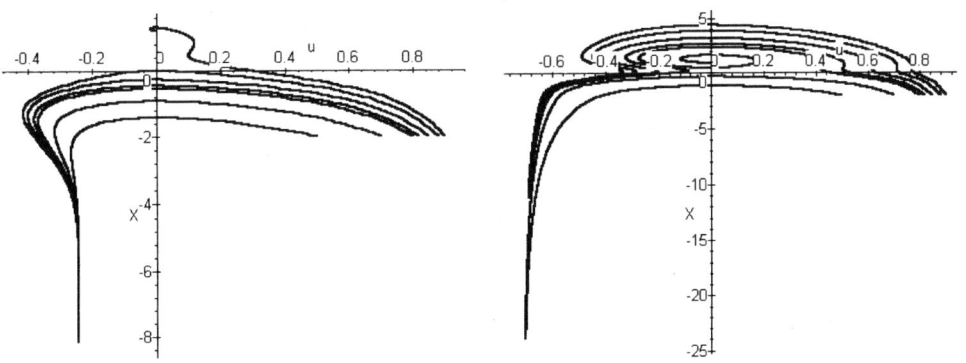

Figure 5. Same as Fig. 1, $W = -0.2\ U$.

Figure 6. Same as Fig. 2, $W = -0.2\ U$.

In the case of one trap (one interchain transition, $i = 1$) X_p is determined by the equilibrium conditions: $du/dt_1=0$, $dX/dt_1=0$. The oscillation frequency can be obtained through the use of assuming the oscillation near the pinning point X_p of the form:

$$u = u_a \exp(-\delta t) \cos(\omega t),$$

$$X = X_p + X_a \exp(-\delta t) \cos(\omega t).$$

For one trap is in $\xi = 0$ solutions X_p and ω of the equation system (6) can be written as:

307

$$X_p = -\text{sech}^{-1}\left(\frac{W}{\mu U}\right)^{\frac{1}{2}} \qquad (8)$$

and
$$\omega^2 = -\frac{\mu}{2}\cdot\left(1+\frac{W\cdot X_p}{2U}\right)\cdot\text{sech}(X_p)\cdot\tanh(X_p). \qquad (9)$$

The system (5) gives the equation for X_p only. X_p is the root of equation:

$$\mp\frac{W}{\mu U}+\text{sech}^2(x-X_p)\cdot\tanh(x-X_p)=0 \qquad (10)$$

and the oscillation frequency is

$$\omega^2 = \frac{\mu}{2}\cdot\left(1\mp\frac{W\cdot(x-X_p)}{2U}\right)\cdot\left[2\cdot\text{sech}^2(x-X_p)\cdot\tanh^2(x-X_p)-\text{sech}^4(x-X_p)\right], (11)$$

where $(x-X_p)$ is a root of equation (10).

Paremeter δ is independent on $F(y)$ and $\delta = \sigma/2$ for both equation systems (5) and (6).

Of course, the dielectric-metal switching can be caused by pulse propagation from one electrode to another only.

If there are a few traps in the channel, the distance between them a highly significant for the shape of phase trajectories.

Such switching description is agree with the experimental fact of current peak dependence on film thickness as d^{-3} [10]. The peak value of current is a sum of currents in all conducting channels. As it follows from analysis of equation systems (5) and (6), some channels do not reach the film surface because of they have not enough length. As channels are quasi-one-dimensional but not straightforward, the thickness dependence of I_{max} is stronger than inverse proportional. It must be $I_{max} \sim d^{-n}$, where n is a function of channels (medium-range configurations) average curvature.

So, the offered analysis of equation systems (5) and (6) gives the conditions of conducting channels formation and destruction in thin amorphous films when $F(y) = \sin y$ and $F(y) = \text{const}$, $\mu_k > 0$, $j = 1$, $k_j = 2$. The same equation systems may be obtained for different $F(y)$ and k_j that is for different types of local structure units and medium-range order configurations. The solutions of equation for conducting state propagation give all types of switching effects: the soliton reflection, capture and propagation. The switching regime is formed by molecular parameters of thin film material.

REFERENCES

1. S.R. Ovshinsky, *U.S. Patent* 3:052:830 (1962); An introduction to ovonic research, *J. Non-Cryst. Solids* 2:99 (1970).
2. A.D. Pearson, W.R. Northover, I.F. Dewald, and W.I. Peck, *Advan. in Glass Technol.*, Plenum. Press, New York (1962).
3. G. Dearnaley, A.M. Stoneham, and D.V. Morgan, Electrical phenomena in amorphous oxide films, *Rept. Progr. Phys.* 33:1129 (1970).
4. H. Pagnia and N. Sotnic, Bistable switching in electroformed metal-insulator-metal devices, *Phys. Stat. Sol. (a)* 108:11 (1988).

5. L.F. Pender and R.J. Fleming, Memory switching in glow discharge polymerized thin films, *J. Appl. Phys*. 46:3426 (1975).
6. Th. Bach, R. Blessing, H. Pagnia, and N. Sotnic, Temperature dependence of the on state regeneration in metal/insulator/metal diodes, *Thin Solid Films* 103:283 (1983).
7. J.G. Simmons and R.R. Verderber, New conduction and reversible memory phenomena in thin insulating films, *Proc. Roy. Soc*. A301:77 (1967).
8. Dearnaley, Electronic conduction through thin unsaturated oxide layers, *Phys. Lett*. A25:760 (1967).
9. C. Barriac, P. Pinard, and F. Danoine, Etude des proprietes electriques des structures Al-Al$_2$O$_3$-metal, *Phys. Stat. Sol*. 34:621 (1969).
10. R.R. Verderber, J.G. Simmons, and B. Eales, Forming process in evaporated SiO thin films, *Philos. Mag*. 16:1049 (1967).
11. M. Kastner, D. Adler, and H. Fritzche, Valence-alternation model for localized gap states in lone-pair semiconductors, *Phys. Rev. Lett*. 37:1504 (1976).
12. A.J. Apling, A.J. Leadbetter, and A.C. Wright, A comparison of the structure of vapour-deposited and bulk arsenic sulphide glasses, *J. Non-Cryst. Solids* 23:369 (1977).
13. E.A. Lebedev and K.D. Tsendin, Switching effect in chalcogenide glassy semiconductors, In *Electronic Switching in Chalcogenide Glassy Semiconductors*, K.D. Tsendin, ed., Nauka, St. Petersburg. (1996).
14. J. Heo and J.D. Mackenzie, Chalcohalide glasses. Vibrational spectra of Ge-S-Br, *J. Non-Cryst. Solids* 113:1 (1989).
15. G. Lucovsky and J.C. Knights, Infrared absorption in bulk amorphous As, *Phys. Rev. B* 10:4324 (1974).
16. K. Murase, K. Yakushiji, and T. Fukunaga, Interaction among clusters in chalcogen-rich glasses of Ge$_{1-x}$(S or Se)$_x$, *J. Non-Cryst. Solids* 59/60:855 (1983).
17. L. Cervinka, Medium-range order in amorphous materials, *J. Non-Cryst. Solids* 106:291 (1988).
18. A.J. Apling, A.J. Leadbetter, and A.C. Wright, A comparison of the structure of vapour-deposited and bulk arsenic sulphide glasses, *J. Non-Cryst. Solids* 23:369 (1977).
19. S.R. Elliot, Origin of the first sharp diffraction peak in the structure factor of covalent glasses, *Phys. Rev. Lett*. 67:711 (1991).
20. S.R. Elliot, A unified model for the low-energy vibrational behaviour of amorphous solids, *Europhys. Lett*. 19:201 (1992).
21. O.A. Ponomarev and E.S. Shikhovtseva, The dielectric-metal transition mechanism for the thread-like structure polymers, *Synth. Metals* 68:99 (1995).
22. O.A. Ponomarev and E.S. Shikhovtseva, Mechanism by which pressure and a field affect the electrical conductivity of conjugated polymers with insulator bridges, *Zh. Eksp. Teor. Fiz*. 107:637 (1995) (*JETP* 80:346 (1995)).
23. B.T. Kolomiets, E.A. Lebedev, and I. A. Taxami, Main parameters of switching on the base of chalcogenide glassy semiconductors, *Fiz. i Tekh. Poluprov*. 3:731 (1969).
24. A.N. Lachinov, A.Yu. Zherebov, and V.M. Kornilov, Influence of uniaxial pressure on conductivity of polydiphenylenephthalide, *Synth. Metals* 44:111 (1991).
25. A.Yu. Zherebov, A.N. Lachinov, V.M. Kornilov, and M.G. Zolotukhin, Metal phase in electroactive polymer induced by uniaxial pressure, *Synth. Metals* 84:735 (1997).
26. P.J. Walsh, J.E. Hall, R. Nicolaides, S. Defeo, P. Callela, J. Kuchmas, and W. Doremus, Experimental results in amorphous semiconductor switching behavior, *J. Non-Cryst. Solids* 2:107 (1970).
27. D.W. McLauglin and A.C. Scott, Perturbation analysis of fluxon dynamics, *Phys. Rev. A* 18:1652 (1978).

28. E.S. Shikhovtseva and O.A. Ponomarev, Stability of the insulator-metal transition in oxygen-containing polymers, *Pis'ma Zh. Eksp. Teor. Fiz.* 64:468 (1996) (*JETP Lett.* 64:509 (1996)).
29. E.S. Shikhovtseva and O.A. Ponomarev, Radiation arisen from the areas of conducting channels formation in the thread-like structure polymers, *Phys. Low-Dim. Struct.* 5/6:43 (1998).

PHENOMENOLOGICAL DESCRIPTION FOR PROCESS OF MULTIPLE DISINTEGRATION* OF SOLIDS UNDER INTENSIVE STRESS ACTION SUCH AS COMPRESSION & SHEAR

Arkadii E. Arinstein

Moscow State University of Technology "STANKIN",
101472 Moscow, Russia
N.N. Semenov Institute of Chemical Physics,
Russian Academy of Sciences, 117334 Moscow, Russia

INTRODUCTION

The effect of multiple disintegration of solids under intensive stress action such as compression & shear (ISACS) was discovered in 70-th years in N.N. Semenov Institute of Chemical Physics of Russian Academy of Sciences at the investigation of mechanochemical reactions in some polymeric and composit materials. The first experimental investigations of this phenomenon were also carried out there (see the review of Nikol'skii and Wol'fson[1]). Initially the interest in physicochemical processes proceeding in polymeric systems subjected to ISACS, was caused by the fact that the velocity of chemical reaction can increase for several orders of magnitude under these conditions and, in some cases, the reaction does not proceed at all within the same temperature range in the absence of ISACS. The experimental data accumulated till now demonstrate, however, that in a solid-state matrix of polymers and polymeric composites subjected to ISACS, not only the mechanochemical processes, but also some other physicochemical ones can proceed in an unconventional way.

Thus, it has been established for composites that under the action of intensive compression and shearing the structure of the solid polymer blends changes irreversibly, namely, the disruption of amorphous regions into microdomains smaller than 100 nanometers in size occurs, their mixing and then the formation of a quasi-homogeneous amorphous phase[2]. It is particularly remarkable that the homogenization takes place also for the blends consisting of partially compatible or even fully incompatible from the thermodynamic point of view polymers.

---

* This term is considered to mean a type of disintegration of a solid which results in its instantaneous break-up into a totality of small fragments, by passing the stage of consecutive disintegration.

Also unusual stands up the abnormally fast diffusion of an impurity in the solid-state polymeric matrix, then the effective value of the diffusion coefficient of the low-molecular admixture in the solid phase increased for 4–5 orders of magnitude observed in numerous experiments under ISACS conditions[3].

Furthermore, in a number of experiments carried out with the Bridgeman anvils, an explosive destruction of inorganic crystals and sheet polymer films subjected at ambient temperatures to shearing and compression under pressures of 50.000–150000 atm. has been found[2-10].

And, finally, it has been disclosed quite recently that the disintegration and dispergation of some polymers heated up to their premelting temperature, can proceed at significantly low compressions, pressure values being no more that 30–200 atm.[4-7]. When it is considered that during the plastic deformation of solid polymers the significant dissipation of energy takes place, whereas their melting point is not too high, it becomes evident that the necessary temperature condition can arise due to the self-heating of the material subjected to ISACS; in result the shear-induced strain disintegration can proceed under rather low pressures.

For all the above phenomena the critical dependence of the intensity of the external action on the effect is the case; thus, if the external action is not strong enough, no anomalies are observed. But if the intensity of this action has attained some critical value, the kinetics of all processes changes crucially.

For the process of shear-indused strain disintegration (the theoretical description of the phenomenon is the subject of the present paper) this is reflected by the fact that if the external action is not strong enough, no destruction of the material is observed at all; as the intensity of the external action attains some critical value, the complete disintegration of a sample occurs almost instantly. This peculiarity of the process of the multiple disintegration under ISACS conditions demonstrates the analogy between the explosive destruction of the strata within the earth's crust and the caving of the mountains slopes observed in nature, and the shear-induced plane polymer films carried out on a laboratory scale; both these phenomena are apparently caused by the same physical reasons.

PHYSICAL MECHANISM FOR DISINTEGRATION UNDER ISACS

Basing on the comprehensive experimental data accumulated till now, it is reasonably safe to suggest that the disintegration of solids and thermoplasts under ISACS condition is a complex physicochemical process. Unfortunately, up to now no workable (adequate) theoretical description of the phenomenon, covering all the anomalies found experimentally does not exist. In the present paper, when describing the regularities and peculiarities of the disintegration of solids under ISACS conditions, we shall estimate on the ingenious model proposed by the author[11, 12].

It is common knowledge that the disintegration (fragmentation) of solids by ordinary crushing proceeds according to the mechanism of "through–the–thickness" crack propagation proposed by Griffits. On the conventional external force application, within the bulk of the material, in its most stressed points, microcracks occur which, in turn, are stress concentrators. As the stress in the vicinity of some microcrack attains its critical value, this microcrack begins to grow rapidly and transforms into a macroscopic one, critical macrocrack threading the whole interior of the sample, which in result breaks apart. Each of the fragments produced destroys according to the same scenario and fine powder results.

Unfortunately, in the context of this scenario for the solids' disintegration one cannot explain adequately a number of regularities, inherent in the disintegration process occurring under ISACS-induced fragmentation. Thus, for instance, the dispersed phase particle size distribution produced by the consecutive fragmentation (as it occurs on conventional crushing) has to be the Poisson one, whereas under the ISACS's conditions it appears to be the different, more narrow one, in particular. As to polymeric materials, during their crushing substantial decrease in the average length of polymeric macromolecules occurs, i.e., their decomposition proceeds, whereas during disintegration under the ISACS's conditions the molecular-weight distribution of polymeric chains remains practically unchanged. Many other unusual facts could be brought forward, but the above two are sufficient to leave no doubt that the disintegration under the ISACS's condition proceeds by a mechanism fundamentally different from that of the "critical crack appearance".

A number of experimental facts' evidences that under the ISACS's conditions the fragmentation of solids does not pass through the stage of consecutive crushing, but at some point in time the specimen instantaneously breaks into a totality of small particles, i.e., the process of the dispersed phase formation is of critical character. This is possible only in the event that the accumulation of microdefects occurs in the material under the ISACS's conditions. Just in the instant the concentration of these defects attains some critical value, the sample is crumbled to a totality of small chips. It has to be noted that the assumption of microdefects accumulating within the bulk of a disintegrating sample, allows to explain one further unusual peculiarity of the process in hand, the mentioned above abnormally rapid diffusion of impurities in solids subjected to ISACS, the diffusion coefficient being increased in this case for many orders of magnitude[13] (see sec. "Diffusion of low-molecular impurities ... under ISACS").

Linking the microdefects with the diffusion of impurities in a polymer matrix, we are able to determine immediately both their nature and the scale, those are the defects of the polymer supramolecular structure with the magnitude on the order of several units of a polymer chain (segment). Thus, a microdefect is considered to mean rupture of any knot of a three-dimensional network of the bonds providing the completeness of a sample as a whole. Whether the bonds in this three-dimensional network are of physical or chemical nature, is of no concern for our consideration. It seems likely that for the linear (polyethylene-like) polymers this network is of physical nature, whereas for the cross-linked (rubber-like) polymers those are chemical bonds. This is to be mentioned that, as to the scale of these microdefects, the microcracks generating the "critical" cracks during the crushing process, are macroscopic defects, and the consideration given below, is unsuitable for their description.

It is apparent that the microdefects considered by us, appear first at the most stressed points of a solid matrix, as is the case for the conventional disintegration of a specimen. However, in contrast to a customary situation, when the microcrack is the stress concentrator, we assume that due to the microdefect formation, the local stressed region of a matrix which induced the formation of this defect, relaxes towards the unstressed state. This is just the assumption that allows to explain why the microdefects accumulating in a system, are uniformly distributed within the bulk of a sample, the latter finally leads to its disintegration without the "through–the–thickness" crack formation. Besides, during relaxation of local stressed regions of a matrix, the elastic strain energy transforms into the thermal one resulting in heating of sample. Thus, the efficient channel for the transformation of the mechanical energy introduced into the system during shearing, into the thermal one, arises in a natural way.

DIFFUSION OF LOW-MOLECULAR IMPURITIES IN POLYMERIC MATRIX UNDER ISACS

In experiments using Bridgeman anvils the homogenization of impurity exhibits at a scale of about several mm (average granule size) during the time $t \sim 10 - 10^2$ sec. In order to explain this effect it is necessary to assume that the effective value of the diffusion coefficient is of the order $D \sim 10^{-5} - 10^{-4}$ cm^2/sec, this for 4–5 orders of magnitude exceeding the specific value of the diffusion coefficient for the low-molecular impurity in a solid phase ($D \sim 10^{-10} - 10^{-9}$ cm^2/sec) for 1–2 orders exceeding the diffusion coefficient for liquids ($D \sim 10^{-6} - 10^{-5}$ cm^2/sec) and only for 4 orders being below the gas coefficient diffusion ($D \sim 10^{-1} - 1$ cm^2/sec). In the present section we shall demonstrate that the dynamics of the microdefects system in a polymer matrix can provide so fast diffusion of an impurity into the solid polymeric matrix under the ISACS's conditions, that, in our opinion, it could serve as a reliable ground for the assumption about the role of microdefects in the shear-induced strain disintegration.

Thus, according to the notion about the mechanism of diffusion of molecules of impurities into the solid-state matrix the following picture is beginning to emerge. In the basic component of the impurity molecules is impracticable; however, if adjacent to the molecule some defect microregion with reduced density (physical or chemical interchain cross-link is disrupted at this point) is available, it can migrate for the distance of the order of this defect "a"; the defects on their own can move randomly, the latter being provided the macroscopic transfer of impurity molecules. Under conventional conditions both concentration of defects and the velocity their movement in such systems are not too high, with the result that the collision rate of defects with the diffusing molecules and the effective value of the diffusion coefficient of the impurity molecules governing by this rate, is also small. However, in result of the external stress action on the substance (especially that of compression combined with shear) the concentration of defects and the velocity of their movement within the bulk of a solid matrix dramatically increases.

It is naturally to assume that at their collisions to one another, the defects shall unite into clusters, wherein the movement of the diffusing molecules proceeds rather rapidly (at least faster than in motionless liquid), with the concentration of the defects C governed by the intensity of the external action, being fixed, the steady-state distribution of clusters over sizes establish within the system rather quickly. In result of acts of attachment to and detachment from the cluster the individual defects, the centre of gravity of each cluster of defects displaces in a random way, i.e., this cluster performs the random walks, it diffuses. The impurity molecules being inside the cluster, diffuse together with them, providing thus the mechanism of fast macroscopic mass transfer of the impurity within the bulk of the solid–phase specimen under the conditions of intensive action of the compression combined with the shear type.

In such a way, out task reduced to the description of the dynamic of a system of defects united into clusters. As the mobility of a cluster as a whole is small as compared with that of individual defects, the dynamics of the cluster system of defects can be described within the framework of the simplest aggregation the particle–cluster process, whose kinetic is described by the following set of equations:

$$\begin{cases} \dot{u}_n = -p_n u_1 u_n + p_{n-1} u_1 u_{n-1} + q_{n+1} u_{n+1} - q_n u_n\,, \quad n \geq 2, \\[6pt] \dot{u}_1 = -u_1 \sum_{n=1}^{\infty} p_n u_n - p_1 u_1^2 + \sum_{n=2}^{\infty} q_n u_n + q_2 u_2 + I - \kappa u_1\,, \end{cases} \quad (1)$$

here u_n is the concentration of clusters consisting of n defects, p_n, q_n are the rate constants for attachment to clusters consisting of n defects, and detachment of individual defect from them, accordingly; I is the velocity of nucleation of new defect, and κ is the rate constant of relaxation disappearance of individual defects.

In dynamic equilibrium the rate of new defects nucleation is equal to that of their disappearance, i.e., $I = \kappa u_1$ is valid, and the set of equations (1) takes the form of equations of the generally recognized Beker-Döring model[14].

On condition that $I = \kappa u_1$ for the kinetic set of equations, the conservation law corresponding to the conservation of the total number of defects in a system, is obeyed:

$$\sum_{n=1}^{\infty} n u_n = C = const. \tag{2}$$

Inasmuch as we assume that in a cluster system the dynamic equilibrium has been already established, their size distribution is described by the stationary solution of equation (1) which is well-known [15]:

$$u_n = u_1^n \prod_{k=1}^{n}(p_k/q_{k+1}), \tag{3}$$

where u_1 is determined from the conservation law (2).

In line with our model a particle penetrated, into the cluster with size n, can freely travel a distance of the order of the size of a cluster $R_n \approx a n^{1/d}$ (limited diffusion), and also travel together with the cluster, the diffusion of the latter being proceeded due to the attachment and detachment of individual defects. We have introduced here a parameter d characterizing the density of defect "packing" inside the cluster (looseness of the cluster). With dense packing, d is coincident with the dimensionality of space ($d = 3$ in a bulk, and in a plane, i.e., in thin films, $d = 2$). Nevertheless, for a wide variety of aggregation processes clusters are the fractal objects with loose packing, and the fractal dimensionality of a cluster d, being the universal characteristic of the process, turns out smaller than the dimensionality of space. For the aggregation considered, the cluster–particle process, it appears that in three-dimensional space $d = 2,46 \pm 0,05$, whereas in the plane $d = 1,68 \pm 0,02$ [16]. For time intervals large enough the displacement of a particle shall be governed by the clusters movement. At each act of attachment or detachment of an individual defect, the center of gravity of a cluster displaces for a value of $\Delta R_n = R_n/n \sim a n^{(1/d-1)}$. Then one could estimate the diffusion coefficient of clusters of the n size and, accordingly, the diffusion coefficient of the particles penetrating into these clusters:

$$D_n = \frac{R_n^2}{\tau_n} \sim \frac{a^2}{\tau_n} n^{2(1-d)/d} \sim a^2 n^{2(1-d)/d}(p_n u_1 + q_n), \tag{4}$$

here $\tau_n = 1/(p_n u_1 + q_n)$ is the mean specific time between the acts of attachment and detachment of defects for the cluster of the n size ($q_1 = 0$).

Since the process of attachment of individual defects to clusters with moderate size is the diffusive–limited one, then for the constant of the rate of attachment p_n, we can use the Smolunovskii formula calculated for the effective rate constant for bimolecular diffusive-limited reaction:

$$p_n = 4\pi D_{def} R_n \sim 4\pi D_{def} a n^{1/d}, \tag{5}$$

here $D_{def} = a^2/\tau_{def}$ is the defects diffusion coefficient, τ_{def} is the specific time for the defect's displacement for the distance of the order of its size, i.e. the specific time of the elementary act of the medium rearrangement.

The rate constant q_n for the detachment of defects from clusters is proportional to the amount of defects located at the clusters surface, i.e. is proportional to the area of its surface. Thus, its can be written that

$$q_n = \frac{4\pi R_n^2}{4\pi a^2} q_0 \sim \frac{n^{2/d}}{\tau_{cl}}, \tag{6}$$

here $\tau_{cl} = 1/q_0$ is the specific time for detachment of a certain defect from the cluster surface.

Since q_n increases faster with n rise than it p_n does, at large n the q_n value makes the major contribution to (4) and the error in the expission (5) for p_n shall be unessential.

The expressions (4)–(6) allow to determine the contribution of particles found thereself inside the clusters with n size, to the effective value of the diffusion coefficient:

$$D_n = D_{def}\left(\frac{\tau_{def}}{\tau_{cl}}\right) n^{-2(d-2)/d}\left(1 + \frac{4\pi a^3 u_1}{n^{1/d}}\frac{\tau_{cl}}{\tau_{def}}\right). \tag{7}$$

Averaging over the sizes of the stationary distribution of clusters (3), we shall obtain the expression for the effective value of the diffusion coefficient of the impurity molecules in a solid-state matrix under the conditions of intensive stress action of the compression combined with shear type:

$$D_{ef} = D_{def}\frac{\tau_{def}}{\tau_{cl}}\frac{\sum_n^\infty D_n u_n}{\sum_n^\infty u_n} = D_{def}\frac{\tau_{def}}{\tau_{cl}}\frac{\sum_n^\infty \frac{1-\delta_{1,n}+\vartheta/n^{1/d}}{n^{2-3/d}(n!)^{1/d}}\vartheta^n}{\sum_n^\infty \frac{\vartheta^n}{(n\cdot n!)^{1/d}}}. \tag{8}$$

The dependence of the effective value of the diffusion coefficient on the defects concentration is given by ϑ parameter, which is specified from the normalization condition (2):

$$\sum_n^\infty \frac{n\vartheta^n}{(n\cdot n!)^{1/d}} = 4\pi a^3 \frac{\tau_{cl}}{\tau_{def}}\ C = \frac{3\tau_{cl}}{\tau_{def}}\theta,, \tag{9}$$

here θ is the concentration of the total volume of defects, i.e. the fraction of the free volume in a medium

The calculation of the $D_{ef}(\theta)$ dependence at the prescribed value of the relation of specific time values τ_{cl}/τ_{def} by the formulae (8) and (9) has been fulfilled numerically.

The dependencies $D_{ef}(\theta)/D_{def}$ for the case of dense packing of defects in clusters ($d=3$) at $\tau_{cl}/\tau_{def}=3$ (1), 5 (2), and 10 (3) are given in Fig. 1. For the case of loose packing of clusters (fractal dimensionality $d=2,46$) these dependencies are essentially the same. Fig. 1 evidences, in particular, that when the relation τ_{cl}/τ_{def} is reasonably large, the dependence $D_{ef}(\theta)$ possesses the weakly-defined maximum, the latter can be easily explained by the fact that as the amount of defects increases, the more important part play the large clusters, whose mobility is small.

Inasmuch as in the proposed model we have not took into account the possibility of formulation of macroscopic cluster of defects occupying the whole bulk of a system (gel), which occurs in the percolation point (at the volume fraction of defects concentration $\theta_{cr} \approx 0,25$), we thus entitled formally to use the initial portions of the revealed dependencies at θ altering form zero to $0,25-0,3$ only. (The presented Figure demonstrates that it is just over this values interval of the defects volume fraction, the substantial change of the effective value of the diffusion coefficient takes place).

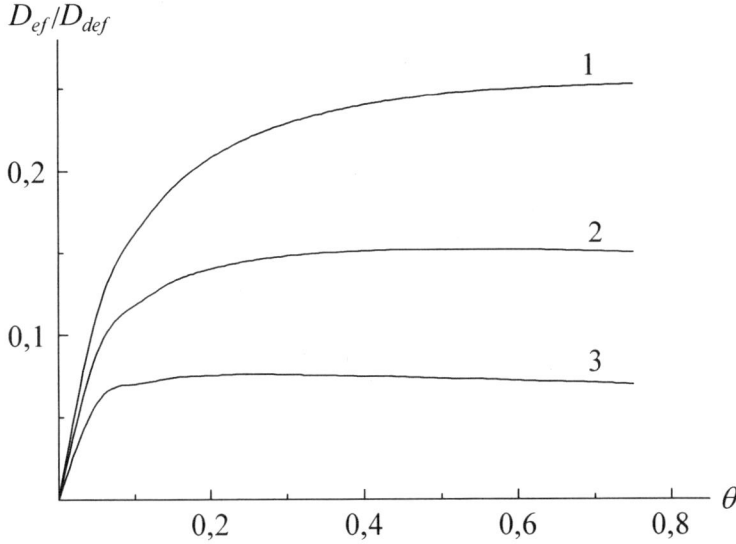

Figure 1. The relation value of the effective diffusion coefficient in there dimensional for dense "packing" of defects into clusters ($d=3$) at $\tau_{cl}/\tau_{def} = 3 - 1, 5 - 2$ and $10 - 3$.

At $\theta > \theta_{cr}$ the mechanism of the macroscopic transfer changes, namely, the particle can diffuse as far as it is wished, being inside of the single, though immobile, cluster of defects, is of the type, which allows to use the obtained dependencies also at θ exceeding θ_{cr}. The matter is that at θ values slightly exceeding θ_{cr}, the macroscopic cluster consists of individual loosely bound conglomerations of defects, and the jump-over of the diffusing particle from whatever conglomeration of defects into some other one shall proceed rather slowly. In result, over the interval of some values of the defects volume fraction exceeding θ_{cr}, the macroscopic transfer shall be effected mainly due to the movement of conglomerations of defects being a constituent of a macroscopic cluster. Once θ is approached to the second critical value $\theta_{cr} = 0,75$, the consolidation of cluster of defects becomes the sufficient one in order the mechanism of the intra-cluster diffusion to prevail. On transition to this transfer mechanism the importance of the effective diffusion coefficient shall radically increase and shall correspond to the intra-cluster transfer rate. However, inasmuch as $\theta_{cr} = 0,75$ corresponds to the disturbance of the completeness of a sample (sample is destroyed), all diffusion processes in a sample have to manage to proceed before θ shall attain θ_{cr} value through the mechanism proposed above.

To estimate quantitatively the diffusion coefficient value within the framework of the proposed mechanism, we assume that $\tau_{cl}/\tau_{def} \sim 0,2$, the diffusion coefficient of defects $D_{def} \sim 5 \times 10^{-3}$ cm^2/sec (under reasonably intensive stress action its value has to exceed substantially the "liquid" one), and the mobility of impurity molecules in the interior of clusters of defects has to approach the "gaseous" value and to be smaller than the latter by an order of magnitude only. It can be seen from the plots that, as θ approaches θ_{cr} $D_{ef} \approx 0,15 D_{def} \sim$ $\sim 0,75 \times 10^{-3}$ cm^2/sec. Such value of D_{ef} may explain entirely the rate of low-molecular impurity homogenization observed experimentally.

MATHEMATICAL MODEL FOR PROCESS OF MULTIPLE DISINTEGRATION

In this section unit we shall write out a set of equations which describes the kinetics of process of multiple disintegration of a solid polymeric matrix under conditions ISACS. We shall analyze also the structure of steady-states of this equations set.

We are assuming thus, that the external stress action gives rise to the formation of some local stressed microregions, inducing the generation of defects in a three-dimensional network of cross–links. Besides, one has to take into account that the spontaneous rupture of bonds caused by the thermal oscillations of molecules, proceeds at the absence of an external stress action, the rate of this spontaneous process being rised sharply as the melting temperature value (T_m) is attained. Considering that the network rupture proceeds by an activation pathway, the rate constant of the process is temperature dependent, according to the Arrhenius law. The destroyed bonds can be restored then but, as the concentration of a network bonding ruptures attains some critical value, at which the material's completeness is disturbed, the rate of the network bonds restoration falls abruptly. Note that the viscosity of the material falls at the same concentration of bonds ruptures, resulting in an abrupt decrease in the stress value within the bulk of a material. The kinetics of the process of the destruction of three-dimensional network can be described by the following equation:

$$\dot{u} = k_T(1-u) + \sigma k_1 \frac{w}{1-w}(1-u) - k_2 u \theta(u_{cr} - u), \qquad (10)$$

where u and w denote the volume fractions of microdefects and local stressed regions, accordingly ($0 \leq u, w \leq 1$); $\sigma = \kappa F \theta(u_{cr} - u)$ is the stress value caused by the external force F; function $\theta(x_0 - x) = 1$ at $x \ll x_0$, $\theta(x_0 - x) = \varepsilon \ll 1$ at $x \ll x_0$, and in the narrow transition region at $x \propto x_0$ the function $\theta(x_0 - x)$ changes smoothly from 1 to ε; $k_T = k_T(T)$ and $k_1 = k_1(T)$ are the rate constants for the spontaneous microdefects appearance and that induced by the external action; $k_2 = k_2(T)$ is the rate constant for "healing" of defects, the expressions $k_T(T)/k_1(T) = k_{T0} + \delta_T(\theta(T_m - T) - \varepsilon)/(1 - \varepsilon)$ and $k_2(T)/k_1(T) = \exp(\Delta E/T)$ being valid in view of the above considerations (here and later the system of units is assumed whereby the Boltzmann constant is taken to be unity).

The denominator in the addend of the right-hand member in equation (10) shows that in an entirely stressed matrix the rate of defects' nucleation in a three-dimensional network of bonds increases indefinitely, i.e., the over-stressed specimen appears to blow up and collapse into small pieces. The step function $\theta(u_{cr} - u)$ takes into account that the viscosity of the material is high, and restoration of bonds proceeds efficiently till the specimen remains intact, i.e. till the it does not collapse into small pieces, as this is the case at $u \approx u_{cr} \approx 0.75$ (the percolation transition point). The third addend of a right-hand member in equation (10) describes the process of restoration of a network of bonds as a reaction of the first order (first degree of the variable u). This is true if we consider a network of physical bonds. For a network of chemical bonds the restoration of bonds is the process of the second order and the first degree u should be replaced with second (u^2). In the analysis reduced below we shall restrict ourselves to reviewing only of first case (reactions of the first order).

The appearance of microdefects in the network causes the relaxation of the local stressed regions induced in the specimen by an external force. The evolution of these local stressed regions is described by the following kinetic equation:

$$\dot{w} = \sigma k_3(1-u)(1-w) - k_4 u w, \qquad (11)$$

where k_3, k_4 are the rate constants of the nucleation and relaxation of the local stressed regions, k_3 having only a weak dependence on the temperature, and this dependence can be neglected, whereas $k_4 = k_4(T)$ increases with the temperature by the exponential law ($k_4(T) = k_{rel}T$), as the relaxation processes in polymeric systems are limited by the polymer chains mobility.

Inasmuch as under the action of a force applied to the specimen, the dissipation of energy proceeds leading to its heating, we have to describe the process of establishment of thermal equilibrium in the system. The heating of a specimen takes place at the account of the energy stored in local stressed regions during their relaxation, however, a portion of this heat is transferred to the environment. Thus we can write the following equation for the temperature balance:

$$\dot{T} = k_5 uw - k_6(T - T_{env}), \qquad (12)$$

where k_5, k_6 are the temperature-independent rate constant for the specimen's heating at the account of relaxation of local stressed regions, and the heat transfer coefficient of the system to the environment, its temperature being equal T_{env}.

The equations (10–12) describe the evolution of the system under investigation subjected to ISACS. To analyze the character of the evolution of the system described by the (10–12) equations, we shall find the steady-state conditions for these equations. The steady-state conditions to be determined are described by the following set of algebraic equations:

$$\begin{cases} (k_T(T)/k_1(T)) + \left[\kappa F \dfrac{w}{1-w}(1-u) - (k_2(T)/k_1(T))u\right]\theta(u_{cr} - u) = 0, & (13a) \\ \kappa F k_3(1-u)(1-w)\theta(u_{cr} - u) - k_{rel}Tuw = 0, & (13b) \\ k_5 uw - k_6(T - T_{env}) = 0. & (13c) \end{cases}$$

Expressing the function w from the equation (13c) and substituting it in the equation (13b), we shall obtain the equation of the second degree for the variable T, its positive root being of the form:

$$T = \dfrac{T_{env}}{2}\left\{\sqrt{(\varphi(u)q+1)^2 + 4\dfrac{k_5}{k_6 T_{env}}u\varphi(u)q} - (\varphi(u)q - 1)\right\}, \qquad (14)$$

in this case the volume fraction of the local stressed regions w in the specimen shall be, correspondingly equal to:

$$w = \dfrac{k_6 T_{env}}{2k_5 u}\left\{\sqrt{(\varphi(u)q+1)^2 + 4\dfrac{k_5}{k_6 T_{env}}u\varphi(u)q} - (\varphi(u)q + 1)\right\}, \qquad (15)$$

here the designations $q = k_3\kappa F/k_{rel}T_{env}$ and $\varphi(u) = \theta(u_{cr} - u)\dfrac{1-u}{u}$ are inserted. Under a weak external action ($q \ll 1$) we shall find that $w \approx \varphi(u)q$ and $T \approx T_{env} + (k_5/k_6)u\varphi(u)q$, whereas under the intense one ($q \gg 1$) – $w \approx 1 - T/\varphi(u)qT_{env}$ and $T \approx T_{env} + uk_5/k_6$.

Substituting the expressions obtained into the equation (13a), we shall obtain a transcendental equation determining the steady-state value of the variable u_{st}. Solving this equation numerically (with fixed parameters), we shall determine the (stationary) steady-state values of the variables u_{st}, the volume fraction of microdefects in a three-dimensional network, w_{st}, the volume fraction of the local stressed regions, and T_{st}, the temperature. The corresponding bifurcation diagrams with governing parameter q, which varies in direct proportion to the intensity of the internal force F ($q = k_3\kappa F/k_{rel} T_{env}$), are given in Figs. 2–4.

It turns out that in the system the two regions of bistability are observed under the certain constants values of the process. Namely, under the low intensity of external stress the system can exist in the only stable state; as the intensity of the external stress increases, there appear two another steady-states, one being the stable, and the other the unstable one. With the further increase in the external stress value, this new unstable state and the initial stable one merge together and disappear. As a result only one steady-state remains in the system.

The behavior of the bistable system can be traced from Fig. 2, for instance. With the low intensity of the external action, the system can exist in a solitary steady-state. As the governing parameter value increases, the state of the system changes, moving along the **AB** curve. The origination of a pair of steady-states (one being stable at the **CD** branch and the other being unstable at the **CB** branch) in the **C** point does not bring about instantaneously drastic changes in the state of the system. The jumping to the another branch of the states shall occur not until **B** point, when the stable branch **AB** shall merge together with the unstable branch **CB**. Once the jumping has occurred, the system is close to collapse state, as the volume fraction of microdefects at the **CD** branch is essentially equal to the critical one, wherein the macroscopic three-dimensional network connectivity, providing the material intact, is destroyed.

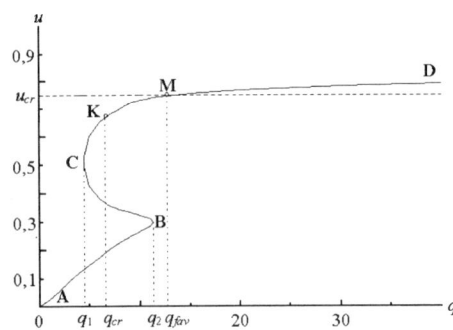 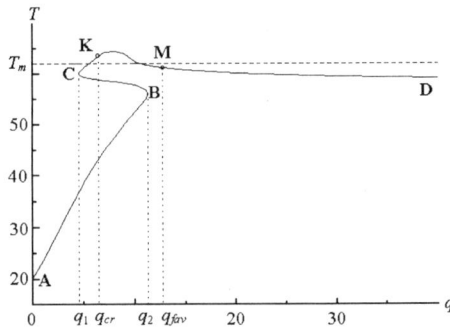

Figure 2. Bifurcation diagram for volume fraction of microdefects in the links network as a function of governing parameter q ($q = k_3\kappa F/k_{rel} T_{env}$). Points **C** and **B** are the bifurcation points and correspond to the critical values of governing parameter q_1 and q_2. In point **K** at $q = q_{cr}$ the branch **CD** loses its steadiness. Point **M** marked off by a circle, corresponds to the value of governing parameter q_{fav}, wherein the multiple disintegration of a material proceeds in the most favoured mode.

Figure 3. Bifurcation diagram for the temperature of a system as a function of governing parameter q ($q = k_3\kappa F/k_{rel} T_{env}$, logarithmic scale). Points **C** and **B** are the bifurcation points and correspond to the critical values of governing parameter q_1 and q_2. In point **K** at $q = q_{cr}$ the branch **CD** loses its steadiness. In point **M** multiple disintegration of a material proceeds in the most favoured mode, the specimen temperature under these conditions being only slightly lower than that of melting point.

After the jumping to the upper branch of states **CD**, if q exceeds slightly q_2, and u is only few less than u_{cr}, the three-dimensional network of the system metamorphoses into set of untied clusters immersed in a slightly coupled infinite percolational cluster, which cannot ensure mechanical solidity and wholeness of a system. Thus after the jumping to the branch **CD** the system finds oneself at once in the highly dispersed phase. With the further increase of the external action (magnification of the parameter q) this macroscopic cluster disappears also.

It seems likely that the multiple disintegration of a substance occurs in the point **M** marked off by a circle at $u_{st} = u_{cr}$. In this particular point the temperature of a substance is close to the melting temperature, but yet remains somewhat below (see Fig. 3). Under these conditions the rupture work value becomes rather small, and the disintegration process proceeds easily, with low energy consumption. Should the temperature of a specimen prove to be somewhat higher than that of a melting point, the melting of the substance occurs, and the polymer melt would appear instead of the highly dispersed phase (the latter is, evidently, possible for the noncross-linked polymers). For the cross-linked polymers (such as, for example, rubber) it is necessary to consider the devulcanization temperature.

Note that it is just the point **M**, wherein the volume fraction of the local stressed regions is relatively small, far smaller than even under very small actions (see Fig. 4). Inasmuch as at the stage of formation a material is subjected, as a rule, to random stress actions, it always comprises some stressed regions. It may happen that in a substance after being subjected to ISACS the portion of local stressed regions can prove to be smaller than before this treatment, with the result that its physical and mechanical properties shall improve.

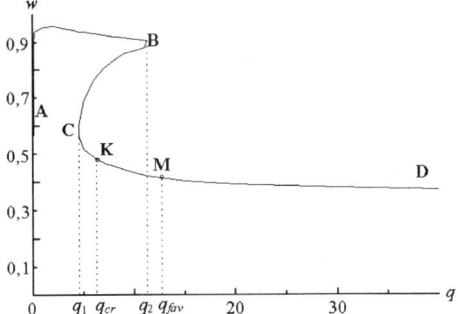

Figure 4. Bifurcation diagram for the volume fraction of a local stressed regions as a function of governing parameter q ($q = k_3 \kappa F / k_{rel} T_{env}$). Points **C** and **B** are the bifurcation points and correspond to the critical values of governing parameter q_1 and q_2. In point **K** at $q = q_{cr}$ the branch **CD** loses its steadiness. In point **M** the multiple disintegration of a material proceeds in the most favoured mode.

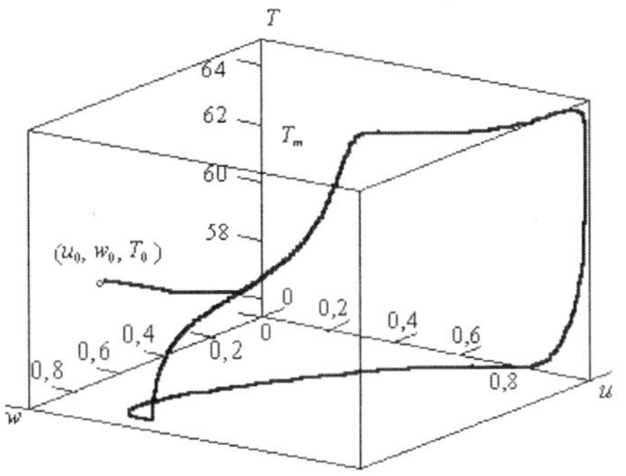

Figure 5. Phase trajectory in the phase space (u, w, T) describing the kinetics of repetitive changes of the system state at the value of governing parameter $q > q_2$.

In conclusion of the present section it has to be emphasized that the bistability occurs in a system only on the condition of the rigid enough choice of the values of the process parameters (constants entering into the set of equations (10)). But even selecting the parameters values necessary for the bistability appearance, we cannot guarantee the "correct" proceeding of the process. The matter is that even with the availability of bistability, after jumping to the branch **CD** at $q > q_2$, if the choice of the parameters would be a poor one, a system can find itself in a state, where the portion of the ruptured bonds u shall be far smaller the critical one, u_{cr}. This means that its compactness in this case shall not disturb, and no multiple disintegration of the specimen does occur. Some other adverse, situation is also possible, namely, the bistability exists in the system and after jumping to the branch **CD** at $q > q_2$ the portion of the ruptured bonds shall be close to the critical one, but the temperature of the system T appears to be higher than T_m, and the melt shall be formed instead the dispersed phase.

By this means the phenomenon of the multiple disintegration of solid polymers under ISACS conditions is very sensitive both to the properties of the material to be dispersed, and to the conditions under which this process proceeds, and for its realization the fulfilment of very rigid limitations is required.

KINETICS OF PROCESS OF MULTIPLE DISINTEGRATION UNDER ISACS

When treating the structure of steady-states of the system, we now turn to the description of the kinetics of multiple disintegration process under ISACS conditions. For this purpose it is necessary to determine the stability of the found steady-states and to calculate the corresponding relaxation times. The knowledge of relaxation times shall allow to determine the specific life time of a system in the neighbourhood of unstable states and specific time of attainment of the stable state, i.e. to determine the time of the transition of a system into the dispersed state (disintegration time).

Since the steady-states have been calculated numerically with fixed values of the process parameters, the study of the stability of these steady-states and calculation of relaxation times has been fulfilled numerically also at the same values of parameters.

The performed analysis of the stability of the determined steady-states shows that the branch **AB** is stable, whereas branch **CB** is the unstable one, which is quite natural. However, the branch **CD**, which is the stable one at the instant of its origination, loses its stability, as the governing parameter attains q_{cr} value ($q_1 < q_{cr} < q_2$). Note that the loss in stability of the branch of states **AB** proceeds without origination of any other steady-states. Though this situation is not encountered too frequently, it is investigated, nevertheless, in the catastrophe theory. The mechanism of the loss of the state stability is the following in this particular case. For each stable stationary state there occurs a certain "attractive region", and if the initial state of a system has thrown into this region, the phase trajectory describing the evolution of the system, shall sooner or later find itself in the stationary point. Those phase trajectory, whose origination have not thrown into the "attractive region", shall, due to evolution of a system, drift apart this stationary point, as though it is unstable. The size of the "attractive region" of the stable stationary point may vary with the change of the governing parameter. If the size of this "attractive region" of the stable stationary point shall decrease with the increase in the governing parameter value, and at some finite quantity of the governing parameter shall become zero, it shall seem to us that the steady-state has lost its stability. Such is indeed the case in the system being studied.

In result at $q_{cr} < q < q_2$ among all the three stationary states of a system the stable shall be the only one, and at $q < q_2$ the sole unstable stationary state shall remain in a system. This means that if the action is the intense enough ($q > q_2$), the state of a system shall constantly vary with time (see Fig. 5), these variations being of a regular cyclic character. The regularity of cycle's appearance has been studied separately, and it has been shown that (with of accuracy of numerical calculations) that the phase trajectory passes in each cycle through the same points as it does in the previous one, i.e. it is not an strange attractor.

Thus, at $q > q_2$ the evolution of a system proceeds in the following way. As the local stressed regions are accumulated, the links providing the completeness of a specimen, begin to rupture. The strain-induced stress relaxes in the neighbourhood of the ruptured links resulting thus in the heating of a system. As the temperature increases, the links are ruptured more and more intensively, whereas the local stressed region progressively runs out, with the resulting formation of a broken up hot unstressed system. Due to the weak consolidation of a system the stresses within its volume are absent and the channel for the transfer of mechanical energy into the thermal one is "switched-off". Giving up the heat to the environment, the system cools down. But in the cold unstressed system the links are re-established quickly and a specimen becomes the monolithic one. In this latter the stresses begin to accumulate once again, resulting in the increase of the process of links rupture, and the cycle begins at the beginning (see Fig. 5).

The described peculiarity of a system allows to explain a number of unusual regularities observed on grinding of polyethylene and rubber in rotor dispersers under ISACS conditions, specifically, periodicity of energy consumption by the installation, delocalization of the grinding zone et al.

CONCLUSION

The mechanism proposed for disintegration of solids under ISACS conditions, differs fundamentally from the universally accepted one for the brittle disintegration based on the notion of the "through–the–thickness" crack propogation. It is just the abandonment of the concept according to which the most stressed points in the matrix, wherein the disintegration occurs, are the stress concentrators, allowed to give the proper description of the abnormal regularities inherent to the disintegration of solids under ISACS conditions.

It is our opinion that of particular interest are two inplications following from the above consideration: disintegration of solids under ISACS conditions belongs to the category of critical phenomena; the dispersing process of solid polymers proceeds by a cyclic process, namely, the disintegration occurs with the subsequent reconstruction of the sample's completeness.

It is apparent that the phenomenological treatment requires an additional validity. For this purpose it is necessary to carry out the microscopic treatment of the dynamics of a system of polymer chains under ISACS, account must be taken of their chemical constitution, rigidity, conformational statistics as well as supermolecular structure. The relevant study is now underway[17–20], but this group of problems is beyond the scope of the phenomenological approach proposed in the present article.

This work is supported by Russian Foundation for Basic Research (grants No 96-03-33237) and by Federal Program "Integration" (project No 43–2.1).

REFERENCES

1. V.G. Nikol'skii, S.A. Wol'fson, *Vysokomol. Soed.* 36B, 6: 1040 (1994) (in Russian).
2. V.A. Zchorin, N.A. Mironov, V.G. Nikol'skii, N.S. Enikolopian, *Doklady Academii Nauk USSR* 244: 1153 (1979) (in Russian).
3. V.A. Zchorin, *Dr. Sci. Thesis.* Institute of Chemical Phys., Moscow (1992) (in Russian).
4. N.S. Enikolopian, V.G. Nikol'skii, E.L. Akopian, *Macromol., Chem. Suppl.* 6: 316 (1984)
5. N.S. Enikolopian, ..., V.G. Nikol'skii, *Vysokomol. Soed.* 30A: 2403 (1988) (in Russian).
6. N.S. Enikolopian, A.Ju. Karmilov, V.G. Nikol'skii, A.M. Ckachatrjan, E.V. Ljadovskaja, *Doklady Academii Nauk USSR* 318, 1: 126 (1991) (in Russian).
7. A.M. Ckachatrjan, A.Ju. Karmilov, V.G. Nikol'skii, E.V. Poddymai, *Vysokomol. Soed.* 33B, 6: 446 (1991) (in Russian).
8. N.S. Enikolopian, ..., V.G. Nikol'skii, *Vysokomol. Soed.* 30A: 2397 (1988) (in Russian).
9. A.K. Karagev,..., D.A. Gor'kov, V.G. Nikol'skii, N.M. Styrikovich, *Vysokomol. Soed.*, 34A: 54 (1992) (in Russian).
10. A.K. Karagev,..., D.A. Gor'kov, V.G. Nikol'skii, N.M. Styrikovich, *Vysokomol. Soed.*, 34A: 98 (1992) (in Russian).
11. A.E. Arinstein, *Doklady Academii Nauk*, 364, 6 (1999) (in Russian).

12. I.I. Akimova, A.E. Arinstein, *Structure and dynamics of molecular systems*. The papers collection. Part 1: 134 (1998) (in Russian).
13. A.E. Arinstein, *Doklady Academii Nauk*, 355, 4: 484 (1997), (in Russian).
14. R. Beker and W. Döring. *Ann. Phys.* (Leipzig) 24B: 719 (1935).
15. P. Dubovskii, *Lecture Notes Series* 23, Seoul, Seoul National University, Research Institute of Mathematics, Global Analysis Research Center: 169 (1994).
16. B.N. Smirnov. *The physics of fractals*: 136, Nauka, Moscow (1991) (in Russian).
17. A.E. Arinstein, L.I. Manivich, S.M. Mezhikovskii, Statistic of oligomeric chain depending on its length, rigidity and temperature. *Preprint. N.N. Semenov Institute of Chemical Phys. RAS*, Chernogolovka (1997) (in Russian).
18. V.I. Irzhak, The role of physical knots during a relaxation oligomeric and polymeric systems. *Preprint. N.N. Semenov Institute of Chemical Phys. RAS*, Chernogolovka. (1997) (in Russian).
19. V.I. Irzhak, *6-th International Conference "Chemistry and Physico-chemistry of oligomers" 1997. Reports Thesises*: 16, Chernogolovka (1997) (in Russian).
20. A.E. Arinstein, *Doklady Academii Nauk*, 358, 3: (1998) (in Russian).

VAPORIZATION AND GROWTH OF AEROSOL PARTICLES, GIVEN INTERNAL HEAT RELEASE AND RADIANT HEAT EXCHANGE

Alexei B. Nadykto[1], Evgenii R. Shchukin[2],

[1]Moscow State Industrial University,
109280 Moscow, Russia

[2]Institute of High Temperature, Russian Academy of Science,
127412 Moscow, Russia

INTRODUCTION

The vaporization and growth processes of drops (solid particles) proceed in a most complex manner when in their volume heat energy[1-5], for example, of electromagnetic nature[4] occur and the process of drop (solid particle) heat exchange with gaseous medium is noticeably influenced by radiant heat exchange.

The major mechanisms of volatile component molecule transport in the vicinity of an immovable particle are molecular diffusion, thermal diffusion and convective radial[3] motion of the gaseous medium. If the molecule transport process is mainly affected by the first two mechanisms, then the vaporization and growth conditions are called diffusive. If the medium convective motion exerts a considerable influence on the vaporization and growth rate, then the conditions are called convective. In convective vaporization conditions the heat and mass transfer process in the drop vicinity is considerably affected by diffusion, thermal diffusion and mass motion of gaseous medium. For the first time the potential existence of gaseous medium hydrodynamical flow near vaporizing body surface directed from the surface was conjectured by Stefan in 1881.

Each particle whose radius is much larger than gaseous mixture molecular free path is surrounded with a thin gas layer (Knudsen layer) in which molecules escaping from the particle surface collide on those flying to the particle surface. If the Knudsen layer effect on the vaporization and growth is minor, then the particle is called large. Otherwise this is referred to as moderately large.

At solution of the large particle vaporization and growth problem the gas temperature near the drop surface can be considered equal to the drop surface temperature T_i and drop material vapor concentration equal to saturated vapor concentration.

When it is necessary to take into account the vaporization growth effect of the Knudsen layer, the problem solution becomes considerably more complicated. At the first phase at solution of volume transport equations relations are determined for distributions of gaseous component relative concentrations c_i and temperature T_e outside the Knudsen layer. Then auxiliary gas kinetics boundary conditions are used to interpolate the relations to the particle surface. For aerosol particles of pure materials the boundary conditions are:

$$(T_e - T_i)_{r=R} = \left(K_T^{(T)} \frac{\partial T_e}{\partial r} + K_T^{(n)} T_i \frac{\partial c_1}{\partial r} \right)_{r=R},$$

$$(c_1 - c_{1S}(T_i))_{r=R} = \left(K_n^{(n)} \frac{\partial c_1}{\partial r} + K_n^{(T)} \frac{1}{T_i} \frac{\partial T_e}{\partial r} \right)_{r=R},$$

where T_i is particle surface temperature; $c_{1S}(T_i) = \left. \frac{n_{1S}(T_i)}{n} \right|_{r=R}$, $n_{1S}(T_i)$ is saturated vapor concentration at temperature T_i, R is particle radius. $(T_e - T_i)|_{r=R}$ and $(c_1 - c_{1S}(T_i))|_{r=R}$ are called temperature and concentration jumps, coefficients $K_T^{(T)}$, $K_T^{(n)}$ and $K_n^{(n)}$, $K_n^{(T)}$ are referred to as gas-kinetic factors of temperature and concentration jumps. Expressions for the gas-kinetic factors are found from Boltzmann equation system solution.

THEORY

Conditions

The particle vaporization and growth occur in two-component gaseous medium. The heat and mass transfer process in the particle vicinity occurs in the quasi-stationary manner. The gravitational convection effect on the particle heat and mass exchange with the environment is minor, as the characteristic sizes of volumes containing a finite quantity of interacting particles are assumed quite small. When the temperature differences in the particle-gaseous medium system are large, it is necessary to take into account the gaseous medium temperature dependence of molecular transport factors. The description of vaporization and growth in gaseous mixtures of dramatically different molecule masses must take into account dependence of molecular transport factors on relative concentrations of gaseous components.

DIFFUSIVE VAPORIZATION AND GROWTH

Equations and Formulas

At large temperature differences in the particle-gaseous medium system it is possible to analytically solve the problem on diffusive vaporization and growth of a single large particle of an arbitrary surface shape. The heat conductivity factor of the particle κ_i is considerably higher than that of gaseous medium κ_e.

If $\kappa_i \gg \kappa_e$, then the temperature distribution across the drop surface can be considered uniform. Here T_e and c_1 distributions in the drop vicinity are described with equation system (1) with boundary conditions (2)

$$\text{div}[nD_{12}(\nabla c_1 + K_T(1/T_e)\nabla T_e)] = 0,$$

$$\mathrm{div}(\kappa_e, \nabla T_e) = 0, \qquad (1)$$

$$T_e|_{S_d} = T_{i0}, \qquad c_1|_{S_d} = c_{1S}(T_{i0}),$$

$$T_e|_\infty = T_{e\infty}, \qquad c_1|_\infty = c_{1\infty} \qquad (2)$$

where T_{i0} is drop surface temperature; $c_{1S}(T_{i0})$ is relative concentration of saturated drop vapors at temperature T_{i0}.

The first two of conditions (2) hold at each drop surface point S_d. The solution to non-linear boundary problem (1) – (2) is

$$c_1 = \left\{ c_{1\infty} + B \int_{T_{e\infty}}^{T_e} \left[F_2 \exp\left(\int_{T_{e\infty}}^{T_e} F_1 dT_e \right) \right] dT_e \exp\left(-\int_{T_{e\infty}}^{T_e} F_1 dT_e \right) \right\}, \qquad (3)$$

$$\int_{T_{e\infty}}^{T_e} \kappa_e dT_e = \int_{T_{e\infty}}^{T_{i0}} \kappa_e dT_e \, G(x_f). \qquad (4)$$

At given T_{i0} the integration constant B appearing in (3) is evaluated through solution to the algebraic equation

$$c_{1S}(T_{i0}) = \left\{ c_{1\infty} + B \int_{T_{e\infty}}^{T_{i0}} \left[F_2 \exp\left(\int_{T_{e\infty}}^{T_e} F_1 dT_e \right) \right] dT_e \right\} \exp\left(-\int_{T_{e\infty}}^{T_{i0}} F_1 dT_e \right). \qquad (5)$$

At integration (4) c_1 appearing in κ_e (on estimation of B with (5)) are found with formula (3). The distribution of function $G(x_f)$ depending solely on spatial coordinates x_f, in the drop vicinity is found from solution to Laplace equation with the first-kind boundary conditions

$$\Delta G = 0,$$

$$G|_S = 1, \qquad G|_\infty = 0. \qquad (6)$$

For drops of spherical, spheroidal and ellipsoidal surface shape $G(x_f)$ are evaluated with analytical expressions.

The flux of drop material molecules Q_1 can be estimated with formula (7),

$$Q_1 = \theta_1(T_{i0}, T_{e\infty}) Q_G, \qquad (7)$$

where

$$Q_G = -\oint_{S_d} \nabla G d\vec{S}.$$

$$\theta_1(T_{i0}, T_{e\infty}) = \left[c_{1S}(T_{i0}) \exp\left(\int_{T_{e\infty}}^{T_{i0}} F_1 dT_e \right) - c_{1\infty} \right] \frac{\int_{T_{e\infty}}^{T_{i0}} \kappa_e dT_e}{\int_{T_{e\infty}}^{T_{i0}} \left[F_2 \exp\left(\int_{T_{e\infty}}^{T_e} F_1 dT_e \right) \right] dT_e}. \tag{8}$$

The integration constant B is estimated basing on condition (5). In Q_G (7) $d\vec{S}_d$ is the drop surface area differential vector element whose direction coincides with that of the outer normal. For drops of spherical, spheroidal and ellipsoidal surface shape the integral coefficients Q_G can be estimated analytically.

The expression for the heat flux $Q_T^{(M)}$, drawn off or to the drop surface due to molecular heat conduction is found from integration of the expression.

$$Q_T^{(M)} = -\oint_{S_d} \kappa_e \vec{\nabla} T_e d\vec{S}_d. \tag{9}$$

Taking into account (4), expression (9) takes the form

$$Q_T^{(M)} = \int_{T_{e\infty}}^{T_{i0}} \kappa_e dT_e Q_G. \tag{10}$$

At steady-state drop vaporization and growth temperature T_{i0} is evaluated with the integral condition of heat conservation

$$Q_W = L_{1S} m_1 Q_1 + Q_T^{(M)} + Q_L, \tag{11}$$

where Q_W is total power of the heat sources; L_{1S} is specific vaporization heat at temperature T_{i0}; Q_L is heat flux due to the heat-exchange radiation equal to

$$Q_L = W(T_{i0}, T_{e\infty}) K_L S_d \tag{12}$$

In (12) S_d is drop surface area; K_L accounts the heat exchange effect of the drop surface shape (in convex particles $K_L = 1$); $W(T_{i0}, T_{e\infty})$ is density of the heat flux due to heat radiation. For quite high temperatures

$$W(T_{i0}, T_{e\infty}) = \sigma(\varepsilon_{i0} T_{i0}^4 - \varepsilon_{ic} T_{e\infty}^4), \tag{13}$$

where σ is Stefan-Boltzmann constant; ε_{i0}, ε_{ic} are drop material radiating and absorption capacity factors.

The area expressions for drops of spherical, spheroidal and ellipsoidal surface shape are

$$S_d = 4\pi R^2;$$

$$S_d = 2\pi \left[b^2 + \frac{a^2 b}{c} \arcsin \frac{c}{a} \right], c = \sqrt{a^2 - b^2}, a > b;$$

$$S_d = 2\pi\left[b^2 + \frac{a^2 b}{2c}\ln\frac{\left(1+\frac{c}{b}\right)}{\left(1-\frac{c}{b}\right)}\right], c = \sqrt{b^2 - a^2}, a < b; \tag{14}$$

$S_d = 8ac\{F_1(e_1, e_2)F_2(e_3, e_4) + F_1(e_3, e_4)F_2(e_1, e_2) - F_2(e_1, e_2)F_2(e_3, e_4)\}; \quad a<b<c;$

where $F_1(k,r) = \int_0^{\pi/2}\left[(1+k\sin^2\alpha)(1+r\sin^2\alpha)\right]^{1/2} d\alpha,$

$$F_2(k,r) = \int_0^{\pi/2}\left[\frac{(1+k\sin^2\alpha)}{(1+r\sin^2\alpha)}\right]^{1/2} d\alpha, \tag{15}$$

$$e_1 = -\frac{a^2-b^2}{a^2}, e_2 = -\frac{a^2-b^2}{a^2-c^2}, e_3 = \frac{b^2-c^2}{c^2}, e_4 = -\frac{b^2-c^2}{a^2-c^2}.$$

Taking into account (6) and (10), condition (11) becomes

$$Q_W = (L_{1S}m_1 B + 1)\int_{T_{e\infty}}^{T_{i0}}\kappa_e dT_e Q_G + W(T_{i0}, T_{e\infty})K_L S_d. \tag{16}$$

Eqs.(11) and (16) should be solved simultaneously with Eq.(5). Here T_{i0} and constant B are simultaneously evaluated.

Having estimated B with (5) and (16), the variation rate of drop mass M_d at the time under consideration (at a given drop surface shape) can be found by formula

$$\frac{dM_d}{dt} = -m_1 Q_1. \tag{17}$$

If the drop volume and shape are related with one-to-one correspondence, in this case at steady-state drop vaporization and growth the drop volume variation with time can be estimated by integration of Eq.(17) simultaneous with solution of algebraic equation system (5), (16).

When drop volume V and integral Q_G are functions of one variable Δ, the drop size variations can be estimated by integration of equation

$$\frac{1}{Q_G}\left(\frac{dV}{d\Delta}\right)\frac{d\Delta}{dt} = -\frac{m_1}{\rho_i} B \int_{T_{e\infty}}^{T_{i0}}\kappa_e dT_e. \tag{18}$$

At the drop transition to the steady state of vaporization or growth the time variation of temperature T_{i0} and drop sizes (at the one-to-one correspondence between the drop volume and surface shape) is found from solving the following differential equation system:

$$Q_W = M_d \alpha_p^{(i)} \frac{dT_{i0}}{dt} + L_{1S} m_1 Q_1 + Q_T^{(M)} + Q_L, \qquad (19)$$

$$\frac{dM_d}{dt} = -m_1 Q_1, \qquad (20)$$

where $\alpha_p^{(i)}$ is drop specific heat capacity. Q_1, $Q_T^{(M)}$ и Q_L are found by formulas (6), (10), (12).

At integration (19) - (20) the coefficient B is evaluated with algebraic equation (5).

Two Particles

When two immovable highly conducting solid particles are heated, the temperature distribution in their vicinity is described by formula

$$\int_{T_{e\infty}}^{T_e} \kappa_e dT_e = \sqrt{2(ch\xi - \cos\eta)} \left[\varphi_1(\xi,\eta) \int_{T_{e\infty}}^{T_{i0}^{(1)}} \kappa_e dT_e + \varphi_2(\xi,\eta) \int_{T_{e\infty}}^{T_{i0}^{(2)}} \kappa_e dT_e \right], \qquad (21)$$

where $T_{i0}^{(1)}$ and $T_{i0}^{(2)}$ are solid particle surface temperatures. The expressions for heat fluxes $Q_T^{(M,1)}$ and $Q_T^{(M,2)}$ (withdrawn from the particle surfaces due to molecular heat conduction) have form (22) and are equal to (23) and (24), respectively,

$$Q_T^{(M,j)} = -\oint_{S_j} \kappa_e \vec{\nabla} T_e d\vec{S}_d, \qquad (22)$$

$$Q_T^{(M,1)} = 4\pi R_1 \left[\psi_1^* \int_{T_{e\infty}}^{T_{i0}^{(1)}} \kappa_e dT_e - \psi_3^* \int_{T_{e\infty}}^{T_{i0}^{(2)}} \kappa_e dT_e \right], \qquad (23)$$

$$Q_T^{(M,2)} = 4\pi R_2 \left[\psi_2^* \int_{T_{e\infty}}^{T_{i0}^{(2)}} \kappa_e dT_e - \psi_4^* \int_{T_{e\infty}}^{T_{i0}^{(1)}} \kappa_e dT_e \right]. \qquad (24)$$

The time dependence of $T_{i0}^{(1)}$ and $T_{i0}^{(2)}$ is found by integration of the equation system

$$M^{(1)} \alpha_p^{(i,1)} \frac{dT_{i0}^{(1)}}{dt} = Q_W^{(1)} - Q_T^{(M,1)} - Q_L^{(1)}, \qquad (25)$$

$$M^{(2)} \alpha_p^{(i,2)} \frac{dT_{i0}^{(2)}}{dt} = Q_W^{(2)} - Q_T^{(M,2)} - Q_L^{(2)},$$

where $\alpha_p^{(i,j)}$ are particle material specific heat capacities; $Q_L^{(j)}$ is heat flux withdrawn from the surface of the j-th particle due to radiant heat withdrawal. At steady-state particle heat exchange with the environment temperatures $T_{i0}^{(j)}$ are found with formulas

$$\int_{T_{e\infty}}^{T_{i0}^{(1)}} \kappa_e dT_e = \frac{1}{4\pi R_1 \left(\psi_1^* \psi_2^* - \psi_3^* \psi_4^*\right)} \left[\psi_2^* \left(Q_W^{(1)} - Q_L^{(1)}\right) + \psi_3^* \frac{R_1}{R_2} \left(Q_W^{(2)} - Q_L^{(2)}\right)\right], \quad (26)$$

$$\int_{T_{e\infty}}^{T_{i0}^{(2)}} \kappa_e dT_e = \frac{1}{4\pi R_2 \left(\psi_1^* \psi_2^* - \psi_3^* \psi_4^*\right)} \left[\psi_1^* \left(Q_W^{(2)} - Q_L^{(2)}\right) + \psi_4^* \frac{R_2}{R_1} \left(Q_W^{(1)} - Q_L^{(1)}\right)\right].$$

When the particles approach each other ($h \to R_1 + R_2$) their temperatures become equal to total temperature T_{i0} evaluated with formula

$$\int_{T_{e\infty}}^{T_{i0}} \kappa_e dT_e = \frac{\left(Q_W^{(1)} + Q_W^{(2)} - Q_L^{(1)} - Q_L^{(2)}\right)}{4\pi \left(R_1 f^{(1)} + R_2 f^{(2)}\right)}. \quad (27)$$

Assembly of N Large Particles

The problem on vaporization of N large highly conducting particles severely heated by internal heat sources can be solved with involvement of analytical formulas, given the distance between surfaces of all the drops is simultaneously much shorter than the radius of the smallest drop. In this case, when making the estimations, the drop surface temperature values can be assumed equal (i.e. $T_{i0}^{(j)} = T_{i0}$). Here the distributions of relative concentration c_1 and temperature T_e in the drop vicinity are described by (3) and (4) with function $G(x_j)$.

The dependence of temperature T_{i0} and drop radii on time t is determined by integration of a system composed of the following $N + 1$ differential equations

$$\sum_{j=1}^{N} Q_W^{(j)} = \sum_{j=1}^{N} \left(\alpha_p^{(i)} M^{(j)} \frac{dT_{i0}}{dt} + L_{1S} m_1 Q_1^{(j)} + Q_T^{(M,j)} + Q_L^{(j)}\right), \quad (28)$$

$$\frac{dM^{(j)}}{dt} = -m_1 Q_1^{(j)}, \quad (29)$$

where $M^{(j)}$ is mass of the j-th drop. The expressions for fluxes $Q_1^{(j)}$ and $Q_T^{(M,j)}$ are

$$Q_1^{(j)} = B \int_{T_{e\infty}}^{T_{i0}} \kappa_e dT_e Q_G^{(j)}, Q_T^{(j,M)} = \int_{T_{e\infty}}^{T_{i0}} \kappa_e dT_e Q_G^{(j)}, Q_G^{(j)} = -\oint_{S_j} \vec{\nabla} G d\vec{S}_j. \quad (30)$$

At integration of system (28) – (29) the B coefficient is evaluated from solution to algebraic system (5). When evaluating T_{i0} at steady-state drop vaporization, in Eq.(28) the derivative dT_{i0}/dt should be assumed zero.

CONVECTIVE VAPORIZATION AND GROWTH OF SPHERICAL PARTICLES

The problem of large and moderately large particle convective vaporization and growth can be analytically solved at small differences of gaseous component concentrations. When the particle vaporization and growth occur at small differences in relative concentration of the second component molecules $|c_2 - c_{2\infty}|/c_{2\infty} \ll 1$, the expressions for the molecule and heat flux densities become

$$q_1 = -\frac{nD_{12}^{(0)}}{c_{2\infty}}\left[\frac{dc_1}{dr} + \tilde{K}_T \frac{1}{T_e}\frac{dT_e}{dr}\right], \tag{31}$$

$$q_2 = 0, \tag{32}$$

$$q_T = (m_1 h_1 + \Phi_e X^{(0)})q_1 - \kappa_e dT_e/dr \tag{33}$$

where

$$D_{12}^{(0)} = D_{12}\big|_{c_1 = c_{1\infty}},\ \tilde{K}_T = \left(K_T^{(0)} + K_T^{(1)}\Delta c_1\right), \kappa_e = \kappa_e^{(0)} + \kappa_e^{(1)}\Delta c_1. \tag{34}$$

In (31) – (32)

$$\Delta c_1 = c_1 - c_{1\infty},\ D_{12}^{(1)} = \frac{1}{D_{12}^{(0)}}\frac{\partial D_{12}}{\partial c_1}\bigg|_{c_1=c_{1\infty}},\ K_T^{(0)} = K_T\big|_{c_1=c_{1\infty}}, \tag{35}$$

$$K_T^{(1)} = \left[K_T^{(0)}\left(\frac{1}{c_{2\infty}} + D_{12}^{(1)}\right) + \frac{\partial K_T}{\partial c_1}\right]_{c_1=c_{1\infty}},\ X^{(0)} = \frac{K_T}{c_1}\bigg|_{c_1=c_{1\infty}},\ \kappa_e^{(0)} = \kappa_e\big|_{c_1=c_{1\infty}},\ \kappa_e^{(1)} = \frac{\partial \kappa_e}{\partial c_1}\bigg|_{c_1=c_{1\infty}},$$

$$p = nkT.$$

Functions h_1, Φ_e and coefficients $D_{12}^{(0)}, D_{12}^{(1)}, K_T^{(0)}, K_T^{(1)}, X^{(0)}, \kappa_e^{(0)}, \kappa_e^{(1)}$ are arbitrarily dependent on temperature T_e. The terms appearing in q_T and proportional to Φ_e are due to the Dufour effect, i.e. thermal energy release due to gaseous component concentration gradients.

The quasi-stationary equations of heat and mass conservation are of form (36)

$$\frac{d}{dr}r^2 q_1 = 0,\ \frac{d}{dr}r^2 q_T = 0. \tag{36}$$

For a large particle equation system (36) should be solved simultaneously with boundary conditions (37)

$$T_e|_{r=R} = T_{i0},\ c_1|_{r=R} = c_{1S}(T_{i0}),$$

$$T_e|_{r\to\infty} = T_{e\infty},\ c_1|_{r\to\infty} = c_{1\infty}, \tag{37}$$

where $c_{1S}(T_{i0}) = n_{1S}(T_{i0})/n|_{r=R}$, $n_{1S}(T_{i0})$ is particle material saturated vapor concentration at temperature T_{i0}.

When determining distributions c_1 and T_e in the vicinity of a moderately large particle, one should take into consideration boundary conditions (38)

$$T_{e0} - T_{i0} = \left(K_T^{(T)} \frac{dT_e}{dr} + T_{i0} K_T^{(n)} \frac{dc_1}{dr} \right)\bigg|_{r=R}, \qquad (38)$$

$$c_{10} - c_{1S} = \left(K_n^{(n)} \frac{dc_1}{dr} + \frac{K_n^{(T)}}{T_{io}} \frac{dT_e}{dr} \right)\bigg|_{r=R},$$

$$T_e|_{r \to \infty} = T_{e\infty}, \ c_1|_{r \to \infty} = c_{1\infty}, \ T_{e0} = T_e|_{r=R}, \ c_{10} = c_1|_{r=R},$$

where $c_{1S} = (T_{i0}) = n_{1S}(T_{i0})/n_0$, $n_0 = n|_{T_e = T_{e0}}$; $K_T^{(T)}$, $K_T^{(n)}$, $K_n^{(n)}$, $K_n^{(T)}$ are gas-kinetic factors of temperature and concentration jumps.

Integrate system (36) to obtain the analytical expression for dependence of concentration c_1 on temperature T_e

$$c_1 = \left\{ c_{1\infty} + \int_{T_{e\infty}}^{T_e} \left[\varphi_2^{(0)} \exp\left(\int_{T_{e\infty}}^{T_e} \varphi_1^{(0)} dT_e \right) \right] dT_e \right\} \exp\left(-\int_{T_{e\infty}}^{T_e} \varphi_1^{(0)} dT_e \right) \qquad (39)$$

c_1 is evaluated by solution of Eq.(40)

$$\frac{dc_1}{dT_e} + \varphi_1^{(0)} c_1 = \varphi_2^{(0)}, \qquad (40)$$

where

$$\varphi_1^{(0)} = \left\{ K_T^{(1)} \frac{1}{T_e} - B\left(\frac{\kappa_e^{(1)} c_{2\infty}}{nD_{12}^{(0)}} \right) \frac{1}{\left[1 - B\left(m_1 h_1 + \Phi_e X^{(0)} \right) \right]} \right\}, \qquad (41)$$

$$\varphi_2^{(0)} = \left\{ B \frac{\left(\kappa_e^{(0)} - \kappa_e^{(1)} c_{1\infty} \right) c_{2\infty}}{nD_{12}^{(0)} \left[1 - B\left(m_1 h_1 + \Phi_e X^{(0)} \right) \right]} - \left(K_T^{(0)} - K_T^{(1)} c_{1\infty} \right) \frac{1}{T_e} \right\}. \qquad (42)$$

Distribution of temperature T_e in the vicinity of a large and moderately large particle is determined by relations (43) and (44), respectively,

$$\int_{T_{e\infty}}^{T_e} \varphi_2(T_e, c_1) dT_e = \frac{R}{r} \int_{T_{e\infty}}^{T_{i0}} \varphi_2(T_e, c_1) dT_e, \qquad (43)$$

$$\int_{T_{e\infty}}^{T_e} \varphi_2(T_e, c_1) dT_e = \frac{R}{r} \int_{T_{e\infty}}^{T_{e0}} \varphi_2(T_e, c_1) dT_e, \qquad (44)$$

where

$$\varphi_1 = B \frac{c_{2\infty}}{n D_{12}^{(0)} \left[1 - B\left(m_1 h_1 + \Phi_e X^{(0)}\right)\right]} - \left[K_T^{(0)} + K_T^{(1)}\left(c_1 - c_{1\infty}\right)\right]\frac{1}{T_e}$$

$$\varphi_2 = \frac{\kappa_e}{\left[1 - B\left(m_1 h_1 + \Phi_e X^{(0)}\right)\right]}. \tag{45}$$

The expressions for fluxes of molecules Q_1 and heat $Q_T^{(M)}$ and Q_T for large and moderately large particles are (46) – (48) and (49) – (51)

$$Q_1 = 4\pi RB \int_{T_{e\infty}}^{T_{i0}} \varphi_2 dT_e, \tag{46}$$

$$Q_T^{(M)} = 4\pi R\left\{1 - B\left[m_1 h_{1S} + \Phi_{eS} X_S^{(0)}\right]\right\} \int_{T_{e\infty}}^{T_{i0}} \varphi_2 dT_e, \tag{47}$$

$$Q_T = 4\pi R \int_{T_{e\infty}}^{T_{i0}} \varphi_2 dT_e, \tag{48}$$

$$Q_1 = 4\pi RB \int_{T_{e\infty}}^{T_{e0}} \varphi_2 dT_e, \tag{49}$$

$$Q_T^{(M)} = 4\pi R\left\{1 - B\left[m_1 h_{10} + \Phi_{e0} X_0^{(0)}\right]\right\} \int_{T_{e\infty}}^{T_{e0}} \varphi_2 dT_e, \tag{50}$$

$$Q_T = 4\pi R \int_{T_{e\infty}}^{T_{e0}} \varphi_2 dT_e, \tag{51}$$

where $Q_T^{(M)}$ is heat flux due to molecular heat conduction, Q_T is total heat flux, subscripts "S" and "0" denote the physical values at $T_e = T_{i0}$ and $T_e = T_{e0}$, respectively.
For large particles B is determined by Eq.(52)

$$c_{lS}(T_{i0}) = c_1(T_{i0}, B) \tag{52}$$

At vaporization of moderately large particles the functional dependence of constant B and temperature T_{e0} on temperature T_i and radius R is found by solution of algebraic equation system (53)

$$T_{e0} - T_{i0} = -\left[K_T^{(T)} + T_{i0} K_T^{(n)} \varphi_1(T_{e0}, c_{10})\right] \frac{1}{R\varphi_2} \int_{T_{e\infty}}^{T_{e0}} \varphi_2 dT_e \bigg|_{T_e = T_{e0}}, \tag{53}$$

$$c_{10} - c_{1S} = -\left[K_n^{(n)}\varphi_1(T_{e0},c_{10}) + K_n^{(T)}\frac{1}{T_{i0}} \right]\frac{1}{R\varphi_2}\int_{T_{e\infty}}^{T_{i0}}\varphi_2 dT_e \Bigg|_{T_e=T_{e0}},$$

Temperature T_{i0} for large and moderately large particles are determined by relations (54) and (55), respectively:

$$\frac{Q_W}{4\pi R} = \left\{1 + B\left[L_{1S}m_1 - \left(m_1 h_{1S} + \Phi_{eS}X_S^{(0)}\right)\right]\right\}\int_{T_{e\infty}}^{T_{i0}}\varphi_2 dT_e + RW(T_{i0}, T_{e\infty}), \quad (54)$$

$$\frac{Q_W}{4\pi R} = \left\{1 + B\left[L_{10}m_1 - \left(m_1 h_{10} + \Phi_{e0}X_0^{(0)}\right)\right]\right\}\int_{T_{e\infty}}^{T_{e0}}\varphi_2 dT_e + RW(T_{i0}, T_{e\infty}), \quad (55)$$

where $W(T_{i0},T_{e\infty}) = Q_L/4\pi R^2$ is heat flux density due to heat radiation.
Variation in the particle radius R can be found by integration of equations (56) (large particles) and (57) (moderately large particles)

$$R\frac{dR}{dt} = -\frac{m_1}{\rho_i} B \int_{T_{e\infty}}^{T_{i0}}\varphi_2 dT_e, \quad (56)$$

$$R\frac{dR}{dt} = -\frac{m_1}{\rho_1} B \int_{T_{e\infty}}^{T_{e0}}\varphi_2 dT_e, \quad (57)$$

where ρ_i is particle material density.
At transition to the steady-state vaporization or growth the dependence of of particle radius R and temperature T_{i0} on time t is determined by simultaneous solution of differential equations (52),(56) and (58) (for large particles) and, respectively, differential equations (53),(57) and (59) (for moderately large particles)

$$\alpha_p^{(i)}\rho_i \frac{R^2}{3}\frac{dT_{i0}}{dt} = \frac{Q_W}{4\pi R} - \left\{1 + B\left[L_{1S}m_1 - \left(m_1 h_{1S} + \Phi_{eS}X_S^{(0)}\right)\right]\right\}\int_{T_{e\infty}}^{T_{i0}}\varphi_2 dT_e - RW(T_{i0}, T_{e\infty}), \quad (58)$$

$$\alpha_p^{(i)}\rho_i \frac{R^2}{3}\frac{dT_{i0}}{dt} = \frac{Q_W}{4\pi R} - \left\{1 + B\left[L_{1S}m_1 - \left(m_1 h_{10} + \Phi_{e0}X_0^{(0)}\right)\right]\right\}\int_{T_{e\infty}}^{T_{e0}}\varphi_2 dT_e - RW(T_{i0}, T_{e\infty}), \quad (59)$$

where $\alpha_p^{(i)}$ is specific heat capacity of particle material.

Analysis

At $c_1 \ll 1$ the expressions of section 1 can be used to describe vaporization and growth of large and moderately large particles spherical in shape occurring in the diffusive conditions. Distribution of temperature T_e in the vicinity of a large particle is determined by relation (53). For large particles the expressions for fluxes of molecules Q_1 and heat $Q_T^{(M)}$ and Q_T are of form (46)-(48). The value B is determined by Eq.(52). T_{i0} for large particles is determined by relation (54). The variation in the large particle radius R can be found by integration of Eq.(56). At transition to the steady-state vaporization or growth the dependence of large particle radius R and temperature T_{i0} on time t is determined by simultaneous solution of differential equations (52),(56) and (58).

The distribution of temperature T_e in the vicinity of a moderately large particle is determined by relation (44). For moderately large particles the expressions for fluxes of molecules Q_1 and heat $Q_T^{(M)}$ and Q_T are (49) – (51). The value B is estimated by Eq.(53). T_{io} for moderately large particles is determined by relation (55). The variation in the moderately large particle radius R can be found by integration of Eq.(57). At transition to the steady-state vaporization or growth the dependence of moderately large particle radius R and temperature T_{io} on time t is determined by simultaneous solution of differential equations (57) and (59), conducted simultaneously with conditions (53).

The coefficients $\varphi_1^{(0)}$, $\varphi_2^{(0)}$, φ_1, φ_2 take form (60),(61),(62) which is simpler than (41), (42), (45):

$$\varphi_1^{(0)} = \left\{ K_T^{(1)} \frac{1}{T_e} - B\left(\frac{\kappa_e^{(1)}}{nD_{12}^{(0)}}\right) \frac{1}{\left[1 - B\left(m_1 h_1 + \Phi_e X^{(0)}\right)\right]} \right\}, \quad (60)$$

$$\varphi_2^{(0)} = \left\{ B \frac{\left(\kappa_e^{(0)} - \kappa_e^{(1)} c_{1\infty}\right)}{nD_{12}^{(0)}\left[1 - B\left(m_1 h_1 + \Phi_e X^{(0)}\right)\right]} - \left(K_T^{(0)} - K_T^{(1)} c_{1\infty}\right) \frac{1}{T_e} \right\} \quad (61)$$

$$\varphi_1 = B \frac{1}{nD_{12}^{(0)}\left[1 - B\left(m_1 h_1 + \Phi_e X^{(0)}\right)\right]} \kappa_e - \frac{K_T^{(0)}}{T_e},$$

$$\varphi_2 = \frac{\kappa_e}{\left[1 - B\left(m_1 h_1 + \Phi_e X^{(0)}\right)\right]}. \quad (62)$$

At practical engineering computations the thermal diffusion term in the expression for φ_B can be neglected.

The formulas can be used to describe free (proceeding without internal heat release and radiant heat exchange) vaporization and growth of large and moderately large aerosol spherical particles.

At the free vaporization and growth the coefficient B and functions $\varphi_1^{(0)}$, $\varphi_2^{(0)}$, φ_1, φ_2 become equal to (63) – (68):

$$B = -\frac{1}{\left[L_{1S} m_1 - \left(h_{1S} m_1 + \Phi_{eS} X_S^{(0)}\right)\right]}, \quad (63)$$

$$\varphi_1^{(0)} = \left\{ K_T^{(1)} \frac{1}{T_e} + \left(\frac{\kappa_e^{(1)} c_{2\infty}}{n D_{12}^{(0)}} \right) \frac{1}{\left[L_{1S} m_1 + m_1 (h_1 - h_{1S}) + \left(\Phi_e X^{(0)} - \Phi_{eS} X_S^{(0)} \right) \right]} \right\}, \quad (64)$$

$$\varphi_2^{(0)} = -\left\{ \frac{\left(\kappa_e^{(0)} - \kappa_e^{(1)} c_{1\infty} \right) c_{2\infty}}{n D_{12}^{(0)} \left[L_{1S} m_1 + m_1 (h_1 - h_{1S}) + \left(\Phi_e X^{(0)} - \Phi_{eS} X_S^{(0)} \right) \right]} + \left(K_T^{(0)} - K_T^{(1)} c_{1\infty} \right) \frac{1}{T_e} \right\}, \quad (65)$$

$$\varphi_1 = -\left\{ \frac{c_{2\infty}}{n D_{12}^{(0)}} \varphi_2^* + \frac{[K_T^{(0)} + K_T^{(0)} (c_1 - c_{1\infty})]}{T_e} \right\} \quad (66)$$

$$\varphi_2 = \left[L_{1S} m_1 - \left(h_{1S} m_1 + \Phi_{eS} X_S^{(0)} \right) \right] \varphi_2^* \quad (67)$$

$$\varphi_2^* = \frac{\kappa_e^{(0)} + \kappa_e^{(1)} (c_1 - c_{1\infty})}{\left[L_{1S} m_1 + m_1 (h_1 - h_{1S}) + \left(\Phi_e X^{(0)} - \Phi_{eS} X_S^{(0)} \right) \right]} \quad (68)$$

The expressions for distribution of temperature T_e, fluxes Q_1 and $Q_T^{(M)}$ take form (69) – (71)

$$\int_{T_{e\infty}}^{T_e} \varphi_2^* dT_e = \frac{R}{r} \int_{T_{e\infty}}^{T_{i0}} \varphi_2^* dT_e, \quad (69)$$

$$Q_1 = -4\pi R \int_{T_{e\infty}}^{T_{i0}} \varphi_2^* dT_e, \quad (70)$$

$$Q_1^{(M)} = 4\pi R L_{1S} m_1 \int_{T_{e\infty}}^{T_{i0}} \varphi_2^* dT_e. \quad (71)$$

CONCLUSION

Analysis of the vaporization and growth shows that:
1. The free vaporization and growth of large particles are characterized with the linear dependence of squared particle radius on vaporization time.
2. The steady-state vaporization or growth temperature is independent on particle radius (for large particles).
3. Under free conditions the thermal diffusion and Dufour effect exert no considerable influence on the vaporization and growth processes.
4. The form of the dependency of the molecular transport coefficients D_{12}, κ_e on gaseous medium temperature T_e and gaseous component concentrations can noticeably influence the vaporization or growth temperature, rate and time.

REFERENCES

1. R.L. Armstrong, Aerosol heating vaporization by pulsed light beams, *Applied Optics*, 23, 1: 148 (1984).
2. U.S. Bennet, G.J. Rosasco, Heating microscopic particles with laser beams, *J. Appl. Phys.*, 49,2: 640 (1978).
3. N.A. Fuks. Drop vaporization and growth in gaseous medium, USSR Academy of Sciences Publishing House, Moscow, 91 (1958).
4. E.R. Shchukin, A.B. Nadykto, Kinetics of large particle vaporization processes, given internal heat release, *VANT, Th. and Appl. Phys.*, 1: 39 (1998).
5. F.A. Williams, On vaporisation for mist by radiation, *International Journal of Heat and Mass Transfer*, 8: 575 (1965).

VAPORIZATION AND GROWTH OF LARGE AND MODERATELY LARGE PARTICLES AT CONSIDERABLE DIFFERENCES OF GASEOUS COMPONENT CONCENTRATIONS

Evgenii R. Shchukin[1], Alexei B. Nadykto[2],

[1]Institute of High Temperature, Russian Academy of Science,
127412 Moscow, Russia

[2]Moscow State Industrial University,
109280 Moscow, Russia

INTRODUCTION

The paper is devoted to theoretical description of the convective (Stefan)[1] vaporization and growth of large and moderately large particles spherical in shape at large relative temperature differences in a particle-gaseous medium system and considerable concentration differences of gaseous components. Vaporization and growth processes in binary gaseous mixtures occurring both without[2] and with internal heat release[3-7] and radiant heat exchange are discussed. Besides the molecular diffusion, thermal diffusion and Dufour effect, i.e. thermal energy release due to gaseous component concentration gradients, are taken into account. Solution of the non-linear transport equation system makes allowance for the arbitrary dependence of the molecular transport factors and thermal diffusion on gaseous medium temperature and gaseous component concentrations.

THEORY OF VAPORIZATION AND GROWTH

Conditions

A large and moderately large pure material particle suspended in two-component gaseous mixture is considered. The particle vaporization and growth occur at Mach numbers $M \ll 1$. By virtue of short times of thermal and diffusive relaxations the particle-gaseous medium system heat and mass exchange occur in the quasi-stationary manner. The particle radius is sufficiently small to neglect the effect of the gravitational convection on the processes of particle heat and mass exchange with gaseous medium. The particle material heat conductivity factor κ_i is much higher in its magnitude than the gas heat conductivity factor κ_e. The temperature distribution across the particle surface is assumed close to uniform even when thermal energy, e.g., electromagnetic in nature, is released inside the drop.

Equations

The distributions of gaseous component temperature and relative concentrations in the particle vicinity are described with the non-linear equation system of molecule and heat conservation:

$$\frac{d}{dr}r^2 q_1 = 0, \quad \frac{d}{dr}r^2 q_2 = 0, \quad \frac{d}{dr}r^2 q_T = 0. \tag{1}$$

where r is radial coordinate; q_1, q_2 and q_3 are radial projections of first and second kind molecule flux densities and heat equal to:

$$q_1 = n_1 v_r - \frac{n^2}{\rho_e} m_2 D_{12}\left[\frac{dc_1}{dr} + K_T \frac{1}{T_e}\frac{dT_e}{dr}\right],$$

$$q_2 = n_2 v_r + \frac{n^2}{\rho_e} m_1 D_{12}\left[\frac{dc_1}{dr} + K_T \frac{1}{T_e}\frac{dT_e}{dr}\right], \tag{2}$$

$$q_T = [m_1 h_1 + \Phi_e(K_T/c_1)]q_1 + [m_2 h_2 + \Phi_e(K_T/c_2)]q_2 - \kappa_e \frac{dT_e}{dr}.$$

In (2) v_r is mass velocity; $c_1 = n_1/n$, $c_2 = n_2/n$, $n = n_1 + n_2$; $\rho_e = m_1 n_1 + m_2 n_2$; n_1, n_2 and m_1, m_2 are concentrations and molecule masses, T_e is gas temperature, D_{12} and κ_e are diffusion and heat conductivity factors; K_T is thermodiffusive coefficient, h1 and h2 are specific enthalpies. The terms appearing in q_T and proportional to Φ_e are due to the Dufour effect, i.e. thermal energy release due to gaseous component concentration gradients.

The second component molecules experience no phase transition on the particle surface. Therefore, q_2 is zero. System (1) is simplified to

$$\frac{d}{dr}r^2 q_1 = 0, \quad \frac{d}{dr}r^2 q_T = 0, \tag{3}$$

where

$$q_1 = -\frac{n}{c_2} D_{12}\left[\frac{dc_1}{dr} + K_T \frac{1}{T_e}\frac{dT_e}{dr}\right], \quad q_T = [m_1 h_1 + \Phi_e(K_T/c_1)]q_1 - \kappa_e \frac{dT_e}{dr}. \tag{4}$$

In the general case the coefficients D_{12}, k_e and K_T appearing in (4) depend on T_e and c_1 and functions h_l and Φ_e on temperature T_e. Because of small pressure differences in the particle vicinity the molecule concentrations n can be found by formula

$$n = p_\infty / kT_e \tag{5}$$

where p_∞ is gaseous medium pressure, k is Boltzmann constant. At a given particle surface temperature T_{i0} in the case of a large particle system (3) is solved simultaneously with boundary conditions (6)–(7)

$$T_e|_{r=R} = T_{i0}, \quad c_1|_{r=R} = c_{1S}(T_{i0}), \tag{6}$$

$$T_e|_{r\to\infty} = T_{e\infty}, \quad c_1|_{r\to\infty} = c_{1\infty}, \tag{7}$$

where $c_{1S}(T_{i0}) = n_{1S}(T_{i0})/n|_{r=R}$, $n_{1S}(T_{i0})$ is particle material saturated vapor concentration at temperature T_{i0}.

When determining the distributions c_1 and T_e in the vicinity of a moderately large particle, it is necessary to take into account boundary conditions (8)–(9)

$$T_{e0} - T_{i0} = \left(K_T^{(T)} \frac{dT_e}{dr} + T_{i0} K_T^{(n)} \frac{dc_1}{dr} \right)\bigg|_{r=R},$$

$$c_{10} - c_{1S} = \left(K_n^{(n)} \frac{dc_1}{dr} + K_n^{(T)} \frac{1}{T_{i0}} \frac{dT_e}{dr} \right)\bigg|_{r=R}, \qquad (8)$$

$$T_e|_{r \to \infty} = T_{e\infty}, \quad c_1|_{r \to \infty} = c_{1\infty}, \qquad (9)$$

where $T_{e0}=T_e|_{r=R}$, $c_{10}=c_1|_{r=R}$, $c_{1S}(T_{i0}) = n_{1S}(T_{i0})/n_0$, $n_0=n|_{Te=Te0}$; $K_T^{(T)}$, $K_T^{(n)}$, $K_n^{(n)}$, $K_n^{(T)}$ are gas-kinetic factors of the temperature and concentration jumps[2,3].

Having integrated (3) with respect to variable r, transfer from the second-order non-linear differential equation system to a simpler first-order non-linear differential equation system:

$$\frac{n}{c_2} D_{12} \left(\frac{dc_1}{dr} + K_T \frac{1}{T_e} \frac{dT_e}{dr} \right) = -\frac{Q_1}{4\pi r^2}, \qquad (10)$$

$$[m_1 h_1 + \Phi_e(K_T/c_1)] \frac{n}{c_2} D_{12} \frac{dc_1}{dr} + \left\{ [m_1 h_1 + \Phi_e(K_T/c_1)] \frac{n}{c_2} D_{12} K_T \frac{1}{T_e} + \kappa_e \right\} \frac{dT_e}{dr} = -\frac{Q_T}{4\pi r^2},$$

where Q_I and Q_T are fluxes of the first component molecules and heat drawn toward or outward the molecule. The factors before the derivatives $\frac{dT_e}{dr}$ and $\frac{dc_1}{dr}$ are continuously differentiable limited functions.

The solution to equation system (10) becomes simpler, if, taking into account the one-to-one r dependence of c_1 and T_e, the relative concentration c_1 is assumed a function immediately depending on T_e. The solution to boundary problem (10), (6)–(7) reduces to solution of two simpler boundary problems (11)–(12) and (13)–(14)

$$\frac{dc_1}{dT_e} = \varphi_1(T_e, c_1), \qquad (11)$$

$$c_1|_{T_e=T_\infty} = c_{1\infty}, \quad c_1|_{T_e=T_{i0}} = c_{1S}(T_{i0}), \qquad (12)$$

$$\varphi_2(T_e, c_1) \frac{dT_e}{dr} = -\frac{Q_T}{4\pi r^2}, \qquad (13)$$

$$T_e|_{r \to \infty} = T_{e\infty}, \quad T_e|_{r=R} = T_{i0},$$

where $\quad c_{1S}(T_{i0}) = n_{1S}(T_{i0})/n_S, \quad n_S = p_\infty/kT_{i0}; \quad B = Q_1/Q_T, \qquad (14)$

$$\varphi_1(T_e, c_1) = \left\{ B\left(\frac{\kappa_e}{nD_{12}}\right) \frac{1}{[1 - B(m_1 h_1 + \Phi_e(K_T/c_1))]} c_2 - K_T \frac{1}{T_e} \right\}, \qquad (15)$$

$$\varphi_2(T_e,c_1) = \frac{K_e}{[1 - B(m_1 h_1 + \Phi_e(K_T/c_1))]}, \qquad (16)$$

In the case of moderately large particles equations (11) and (13) must be solved simultaneously with boundary conditions (17)–(18)

$$T_{e0} - T_{i0} = [K_T^{(T)} + T_{i0} K_T^{(n)} \varphi_1(T_{e0},c_{10})]\frac{dT_e}{dr}\bigg|_{r=R},$$

$$c_{10} - c_{1S} = \left[K_n^{(n)} \varphi_1(T_{e0},c_{10}) + K_n^{(T)} \frac{1}{T_{i0}}\right]\frac{dT_e}{dr}\bigg|_{r=R} \qquad (17)$$

$$c_1|_{T_e = T_\infty} = c_{1\infty}, \quad T_e|_{r\to\infty} = T_{e\infty}, \qquad (18)$$

where $T_{e0} = T_e|_{r=R}$, $c_{10} = c_1|_{r=R}$, $c_{1S}(T_{i0}) = n_{1S}(T_{i0})/n_0$, $n_0 = p_\infty/kT_{e0}$.

Eq.(11) is integrated independently on Eq.(13). The distributions c_1 and T_e are determined by the below-described algorithm. First integrate Eq.(11) simultaneously with the initial condition $c_1|_{T_e = T_\infty} = c_{1\infty}$ to find the dependencies of relative concentration c_1 on temperature T_e at given variable values of the parameter B. The obtained dependencies are used to compute the integrals appearing in expressions (19) and (20) for temperature distributions in the vicinity of large and moderately large particles

$$\int_{T_{e\infty}}^{T_e} \varphi_2(T_e,c_1)dT_e = \frac{R}{r}\int_{T_{ex}}^{T_{i0}} \varphi_2(T_e,c_1)dT_e, \qquad (19)$$

$$\int_{T_{e\infty}}^{T_e} \varphi_2(T_e,c_1)dT_e = \frac{R}{r}\int_{T_{ex}}^{T_{e0}} \varphi_2(T_e,c_1)dT_e. \qquad (20)$$

Expressions (19) and (20) were found at integration of Eq.(13). The distributions c_1 and T_e depend on parameter B. On determination (basing on Eq.(11)) of the functional relation between c_1 and temperature T_e and parameter $B(c_1=c_1(T_e, B))$ the dependence of B on large particle surface temperature T_{i0} is found by solution of algebraic equation (21)

$$c_{1S}(T_{i0}) = c_1(T_{i0}, B) \qquad (21)$$

Eq.(21) was found basing on the second condition of (12). The particle radius does not appear in (11) and (21). Hence, the parameter B and functional relation between c_1 and T_e do not depend on the large particle radius R.

For moderately large particles the functional dependence of constant B and, simultaneously, temperature T_{e0} on T_{i0} and R are now found by solution of algebraic equation system (22)

$$T_{e0} - T_{i0} = -[K_T^{(T)} + T_{i0} K_T^{(n)} \varphi_1(T_{e0},c_{10})]\frac{1}{R\varphi_2}\int_{T_{ex}}^{T_{e0}}\varphi_2 dT_e\bigg|_{T_e=T_{e0}}, \qquad (22)$$

$$c_{10} - c_{1S} = -\left[K_n^{(n)}\varphi_1(T_{e0},c_{10}) + K_n^{(T)}\frac{1}{T_{i0}}\right]\frac{1}{R\varphi_2}\int_{T_{ex}}^{T_{e0}}\varphi_2 dT_e\bigg|_{T_e=T_{e0}}.$$

The expressions for fluxed of molecules Q_I and heat $Q_T^{(M)}$ and Q_T found with taking into account (19) and (20) in the case of large and moderately large particles are of form (23)–(24) and (25)–(26), respectively:

$$Q_1 = 4\pi R B \int_{T_{ex}}^{T_{i0}} \varphi_2 dT_e, \qquad (23)$$

$$Q_T^{(M)} = 4\pi R \{1 - B[m_1 h_{1S} + \Phi_{eS}(K_T\ c_1)_S]\} \int_{T_{ex}}^{T_{i0}} \varphi_2 dT_e, \qquad (24)$$

$$Q_T = 4\pi R \int_{T_{ex}}^{T_{i0}} \varphi_2 dT_e, \qquad (24)$$

$$Q_1 = 4\pi R B \int_{T_{ex}}^{T_{e0}} \varphi_2 dT_e, \qquad (25)$$

$$Q_T^{(M)} = 4\pi R \{1 - B[m_1 h_{10} + \Phi_{e0}(K_T/c_1)_0]\} \int_{T_{ex}}^{T_{e0}} \varphi_2 dT_e, \qquad $$

$$Q_T = 4\pi R \int_{T_{ex}}^{T_{e0}} \varphi_2 dT_e, \qquad (26)$$

where $Q_T^{(M)}$ is heat flux due to molecular heat conduction, Q_T is total heat flux; subscripts "S" and "0" denote the physical values at $T_e = T_{i0}$ and $T_e = T_{e0}$, respectively. Knowing the T_{i0} dependence of Q_1 and $Q_T^{(M)}$ fluxes, in the case of steady-state particle vaporization and growth, T_{i0} can be determined basing on thermal energy conservation condition (27)

$$Q_W = L_{1S} m_1 Q_1 + Q_T^{(M)} + Q_L, \qquad (27)$$

where L_{1S} is specific vaporization heat at temperature T_{i0}, Q_L is heat flux due to radiant heat exchange; Q_W is total power of hear sources. Substitute (23)–(24) and (25)–(26) to (27) to arrive at the following algebraic equations allowing to estimate T_{i0} of large and moderately large particles, respectively:

$$\frac{Q_W}{4\pi R} = \{1 + B[L_{1S} m_1 - (m_1 h_{1S} + \Phi_{eS}(K_T\ c_1)_S)]\} \int_{T_{ex}}^{T_{i0}} \varphi_2 dT_e + RW(T_{i0}, T_{e\infty}), \qquad (28)$$

$$\frac{Q_W}{4\pi R} = \{1 + B[L_{1S} m_1 - (m_1 h_{10} + \Phi_{e0}(K_T/c_1)_0)]\} \int_{T_{ex}}^{T_{e0}} \varphi_2 dT_e + RW(T_{i0}, T_{e\infty}), \qquad (29)$$

where $W(T_{i0}, T_{e\infty}) = Q_L / 4\pi R^2$ is density of heat flux due to heat radiation.

From (28) it follows that at free vaporization and growth of large particles

$$B = -1/[L_{1S} m_1 - (m_1 h_{1S} + \hat{O}_{eS}(K_T/c_{1S}))]. \qquad (30)$$

The free particle vaporization and growth occur when there is no internal heat release at negligible radiant heat exchange. If B is equal to (30), the B factor value only depends on the drop surface temperature T_{i0}. T_{i0} is found on substitution of (30) into algebraic equation (21). Hence, at given $c_{1\infty}$ and $T_{e\infty}$ the free steady-state vaporization and growth of large particles, irrespective of their radius, occur at one and the same surface temperature T_{i0}. Substitute (30) into expressions for φ_1 (15) and φ_2 (16) to arrive at

$$\varphi_1(T_e,c_1) = \left\{ \left(\frac{\kappa_e}{nD_{12}}\right) \frac{c_2}{[L_{1S}m_1 + m_1(h_1 - h_{1s}) + \Phi_e(K_T/c_1) - \Phi_{eS}(K_T/c_1)_s]} + K_T \frac{1}{T_e} \right\}, \quad (31)$$

$$\varphi_2(T_e,c_1) = [L_{1S}m_1 - (m_1 h_{1S} + \Phi_{eS}(K_T/c_1)_s)]\varphi_2^*(T_e,c_1) \quad (32)$$

$$\varphi_2^*(T_e,c_1) = \frac{\kappa_e}{[L_{1S}m_1 + m_1(h_1 - h_{1s}) + \Phi_e(K_T/c_1) - \Phi_{eS}(K_T/c_1)_s]}. \quad (33)$$

Taking into account (32), the formulas for distribution of T_e (19) and fluxes of molecules Q_1 and heat $Q_T^{(M)}$ become

$$\int_{T_{e\infty}}^{T_e} \varphi_2^* dT_e = \frac{R}{r} \int_{T_{e\infty}}^{T_{i0}} \varphi_2^* dT_e, \quad Q_1 = -4\pi R \int_{T_{e\infty}}^{T_{i0}} \varphi_2^* dT_e, \quad Q_1 = 4\pi R L_{1S} m_1 \int_{T_{e\infty}}^{T_{i0}} \varphi_2^* dT_e. \quad (34)$$

Having determined the dependence of B and temperatures T_{i0} and T_{e0} on R with algebraic equation systems (21), (28) and (22), (29), respectively, one can find the time variation of particle radius R through the equation integration

$$R \frac{dR}{dt} = -\frac{m_1}{\rho_i} B \int_{T_{e\infty}}^{T_{i0}} \varphi_2 dT_e, \quad (35)$$

$$R \frac{dR}{dt} = -\frac{m_1}{\rho_1} B \int_{T_{e\infty}}^{T_{e0}} \varphi_2 dT_e, \quad (36)$$

where ρ_i is drop material density. Large particle free vaporization and growth occur at constant temperature of the drop surfaces T_{i0}. The particle radius variation can be therewith determined with formula

$$R^2 = R_b^2 + \frac{2m_1}{\rho_i} t \int_{T_{e\infty}}^{T_{i0}} \varphi_2^* dT_e, \quad (37)$$

where R_b is the initial radius of the particle.

At the drop transfer to the steady-state vaporization or growth the dependence of particle radius R and temperature T_{i0} on time t is found though simultaneous solution of differential equations (35) and (36) and, accordingly, differential equations (38) and (39)

$$\alpha_p^{(i)} \rho_i \frac{R^2}{3} \frac{dT_{i0}}{dt} = \frac{Q_W}{4\pi R} - \{1 + B[L_{1S}m_1 - (m_1 h_{1S} + \Phi_{eS}(K_T/c_1)_s)]\} \int_{T_{e\infty}}^{T_{i0}} \varphi_2 dT_e + RW(T_{i0}, T_{e\infty}), \quad (38)$$

$$\alpha_p^{(i)} \rho_i \frac{R^2}{3} \frac{dT_{i0}}{dt} = \frac{Q_W}{4\pi R} - \left\{1 + B\left[L_{1S}m_1 - \left(m_1h_{10} + \Phi_{e0}(K_T/c_1)_0\right)\right]\right\} \int_{T_{e\infty}}^{T_{ei0}} \varphi_2 dT_e + RW(T_{i0}, T_{e\infty}). \quad (39)$$

In (38) and (39) $\alpha_p^{(i)}$ is specific heat capacity of particle material. Differential equations (35), (38) and (36), (39) are solved simultaneously with algebraic equation (21) and algebraic equation system (22).

Analytical Formulas

Non-linear equation (11) can be integrated in quadratures, when the function $\varphi_1(c_1, T_e)$ can be represented in the following two forms at determination of the T_e dependence of c_1:

$$\varphi_1(c_1, T_e) = \varphi_1^{(T)}(T_e)\varphi_1^{(c)}(c_1)$$

$$\varphi_1(c_1, T_e) = F_1(T_e)c_2 - F_2(T_e)c_2^{1+\nu}, \quad (40)$$

where $\varphi_1^{(T)}$, $F_1(T_e)$, $F_2(T_e)$ are functions depending on T_e; $c_2 = 1 - c_1$; ν is a constant coefficient. Substitute expressions (40) into equation (11), then integrate this with taking into account the first of conditions (12) and (18) to obtain

$$\int_{c_{1\infty}}^{c_1} \frac{1}{\varphi_1^{(c)}(c_1)} dc_1 = \int_{T_{e\infty}}^{T_e} \varphi_1^{(T)}(T_e) dT_e, \quad (41)$$

$$c_1 = 1 - \left\{ c_{2\infty}^{-\nu} - \nu \int_{T_{e\infty}}^{T_e} \left[F_2 \exp\left(-\nu \int_{T_{e\infty}}^{T_e} F_1 dT_e\right)\right] dT_e \right\}^{-1/\nu} \exp\left(-\int_{T_{e\infty}}^{T_e} F_1 dT_e\right). \quad (42)$$

If thermal diffusion negligibly affects the process of heat and mass exchange with gaseous medium and ratios (k_e/D_{12}) can be estimated with one of the following formulas:

$$(\kappa_e/D_{12}) = (\kappa_{e\infty}/D_{12\infty})f^{(T)}(T_e)f^{(c)}(c_1), \quad (43)$$

$$(\kappa_e/D_{12}) = (\kappa_{e\infty}/D_{12\infty})[\Delta_1(T_e) - \Delta_2(T_e)c_2^\nu], \quad (44)$$

where functions $f^{(T)}(T_e)$, $\Delta_1(T_e)$, $\Delta_2(T_e)$ and $f^{(c)}(c_1)$, depend on T_e and c_1, respectively. Then

$$\varphi_1^{(T)} = B\left(\frac{\kappa_{e\infty}}{n_\infty D_{12\infty}}\right) \frac{T_e}{T_{e\infty}} f^{(T)}(T_e) \frac{1}{[1 - Bm_1h_1]}, \quad (45)$$

$$\varphi_1^{(c)}(c_1) = c_2 f^{(c)}(c_1), \quad (46)$$

$$F_1(T_e) = B\left(\frac{\kappa_{e\infty}}{n_\infty D_{12\infty}}\right) \frac{T_e}{T_{e\infty}} \Delta_1(T_e) \frac{1}{[1 - Bm_1h_1]}, \quad (47)$$

$$F_2(T_e) = \frac{\Delta_2(T_e)}{\Delta_1(T_e)} F_1(T_e). \quad (48)$$

If the estimations the term $\Phi_e(K_T/c_1)$ can be neglected in expression (4) for flux density q_T, then, if the ratio (k_e/D_{12}) is (44) and

$$K_T = (1 - c_2^\nu)c_2 \Delta_3(T_e), \tag{49}$$

the expressions for the coefficients F_2 and F_2 appearing in (42) take the following form:

$$F_1(T_e) = B\left(\frac{\kappa_{e\infty}}{n_\infty D_{12\infty}}\right) \frac{T_e}{T_{e\infty}} \Delta_1(T_e) \frac{1}{[1 - Bm_1 h_1]} - \Delta_3(T_e) \frac{1}{T_e}, \tag{50}$$

$$F_1(T_e) = B\left(\frac{\kappa_{e\infty}}{n_\infty D_{12\infty}}\right) \frac{T_e}{T_{e\infty}} \Delta_2(T_e) \frac{1}{[1 - Bm_1 h_1]} - \Delta_3(T_e) \frac{1}{T_e}. \tag{51}$$

It should be noted that the ratio $\Phi_e(K_T/c_1)/h_1 m_1$ is no higher than 16 % even in gases of severely differing molecule masses. Therefore, at the estimations the term $\Phi_e(K_T/c_1)$ can be neglected, as a rule, in coefficients φ_1 (15) and φ_2 (16). If the functional relation between c_1 and T_e is determined by formulas (41) and (42), then for large particles the T_{i0} dependence of B is determined by solution of the algebraic equations which transfer to formulas (41) and (42) on substitution of c_1 and T_e for relative saturated vapor concentration $c_{IS}(T_{i0})$ and temperature T_{i0}, respectively, in their upper integration limits. At free vaporization and growth of large drops the expressions for coefficients $\varphi_1^{(T)}(T_e)$, $F_1(T_e)$, $F_2(T_e)$, (45)–(48), (50), (51) taking into account (30), take a simpler form:

$$\varphi_1^{(T)}(T_e) = -\left(\frac{\kappa_{e\infty}}{n_\infty D_{12\infty}}\right) \frac{T_e}{T_{e\infty}} f^{(T)}(T_e) \frac{1}{[L_{1S} m_1 + m_1(h_1 - h_{1S})]}, \tag{52}$$

$$F_1(T_e) = -\left(\frac{\kappa_{e\infty}}{n_\infty D_{12\infty}}\right) \frac{T_e}{T_{e\infty}} \Delta_1(T_e) \frac{1}{[L_{1S} m_1 + m_1(h_1 - h_{1S})]}, \tag{53}$$

$$F_2(T_e) = \frac{\Delta_2(T_e)}{\Delta_1(T_e)} F_1(T_e), \tag{54}$$

$$F_1(T_e) = -\left(\frac{\kappa_{e\infty}}{n_\infty D_{12\infty}}\right) \frac{T_e}{T_{e\infty}} \Delta_1(T_e) \frac{1}{[L_{1S} m_1 + m_1(h_1 - h_{1S})]} - \Delta_3(T_e) \frac{1}{T_e}, \tag{55}$$

$$F_2(T_e) = -\left(\frac{\kappa_{e\infty}}{n_\infty D_{12\infty}}\right) \frac{T_e}{T_{e\infty}} \Delta_2(T_e) \frac{1}{[L_{1S} m_1 + m_1(h_1 - h_{1S})]} - \Delta_3(T_e) \frac{1}{T_e}. \tag{56}$$

At the estimations expressions (43), (44) for the ratio (k_e/D_{12}) and (47) for the coefficient K_T can be used in the case of gas mixtures with molecules both close and severely differing in mass.

It should be noted that in some cases important for practical applications formulas (41), (42) for the T_e dependence of c_1 allow an explicit representation or can be significantly simplified. Thus, for example, if at the estimations the dependence of (k_e/D_{12}) ratio (42) and (44) on component concentrations can be neglected (here in (43), (44) $f^{(c)}(c_1) = 1$, $\Delta_2(T_e) = 0$), then formulas (41) and (42) become

$$c_2 = c_{2\infty} \exp\left(-\int_{T_{e\infty}}^{T_e} \varphi_1^{(T)} dT_e\right), \tag{57}$$

$$c_2 = c_{2\infty} \exp\left(-\int_{T_{e\infty}}^{T_e} F_1 dT_e\right)^*. \tag{58}$$

At an identical dependence of coefficients $\Delta_1(T_e)$ and $\Delta_2(T_e)$ on T_e (44) when

$$\Delta_2(T_e) = a_1 \Delta_1(T_e), \quad a_1 = const \tag{59}$$

and $K_T = 0$, take the inner integral in (42) to obtain

$$c_1 = 1 - \left\{c_{2\infty}^{-\nu} + a_1 \left[\exp\left(-\nu \int_{T_{e\infty}}^{T_e} F_1 dT_e\right) - 1\right]\right\}^{-\frac{1}{\nu}} \exp\left(-\int_{T_{e\infty}}^{T_e} F_1 dT_e\right). \tag{60}$$

At low differences of gaseous component concentrations

$$c_1 = \left\{c_{1\infty} + \int_{T_{e\infty}}^{T_e} \left[\varphi_2^{(0)} \exp\left(\int_{T_{e\infty}}^{T_e} \varphi_1^{(0)} dT_e\right)\right] dT_e\right\} \exp\left(-\int_{T_{e\infty}}^{T_e} \varphi_1^{(0)} dT_e\right) \tag{61}$$

where

$$\varphi_1^{(0)} = \left\{K_T^{(1)} \frac{1}{T_e} - B\left(\frac{\kappa_e^{(1)} c_{2\infty}}{nD_{12}^{(0)}}\right) \frac{1}{\left[1 - B\left(m_1 h_1 + \Phi_e X^{(0)}\right)\right]}\right\},$$

$$\varphi_2^{(0)} = \left\{B \frac{\left(\kappa_e^{(0)} - \kappa_e^{(1)} c_{1\infty}\right) c_{2\infty}}{nD_{12}^{(0)} \left[1 - B\left(m_1 h_1 + \Phi_e X^{(0)}\right)\right]} - \left(K_T^{(0)} - K_T^{(1)} c_{1\infty}\right) \frac{1}{T_e}\right\}.$$

If enthalpy h_1 is linearly dependent on T_e and the formulas for the coefficients $f^{(T)}(T_e)$, $f^{(c)}(c_1)$, $\Delta_1(T_e)$, $\Delta_2(T_e)$ and ratios $\left(\kappa_e^{(0)} / nD_{12}^{(0)}\right)$, $\left(\kappa_e^{(1)} / nD_{12}^{(0)}\right)$ can be represented as

$$f^{(T)}(T_e) = \frac{T_{e\infty}}{T_e} \quad \Delta_1(T_e) = a_2\left(\frac{T_{e\infty}}{T_e}\right), \quad \Delta_2(T_e) = a_1 a_2\left(\frac{T_{e\infty}}{T_e}\right), \tag{62}$$

$$\frac{\kappa_e^{(0)}}{nD_{12}^{(0)}} = a_3, \quad \frac{\kappa_e^{(1)}}{nD_{12}^{(0)}} = a_4, \tag{63}$$

then on the integration expressions (41), (42) and (61) for c_1 and concentration distribution become, respectively,

$$T_e = T_{e\infty} + (T_{e0} - T_{e\infty})\frac{[1-\theta_1(c_1,c_{1\infty})]}{[1-\theta_1(c_{10},c_{1\infty})]}, \quad B = \frac{1}{m_1\left\{h_{1\infty} + \dfrac{\alpha_{pl}(T_{e0}-T_{e\infty})}{[1-\theta_1(c_{10},c_{1\infty})]}\right\}}, \quad (64)$$

$$c_1 = 1 - \left\{c_{2\infty}^{-\nu} + a_1\left[\theta_2^{A_3}(T_e,T_{e\infty}) - 1\right]\right\}^{-\frac{1}{\nu}}\theta_2^{A_2}(T_e,T_{e\infty}),$$

$$B = \frac{1}{m_1\left\{h_{1\infty} + \dfrac{\alpha_{pl}(T_{e0}-T_{e\infty})}{[1-\theta_2(T_{e0},T_{e\infty})]}\right\}}, \quad (65)$$

$$c_1 = \left\{\left[c_{1\infty} + \left(\frac{a_3}{a_4} - c_{1\infty}\right)c_{2\infty}\right]\frac{1}{\theta_3^{A_4}(T_e,T_{e\infty})} - \left(\frac{a_3}{a_4} - c_{1\infty}\right)c_{2\infty}\right\},$$

$$B = \frac{1}{m_1\left\{h_{1\infty} + \dfrac{\alpha_{pl}(T_{e0}-T_{e\infty})}{[1-\theta_3(T_{e0},T_{e\infty})]}\right\}}, \quad (66)$$

where $T_{e0} = T_e|_{r=R}$, $c_{10} = c_1|_{r=R}$ (for large drops $T_{e0} = T_{i0}$, $c_{10} = c_{1S}(T_{i0})$); $a_1, a_2, a_3, a_4, \alpha_{pl}$ are constant coefficients; $h_1 = h_{1\infty} + \alpha_{pl}(T_e - T_{e\infty})$;

$$\theta_1(c_1,c_{1\infty}) = \exp\left(-A_1\int_{c_{1\infty}}^{c_1}\frac{1}{\varphi_1^{(c)}(c_1)}dc_1\right), \quad (67)$$

$$\theta_2(T_e,T_{e\infty}) = 1 + \left[\left(\frac{c_{20}}{c_{2\infty}}\right)^{\frac{1}{A_2}}\left(\frac{1-a_1c_{2\infty}^\nu}{1-a_1c_{20}^\nu}\right)^{\frac{1}{A_3}} - 1\right]\frac{(T_e-T_{e\infty})}{(T_{e0}-T_{e\infty})}, \quad (68)$$

$$\theta_3(T_e,T_{e\infty}) = 1 + \left\{\left[\frac{c_{1\infty} + \left(\dfrac{a_3}{a_4} - c_{1\infty}\right)c_{2\infty}}{c_{10} + \left(\dfrac{a_3}{a_4} - c_{1\infty}\right)c_{2\infty}}\right]^{\frac{1}{A_4}} - 1\right\}\frac{(T_e-T_{e\infty})}{(T_{e0}-T_{e\infty})}, \quad (69)$$

$$A_1 = \frac{D_{12\infty}}{(\kappa_{e\infty}/m_1\alpha_{pl}n_\infty)},$$

$$A_2 = \frac{a_2}{A_1}, \quad A_3 = \nu A_2, \quad A_4 = \frac{a_4 c_{2\infty}}{\alpha_{pl}m_1}. \quad (70)$$

If $a_4 = 0$, then formulas (66) become

$$T_e = T_{e\infty} + (T_{e0} - T_{e\infty})\frac{[1-\theta_4(c_1,c_{1\infty})]}{[1-\theta_4(c_{10},c_{1\infty})]}, \qquad B = \frac{1}{m_1\left\{h_{1\infty} + \dfrac{\alpha_{pl}(T_{e0} - T_{e\infty})}{[1-\theta_4(c_{10},c_{1\infty})]}\right\}}. \qquad (71)$$

In formulas (71)

$$\theta_4(c_1,c_{1\infty}) = \exp\left[-\frac{A_5(c_1 - c_{1\infty})}{c_{2\infty}}\right], \qquad A_5 = \frac{\alpha_{pl} m_1}{a_3}. \qquad (72)$$

If in the T_e variation range under consideration D_{12} and k_e can be found with formulas

$$D_{12} = D_{12\infty}\left[1 + S_g\left(\frac{T_e - T_{e\infty}}{T_{e\infty}}\right)^{A_6}\right]\left(\frac{T_e}{T_{e\infty}}\right), \quad \kappa_e = \kappa_{e\infty}\left[1 + S_g\left(\frac{T_e - T_{e\infty}}{T_{e\infty}}\right)^{A_6}\right], \qquad (73)$$

then make integration in (25) and (26) with taking into account (64) and (71) to obtain the following expressions for fluxes Q_1 and $Q_T^{(M)}$, respectively:

$$Q_1 = 4\pi R n_\infty D_{12\infty}\left\{\ln\frac{c_{2\infty}}{c_{20}} - S_g\left(\frac{T_{e0} - T_{e\infty}}{T_{e\infty}}\right)^{A_6}\int_{c_{2\infty}}^{c_{20}}\frac{1}{c_2}\left[\frac{1-\theta_1(c_1,c_{1\infty})}{1-\theta_1(c_{10},c_{1\infty})}\right]^{A_6} dc_2\right\}, \qquad (74)$$

$$Q_T^{(M)} = \frac{m_1\alpha_{pl}(T_{e0} - T_{e\infty})\theta_1(c_{10},c_{1\infty})}{[1-\theta_1(c_{10},c_{1\infty})]}Q_1, \qquad (75)$$

$$Q_1 = 4\pi R n_\infty D_{12\infty}\frac{1}{c_{2\infty}}\left\{(c_{10} - c_{1\infty}) + S_g\left(\frac{T_{e0} - T_{e\infty}}{T_{e\infty}}\right)^{A_6}\int_{c_{1\infty}}^{c_{10}}\frac{1}{c_2}\left[\frac{1-\theta_4(c_1,c_{1\infty})}{1-\theta_4(c_{10},c_{1\infty})}\right]^{A_6} dc_1\right\}, (76)$$

$$Q_T^{(M)} = \alpha_{pl} m_1 \frac{\theta_4(c_{10},c_{1\infty})(T_{e0} - T_{e\infty})}{[1-\theta_4(c_{10},c_{1\infty})]}Q_1, \qquad (77)$$

where S_g and A_6 are constant factors. At $f^{(c)}(c_1) = 1$ and $A_6 = 1$ formulas (74) and (76) for Q_1 become

$$Q_1 = 4\pi R n_\infty D_{12\infty}\left\{\ln\frac{c_{2\infty}}{c_{20}} - S_g\left(\frac{T_{e0} - T_{e\infty}}{T_{e\infty}}\right)\cdot\left[\frac{1}{(1-\theta_1(c_{10},c_{1\infty}))}\ln\frac{c_{2\infty}}{c_{20}} - \frac{1}{A_1}\right]\right\}, \qquad (78)$$

$$Q_1 = 4\pi R n_\infty D_{12\infty}\frac{1}{c_{2\infty}}\left\{(c_{10} - c_{1\infty}) + S_g\left(\frac{T_{e0} - T_{e\infty}}{T_{e\infty}}\right)\cdot\left[\frac{(c_{10} - c_{1\infty})}{(1-\theta_4(c_{10},c_{1\infty}))} - \frac{c_{2\infty}}{A_5}\right]\right\}. \qquad (79)$$

where $\theta_1(c_{10}, c_{1\infty}) = \exp\left(A_1 \ln \frac{c_{20}}{c_{2\infty}}\right)$.

At constant factors $\Delta_3(T_e)$ (47) and $K_T^{(0)}$, $K_T^{(1)}$ of expression (42) and (61) take the form

$$c_1 = 1 - \left\{ c_{2\infty}^{-\nu} - \nu \int_{T_{e\infty}}^{T_e} \left[F_4 \exp\left(-\nu \int_{T_{e\infty}}^{T_e} F_3 dT_e\right)\right] dT_e \right\}^{-\frac{1}{\nu}} \left(\frac{T_e}{T_{e\infty}}\right)^{\Delta_3} \exp\left(-\int_{T_{e\infty}}^{T_e} F_3 dT_e\right), \quad (80)$$

$$c_1 = \left\{ c_{1\infty} + \int_{T_{e\infty}}^{T_e} \left[\varphi_6 \exp\left(\int_{T_{e\infty}}^{T_e} \varphi_5 dT_e\right)\right] dT_e \right\} \left(\frac{T_{e\infty}}{T_e}\right)^{K_T^{(1)}} \exp\left(-\int_{T_{e\infty}}^{T_e} \varphi_5 dT_e\right), \quad (81)$$

where

$$F_3 = B\left(\frac{\kappa_{e\infty}}{n_\infty D_{12\infty}}\right) \frac{T_e}{T_{e\infty}} \Delta_1(T_e) \frac{1}{(1 - Bm_1 h_1)}, \quad (82)$$

$$F_4 = \left[B\left(\frac{\kappa_{e\infty}}{n_\infty D_{12\infty}}\right) \frac{T_e}{T_{e\infty}} \Delta_2(T_e) \frac{1}{(1 - Bm_1 h_1)} - \Delta_3 \frac{1}{T_e} \right] \left(\frac{T_e}{T_{e\infty}}\right)^{\nu \Delta_3}, \quad (83)$$

$$\varphi_5 = -B\left(\frac{\kappa_e^{(1)} c_{2\infty}}{nD_{12}^{(0)}}\right) \frac{1}{\left[1 - B(m_1 h_1 + \Phi_e X^{(0)})\right]}, \quad (84)$$

$$\varphi_6 = \left\{ B \frac{(\kappa_e^{(0)} - \kappa_e^{(1)} c_{1\infty}) c_{2\infty}}{nD_{12}^{(0)} \left[1 - B(m_1 h_1 + \Phi_e X^{(0)})\right]} - (K_T^{(0)} - K_T^{(1)} c_{1\infty}) \frac{1}{T_e} \right\} \left(\frac{T_e}{T_{e\infty}}\right)^{K_T^{(1)}}. \quad (85)$$

$$D_{12}^{(1)} = \frac{1}{D_{12}^{(0)}} \frac{\partial D_{12}}{\partial c_1}\bigg|_{C_1 = C_{1\infty}}, \quad K_T^{(0)} = K_T\big|_{C_1 = C_{1\infty}},$$

$$K_T^{(1)} = \left[K_T^{(0)} \left(\frac{1}{c_{2\infty}} + D_{12}^{(1)}\right) + \frac{\partial K_T}{\partial c_1}\right]_{C_1 = C_{1\infty}}, \quad X^{(0)} = \frac{K_T}{c_1}\bigg|_{C_1 = C_{1\infty}}, \quad \kappa_e^{(0)} = \kappa_e\big|_{C_1 = C_{1\infty}}, \quad \kappa_e^{(1)} = \frac{\partial \kappa_e}{\partial c_1}\bigg|_{C_1 = C_{1\infty}}.$$

At $\Delta_1(T_e)$, $\Delta_2(T_e)$, $(\kappa_e^{(0)}/nD_{12}^{(0)})$, $(\kappa_e^{(1)}/nD_{12}^{(1)})$ and h_1 according to (62), (63) and linear dependence of the function Φ_e on T_e we take integrals of F_3 and φ_5 to obtain

$$c_1 = 1 - \left\{ c_{2\infty}^{-\upsilon} - \upsilon \int_{T_{e\infty}}^{T_e} F_5 dT_e \right\}^{-\frac{1}{\upsilon}} \left(\frac{T_e}{T_{e\infty}}\right)^{\Delta_3} \theta_4^{A_2}(T_e, T_{e\infty}), \quad (86)$$

$$c_1 = \left\{ c_{1\infty} + \int_{T_{e\infty}}^{T_e} \varphi_7 dT_e \right\} \left(\frac{T_{e\infty}}{T_e} \right)^{K_T^{(1)}} \theta_5^{-A_4}(T_e, T_{e\infty}), \tag{87}$$

where

$$F_5 = F_4 \theta_4^{A_3}(T_e, T_{e\infty}), \quad \varphi_7 = \varphi_6 \theta_5^{A_4}(T_e, T_{e\infty}), \tag{88}$$

$$\theta_4(T_e, T_{e\infty}) = 1 - \frac{Bm_1 \alpha_{pl}(T_e - T_{e\infty})}{[1 - Bm_1 h_{l\infty}]}, \tag{89}$$

$$\theta_5(T_e, T_{e\infty}) = 1 - \frac{B(m_1 \alpha_{pl} + \beta_{pl})(T_e - T_{e\infty})}{\left[1 - B\left(m_1 h_{l\infty} + \Phi_{e\infty} X_\infty^{(0)}\right)\right]}, \tag{90}$$

$$\Phi_e X^{(0)} = \Phi_{e\infty} X_\infty^{(0)} + \beta_{pl}(T_e - T_{e\infty}). \tag{91}$$

Coefficients A_2, A_3, A_4 are found with formulas (70).

Integration (41), (42) becomes considerably simpler when the convective transversal heat transfer weakly affects the distribution of molecule temperature and concentrations, hence, the rate and time of drop vaporization and growth. Here it is not necessary to take into account terms proportional to h_1 and Φ_e in the coefficients appearing in the transport equation and expressions for c_1.

Calculation and Analysis

Fig. 1 presents the curve of ratio $\gamma_1 = Q_1^{(D)}/Q_1^{(K)}$ vs. surface temperature T_{i0} for large water drops in air of $T_{e\infty} = 293$ K, $c_{1\infty} = 0$ and pressure $p = 101325$ Pa. $Q_1^{(D)}$ and $Q_1^{(K)}$ are molecule fluxes withdrawn from the drop surface due to the diffusive and combined diffusive and convective mechanisms, respectively. The values of $Q_l^{(D)}$ and $Q_l^{(K)}$ were found with formulas (74) and (79), respectively, at $A_6 = 1$ and $S_g = 0.8$. The values of the factor $f^{(c)}(c_1)$ were taken from[4].

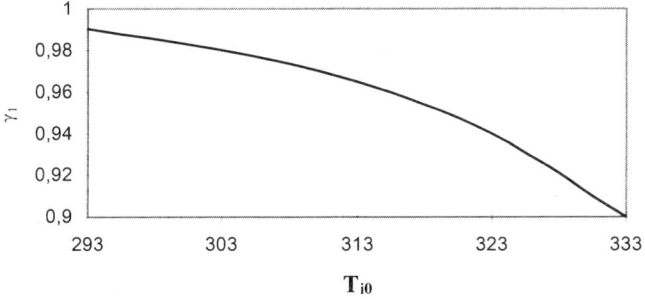

Figure 1. Factor γ_1 as a function of surface temperature T_{i0} for large water drops in air of $T_{e\infty} = 293$ K, $c_{1\infty} = 0$ and pressure $p = 101325$ Pa.

The dependence of the molecular transport factors on T_e can exert a considerable effect on the particle vaporization and growth processes. Fig. 2 presents the curve for temperature $T_{e\infty}$ of flux Q_1^* made dimensionless

$$Q_1^* = Q_1/4\pi R n^* D^*_{12}. \tag{92}$$

At construction of curves 1 the temperature dependence of total molecule concentration and heat conduction and diffusion factors was taken into account. The coordinates of curve 2 were found at constant molecule concentrations and molecular transport factors equal to these variables at large distances from the drop surface. The computations used expression (64) for T_e. The dependencies of molecular transport factors on T_e and the values of α_{pl} were taken from[4]. The asterisks denote n and D_{12} at temperature T_e=272.15 K.

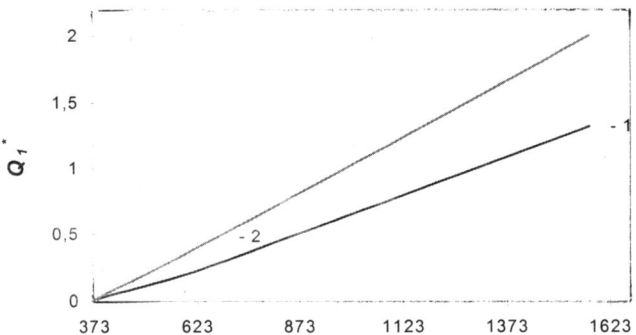

Figure 2. Flux Q_1^* as a function of the temperature $T_{e\infty}$.

Fig. 3 presents the curves for fluxes Q_1^* made dimensionless of potassium atoms withdrawn from a large potassium drop surface vs. ratio $Q_W\ 4\pi R \kappa_{e\infty} T_{e\infty}$. The estimations assumed that the potassium drop vaporization occurs in gaseous medium composed of K and He atoms with $T_{e\infty}$ = 200 K, $c_{1\infty}$ = 0, p_∞=101325 Pa. The values of Q_1^* were found at $n^* = n_\infty$ и $D_{12}^* = D_{12\infty}$, $K_T = 0.3 c_1$ (curve 1) , $K_T = 0$ (curve 2) and the following power dependence of heat conductivity and diffusion factors on temperature

$$\kappa_e = \kappa_{e\infty}(T_e\ T_{e\infty})^\alpha, \quad D_{12} = D_{12\infty}(T_e\ T_{e\infty})^{1+\alpha}, \quad \alpha=0.7 \tag{93}$$

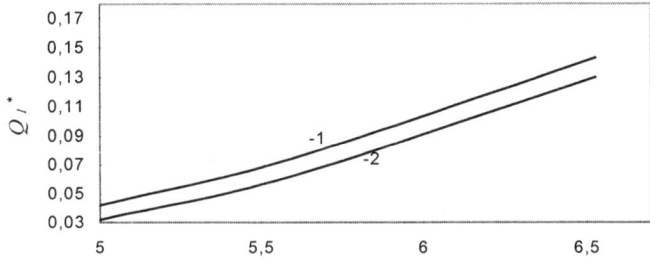

Figure 3. Fluxes Q_1^* as a function of the ratio $Q_W\ 4\pi R \kappa_{e\infty} T_{e\infty}$.

At fixed surface temperature T_{i0} vaporization of drops squeezed out of capillaries can take place. The curves in Fig. 4 show ratio $\gamma_2 = Q_1/4\pi R n^{(S)} D_{12}^{(S)} c_{1S}(T_{i0})$ vs. temperature $T_{e\infty}$ at fixed large water drop surface temperature $T_{i0}=30°C$. The water drop vaporization occurs in gaseous mixture composed of H_2O molecules and He atoms at $p_\infty = 101325$ Pa. The values of Q_1 were found at $K_T=0$ (curve 1), $K_T=0.3c_1$ (curve 2) and power dependence (93) of coefficients k_e and D_{12} on temperature T_e. The subscript "S" denotes the values of n and D_{12} at $T_e = T_{i0}$.

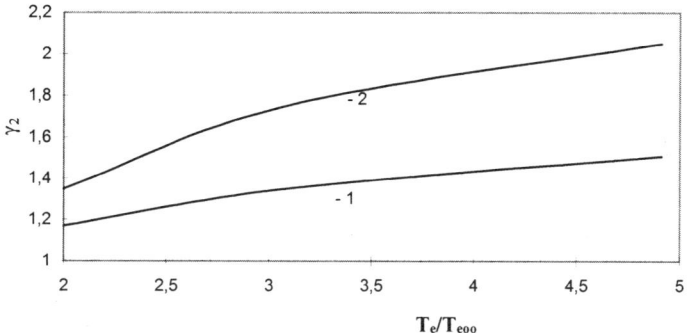

Figure 4. Factor $\gamma_2 = Q_1/4\pi R n^{(S)} D_{12}^{(S)} c_{1S}(T_{i0})$ as a function of the temperature $T_{e\infty}$ at fixed large water drop surface temperature $T_{i0}=30°C$.

Experiment

Vaporization of water drops 14–70 mkm in diameter freely falling in the continuos radiation field of the CO_2 laser of $I = 930$–2200 W/cm² intensity was experimentally studied in[5]. The theoretical results are compared with the experiments in Fig. 5 and Fig. 6. Axis X is time in 10^3 s, axis Y is drop radius in microns.

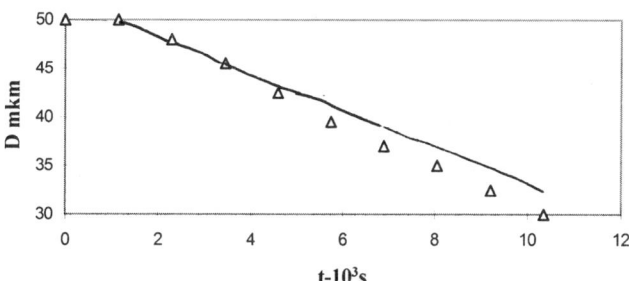

Figure 5. Radius R as a function of the vaporization time for water drops 50 μm in diameter freely falling in the continuos radiation field of the CO_2 laser of $I = 930$ W/cm².

CONCLUSION

The theoretical description found out that convective processes of large particle vaporization and growth occurring when there are no internal heat sources and radiant heat exchange are characterized with a linear time dependence of the squared radius of the particle both at small and considerable temperature and concentration differences of molecules in the particle vicinity and at an arbitrary dependence of the molecular transport factors on medium parameters. The surface temperature of the large particle is independent on its radius.

The effect of the transversal convective heat transfer on the distribution of molecule temperature and concentrations in the vicinity of particles can be neglected if conditions (94) are met:

$$|B[m_1h_1 + \Phi_e(K_T/C_1)]| \ll 1, \qquad [m_1h_{10} + \Phi_{e0}(K_T/C_1)_0] \ll L_{1S}m_1. \qquad (94)$$

At large particle vaporization and growth occurring when there are no internal heat sources and radiant heat exchange the convective heat transfer may be not taken into account if condition (95) is met:

$$|m_1(h_1 - h_{1S}) + \Phi_e(K_T/c_1) - \Phi_{eS}(K_T/c_1)_S| \ll m_1L_{1S}. \qquad (95)$$

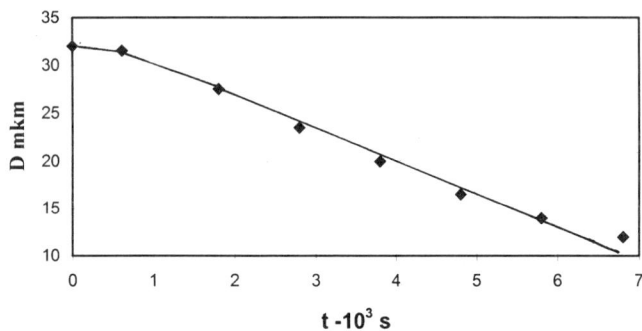

Figure 6. Radius R as a function of the vaporization time for water drops 32 μm in diameter freely falling in the continuos radiation field of the CO_2 laser of $I = 2200$ W/cm^2.

The results of this paper can be used to determine extremes of possible employment of the formulas obtained in the diffusive approximation. The estimations showed that the formulas found in the diffusive approximation, for example[3], allow to estimate the evaporation rate of water drops heated by internal heat sources within 10 % accuracy at relative concentration of water molecules $c_l \le 0.2$.

The obtained results allow to estimate the particle vaporization rate and time with allowance for the thermal diffusion effect. The estimations showed that even in gaseous media with molecules dramatically differing in mass the thermal diffusion exerts no considerable effect on the drop vaporization and growth rate.

REFERENCES

1. N.A. Fuks. Drop vaporization and growth in gaseous medium, USSR Academy of Sciences Publishing House, Moscow, 91 (1958).
2. J.C. Maxwell, Collected Scientific Papers, Cambridge, 11: 625 (1890).
3. E.R. Shchukin, A.B. Nadykto, Kinetics of large particle vaporization processes, given internal heat release , *VANT, Th. and Appl. Phys.*, 1: 39 (1998).
4. N.B. Vargaftick, Handbook, *Thermal and heat qualities of gases and liquids*, Nauka Publishing House, Moscow, 721 (1972).
5. R.V. Ivanov, V.Ya. Korovin, Vaporization of water drops, *I.F.J.*, 34, 5: 807 (1978).
6. R.L. Armstrong, Aerosol heating vaporization by pulsed light beams, *Applied Optics*, 23, 1: 148 (1984).
7. F.A. Williams, On vaporisation for mist by radiation, *International Journal of Heat and Mass Transfer*, 8: 575 (1965).

DIFFUSIVE VAPORIZATION AND GROWTH OF ASSEMBLY OF N-LARGE PARTICLES

Evgenii R. Shchukin[1], Alexei B. Nadykto[2]

[1]Institute of High Temperature, Russian Academy of Science,
127412 Moscow, Russia

[2]Moscow State Industrial University,
109280 Moscow, Russia

INTRODUCTION

The paper discusses the theory of vaporization and growth of an assembly of N-large interacting particles in two-component gaseous medium at high relative temperature differences in the particle vicinity. Vaporization and growth diffusive processes at low particle material concentrations are considered. Major mechanisms of volatile component molecule transport in the vicinity of an immovable aerosol particle are molecular diffusion and convective radial (Stefan) motion of gaseous medium. When the particle vaporization and growth occur at low material vapor concentrations, the major molecule transport mechanisms are diffusion and thermal diffusion. These vaporization and growth conditions are referred to as diffusive.

In real aerodisperse systems particles are arbitrarily apart. When the distances between the particles are comparable with their sizes, their approach to each other exerts a noticeable influence on the vaporization and growth processes. Drop pairs draw close together most frequently[2-6]. Refs.[2-3] discuss vaporization of large pure material drops of identical radius suspended in binary mixture under diffusive conditions at relatively small temperature differences. Ref.[2] found simultaneous solutions to the equations of heat transfer and volatile component molecule transport. The authors of[2] found values of indefinite integration constants at the numerical solution of the algebraic equation system, although the boundary problem allowed for an exact solution. Ref.[2] shows by example of water drops growing in air supersaturated with vapors that temperature and concentration space re-distribution among drops considerably depends on the distance between the drop centers and that the growth rate of each of the interacting drops decreases as they approach each other. Ref.[3] obtained analytical expressions for distributions of volatile component temperature and concentration in the drop vicinity. A detailed theoretical description of the diffusive vaporization of two large drops arbitrarily apart from each other can be found in ref.[4].

Refs.[5, 6] present numerical solution to the problem on vaporization of an assembly of interacting drops spherical and non-spherical in surface shape under diffusive conditions at small temperature differences. At the problem solution the vapor concentrations were assumed uniformly distributed by surface of each drop.

THEORY OF THE VAPORIZATION AND GROWTH

Conditions

The particle vaporization and growth occur in two-component gaseous medium. The heat and mass transfer process in the particle vicinity occurs in the quasi-stationary manner. The gravitational convection effect on the heat and mass exchange with the environment is minor, as the characteristics sizes of the volumes containing a finite number of interacting particles are assumed quite small. It is suggested that when there is no internal heat release and radiant heat exchange the effect of the macroscopic diffusive transport on the particle heat exchange with gaseous medium can be neglected. The effect of macroscopic (in our case diffusive) transport on heat exchange of particles with gaseous medium is minor.

At large temperature differences in the particle-gaseous medium system it is necessary to take into consideration the gaseous medium temperature dependence of the molecular transport factors. When describing the vaporization and growth in gaseous mixtures of severely differing molecule masses, it is necessary to take into account the dependence of the molecular transport factors on relative concentrations of gaseous components.

The vaporization and growth processes of an assembly of N-large interacting particles of arbitrary surface shape occurring without internal heat release and radiant heat exchange are considered.

Equations and Formulas

The distribution of temperatures T_e, $T_i^{(j)}$ and relative concentration c_1 of the particle material molecules is described with equation system (1)–(3) with boundary conditions (4)–(6):

$$\mathrm{div}[nD_{12}(\nabla c_1 + K_T(1/T_e)\nabla T_e)] = 0, \tag{1}$$

$$\mathrm{div}(\kappa_e \nabla T_e) = 0, \tag{2}$$

$$\mathrm{div}\left(\kappa_i^{(j)} \nabla T_e^{(j)}\right) = 0, \tag{3}$$

$$T_e|_{S_j} = T_i^{(j)}|_{S_j}, \qquad c_1|_{S_j} = c_{1S}(T_i^{(j)})|_{S_j}, \tag{4}$$

$$-\kappa_i^{(j)} \nabla_\perp T_i^{(j)}\bigg|_{S_j} = -\kappa_e \nabla T_e - L_1 m_1 n D_{12}\left(\nabla_\perp c_1 + K_T \frac{1}{T_e}\nabla_\perp T_e\right)\bigg|_{S_j}, \tag{5}$$

$$c_1|_\infty = c_{1\infty}, \qquad T_e|_\infty = T_{e\infty}, \tag{6}$$

where $\kappa_i^{(j)}$ and $T_i^{(j)}$ are heat conductivity and temperature factors for the j-th particle; $j=1, 2, ..., N$ are the numbers of the particles; $c_{1S}(T_i^{(j)}) = n_{1S}(T_i^{(j)})/n_S$, $n_{1S}(T_i^{(j)})$ is saturated vapor molecule concentration at temperature $T_i^{(j)}$, $n_S = n|_{S_j} = n_\infty(T_{e\infty}/T_i^{(j)})$ is total molecule concentration of the gaseous medium near the particle surface; L_1 is evaporation heat; subscript "$\perp$" denotes projections of gradients T_e and c_1 normal to the surface S_j.

The obtained system is a second-order non-linear equation system. The factors of molecular transport D_{12}, κ_e and thermal diffusion factor K_T are of an arbitrary form depending on gaseous medium temperature T_e and depend linearly on gaseous component concentrations (7):

$$D_{12} = D_{12}^{(0)}, \kappa_e = \kappa_e^{(0)} + \kappa_e^{(1)}(c_1 - c_{1\infty}), K_T = K_T^{(1)} c_1, \tag{7}$$

where

$$D_{12}^{(0)} = D_{12}\big|_{c_1=c_{1\infty}}, \kappa_e^{(0)} = \kappa_e\big|_{c_1=c_{1\infty}}, \kappa_e^{(1)} = \frac{\partial \kappa_e}{\partial c_1}\bigg|_{c_1=c_{1\infty}}, K_T^{(1)} = \frac{\partial K_T}{\partial c_1}\bigg|_{c_1=0} \tag{8}$$

The solution of the system (1)–(3) with boundary conditions (4)–(6) is:

$$c_1 = \left\{ c_{1\infty} + B \int_{T_{e\infty}}^{T_e} \left[F_2 \exp\left(\int_{T_{e\infty}}^{T_e} F_1 dT_e \right) \right] dT_e \right\} \exp\left(-\int_{T_{e\infty}}^{T_e} F_1 dT_e \right), \tag{9}$$

$$\int_{T_{e\infty}}^{T_e} dT_e = \int_{T_{e\infty}}^{T_{i0}} \kappa_e dT_e G(x_f), T_i^{(j)} = T_{i0}, \tag{10}$$

where x_f are coordinates of space points: L_{1S} is the value of L_1 at temperature T_{i0};

$$F_1 = \left(K_T^{(1)} \frac{1}{T_e} - B \frac{\kappa_e^{(1)}}{nD_{12}^{(0)}} \right), F_2 = \left(\kappa_e^{(0)} - \kappa_e^{(1)} c_{1\infty} \right) \frac{1}{nD_{12}^{(0)}};$$

$$\kappa_e = \left(\kappa_e^{(0)} + \kappa_e^{(1)}(c_1 - c_{1\infty}) \right), B = -\frac{1}{L_{1S} m_1}.$$

The value of T_{i0} is found from solution to algebraic equation

$$\int_{T_{e\infty}}^{T_e} \left[F_2 \exp\left(\int_{T_{e\infty}}^{T_e} F_1 dT_e \right) \right] dT_e + L_{1S} m_1 \left[c_{1S}(T_{i0}) \exp\left(\int_{T_{e\infty}}^{T_{i0}} F_1 dT_e \right) - c_{1\infty} \right] = 0. \tag{11}$$

Function $G(x_f)$ is completely characterized by the geometry properties of the particle system. Function $G(x_f)$ is determined from solution to the linear Laplace equation in the N-connected domain with first-kind boundary conditions (12)

$$\Delta G = 0,$$

$$G|_{S_j} = 1, \qquad G|_\infty = 0. \tag{12}$$

The first of conditions (12) should be met at each point of the surface S_j.

Computation of the function $G(x_f)$ is an independent problem of applied mathematics which can be solved numerically in the case of an arbitrary particle system. In some special cases the form-factor $G(x_f)$ can be found analytically. Thus, when the system is composed of one spherical surface of radius R, the form-factor $G(x_f) = G(r, \theta, \varphi)$ is of form (13)

$$G(r) = R/r, \tag{13}$$

where r is the radial coordinate in the spherical coordinate system whose origin coincides with the sphere center, r, θ, φ are spherical coordinates. At vaporization and growth of a system composed of a single spheroidal particle $G = G(\xi, \eta, \varphi)$ is represented as (14) (prolate spheroid) and (15) (flattened spheroid):

$$G(\xi) = \left[\ln\frac{(ch\xi+1)}{(ch\xi-1)}\right] \Big/ \left[\ln\frac{(ch\xi_0+1)}{(ch\xi_0-1)}\right],$$

$$a = c\, ch\, \xi_0, \quad b = c\, ch\, \xi_0, \quad \xi_0 = \frac{1}{2}\ln\frac{(a+b)}{(a-b)}, \quad c = \sqrt{a^2-b^2}, \quad a > b; \tag{14}$$

$$G(\xi) = \left[arcctg(sh\xi)\right] \Big/ \left[arcctg(sh\xi_0)\right],$$

$$a = c\, sh\, \xi_0, \quad b = c\, sh\, \xi_0, \quad \xi_0 = \frac{1}{2}\ln\frac{(b+a)}{(b-a)}, \quad c = \sqrt{b^2-a^2}, \quad a < b; \tag{15}$$

where a and b are spheroid semiaxes, ξ, η, φ are spheroidal coordinates.

In the case of vaporization of a single ellipsoidal particle $G = G(\lambda, \mu, \nu)$ is

$$G(\lambda) = \left[1 - \left(\int_0^\lambda \frac{1}{R(\lambda)}d\lambda \Big/ \int_0^\infty \frac{1}{R(\lambda)}d\lambda\right)\right],$$

$$R(\lambda) = \sqrt{(\lambda+a^2)(\lambda+b^2)(\lambda+c^2)}, \tag{16}$$

where $a > b > c$ are lengths of the ellipsoid semiaxes, λ, μ, ν are ellipsoidal coordinates.

For a system composed of two spherical particles the form-factor $G(\xi, \eta, \varphi)$ is also estimated analytically:

$$G = \sqrt{2(ch\xi - \cos\eta)}\left[\varphi_1(\xi,\eta) + \varphi_2(\xi,\eta)\right], \tag{17}$$

where

$$\varphi_1(\xi,\eta) = \sum_{n=0}^{\infty} \left\{ \frac{sh\left[\left(n+\frac{1}{2}\right)(\xi+|\xi_2|)\right]}{sh\left[\left(n+\frac{1}{2}\right)(\xi_1+|\xi_2|)\right]} exp\left[-\left(n+\frac{1}{2}\right)\xi_1\right] \right\} P_n(\cos\eta), \qquad (18)$$

$$\varphi_2(\xi,\eta) = \sum_{n=0}^{\infty} \left\{ \frac{sh\left[\left(n+\frac{1}{2}\right)(\xi_1-\xi)\right]}{sh\left[\left(n+\frac{1}{2}\right)(\xi_1+|\xi_2|)\right]} exp\left[-\left(n+\frac{1}{2}\right)|\xi_2|\right] \right\} P_n(\cos\eta), \qquad (19)$$

ξ, η, φ are bispherical coordinates.

At determination of the form-factor for a system composed of three or more spherical particles the solution of boundary problem (12) can be conducted either numerically, or with quite complex analytical methods (for example, with the Cartesian prefix method).

To compute particle vaporization and growth rate, it is necessary to know expressions for integral fluxes $Q_1^{(j)}$ of the particle material molecules drawn away or to the particle surfaces

$$Q_1^{(j)} = \oint_{S_j} \vec{q}_1 d\vec{S}_j, \qquad (20)$$

where $\vec{q}_1$ is particle material molecule flux density; $d\vec{S}_j$ is the differential vector element of the j-th particle surface whose direction coincides with that of the outer normal.

Transform (20) taking into account (9) and (10) to obtain

$$Q_1^{(j)} = \theta_1(T_{i0}, T_{e\infty}) Q_G^{(j)}. \qquad (21)$$

The function $\theta_1(T_{i0}, T_{e\infty})$ is determined by one of formulas (22), (23):

$$\theta_1(T_{i0}, T_{e\infty}) = B \int_{T_{e\infty}}^{T_{i0}} \kappa_e dT_e, \qquad (22)$$

$$\theta_1(T_{i0}, T_{e\infty}) = \left[c_{1S}(T_{i0}) exp\left(\int_{T_{e\infty}}^{T_{i0}} F_1 dT_e \right) - c_{1\infty} \right] \frac{\int_{T_{e\infty}}^{T_{i0}} \kappa_e dT_e}{\int_{T_{e\infty}}^{T_{i0}} \left[F_2 \left(exp \int_{T_{e\infty}}^{T_e} F_1 dT_e \right) \right] dT_e}, \qquad (23)$$

where

$$Q_G^{(j)} = -\oint_{S_j} \nabla G d\vec{S}_j, \qquad (24)$$

In the case of a single spherical particle of radius R

$$Q_G = 4\pi R. \tag{25}$$

For spheroidal particles

$$Q_G = 4\pi c I(\xi_0), \tag{26}$$

where

$$I(\xi_0) = 2 \Big/ \ln\left(\frac{ch\xi_0 + 1}{ch\xi_0 - 1}\right) = 2 \Big/ \ln\left[\frac{1 + \frac{\sqrt{a^2 - b^2}}{a}}{1 - \frac{\sqrt{a^2 - b^2}}{a}}\right], a > b; \tag{27}$$

$$I(\xi_0) = 1\big/ arcctg(sh\xi_0) = 1\Big/ arctg\frac{\sqrt{b^2 - a^2}}{a}, a < b.$$

At the ellipsoidal particle shape

$$Q_G = 8\sqrt{a^2 - c^2}\left[E(P_1, \alpha_1)K(P_2, \alpha_1) + E(P_2, \alpha_1)K(P_1, \alpha_2) - K(P_1, \alpha_1)K(P_2, \alpha_2)\right], \tag{28}$$

where $K(r, \alpha)$ and $E(r, \alpha)$ are the first and second kind elliptic integrals

$$K(r, \alpha) = \int_0^\alpha \frac{1}{\sqrt{1 - r^2 \sin^2 \alpha}} d\alpha, E(r, \alpha) = \int_0^\alpha \sqrt{1 - r^2 \sin^2 \alpha} d\alpha,$$

$$P_1 = \sqrt{\frac{a^2 - b^2}{a^2 - c^2}}, P_2 = \sqrt{\frac{b^2 - c^2}{a^2 - c^2}}, \alpha_1 = \frac{\pi}{2}, \alpha_2 = \arcsin\sqrt{\frac{a^2 - c^2}{a^2}}, a > b > c$$

For a system composed of two spherical particles $Q_G^{(1)}$ и $Q_G^{(2)}$ is determined by analytical relations:

$$Q_G^{(1)} = 4\pi a(\psi_1 - \psi_3), \qquad Q_G^{(2)} = 4\pi a(\psi_2 - \psi_3), \tag{29}$$

where

$$\psi_1 = \sum_{n=0}^\infty \frac{\exp\left[\left(n + \frac{1}{2}\right)(|\xi_2| - \xi_1)\right]}{sh\left[\left(n + \frac{1}{2}\right)(\xi_1 + |\xi_2|)\right]}, \psi_2 = \sum_{n=0}^\infty \frac{\exp\left[\left(n + \frac{1}{2}\right)(\xi_1 - |\xi_2|)\right]}{sh\left[\left(n + \frac{1}{2}\right)(\xi_1 + |\xi_2|)\right]},$$

$$\psi_3 = \sum_{n=0}^\infty \frac{\exp\left[-\left(n + \frac{1}{2}\right)(\xi_1 + |\xi_2|)\right]}{sh\left[\left(n + \frac{1}{2}\right)(\xi_1 + |\xi_2|)\right]}.$$

In steady-state the evaporation or growth rate of the j-th assembly particle is determined by expression (30):

$$\frac{dM^{(j)}}{dt} = -m_1\theta_1(T_{i0},T_{e\infty})Q_G^{(j)}. \tag{30}$$

where $M^{(j)}$ is mass of the j-th particle, t is time.

The mass variation rate of an assembly of N particles in the vaporization and growth steady state is determined with relation (31):

$$\frac{dM}{dt} = -m_1\theta_1(T_{i0},T_{e\infty})\sum_{j=1}^{N}Q_G^{(j)} = -m_1\theta_1(T_{i0},T_{e\infty})\sum_{j=1}^{N}\oint_{S_j}\vec{\nabla}G d\vec{S}_j. \tag{31}$$

where M is the total mass of all the assembly particles.

If particle volumes $V^{(j)}$ and surface shape are related with one-to-one correspondence, then one can make integration (30) to determine the particle vaporization and growth time. Equation system (30) allows for integration in quadratures when each volume $V^{(j)}$ and integral factor $Q_G^{(j)}$ is a function of a single independent variable Δ. common to all the particles. This condition is met, for example, at vaporization and growth in infinite media of separate particles of spherical ($\Delta = R$), spheroidal ($\Delta = b$, $a > b$, $a = const$; $\Delta = a$, $a < b$, $b = const$;) and ellipsoidal ($\Delta = c$, $a > b > c$, $a = const$, $b = const$) surface shape. Volumes $V^{(j)}$ and integrals $Q_G^{(j)}$ of N large particles identical in shape and equal in volume, symmetrically located at vertices of regular polygons and polyhedra depend on one parameter Δ.

In such systems

$$V^{(1)} = V^{(2)} = \ldots V^{(N)} = V,$$

$$Q_G^{(1)} = Q_G^{(2)} = \ldots Q_G^{(N)} = Q_G.$$

If volumes $V^{(j)}$ and integrals $Q_G^{(j)}$ depend only on one variable Δ, then system (30) reduces to equation (32)

$$\frac{dV}{dt} = -m_1\theta_1(T_{i0},T_{e\infty})Q_G/\rho_i, \tag{32}$$

where ρ_i is particle material density. Integrate (32) to obtain expression (33) allowing to find time variation of the particle sizes

$$\Phi(\Delta,\Delta_0) = -m_1\theta_1(T_{i0},T_{e\infty})t/\rho_i, \tag{33}$$

where

$$\Phi(\Delta,\Delta_0) = \int_{\Delta_0}^{\Delta}\frac{1}{Q_G}\left(\frac{dV}{d\Delta}\right)d\Delta. \tag{34}$$

In (33), (34) Δ and Δ_0 are values of the variable Δ at current, t, and initial, $t = 0$, times, respectively.

For systems composed single spherical, spheroidal, ellipsoidal particles as well as for a system of two spherical particles of identical radii the vaporization and growth rate is described with analytical relations.

For a spherical particle of radius R

$$\Delta = R, \quad V = \frac{4}{3}\pi R^3. \tag{35}$$

Integrate (34) taking into account (35) and (25) to obtain

$$\Phi(R, R_0) = \frac{1}{2}\left(R^2 - R_0^2\right). \tag{36}$$

In case of prolate and oblate spheroidal particles

$$\Delta = b, \quad V = \frac{4}{3}\pi a^2 b, \quad a > b, \quad a = const;$$

$$\Delta = a, \quad V = \frac{4}{3}\pi ab^2, \quad a < b, \quad b = const. \tag{37}$$

Integrate (34) taking into account (26) and (37) to obtain

$$\Phi(b, b_0) = -\frac{2}{3}a^2\left\{\left[\left(1 - \sqrt{1 - \left(\frac{b}{a}\right)^2}\right)\ln\left(\frac{b}{a}\right) + \sqrt{1 - \left(\frac{b}{a}\right)^2}\ln\left(1 + \sqrt{1 - \left(\frac{b}{a}\right)^2}\right)\right] - \right.$$
$$\left. - \left[\left(1 - \sqrt{1 - \left(\frac{b_0}{a}\right)^2}\right)\ln\left(\frac{b_0}{a}\right) + \sqrt{1 - \left(\frac{b_0}{a}\right)^2}\ln\left(1 + \sqrt{1 - \left(\frac{b_0}{a}\right)^2}\right)\right]\right\}, \quad a > b;$$

$$\Phi(a, a_0) = -\frac{b^2}{6}\left[arccos^2\left(\frac{a}{b}\right) - arccos\left(\frac{a_0}{b}\right)\right], \quad a < b, b = const. \tag{38}$$

At vaporization and growth of an ellipsoidal drop with semiaxes $a > b > c$, when lengths a and b remain constant,

$$\Delta = c, \quad V = \frac{4}{3}\pi abc. \tag{39}$$

Transform (34) taking into account (28) and (39) to obtain

$$\Phi(c, c_0) = ab\frac{\pi}{6}\int_{c_0}^{c}\frac{K(P_1, \alpha_2)dc}{\sqrt{a^2 - c^2}\left[E(P_1, \alpha_1)K(P_2, \alpha_1) + E(P_2, \alpha_1)K(P_1, \alpha_1) - K(P_1, \alpha_1)K(P_2, \alpha_1)\right]}. \tag{40}$$

When the system consists of two spherical particles of identical radius, the vaporization and growth rate can be described with the analytical relation. In this case

$$\Delta = \xi_1, ch\xi_1 = \frac{h}{2R}, V = \frac{\pi}{G}\frac{h^3}{ch^3\xi_1}, h = const. \tag{41}$$

Transform (34) taking into account (29) and (41) to obtain relation (42)

$$\Phi(\xi_1,\xi_2) = -\frac{h^2}{8}\int_{\xi_{10}}^{\xi_1}\frac{d\xi_1}{ch^3\xi_1\sum_{n=0}^{\infty}\frac{exp[-(2n+1)\xi_1]}{(1+exp[-(2n+1)\xi_1])}}. \tag{42}$$

Calculation

Earlier it was shown that the particle vaporization and growth rate depends on material distribution inside the system (determined by the function $G(x_i)$). At vaporization and growth in gaseous medium with given $T_{e\infty}$, $c_{1\infty}$ and p_∞ of identical-volume particles the difference in the process kinetics depends on different particle shapes. For a system composed of a single particle $G(x_i)$ depends on parameters of the particle itself. The estimations showed that under identical conditions the vaporization and growth of spherical particles occurred with the lowest rate. The data comparison showed that the differences in vaporization and growth kinetics for particles of different surface shape can be quite considerable[7].

The drawing of the particles closer together can also lead to a considerable change in the vaporization or growth time. At vaporization of a particle pair, for example, presence of a large particle considerably affects the vaporization rate of the smaller one. Drawing together to short distances exerts a considerable influence on vaporization of close geometry particles. Fig.1 presents the curves of the ratio R/R_0 vs. dimensionless time τ for vaporization of a system of two large drops of identical initial radii $R_0 = 5$ μm located at distance $h = 10,6$ μm (continuous curve) and $h = \infty$ (dotted curve). τ is determined as

$$\tau = \frac{t}{t_d} \tag{43}$$

where t_d is the time of complete vaporization for a single drop of initial radius $R = R_0$.

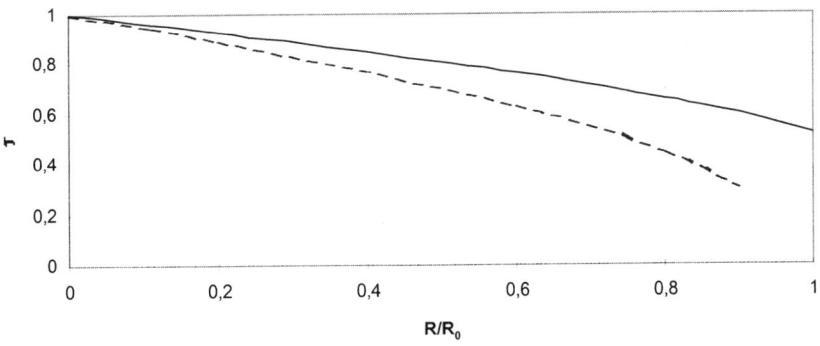

Figure1. Dimensionless time τ as a function of the ratio R/R_0

Analysis

All the formulas describing the vaporization and growth of an assembly of N-large particles are obtained under the assumption that the effect of macroscopic (in our case diffusive) transport on heat exchange of particles with gaseous medium is minor. One can find the extremes of applicability for the obtained expressions, having determined temperature and concentration distributions of gaseous components in the particle-gaseous medium system from solution to non-linear equation system (44)–(46), more complex than (1)–(3):

$$\text{div}\left[nD_{12}\left(\nabla c_1 + K_T \frac{1}{T_e}\nabla T_e\right)\right] = 0, \tag{44}$$

$$\text{div}\left[\left(m_1 h_1 + \Phi_e\left(\frac{K_T}{c_1}\right)\right)nD_{12}\left(\nabla c_1 + K_T \frac{1}{T_e}\nabla T_e\right) + \kappa_e \nabla T_e\right] = 0, \tag{45}$$

$$\text{div}\left(\kappa_i^{(j)} \nabla T_i^{(j)}\right) = 0. \tag{46}$$

The term containing Φ_e relates to the Dufour effect, i.e. heat energy release due to gaseous component concentration gradients. When determining the distributions T_e and c_1, it is necessary to take into account boundary conditions (4)-(6). The molecular transport and thermal diffusion factors are determined by relation (8). As the particle surface is impermeable to the second-kind molecules, $\dot{q}_2$ is considerably less than $\dot{q}_1$ (for spherical particles $\dot{q}_2=0$). In the heat transport equation the convective term proportional to $\dot{q}_2$ makes a noticeably less contribution than that proportional to $\dot{q}_1$. Hence, in Eq. (45) the effect of the term proportional to $\dot{q}_2$ on distribution of temperature T_e in gaseous medium can be neglected. The solution to system (44)–(46) with boundary conditions (4)–(6) is:

$$c_1 = \left\{c_{1\infty} + B\int_{T_{e\infty}}^{T_e}\left[F_2 \exp\left(\int_{T_{e\infty}}^{T_e} F_1 dT_e\right)\right]dT_e\right\}\exp\left(-\int_{T_{e\infty}}^{T_e} F_1 dT_e\right), \tag{47}$$

$$\int_{T_{e\infty}}^{T_e} \Omega dT_e = \int_{T_{e\infty}}^{T_{i0}} \Omega dT_e G(x_i), \quad T_i = T_{i0}, \tag{48}$$

$$F_1 = K_T^{(1)} \frac{1}{T_e} - B\left(\frac{\kappa_e^{(1)}}{nD_{12}^{(0)}}\right)\frac{1}{\left[1 - B\left(m_1 h_1 + \Phi_e K_T^{(1)}\right)\right]}, \tag{49}$$

$$F_2 = \left(\kappa_e^{(0)} - \kappa_e^{(1)} c_{1\infty}\right)\frac{1}{\left(nD_{12}^{(0)}\right)\left[1 - B\left(m_1 h_1 + \Phi_e K_T^{(1)}\right)\right]}, \tag{50}$$

$$\Omega = \frac{\left[\kappa_e^{(0)} + \kappa_e^{(1)}(c_1 - c_{1\infty})\right]}{\left[1 - B\left(m_1 h_1 + \Phi_e K_T^{(1)}\right)\right]}, \quad (51)$$

$$B = -\frac{1}{\left[L_{1S} m_1 - m_1 h_{1S} - \Phi_{eS} K_{TS}^{(1)}\right]}. \quad (52)$$

Here subscript "S" denotes the magnitudes of the values at $T_e = T_{i0}$. Temperature T_{i0} is found by solving algebraic equation (53) with taking into account (52)

$$\left[c_{1S}(T_{i0}) \exp\left(\int_{T_{e\infty}}^{T_{i0}} F_1 dT_e\right) - c_{1\infty}\right] - B \int_{T_{e\infty}}^{T_{i0}} \left[F_2 \exp\left(\int_{T_{e\infty}}^{T_e} F_1 dT_e\right)\right] dT_e = 0. \quad (53)$$

When computing the integrals in expression (48) concentrations c_1 are found by formula (47). Taking into account (47) and (48), the expression for molecule fluxes takes form (54):

$$Q_1^{(j)} = \theta(T_{i0}, T_{e\infty}) Q_G^{(j)}, \quad \theta(T_{i0}, T_{e\infty}) = B \int_{T_{e\infty}}^{T_{i0}} \Omega dT_e, \quad Q_G^{(j)} = -\oint_{S_j} \nabla G d\vec{s}_j. \quad (54)$$

The particle mass and volume variation with time t is determined with relations (30)–(42).

The expressions for coefficients (49)–(51) taking into account (52) take the form:

$$F_1 = K_T^{(1)} \frac{1}{T_e} + \left(\frac{\kappa_e^{(1)}}{nD_{12}}\right) \frac{1}{\left[L_{1S} m_1 + m_1(h_1 - h_{1S}) + \left(\Phi_e K_T^{(1)} - \Phi_{eS} K_{TS}^{(1)}\right)\right]}, \quad (55)$$

$$F_2 = \left[L_{1S} m_1 - m_1 h_{1S} - \Phi_{eS} K_{TS}^{(1)}\right] F_2^*, \quad (56)$$

$$F_2^* = \frac{\left(\kappa_e^{(0)} - \kappa_e^{(1)} c_{1\infty}\right)}{nD_{12}^{(0)} \left[L_{1S} m_1 + m_1(h_1 - h_{1S}) + \left(\Phi_e K_T^{(1)} - \Phi_{eS} K_{TS}^{(1)}\right)\right]}, \quad (57)$$

$$\Omega = \left[L_{1S} m_1 - m_1 h_{1S} - \Phi_{eS} K_{TS}^{(1)}\right] \Omega^*, \quad (58)$$

$$\Omega^* = \frac{\left[\kappa_e^{(0)} + \kappa_e^{(1)}(c_1 - c_{1\infty})\right]}{\left[L_{1S} m_1 + m_1(h_1 - h_{1S}) + \left(\Phi_e K_T^{(1)} - \Phi_{eS} K_{TS}^{(1)}\right)\right]}. \quad (59)$$

On substitution of formulas (55)-(58) to expressions for distributions of concentration c_1 (47) and T_e (48), flux $Q_1^{(j)}$ (54) and equation (53), we arrive at

$$c_1 = \left\{ c_{1\infty} - \int_{T_{e\infty}}^{T_e} \left[F_2^* \exp\left(\int_{T_{e\infty}}^{T_e} F_1 dT_e \right) \right] dT_e \right\} \exp\left(-\int_{T_{e\infty}}^{T_e} F_1 dT_e \right), \quad (60)$$

$$\int_{T_{e\infty}}^{T_e} \Omega^* dT_e = \int_{T_{e\infty}}^{T_{i0}} \Omega^* dT_e G(x_i), \quad (61)$$

$$Q_1^{(j)} = -\int_{T_{e\infty}}^{T_{i0}} \Omega^* dT_e Q_G^{(j)}, \quad (62)$$

$$\left[c_{1S}(T_{i0}) \exp\left(\int_{T_{e\infty}}^{T_{i0}} F_1 dT_e \right) - c_{1\infty} \right] + \int_{T_{e\infty}}^{T_{i0}} \left[F_2^* \exp\left(\int_{T_{e\infty}}^{T_e} F_1 dT_e \right) \right] dT_e = 0. \quad (63)$$

Comparing (9), (10), (21) with (60)–(62) and Eq. (9) with Eq. (63), one can infer that at steady-state vaporization and growth the effect of diffusive transport on heat exchange of the particle with gaseous medium will be minor, when condition (64) is met:

$$\left| m_1(h_1 - h_{1s}) + (\Phi_e K_T^{(1)} - \Phi_{eS} K_{TS}^{(1)}) \right| \ll L_{1S} m_1. \quad (64)$$

EXPERIMENT

In ref.[8] the experiments were conducted under conditions close to free vaporization of immovable props at a small effect of gravitational convection on the vaporization rate. Vaporization of single drops 0.05–0.2 mm in diameter was studied. Vaporization constants were estimated reasoning from the obtained data. According to estimations by the authors, the error in drop diameter estimation was no more than 5 %. The experimental data and theoretical curves of vaporization constant K vs. air temperature $T_{e\infty}$ are presented in Fig. 2 (for water drops) and Fig. 3 (for alcohol drops).

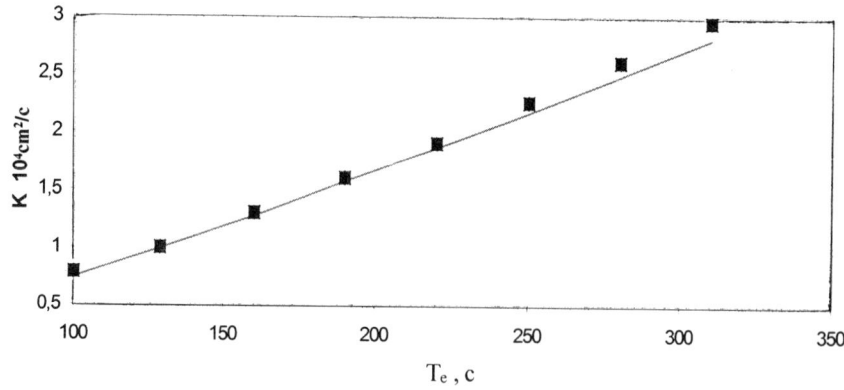

Figure 2. Constant K as a function of temperature $T_{e\infty}$ for water drops.

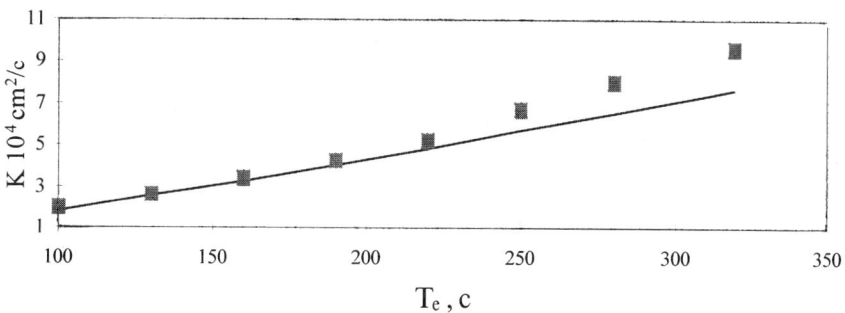

Figure 3. Constant K as a function of temperature $T_{e\infty}$ for alcohol drops.

CONCLUSION

Analysis of analytical solutions to the model problem and data of numerical computations allows to determine some mechanisms of the vaporization and growth processes for an arbitrary assembly of N large particles occurring when there are no internal heat release and radiant heat exchange of particles with gaseous medium:

1. In steady state all the assembly particles either grow or vaporize simultaneously at identical temperature of the surface.

2. The steady-state vaporization or growth temperature is independent on spatial distribution of particle material inside the system. The form of the dependency of the molecular transport coefficients D_{12}, κ_e on gaseous medium temperature T_e and gaseous component concentrations can noticeably influence the vaporization or growth temperature, rate and time.

3. The particle assembly evaporation and growth rate depends on material distribution inside the system. Drawing the particles together can lead to a longer time of vaporization and growth. Vaporization rate and time of a small particle vaporizing or growing at presence of a larger one change more considerably.

4. Comparison of the data on vaporization of large particles of spherical, spheroidal, ellipsoidal and cubic shapes shows that under identical conditions at equal volumes the spherical particles vaporize at the lowest rate.

The obtained results can be used to describe vaporization and growth in combinations not only with close, but also severely different molecule masses. The suggested approach allows to describe vaporization and growth for an arbitrary system of N particles in a fair detail. Alongside relation (31) characterizing the rate of variation in mass of an assembly of N particles, relation (30) is obtained which describes vaporization and growth of each assembly particle.

REFERENCES

1. N.A. Fuks, Drop vaporization and growth in gaseous medium, USSR Academy of Sciences Publishing House, Moscow, 91 (1958).
2. J.C. Carstens, A. Williams, J.T. Zung, Theory of droplet growth in clouds, *J. Atmos. Sci.*, 27, 8: 798 (1970).
3. A. Williams, J.C. Carstens, J. Atmos. Sci., 28: 1298 (1971).

4. Yu.I. Yalamov, E.R. Shchukin, M.F. Barinova, Yu.K. Ostrovsky, Theory of vaporization of two drops located at arbitrary distances from each other, *DAN SSSR, Fizika*, 284, 2: 341 (1985).
5. J.L. Griffin, S.K. Loyalka, Vapor condensation on multiple spheres and spheroids in the near – continuum regime, *J. Aerosol Sci.*, 25, 7: 1271 (1994).
6. J.L. Griffin, S.K. Loyalka, Condensation on aerosol particles: boundary element formulation, *J. Aerosol Sci.*, 27, 1: 3 (1996).
7. E.R. Shchukin, A.B. Nadykto, Kinetics of large particle vaporization, *VANT, Th. and Appl. Phys.*, 1: 43 (1998).
8. M.D. Apashev, R.V. Malov, Vaporization of single particles of different liquids, *I. AN SSSR, Energetics and Automatics*, 2: 185 (1960).

ANALYSIS OF THE PHENOMENON OF AUTOACCELERATION IN FREE RADICAL POLYMERIZATION

Vladimir A. Kaminsky, M.V. Egorov

Karpov Institute of Physical Chemistry
Moscow, Russian

1. INTRODUCTION

Free radical polymerization provides a good example of a process for which non-linear kinetics of a complex system is governed to a great extent by the physical properties of the medium. The kinetic scheme of bulk and solution polymerizations includes the following reactions[1]:

Initiation:
$$I \xrightarrow{k_d} 2R^*$$
$$R^* + M \xrightarrow{k_{pi}} R_1$$

Propagation:
$$R_n + M \xrightarrow{k_p} R_{n+1}$$

Chain transfer:
$$R_n + T \xrightarrow{k_{tr}} P_n + R_1$$

Termination

(a) by combination:
$$R_n + R_m \xrightarrow{(1-\lambda)k_t} P_{n+m}$$

(b) by disproportionation:
$$R_n + R_m \xrightarrow{\lambda k_t} P_n + P_m$$

In the above kinetic scheme initiator of concentration I decomposes with rate coefficient k_d to produce free radicals R^*, a fraction of which f can in turn react with monomer with rate coefficient k_{pi}, the overall rate of production of growing free radicals (initiation rate) R_{in} is therefore $2fk_dI$; M is monomer concentration; k_p is the rate coefficient of propagation, in which a radical of length n (any radical which contains exactly n monomeric units) becomes a free radical of length $n + 1$; k_{tr} is the rate coefficient of chain transfer to chain transfer agent (which can be special substance, monomer, solvent or initiator), in which free radical activity is transferred from a free radical of length n to a chain transfer agent molecule; k_t is the rate coefficient of bimolecular termination, in which mutual annihilation of the radical activity of two free radicals, which contain n and m monomeric units respectively, occurs; λ is the fraction of combination in the overall termination rate. In some cases reactions such as chain transfer to polymer or inhibition also occur.

According to the above kinetic scheme, the rate of production of polymer chains, i.e. polymerization rate, is given by the expression:

$$R_p = -\frac{dM}{dt} = k_p M \left(\frac{R_{in}}{k_t}\right)^{1/2} \quad (1.1)$$

In obtaining Eq. (1.1) the quasi-stationary approximation was used. Also, it was assumed that the termination reaction can be described with a single termination rate coefficient.

A distinguishing feature of polymerization of many monomers is the effect of autoacceleration, or gel effect, which consists in that polymerization rate begins to increase sharply at some monomer conversion. The effect of autoacceleration was first discovered more than 50 years ago[2]. Soon it was realized that the reason for the effect is the diffusion control of the termination reaction[3]. It was understood also that microscopic termination rate coefficients $k_t(n, m)$ must depend not only on conversion but also, in general, on the chain lengths m and n of reacting radical species. So a demand was created for a systematic theory which could explain the phenomenon of autoaccelaration. It was desirable to construct such a theory on the basis of "first principles"[4, 5] considering diffusion-controlled reactions (DCR) in polymer solutions. In addition to some questions associated with the multiparticle character of the problem, which arise also in describing the DCR between small molecules or colloid particles, the reactions between macromolecules have their own distinguishing features, one of which is that the motion of polymer chains in polymer solutions has multimode character so the motion of polymer chains depends not only on polymer concentration but also on the size of the fragment of a polymer chain that takes part in the motion. So experimental data not only on self-diffusion coefficients but also on intramolecular dynamics in polymer solutions are needed. However comprehensive experimental data on the dynamic properties of polymer solutions for real polymerizing systems are not available even now. Therefore the study of the kinetic evolution of polymerizing systems can be interesting from the point of view that it can provide additional information about the mechanism of DCR in polymer solutions.

Of course, further investigation of the dynamic properties of polymer solutions on different time and spatial scales must be done in order to build a comprehensive theory of reactions between macromolecules and to determine individual termination rate coefficients from the "first principles". However, even though the uncertainty of the physical picture of such reactions and of the corresponding parameters does exist, some predictions about free-radical polymerization kinetics can be made on the basis of general concepts of polymer dynamics.

The absence of independent experimental data on physical properties of polymer solutions has led to the creation of a large number of phenomenological models of termination reaction, icluding those that take into account the effect of the dependence of microscopic termination rate coefficients on the lengths of growing radical species, and considered in review papers[6–8]. Once a particular model for microscopic chain length dependent (CLD) termination rate coefficients $k_t(n, m)$ is chosen, then the next step is to calculate the polymerization rate and molecular weight distribution (MWD) for a given polymerizing system and to analyze the influence of different kinetic and physico-chemical parameters on the polymerization kinetics of various monomers. In other words, there exists a "microscopic to macroscopic" aspect of the termination rate problem. There are two approaches (methods) to calculate overall, or effective, termination rate coefficients from some given CLD microscopic termination rate coefficients and to analyze the kinetic evolution of a polymerizing system:

i) A direct method, which is to find solutions to a complete set of differential equations for the time evolution of the individual radical concentrations $R(n, t)$[9, 10]. Once chain length distribution of radicals is known, then the polymerization rate and MWD of the polymer formed, including those for polymerization under non-stationary conditions, can be easily calculated. The main feature of this method, in which an initial set of equations is solved directly, is the fact that different parameters pertaining to initiation, propagation and termination rates are used in the equations independently. This method enables one to perform calculations for any particular polymerizing system. However, a large number of independent parameters makes it difficult to determine governing parameters and to compare the kinetic behavior of different polymerizing systems.

ii) An alternative method, in which a discrete set of differential equations is rolled up to produce an equation for the overall, or effective, termination rate coefficient. In this approach it is assumed that the chain length of macroradicals n can be considered as a continuous variable and that the chain length distribution of radicals is quasi-stationary. This method of calculating overall termination rate coefficients from individual CLD termination rate coefficients was first introduced by Benson and North [3]. In this approach the number of independent dimensionless parameters can be introduced, as was done, for example, in the work of Soh and Sundberg [11], that is, the number of independent parameters is significantly reduced, so the analysis of polymerization kinetics becomes much easier. One of the disadvantages of this method is that a closed equation for the overall termination rate coefficient can be obtained for a few specific types of termination rate models only, e.g. for an additive termination rate model[3]. However the additive termination rate model seems to be the most realistic.

In this paper an attempt is made to consider some physical aspects of the termination rate problem, to determine the parameters that govern the kinetics of free-radical polymerizations and to investigate the kinetic evolution of a polymerizing system on the basis of method (ii).

2. KINETIC EQUATION

The kinetic scheme described above gives rise to the following free-radical concentration balance equation for the time evolution of the concentration $R(n, t)$ of radicals of degree of polymerization n:

$$\frac{dR(n,t)}{dt} = \left(R_{in} + \sum_j k_{tr,j} T_j R \right) \delta(n) - k_p M \frac{\partial R(n,t)}{\partial n} - \sum_j k_{tr,j} T_j R(n,t) - \int_0^\infty k_t(n,m) R(n,t) R(m,t) dm \qquad (2.1)$$

In the above equation $R(t) = \int_0^\infty R(n,t)dn$ is the total concentration of radicals; R_{in} denotes initiation rate ($R_{in} = 2fk_dI$); T_j is the concentration of added chain transfer agent j, M is monomer concentration; the formation of short radicals due to initiation and chain transfer is represented by Dirac's δ-function. Hereafter the long chain approximation will be used and always an integration over chain length will replace the summation. Because the lifetime of macroradicals is much less then the duration of polymerization, steady-state approximation is valid, even in the stage of a strong gel effect. It should be noted that the effect of conversion induced volume contraction, which, as can be easily shown, plays a truly insignificant role in the calculation of $R(n, t)$, is neglected in calculations, both here and in all that follows. Of course, the use of the partial derivatives (with respect to n) and the replacement of the summation with an integration over chain length in Eq. (2.1) do need some justification, however this is here omitted, and so Eq. (2.1) is taken for granted.

Eq. (2.1) is equivalent to an infinite set of differential equations, which are considered in method (i). In method (ii) Eq. (2.1) is integrated over chain length using steady state approximation. This yields the following equation for the overall termination rate coefficient $k_{t,eff} = k_t(1, 1)F$:

$$\frac{v}{1+vc} = \int_0^\infty \exp\left(-s \cdot c - \frac{\varphi(s)}{vF}\right) ds \tag{2.2}$$

$$\varphi(s) = \int_0^s dn \int_0^\infty dm\, k_t(n,m) R(m) \frac{1}{k_t(1,1) R}$$

In the above equation $c = \sum_j k_{tr,j} T_j / (k_p M)$ denotes the overall (dimensionless) constant of chain transfer; $k_t(1, 1)$ accounts for the dependence of the termination rate coefficients of "short" radicals on conversion; while F signifies the parameter which is defined by the following equation

$$k_{t,ef} = \frac{1}{R^2} \int_0^\infty \int_0^\infty dm\,dn\, k_t(n,m) R(m) R(n) = k_t(1,1) F \tag{2.3}$$

As can be seen from equation (2.3), the parameter F has the meaning of the fraction of short, mobile radicals and it shows to what extent the effective, or overall, termination rate coefficient differs from $k_t(1, 1)$. Also, v is the kinetic chain length (number of monomeric units polymerized per one radical), i.e. $v = k_p M (R_{in} k_{t,eff})^{-1/2} = v_0 F^{-1/2}$, where v_0 is the kinetic chain length calculated under the assumption that microscopic termination rate coefficients are independent of radical lengths.

It is important to note that equation (2.2) can be used for any termination model, that is, it is in no way dependent on which model is actually employed. In other words, this equation is valid for any assumptions about the dependence of individual (microscopic) termination rate coefficients $k_t(n, m)$ on macroradical chain lengths. However it needs to be pointed out that Eq. (2.2) is unclosed and can be solved only together with Eq. (2.1). It becomes a closed equation for F, and therefore can be solved on its own, for a few particular types of termination models only. In the next section one of such termination models, which is actually used in this paper, is outlined. So instead of a complete set of equations for $R(n, t)$ a formal solution to Eq. (2.1) for an arbitrary n, i.e. Eq. (2.2), which contains only one unknown parameter F, can be used. Once factor F is known, one can calculate the polymerization rate and statistical moments of MWD of polymer formed.

3. DIFFUSION–CONTROLLED TERMINATION REACTION MODEL

Chain termination in free-radical polymerization occurs as a result of interaction between the active terminal units of macroradicals, that is, the true radius of the reaction is of the order of the size of a monomer unit a. One main feature of the reactions involving macromolecules is the existence of a spectrum of different motions from small-scale segmental motions to the motion of a polymer coil as a whole. As a consequence, even the reaction between two isolated coils cannot be described with the use of a single constant diffusion coefficient. This is why even in the first papers devoted to the study of DCR between macromolecules, the effects of translational and segemntal diffusion were studied separately[3]. Translational diffusion was thus treated as the relative movement of polymer coils prior to overlapping. However, the tranlational diffusion with the reaction radius equal to the sum of the radii of the reacting coils greatly overestimates the values of the termination rate coefficient at initial conversion; this made Benson and North conlude that segmental diffusion is the slowest process. Moreover, Kozlov et al.[14] supposed that for reaction in dilute solution the relative diffusion coefficient of the active units sharply decreases once the coils have overlapped and that it is similar to the value of the diffusion coefficient in polymer melt which is quite impossible.

The relative contribution of segmental and translational diffusion has been studied in many papers. It should be noted that the treatment of the diffusion process as a separate stage with its own rate constant may lead to certain misunderstanding. In particular, the possibility of a particle returning many times to its initial point excludes the representation of the average time of reaction between initially separated coils as the sum of the time of movement prior to collision and the time of motion of the overlapped coils before reaction. This is explained by the fact that after collision the coils will again be separated for some time with a probability close to unity and will again translate.

At the early stages of the theory of DCR involving macromolecules, in particular, for chain termination reaction in free radical polymerization, the change in the character of motion of polymer coils after they are overlapped was taken into account phenomenologically[15–17]. The development of physically substantiated models of the reactions between macromolecules in polymer solution became possible due to successes in the physics of polymers, including the study of the dynamics in polymer solutions[4, 5]. Unfortunately, hitherto only a few authors have used modern approaches to describe the dynamics of polymer solutions (in particular, the dynamics of intramolecular motions) to develop quantitative theory of intermolecular DCR between macromolecules[11, 18, 19]. The main difficulty lies in the correct description of intramolecular motions resulting in the collision of the active units of macroradicals and the subsequent reaction between them after the coils become overlapped.

First of all, it should be mentioned that, due to the existence of a whole spectrum of relaxation motions of a polymer chain, the description of the motion of a single selected unit by an equation of diffusion type may be done only approximately, the respective diffusion coefficients being determined by the scale on which the motion of the unit is considered. Let us analyze the motion of the terminal units of different macromolecules of the same length n[7]. Prior to the overlap of the polymer coils ($r > 2r_n$), their relative motion is described by a single coefficient of relative diffusion equal to the sum of the diffusion coefficients of each coil, $D = 2D_n$. After the coils become overlapped, one should select from the entire spectrum of motions only those which make the main contribution to the relative motion. For distance r between the active units, the main contribution is made by the terminal segments of the chains of spatial size about r. Assuming Gaussian chains, such segments, on the average, contain $m \approx n(r/r_n)^2 \approx (r/a)^2$ monomer units. As far as the mobility of a segment depends on its length, for any give relative distance beetwent the active units it is possible to introduce an effective diffusion coefficient $D(r)$, which makes them come close together. The dependence of D on r can be obtained if the relation between the mobility of the segment and the number of monomer units in it is known. If this dependence is a power law[5], i.e., $D = D_0 m^{-\alpha}$, then

$$D(r) = 2D_m = 2D_0(a/r)^{2\alpha} \qquad (3.1)$$

To determine the slowest stage of the reaction between two macroradicals and to estimate the relative role of both the translational diffusion and intramolecular motions after the overlap, one should calculate the probability $u_{nn}(2r_n)$ that after collision polymer coils containing n monomer units each will enter the reaction over infinite time. The value $u_{nn}(2r_n)$ is related to $D(r)$ via

$$u_{nn}(2r_n) = \int_{2r_n}^{\infty} \frac{dr}{r^2 D(r)} \Big/ \int_{a}^{\infty} \frac{dr}{r^2 D(r)} \tag{3.2}$$

The termination rate coefficient $k(n, n)$, being the diffusion flux of sinking particles at $r = a$ per unit concentration of the particles, is given by

$$k(n,n) = 16\pi D_n r_n u_{nn}(2r_n) \tag{3.3}$$

If $u_{nn}(2r_n)$ is close to unity, the slowest stage is translational diffusion, and after the overlap the coils move in the regime of compact exploration[20]. To obtain quantitative estimates for the reaction probability $u_{nn}(2r_n)$ and, consequently, the rate coefficient of intermolecular reaction, the knowledge of the intramolecular dynamics for all possible scales of motion is needed. The finite flexibility of chains should also be taken into account. The point is that the increasing length of the Kuhn segment ξ_K increases the volume of the region over which the regime of compact exploration does not hold ($r \approx \xi_K$) and, as a result, the probability $u_{nn}(2r_n)$ may decrease.

Although chain termination reaction seems to be diffusion-controlled from the very beginning of polymerization, a comparison of Eq. (3.3) with the value obtained from the initial rate of polymerization shows that at low conversions $u_{nn}(2r_n) \ll 1$. This means that the effective radius of reaction defined as $\rho_{eff} = 2r_n u_{nn}(2r_n)$ is much less that the size of a polymer coil. As conversion increases, the conditions affecting both the intramolecular dynamics and the translational mobility of polymer chains change, large-scale motions decelerating faster than small-scale. This results in an increase in the effective radius. These facts explain why $D\rho_{eff}$ does not change significantly with conversion at the initial stage of polymerization and so the termination reaction may be described with a single effective termination rate coefficient that is only weakly dependent on conversion, even though the change of the diffusion coefficients of polymer coils over this range of conversions is significant. After the motion of the chains becomes repetitive due to the formation of a physical network of entanglements [4], the coefficient of tranlslational diffusion falls sharply whereas the effective radius of reaction becomes equal to the radius of a coil, and a regime of compact exploration is expected for which the following relation can easily be obtained:

$$k(n,n) \approx 16\pi D_n r_n \tag{3.4}$$

Thus, in the stage of autoacceleration when $u_{nn}(2r_n) \approx 1$, the main laws of polymerization kinetics are governed by the dependence of the translational diffusion coefficients of macroradicals on their lengths and the concentration of polymer solution.

Expression (3.4) is derived for the reaction between polymer coils of equal length. For an arbitrary relationship between the lengths, the following expression that is similar to (but not identical with!) the Smoluchowski equation can be obtained:

$$k_t(n, m) = 8\pi(D_n r_n + D_m r_m) + k_{tp} \tag{3.5}$$

In the equation above D_n and D_m denote translation diffusion coefficients of macroradicals of degree of polymerization n and m; r_n and r_n their effective radii, respectively. Equation (3.5) recognizes (through the term k_{pt}) that a radical chain end diffuses not just as a result of center-of-mass diffusion of a macroradical as a whole, but also as a result of so-called 'diffusion via propagation' (some authors use the term 'reaction diffusion'): as propagation occurs at a growing chain end, the active unit of a macroradical itself moves.

Eq. (3.5) was used by many authors. However, so far no satisfactory physical justification of this equation has been given. So it seems quite reasonable to discuss it here in more detail. One could argue that Eq. (3.5) is a good approximation of the reality because it holds true both for macroradicals considerably different in length ($n \ll m$) and for macroradicals of equal degree of polymerization ($n = m$). Indeed, as far as the diffusion coefficient of a macroradical decreases sharply with increasing chain length ($D_n \propto n^{-2}$ for the reptation model), one may take approximately that for $r > 2r_n$ ($n \ll m$) the approach of the active ends of diffusing species to each other is determined mainly by the mobility of the shorter chain, whereas for $r < 2r_n$ the intramolecular motions of both chains exert an equal effect. For the regime of compact exploration[4] this gives $k_t(n, m) \cong 8\pi D_n r_n$ for $n \ll m$, which is in accord with Eq. (3.5). In the other limiting case ($n = m$) one can obtain for a microscopic termination rate coefficient $k_t(n, m) \cong 16\pi D_n r_n$, which is also in agreement with Eq. (3.5). In case of an arbitrary relationship between n and m it seems reasonable to accept an additive representation for $k_t(n, m)$ in the form of Eq. (3.5), which correctly describes the reality in each of the two limiting cases. Eq. (3.5) can be expressed in equivalent terms in the form:

$$k_t(n,m) = k_t(1,1)(\omega(n) + \omega(m))/2 + k_{tp} \tag{3.6}$$

The function $\omega(n)$ in Eq. (3.6) accounts for the decrease in $k_t(n, m)$ with increasing chain length; while $k_t(1, 1)$ accounts for the dependence of the termination rate coefficients of short radicals on conversion. The dependence of $\omega(n)$ on n can be found if the effective radius r_n is equal to the radius of the polymer coil and the dependence of D_n on n is known. The diffusion coefficient D_n of a macroradical of degree of polymerization n has largely different values for short and long chains, the boundary between which is defined by the value of entanglement scale n_e. In semidilute and concentrated solutions a physical network of entanglements is formed, which imposes physical restrictions on the lateral movement of long ($n > n_e$) chains. The motion of short radicals, i.e. radicals of degree of polymerization $n < n_e$, on the other hand, is not impeded by the entanglements (according to the definition, "short" radicals can be not so short, on the contrary they can be quite long at the early stages of polymerization, when n_e is large; the only thing that really matters is whether their degree of polymerization is less than the scale of entanglements. So by short radicals we mean simply radicals for which $n < n_e$). The product $D_n r_n$ for short radicals is only weakly dependent on macroradical chain length, so $\omega(n)$ for short radicals may be taken as being chain-length independent and equal to a constant, whereas for long radicals the center-of-mass diffusion coefficient D_n is proportional to n^{-2} according to the widely known reptation theory of polymer diffusion[4,5], and as the radius of a macroradical of degree of polymerization n is proportional to $n^{-1/2}$, one obtains the following equation for $\omega(n)$:

$$\omega(n) = \begin{cases} 1, & n < n_e \\ (n_e/n)^{3/2}, & n > n_e \end{cases} \tag{3.7}$$

Sometimes another value for the exponent in Eq. (3.7) is chosen. For instance, in[11] it is taken to be equal to 2. The sensitivity of the final results to the choice of a particular value for the exponent is discussed in[7], in which the solutions for different models are compared with an asymptotic one.

In order to take into account the dependence of diffusion coefficients of small molecules on conversion at the initial stage of polymerization, the free volume theory [21] was used, which yields the following expression for $k_t(1,1)$:

$$k_t(1,1) = k_{t0} \exp\left(\frac{B}{V_m} - \frac{B}{V_f}\right) \tag{3.8}$$

In the equation above k_{t0} is the value of the termination rate coefficient at zero conversion; B is a coefficient of the order of unity; while V_f is an additive function of the fractional volumes of monomer, polymer and solvent:

$$V_f = V_m \phi_m + V_p \phi_p + V_s \phi_s \tag{3.9}$$

V_p, V_m and V_s being the fractional free volumes of polymer, monomer and solvent respectively; ϕ_p, ϕ_m and ϕ_s are the volume fractions of polymer, monomer and solvent respectively; V_{f0} is the free volume of the system at the beginning of polymerization, i.e. at $\phi_p = 0$.

The last term in the right-hand side of Eq. (3.6), which accounts for the diffusion of the growing chain end of a macroradical due to propagation, is given by[7]:

$$k_{tp} = \frac{8\pi}{3} \beta k_p M a^3 n_e^{1/2} \tag{3.10}$$

In the equation above β is a parameter of the order of unity, which accounts for the effect of finite flexibility on the effective radius of reaction; while n_e is the scale of entanglement, and which according to[5] is a function of ϕ_p:

$$n_e = n_{e0} \phi_p^2 \tag{3.11}$$

If no solvent is present in the system, then Eqs. (3.9) and (3.11) become

$$V_f = V_m(1-x) + V_p x \tag{3.9'}$$

$$n_e = n_{e0}/x^2 \tag{3.11'}$$

where x is conversion.

In case of solution polymerization (the effect of volume contraction being neglected in all calculations that follow) one may write approximately:

$$\phi_m \cong \frac{M}{M_0 + S} = \frac{1-x}{1+b}$$

$$\phi_p \cong \frac{P}{M_0 + S} = \frac{x}{1+b}$$

$$\phi_s \cong \frac{S}{M_0 + S} = \frac{b}{1+b}.$$

$$V_f \cong V_m(\phi_m + \phi_s) + V_p\phi_p = V_m(1-y) + V_p y$$

where $y = x/(1-b)$; $S = M_0/$; S, P and M are the concentrations of solvent, polymer and monomer respectively; while M_0 denotes monomer concentration at the beginning of polymerization. So in case of solution polymerization Eq. (3.9') and (3.11') are still valid with x replaced by $y = x/(1+b)$:

$$V_f = V_m(1-y) + V_p y \tag{3.9''}$$

$$n_e = n_{e0}' y^2 \tag{3.11''}$$

All other equations remain unchanged.

4. CALCULATIONS AND DISCUSSION

Given a termination model, one can calculate, using Eq. (2.2) or Eq. (2.1), the factor $F = k_{t,eff}/k_t(1, 1)$ and, correspondingly, the value of an overall termination rate coefficient from microscopic chain length dependent termination rate coefficients for any given chemical composition of the system (e.g., initiator concentration) and at any specific physical conditions of polymerization (e.g., at any given temperature or pressure), as is done in method (i). However, such method of calculating the overall termination rate coefficient seems to be not so productive, because it tells us nothing about what we might expect at other conditions of polymerization. In other words, the number of the parameters employed seems to be too large to get a clear picture of what might happen should the conditions of polymerization be changed (in order to do this one would have to perform calculations for the new conditions of polymerization). Method (ii), on the contrary, makes it possible to evaluate the conditions for which the employment of simpler models, which use chain length independent model termination rate coefficient as an overall termination rate coefficient, is justifiable, and to predict the kinetic behavior of a polymerizing system for an arbitrary set of initial conditions in a simple and effective way.

First of all, let's turn to Eq. (2.2). For the additive termination model, which is used here (Eqs. (3.6)–(3.11)), Eq. (2.2) becomes:

$$\left(c + \frac{F^{1/2}}{v_0}\right)\int_0^\infty e^{-G(n)} dn = 1, \tag{4.1}$$

$$G(n) = cn + \frac{F^{1/2}}{2v_0} n + \frac{Q(n)}{2v_0 F^{1/2}}, \quad Q(n) = \int_0^n \bar{\omega}(s)ds, \quad \bar{\omega}(s) = \omega(s) + \frac{k_{tp}}{k_t(1,1)}$$

As can be seen from the above equation, F is a function of four parameters: v_0, c, n_e and $k_{tp}/k_t(1, 1)$. Eq. (4.1) can be simplified by introducing a new integration variable defined by the expression $u = n/v_R$, in which $v_R = v_0/(1 + cv_0)$, signifies the average length of macroradicals, $v_0 = vk_p M/(R_{in}k_t(1, 1))^{1/2}$ being the kinetic length in the absence of the dependence of the termination rate coefficients on radical lengths. With the introduction of the new integration variable the number of the parameters on which F depends is reduced so that Eq. (4.1) becomes:

$$\left(q+(1-q)F^{1\cdot 2}\right)\int_0^\infty e^{-G(u)}\,du = 1 \tag{4.2}$$

$$G(u) = qu + \frac{(1-q)F^{1\cdot 2}}{2}u + \frac{(1-q)Q(u)}{2F^{1\cdot 2}}, \quad Q(u) = \begin{cases} u, & u < p \\ \left(3-2(p/u)^{1\cdot 2}\right)p + r, & u > p \end{cases}$$

In the above equation the following notations were used:

$$p = \frac{n_e}{v_R}, \quad q = cv_R, \quad r = \frac{k_{tp}}{k_t(1,1)}. \tag{4.3}$$

As Eq. (4.2) reveals, with the introduction of a new variable u, F has become a function of only three dimensionless conversion dependent parameters p, q and r.

It would be interesting to compare the number of the parameters used in method (ii) with that used in method (i) (such as in the work of Russell [10]), in which a solution to an infinite set of radical concentration balance equations similar to Eq. (2.1) is sought. In method (i), for the same termination model, there are seven independent parameters in the initial set of equations similar to Eq. (2.1) (using the same notations):

$$R_{in}, \quad c_{tr}, \quad k_p, \quad M, \quad n_e, \quad k_{tp}, \quad k_t(1,1)$$

It should be noted that initiation rate is considered as a single parameter, even though it includes three different parameters: initiator concentration, decomposition rate constant and initiation efficiency. The same can be said about the constant of chain transfer c_{tr}, k_{pt} and $k_t(1,1)$. In method (ii), on the contrary, the number of independent parameters on which an overall termination rate coefficient depends is reduced to three (Eq. (4.3)). It should be emphasized that in the work of Soh and Sundberg [11] the number of model parameters was also reduced, however the parameters used here (Eq. (4.3)) are more convenient because F varies mainly with p and r as will be shown later. So the analysis of polymerization kinetics for any monomer can be done with the use of these two parameters. This is so because the constant of chain transfer is included not only in q but also in p.

As follows from Eqs. (4.2)–(4.3), the problem of calculating an effective termination rate coefficient for a concrete system can be divided into two parts: (i) the analysis of the factor F as a function of the parameters p, q and r and (ii) the analysis of the variation of these parameters with conversion.

Prior to proceeding with such analysis, let us discuss briefly what physical meaning the parameters introduced have. The parameter p is the ratio n_e/v_R which characterizes a change to the reptation regime due to the formation of entanglement network in polymer solution. In this ratio n_e is a hypothetical entanglement length corresponding to a given polymer concentration, so if the average length of macroradicals v_R is less than n_e, then there are no entanglements. In the opposite case ($p \ll 1$, that is, $n_e \ll v_R$) the motion of macroradicals in the entangled polymer network is reptative. Therefore the equality $p = 1$ can be considered as a condition for the formation of an entangled polymer network and, respectively, as a condition for the onset of gel effect in a given polymerizing system. The parameter q characterizes a polymer fraction which is formed as a result of chain transfer; it equals zero if no chain transfer takes place and approaches unity when the reaction of chain transfer to monomer or added chain transfer agent becomes predominant (over the termination reaction). The parameter r may be regarded as a dimensionless residual termination rate constant: it denotes the minimal value of $k_{tp}/k_t(1,1)$ as both n and m

approach infinity. In other words, this parameter accounts for the fact that the diffusion of the chain ends of macroradicals of significantly large degree of polymerization proceeds mainly through propagation because such macroradicals are practically immobile (their center-of-mass diffusion coefficient equals zero).

Let us now turn to the question about the dependence of F on the parameters p, q and r. For some limiting cases the following asymptotic solutions to Eq. (4.2) can easily be obtained:

$$q \to 1: \quad F = F_0 + r, F_0 = 1 - e^{-p} + 2pe^{-p} - 2p^{3/2}\sqrt{\pi}\left(1 - erf(\sqrt{p})\right) \quad (4.4)$$

$$q = 0: \quad F \cong F_0 + r, F_0 \cong \left(\frac{3}{2\ln 2}p\right)^2, \quad p \ll 1 \quad (4.5)$$

where $erf\, z = \frac{2}{\sqrt{\pi}} \int_0^z e^{-t^2} dt$ is the error function. The solution (4.4) was obtained for the termination model used here (Eqs. (3.6)–(3.9)). If $\omega(n)$ decreases with n according to the law $\omega(n) \propto n^{-\alpha}$, $\alpha > 1$, than one can obtain:

$$q \to 1: \quad F = F_0 + r, F_0 = \int_0^\infty e^{-u} Q(u) du \quad (4.6)$$

$$Q(u) = \int_0^u \omega(s) ds, \quad \omega(s) = \begin{cases} 1, & s < p \\ (p/s)^\alpha, & s > p \end{cases}$$

Eq. (4.4) is obviously a specific case of Eq. (4.6): one can obtain it by substituting Eq. (3.7) into Eq. (4.6).

In case of an arbitrary relationship between p, q and r, a numerical solution to Eq. (4.2) can be found. Calculations show that, for any p, q and r, the factor F can be approximately represented as the sum of F_0 and r, in which function $F_0(p, q)$ means $F_0(p, q, r)$ at $r = 0$. That is, the following approximate expression is valid:

$$F \cong F_0 + r \quad (4.7)$$

The results of the calculations of factor F_0 are represented in Fig. 1. In the picture the dependence of factor $F_0^{1/2}$ on parameter p for three different values of q – the lowest one ($q = 0$), the highest one ($q \to 0$) and an intermediate one ($q = 0.5$) – is shown. As can be seen from the picture, $F_0^{1/2}$ begins to fall sharply with p as p approaches zero at a point $p = 1$. Therefore, as was already mentioned, the equality

$$p(x) = 1 \quad (4.8)$$

can be taken as a condition for the onset of gel effect. From Eqs. (4.3) and (4.8) one can find critical value of conversion x_c corresponding to the onset of gel effect. As follows from Eqs. (4.3) and (4.8) the main parameter on which x_c depends is the kinetic length of macroradicals v_0 at the initial stage of polymerization (at $x = 0$) because the dimensionless constant of chain transfer c does not change significantly with temperature and conversion (Fig. 2a). Another observation is that the dependence of F on q is not so significant in comparison with its dependence on p, as Fig. 1 reveals: any real curve will surely lie somewhere between curves 1 and 3 in Fig. 1, so it will not differ much from either curve 1 or curve 3.

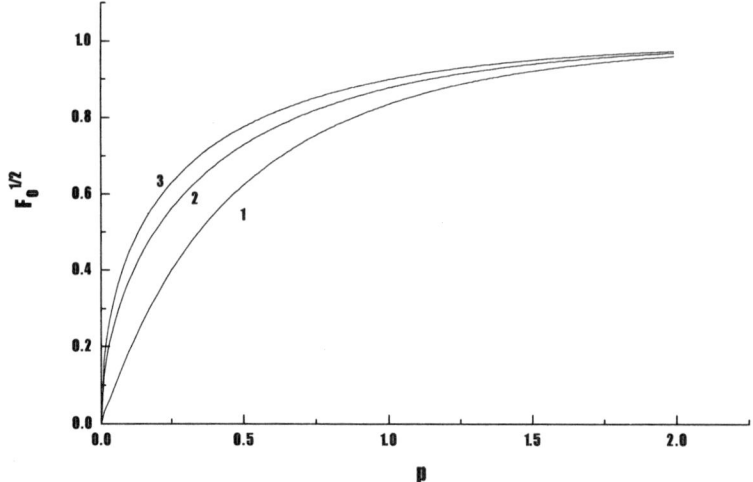

Figure 1. Variation of factor $F_0^{1/2}$ with p. $1 - q = 0$; $2 - q = 0.5$; $3 - q = 1$.

Once F_0 is calculated, one can determine the range of parameter values, in which chain length dependence should be taken into account. As follows from Eq. (4.7) if $r \ll F_0 \ll 1$ then the chain length dependence in termination reaction is really important. If $F_0 \leq r$, the value of the factor F can be much less than unity, however termination reaction can be described with a single termination rate coefficient. Therefore the following factor for evaluating the range of parameter values in which chain length dependence of the termination rate coefficient is significant can be proposed:

$$\Phi = \frac{(F - F_0)/F_0 + F_0}{(F - F_0)/F_0 + 1} \quad (4.9)$$

In view of Eq. (4.7) the above equation can be rewritten as follows:

$$\Phi = \frac{r/F_0 + F_0}{r/F_0 + 1} \quad (4.10)$$

As can be seen from Eq. (4.10), $\Phi \ll 1$ when both r/F_0 and F_0 are much less than unity, while $\Phi \cong 1$ when either $F_0 \cong 1$ or $r/F_0 \geq 1$. The former conditions mean that the role of 'reaction diffusion' is truly insignificant and that the fraction of short radicals, which equals F_0, is small. In the opposite case ($\Phi \cong 1$) either the fraction of short radicals is large ($F_0 \cong 1$) or the diffusion of the chain ends of macroradicals proceeds mainly through propagation. So if $\Phi \cong 1$ the use of simplified models with a single chain length independent termination rate coefficient is undoubtedly justifiable.

In order to give some further illustrations to the method presented here the calculations that follow were performed using parameter values typical for methyl methacrylate (MMA) bulk polymerization with 2,2'-azoisobutyronitrile (AIBN) as initiator.

Also, initiation efficiency and propagation rate coefficient are assumed to be independent of conversion, so the calculations that follow do not cover very high conversions ($x \geq 0.8$) where both initiation efficiency and propagation rate coefficient become dependent on conversion. A set of parameter values held to be appropriate for MMA polymerization is given in ref.[11,22].

In Fig. 2, 3 the variation of the parameters p, q and r with conversion is presented. The curves were drawn using free volume parameter values typical for MMA. However, the sensitivity to them is not so strong. As can bee seen from Fig. 2a, the critical value of conversion x_c is increased with decreasing value of v_0 (at $x = 0$), and thus the onset of the gel effect shifts to higher conversion as R_{in}, or initiation concentration I, is increased in view of $v_0 = k_p M/(R_{in} k_t(1, 1))^{1/2}$. Another point that should be made is that F depends only on p and q at lower conversions as Eq. (4.7) and Figs. 1–3 reveal, while at higher conversions F ceases to depend on p and q and depends only on r, as the pictures imply. That is, at higher conversions an effective, or overall, termination rate coefficient becomes equal to one chain length independent coefficient k_{tp}, as could be anticipated.

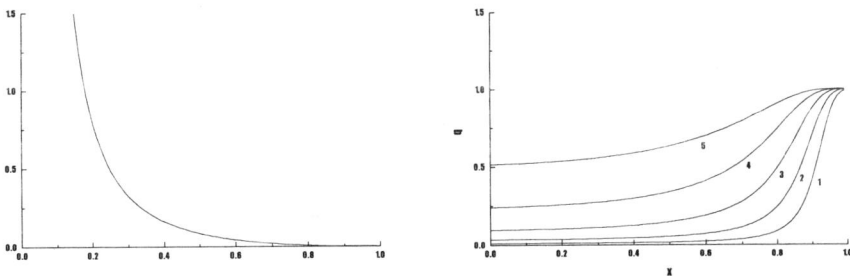

Figure 2. Variation of parameters p and q with conversion x (Fig. 2a and 2b respectively). Constant of chain transfer to monomer $c = 10^{-5}$: 1 – $v_0(x = 0) = 1.06 \cdot 10^3$; 2 – $v_0 = 3.5 \cdot 10^3$; 3 – $v_0 = 1.06 \cdot 10^4$; 4 – $v_0 = 3.18 \cdot 10^4$; 5 – $v_0 = 1.06 \cdot 10^5$.

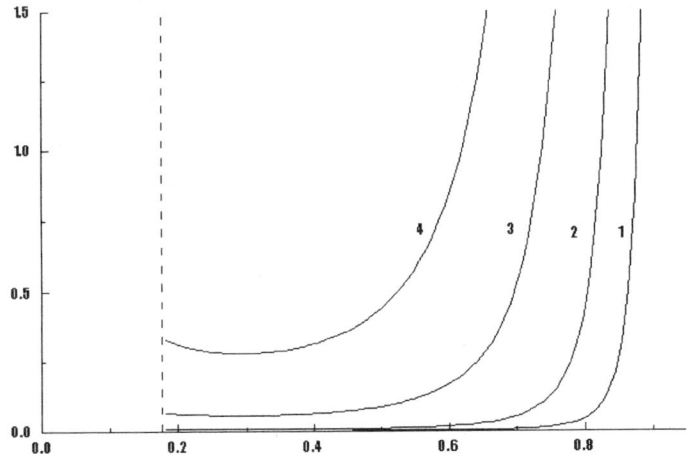

Figure 3. Variation of parameter r with conversion x. 1 – $(k_p/k_t)_0 = 2 \cdot 10^{-6}$; 2 – $(k_p/k_t)_0 = 2 \cdot 10^{-5}$; 3 – $(k_p/k_t)_0 = 2 \cdot 10^{-4}$; 4 – $(k_p/k_t)_0 = 2 \cdot 10^{-3}$.

Variations of Φ and of the respective overall termination rate coefficient with conversion for MMA polymerization for different values of v_0 or different initiation rates are shown in Figs. 4a,b. As can be seen from the pictures, at low conversions, when $\Phi \cong 1$, the k_t curves coincide, at intermediate conversions ($0.2 \leq x \leq 0.7$) they diverge, and then converge again at a point $x \cong 0.7$ as Φ becomes close to unity. For the system in question (MMA-AIBN), therefore, microscopic termination rate coefficients are essentially chain length dependent in the range of conversions $0.2 \leq x \leq 0.7$, in which Φ is significantly less than unity, so the use of simplified models, in which such dependence is neglected, is unjustifiable. In this range of conversions the overall termination rate coefficient will depend on any factors, which influence chain length distribution of macroradicals (e.g., rates of initiation, propagation and chain transfer). It is remarkable to note that $k_{t,\mathit{eff}}$ will depend on these factors - say on the rate of initiation or chain transfer – through only three independent parameters: v_0 (at $x = 0$), c and $(k_p/k_t)_0$ as Eq. (4.3) implies.

Thus the analysis of a particular polymerizing system can be done in the following way. One should draw curves similar to those shown in Fig. 2, 3 for given values of v_0 (at $x = 0$) and $(k_p/k_t)_0$ using Eq. (4.3). Then, using the dependences $p = p(x)$, $q = q(x)$, and $r = r(x)$, one should determine the critical value of conversion corresponding to the onset of gel effect and the values of the factors F and Φ.

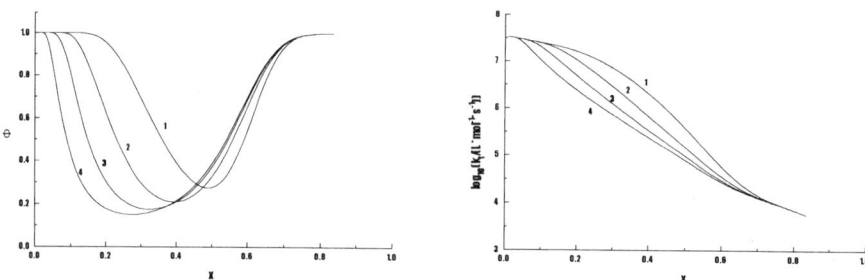

Figure 4. Variation of factor Φ (Fig. 4a) and of effective termination rate coefficient k_t (Fig. 4b) with conversion x for MMA. Constant of chain transfer to monomer $c = 1.78 \cdot 10^{-5}$;; $(k_p/k_t)_0 = 2.0 \cdot 10^{-5}$: $1 - v_0(x = 0) = 1.12 \cdot 10^3$; $2 - v_0 = 3.54 \cdot 10^3$; $3 - v_0 = 1.12 \cdot 10^4$; $4 - v_0 = 1.12 \cdot 10^{-5}$.

The kinetic behavior of a particular polymerizing system can be described more informatively by using phase curves in p, q, r - space. To each point on a phase curve corresponds a particular value of the factor F. Since the dependence of factor F on r is rather simple according to Eq. (4.7), it is quite sufficient to show a phase curve on the plane (p, q) and to find the change of factor F_0 along this curve. The values of factor F_0 for any given phase curve lie on a surface as shown in Fig. 5. This surface is universal, as it describes the variation of factor F_0 for any given monomer and any given polymerization conditions, once the dependence of $\omega(n)$ on n (Eq. (3.7)) is specified. For other dependences of $\omega(n)$ on n the form of the surface will not change significantly. As an example, in Fig. 5 is shown the variation of F_0 along the phase trajectory which corresponds to curves 3 in Fig. 2a and Fig. 2b. The curve in Fig. 5 can be construed as that pertaining to MMA polymerization.

5. CONCLUSION

The analysis made has shown the efficiency of the modeling of termination reaction with reduced number of governing parameters for describing the kinetics of free-radical polymerizations. Such modeling makes it possible to determine the range of parameter values in which the chain length dependence of termination rate coefficients is significant. In order to model termination reaction for particular systems additional experimental data about the dependence of translational diffusion coefficients of polymer chains on chain lengths and polymer concentration are needed. Experimental data on intramolecular dynamics should also be obtained in order to determine the effective radius of termination reaction, both when the reaction is chain length dependent and when the random motion of the active unit of a macroradical proceeds mainly through propagation so that this reaction is described with the use of a single chain length independent coefficient.

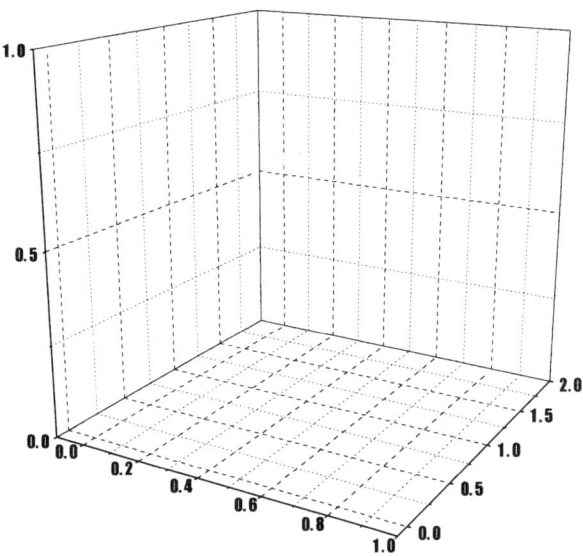

Figure 5. Variation of factor $F_0^{1/2}$ with p and q (the surface). Any real curve will lie on the surface as shown in the picture.

REFERENCES

1. C.H. Bamford, Radical Polymerization, in: Encyclopedia of Polymer Science and Engineering, 2-nd ed., J. Wiley & Sons, N.Y. 13: 708 (1989).
2. R.G.W. Norrish and R.R. Smith, *Nature*, 150: 336 (1942).
3. S.W. Benson, *North A. M. J. Am. Chem. Soc.*, 84: 935 (1962)/
4. P.G. De Gennes, Scaling Concepts in Polymer Physics, Cornell University Press, Ithaca, NY 1979.
5. M. Doi, S.F. Endwards, The Theory of Polymer Dynamics, Oxford Univ. Press., New York (1986).
6. I. Mita and K.J. Horie, *Macromol. Sci. -Macrom. Chem. Phys. C*, 27: 91 (1987).
7. G.I. Litvinenko, V.A. Kaminsky, *Prog. React. Kinetics.*, 19: 139 (1994).
8. J. Gao and A.J. Penlidis, *Macromol. Sci. -Revs. Macromol. Chem. Phys. C*, 36: 199 (1996).
9. M. Wulkow, *Macormol. Theory Simul.*, 5: 393 (1996).
10. G.T. Russell, *Macromol. Theory Simul.*, 3: 439 (1994).
11. S.K. Soh, D.C. Sundberg, *J. Polymer Sci., Polym. Chem.*, 20: 1299, 1315, 1331, 1345 (1982)
12. J.N. Cardenas, K.F. O'Driscoll, *J. Polymer Sci., Polym. Chem.*, 14: 883 (1976).
13. J.N. Cardenas, K.F. O'Driscoll, *J. Polymer Sci., Polym. Chem.*, 15: 2097 (1977).

14. S.V. Kozlov, S.L. Kamenomostskaya, A.A. Ovchinnikov, N.S. Enikolopyan, *Dokl. Akad. Nauk SSSR*, 191: 1063 (1970).
15. P.E.M. Allen and C.R. Patrick, *Macrom. Chem.*, 72: 106 (1964).
16. R.D. Burkhart, *J. Polym. Sci.*, 3: 883 (1965).
17. R.D. Burkhart, North A.M. *Macrom. Chem.*, 83: 15 (1965).
18. T.G. Tulig, T. Tirrel, *Macromolecules*, 14: 1501 (1981).
19. H.M. Boots6 *J. Polym. Sci., Polym. Phys.*, 20: 1695 (1982).
20. P.G. De Gennes6 *J. Chem. Phys.*, 76: 3316, 3322 (1982).
21. H. Fujita, A. Kishimoto, *J. Chem. Phys.*, 34: 393 (1961).
22. V.A. Kaminsky, V.A. Ivanov, E.B. Brun, *Vysokomol. Soedin. B.*, 34: 14 (1992).
23. V.A. Ivanov, A.F. Kamenshchikov, V.F. Gromov, V.A. Kaminsky, E.B. Bune, *Vysocomolec. Soed.*, 34: 15 (1992).
24. V.F. Gromov, N.I. Galperina, T.O. Osmanov, P.M. Khomikovskii, A.D. Abkin, *Eur. Polym. J.*, 16: 529 (1980).

QUANTUM-CHEMICAL MODELS OF THE ACTIVE CENTRES OF TRANSITION METALS BIOLOGICALLY ACTIVE COMPLEXES. INTERCONNECTION OF THE ACTIVE CENTRES STRUCTURE AND THE FUNCTIONS.

Ludmila Yu. Vasil'eva

Tver' State University,
170002 Tver, Russia

INTRODUCTION

Biologically active complexes containing transition metal ions take part in the metabolism processes. The following conclusions were arisen from the structure analysis of the simple metalloenzymes and more complicated complexes: 1) such systems have much in common in the active centres structure though performing different functions; 2) the acctive centres stereochemistry is determined by the electronic configuration of the central ions and the nature of the ligands; 3) interactions between the central ions and the ligands have a donor-acceptor nature: the metal ions give for the bond the vacant binding sites (unoccupied orbitals) and the peptide nitrogen or oxygen – the lone electron pairs.

In setting up the model of the active centres is seemed possible to use the general concepts of the theory of the crystalline field and ligand field theory [1]. The structure of some metalloenzymes are come to know very well but their functioning mechanism is still unclear [2,3].

According to the theory of the crystalline field the d-levels of the central ions are splitting in the ligand invirons. The character of splitting is determined by the transition ion type. The ligand field theory permits to combine the ligand splitting d–orbitals of the central metal ions and the atomic orbitals setting up the molecular orbitals (MO) of the complex. As a result it is the quantum chemical model of active center. These models are called the correlation diagrams. The approach of such kind allows to represent the electronic structure of the active centres is qualitative. As for the quantitative calculations the strict mathematical calculations are inapplicable.

The functioning mechanism may be explained by the electronic structure of the active centres. On the basis of the theory of the crystalline field and the ligand field theory was worked out the algorithm of the electronic structure construction for the mono – and polynuclear biologically active complexes of the transition metals. It is the following summery of the algorithm: 1) first of all, it is necessary to attribute the atomic orbitals of central ions to either type of symmetry; 2) secondly, to obtain the type of symmetry of the ligands hybrid orbitals; 3) then, as molecular orbitals are the linear combination of the central ion atomic orbitals and the ligands hybrid orbitals, it needs to combine the atomic and hybrid orbitals belonging to the some type of symmetry.

As for the accuracy of computational procedures it is impossible to take into account all structural parameters of a protein macromolecule that contains hundreds of amino acid residues. So the desire to increase the accuracy of calculations can lead to the opposite effect and distort the true picture because of the use of too many approximations. It is more important to obtain qualitative results, which explain the mechanism of functioning of the active centres of the transition metal complexes than to get any quantitative information. This approach to the qualitative and quantitative consideration of the structure of protein complexes provides a means to develop their quantum – chemical models.

It is of interest to take advantage of the algorithm for structural modeling of the protein manganese complex active centres participating in photosinthetic production of oxygen.

STRUCTURE AND FUNCTION OF MANGANESE–PROTEIN COMPLEX IN PHOTOSYNTHETIC OXYGEN EVOLUTION

The first model of the oxygen evolution process is considered well known Kok's four steps model. The main states of Kok model are: 1. for the molecular oxygen evolution it is necessary four oxidizing states; 2. four subsequent flashes induce four oxidizing states; 3. one positive charge is generated by each flash; 4. liberation of oxygen requires the cooperation of four positive charges in the reaction center. The accumulation of four positive charges on the reaction center leads to evolution of one molecular oxygen. The mechanism of the process and the damping of the O_2 evolution (the misses and double hits) have not been explained by Kok's model.

All following works may be divided into three groups. First group is made of experimental and theoretical investigations, having for an object the corroboration of Kok's model, the misses and double hits[5, 6, 7].

Kutjurin's works are related to the second group[8, 9]. The main point of Kutjurin's model is that for the oxygen evolution it is necessary the spacing conjugation of the photochemical centres with the catalytical system of O_2 evolution and the simultaneously participation of the chlorophyll molecule of the Photosystem-2 (PS2) reaction center in the electron transfer process. To produce one molecule O_2 from water requires the removal of four electrons from two H_2O molecules. The water–splitting reaction is realized by oxidized chlorophyll. It has not been confirmed with the experimental dates.

The third group of works make use of the fact that manganese is required for oxygen evolution. It has been shown that Mn–deficient algae and spinach has reduced the intensity of oxygen evolution about one third of the standard[8, 9]. It has been proposed that manganese complex and the water oxidation site of the reaction center took place in the direct interaction of manganese with the electron carriers[10, 11, 12]. It is postulated that manganese is oxidized in light and reduced in dark, and that catalyticaly active manganese is strongly bound with proteins and lipids[13, 14]. Renger and co–workers investigate the role of manganese in the photosynthetical O_2 evolution[15, 16]. According to the foursteps model, developed by Kok, the process of photolysis of water is divided in five stages: 1. the formation of positive charges by light on a primary electron donor of PS2; 2. the formation of the free "holes"; 3. transformation of these "holes" into "traps"; 4. cooperative fast reaction of "holes" with water and the following O_2 evolution; 5. the reaction of deactivation or the discharge of "holes". Manganese plays a role of "trap" on the third state accumulation of four positive charges ("holes") and so takes part in watersplitting reaction. The model for oxygen evolution proposed by Renger was essentially the first structural model of metal-containing protein complex taking part in the photosynthetic water cleavage. Shortcoming of this model is its descriptive character as there is not an idea about the structures of primary and secondary electron donors and true mechanism of manganese ion action in water-splitting system.

During the last few years the role of manganese in the process of O_2 evolution has been extensively studied by a variety of physical and chemical methods[14, 17, 18, 19]. According to recent data it may come to the following conclusions: the process of Mn^{2+} oxidized to Mn^{3+} or to higher valence states[20, 21, 22]. It has been shown that manganese was complexed with protein and involved into the water splitting enzyme system as secondary electron donor of PS2. Several models were proposed to explain the involvement of manganese in process. All of them make use of the fact that manganese possesses several oxidation states and, therefore, can donate and accept electrons in the photosynthetic cycle; so manganese may be a likely candidate for the positive charge accumulator (Si-state of Kok's scheme) in the watersplitting enzyme.

All these models are joined with the idea of direct participation of one or several manganese ions in initial valence state Mn^{2+} and certain ligands. Under the action of light each flash transfers manganese into the next oxidizing state according to Kok,s scheme. On the fourth flash manganese ion returns to the initial. S_0 state produsing O_2 evolution. All proposed models of O_2 evolution process are distinguished with amount of manganese ions and the nature of the ligands surrounding of the manganese [21, 23, 24, 25].

MODEL OF THE ACTIVE CENTER OF THE PROTEIN MANGANESE COMPLEX

To set up a structural model of the manganese complex active center and to describe the mechanism of its action, it is necessary to view the problem of oxygen evolution in photosynthesis systems as a whole. It then becomes clear that the process must be examined from several angles: the chemical aspect of the process: the formation of the oxygen molecule accompanied by the formation of radicals; the physical problem: electron transport and removal of radicals from the "site of action"; the problems of the structure of the oxygen evolving system and the kinetic aspect of the process: the strict sequence of all the stages with time, i.e. the operation of a well–adjusted biological mechanism. Models to date of the oxygen–evolving system have referred either only to the chemical or to the structural problem. To reconstruct a more or less complete picture, it is necessary to allow for the physical basis of the process, for which it is necessary to resort to quantum-chemical ideas and to use the methods of quantum chemistry. For this purpose the setting up of a model by analogy with the biologically active complexes of the transition metals already known is structural and functional terms seems most realistic. It is unlikely that nature created for the process of photosynthesis an entirely unique complex which does not have an analogue in biosystems.

The basic points of the discussed model are the following: 1) for the mononuclear complex active centre appears to contain only one manganese ion Mn^{2+}; 2) the ligand is a fragment of the polypeptide chain. It is assumed that the complex is formed directly by the skeleton of the chain, with the formation of the donor–acceptor bond of the metal ion with four peptide nitrogens or oxygen's of $d2sp3$–hybridization and two bonding sites are for the complex functioning. The hinge rod structural model thus constructed as an octahedral complex with three- five member chelate rings (Fig. 1).

The model was constructed with reference to the fact that two bonds of the central ion must take part in the functioning of the complex and are not incorporated into the coordination bond with a protein ligand. A necessary condition for the efficient functioning of the complex is the constant maintenance of its symmetry, so the two accepting binding sites remain free but, at transient stages of functioning, they are blocked by CL-ions, which stabilizes the structural stability of the active center.

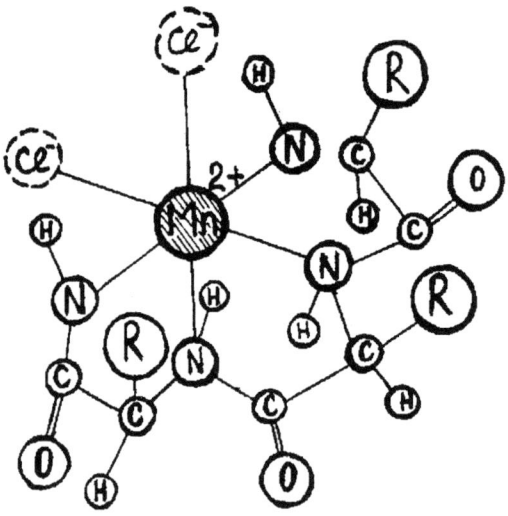

Figure 1. Schematic representation of the structure of the active centre of the protein manganese complex.

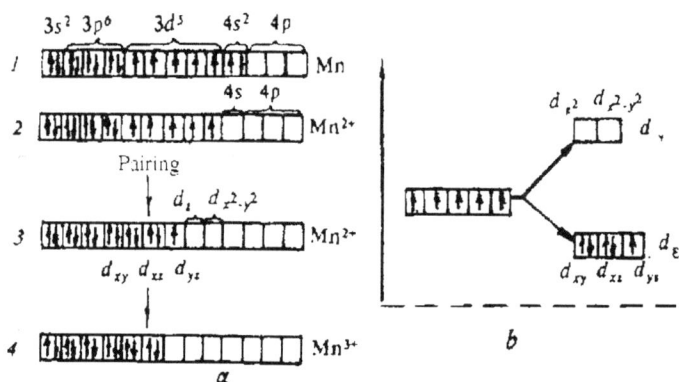

Figure 2. Electron configuration of the Mn: atom (1), the Mn^{2+} ion (2) in the free state, Mn^{2+} (3) and Mn^{3+} (4) ions in complex (a); scheme of splitting the d-levels of the central ion in the octahedral field of the ligands (b)

The behaviour of the central manganese ion in the ligand environs is shown in Fig. 2. Manganese belongs to the transition metals. Peptide nitrogen or oxygen are a strong field ligand splitting the five fold degenerate d-level of the Mn^{2+} ion into two sublevels, one of which is triply degenerate d_ε and the other doubly degenerate d_γ. With the electron distribution at these sublevels, first the lower orbitals d_ε are filled ,according to the Pauli principle. Five electrons are positioned at the d-level in the free state. In the octahedral field of the ligands these electrons are positioned in the dxy-, dxz- and dyz-orbitals, while the orbitals d_{z^2}, $d_{x^2-y^2}$ remain free. Fig. 3 shows the orientation of these orbitals in the ligand field:

the d_{z^2}- and $d_{x^2-y^2}$-orbitals are directed towards the location of the ligands and the d_{xy}-, d_{xz}- and d_{yz}-orbitas are directed to the ribs of octahedron. The donor–acceptor of the central ion with the ligand is formed as a result of d^2sp^3-hybridization, i.e. the rearrangements of the initial atomic orbitals of different symmetry of the central ion leading to more complete overlapping of the hybrid orbitals of the atoms of the ligand involved in the formation of the bond. Thus, d^2sp^3-hubridization involves four orbitals s- and two d_{z^2}-, $d_{x^2-y^2}$-orbitals, which is characteristic of the octahedral complexes of the coordination compounds of the transition metals. As a result six hybrid d^2sp^3-orbitals form, directed from the nucleus of the atom to the apices of the octahedron at an angle of 90° to each other.

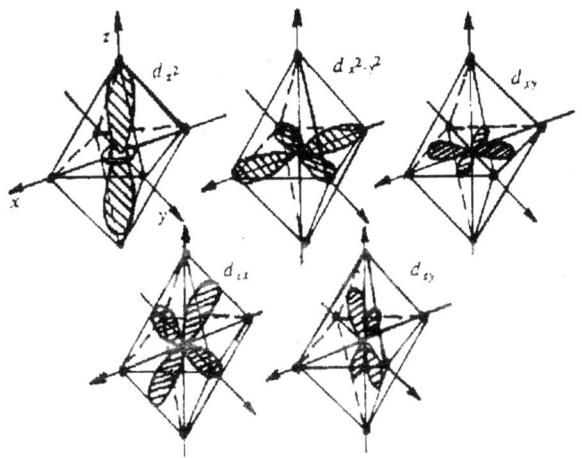

Figure 3. Directivity of the d-orbitals in the octahedral field of the ligands 26.

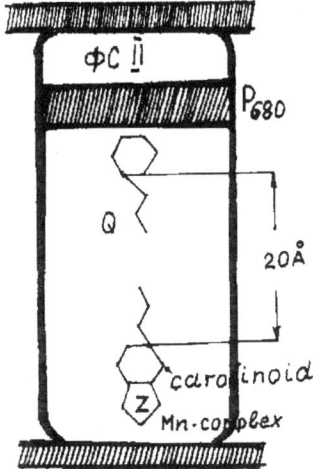

Figure 4. The scheme of the O_2 evolution part structure in PS2.

The functioning of the complex A definite role of carotinoids of the functioning of manganase complex is assumed. The scheme of the PS2 O_2 evolution part structure is given in the Fig. 4[27]. It is supposed that P_{680} is in touch nearby environment with one of the carotinoid iononic rings, the inner side of the thylakoid membrane and is in touch with the protein –manganese complex. The main stages of function are following: 1) generation under the action of light on oxidzed P^+_{680}; 2) translocation of electron from the opposite iononic ring into P^+_{680} by the tunnelling process; 3. interaction between the ionic ring and manganese complex. In discussing model carotinoids are accomplished two function: structural element and conductor of electrons from complex to P^+_{680}.

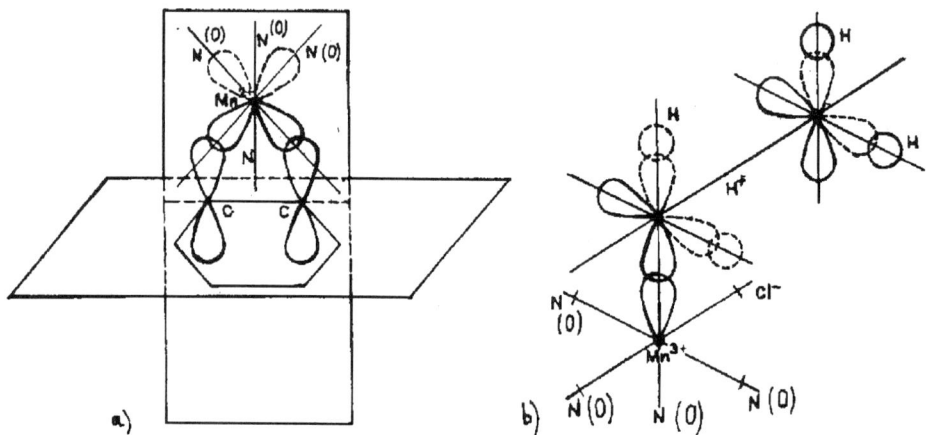

Figure 5. Schematically representation of two functional stages of the complex: a) congigation of Mn–complex and iononic ring; b) incorporation of water molecule into ligand surrounding.

The schematic illustration of key–stages of the complex functioning: conjunction of manganese complex and iononic ring and interaction with H_2O molecule is shown in Fig. 5. The complex mobility is precisely due to the peptide ligand[28]. At first point complex–iononic ring interaction is provided with two T-bonds (two π–hybrid orbits of manganese ion in complex and π–electrons carbon) (Fig.5a). As a rule there is the breaking of double bond in iononic ring and the formation of a united structural system, then the transition of the unpaired manganese electron from manganese ion to $P+680$ along the carotinoidal chain by the tunnelling process and transition $Mn^{2+} \to Mn^{3+}$ correspondingly. The tunnel transport of electron in carotinoids has been investigated [29].

The next point of functioning is water–spliting reaction .We assume, that one water molecule coordinates to the manganese complex (oxygen donates an electron pair to a hybrid orbital of the Mn^{3+} ion) and acts as an electron and proton donor (Fig. 5a).The second H_2O molecule accepts the proton: $H_2O + H^+ \to (H_3O)^+$. The $(H_3O)^+$-ion can transfer the proton to a carrier ,for example, plastoquinon 30, thus yielding the $OH°$-radical and inducing the $Mn^{3+} \to Mn^{2+}$ transition. The complex returns to the initial state and is again ready for binding with the ionone ring. At this transient stage, the binding sites that become free in the complex are blocked be Cl-ions. Water-splitting reaction may be explained with conformation changes of water molecule provided its position between Mn-complex and second water molecule. The molecule orbital (MO) calculations for the hypothetic square planary model of Mn-complex 31 have shown that the donation of electron from $(OH)^-$ to Mn^{3+} in complex is possible because of certain molecule correlation of molecule orbitals, water is is oxidized in one step (first flash) and as result of four steps four hydroxyl radicals $(OH)°$ are accumulated, beeing strong oxidizing agent. More commonplace reaction of hydroxyl radicals is the formation of H_2O_2 hudrogen peroxide and following O_2 evolution: $2H_2O_2 \to 2H_2O + O_2$. But it has been supposed that this way is not agreement with the experimental data on the rate of the oxygen evolution 32.

In the light of the present investigation it can be necessary to suggest the existence of hydroxyl radicals accumulator. We have good reason for suggestion that xanthaphylls involving in PS2 reaction center of plants may be accumulator of hydroxyl radicals.

THE FUNCTIONING OF CAROTINOIDS IN O_2 EVOLUTION.

Carotinoids function on photosynthetic O_2 evolution has been in detail investigated in several reports by the group of Saposhnikov, Hager, Yamamoto[33, 34, 35] and others[36, 37]. All these works are devoted to violaxanthin transformation in dark. It was received information on the function and localization of xanthophylls in the thylacoid membrane. As a whole the process has some intermediates as zeaxanthin, neoxanthin and antheraxanthin.

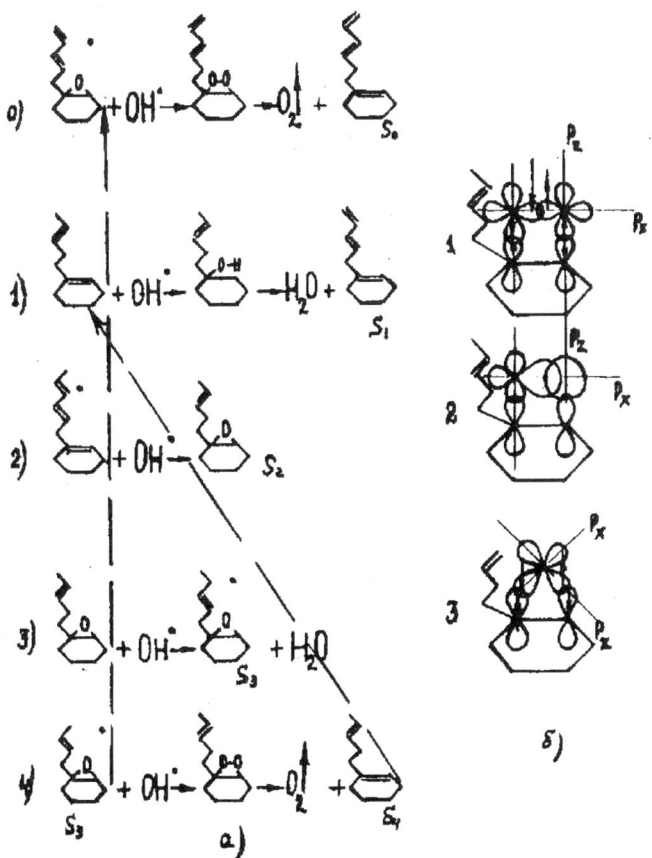

Figure 6. Schematic representation of O_2 evolution process with xanthophyll participation. Stages: 0) preliminary flash (S_0) addition of $OH°$-radical to violaxanthin, and violaxanthin– –lutein (or zeaxanthin) transformation and formation of intermediate $-1b$; 1) the formation of radical state S_1 (antheraxanthin) by way of addition $OH°$-radical, liberating one water molecule and formation of S_2 state by way of addition of the next $OH°$-radical (second flash); 3) the formation of S_3–radical state by way of addition next $OH°$-radical and liberating one water molecule $-3b$; 4) O_2 evolution and returning to initial state 1

We assume it will be interesting to connect the functioning of manganese complex with the O_2 evolution process by violaxanthin cycle. The scheme of the xanthophyll partieipation is given in Fig. 6. Violaxanthin–lutein (zeaxanthin) transformation is induced by preliminary flash by way of intermediate with peroxy group (Fig. 6–1b). The initial state so is violaxanthin. Second flash induced the formation of radical state S_1 accompanied by the liberation of one water molecule. Close location of two neighbouring CH_2-groups in carotinoid chain, as a rule, causes the rotation of covresponding part of chain round C–C-bond. As a result of stereochemical interaction between nearest CH_2-group and the chain OH–group (Fig. 6–2b) there is homolytical isolation of $H°$ from CH_2-group and then combination of $H°$ and chain OH-group, then water molecule evolution. In this way is former the intermediate radical state S_1 (lutein or zeaxanthin). The next flashes are rising the combination of hydroxyl radicals and formation of S_2, S_3 and S_4 states and on the fourth flash O_2 evolution. The schematical representation of combination of the oxygen and xanthophyll iononic ring is shown on Fig. 6 (3b). All radical states are metastable. Discussed scheme of protein–manganese complex may have two reaction centres: two synchronous functioning complex, two xanthophylls, performing violaxanthin cycle and two carotinoid connecting complex and P_{680}.

It must be noted that one of four stages the radical state of the xanthophylls formats or disappears interacting with the hydroxyiradicals. The formation and disappearances of the radical states is accompanied by the bond alteration in the xanthhophyll polyene chain: uncompensated bond (radical) is moved away at the length of the polyene chain from the reaction place and besides it is become localized, i.e. distributing among some bonds, migrating between them. It is important for the biological systems. The radical states are reactivity as they may destroy the system by action of the external chemical agents. The radical moving away deep into the membrane and it delocalization lowers the danger of the destruction. Then in discussed scheme the xanthophylls carry out the protective role. The bond migration along the polyene chain is the spreading of the soliton with physical point of view.

According to the present notion about the photosynthetic apparatus structure it is not exeluded possibility of the presence in PS2 of two reaction centers, so for the discussed model it is means that the active centers two mononuclear manganese complexes functioning synchronically or one complex, containing two mononuclear active center.

Some authors connect the oxygen evolution with the participation of the plastoquinons or amino acids. According to the discussed model only quinons having the $-CH_2-CH_2-$ groups may interact with $OH°$-radicals. The base scheme expaining the mechanism of the interaction $OH°$-radicals with any organic, capable of formation the epoxyde groups is shown in Fig. 7. The total scheme of the oxygen evolution process, including all participators is shown in Fig. 8.

During last 20 years it is discussed the amount of the manganese ions in the active centre. Many authors making use of EXAFS method [38,39] suppose the complex active centre is polynuclear, having four manganese ions: $Mn(2)$ $Mn(3)Mn(3)Mn(4)$. The scheme of such active center structure is shown in Fig. 9 (a) [40, 41]. The model of the cluster $Mn(2)Mn(3)Mn(3)Mn(4)$ constructing on the basis of the theory of the crystalline field in Fig. 9 (b). It is proposed that the active center functioning is accompanied by the transition $Mn(2) \rightarrow Mn(3)$, $Mn(3) \rightarrow Mn(4)$ according to the Kok's S_i-oxidation states. Discussed mononuclear manganese model may be considered as the fragment of the polynuclear complex.

The advantages of four manganese ions model are the following: the decomposition of H_2O molecule is not accompanied by formation of $OH°$-radicals and the oxygen evolution is the result of MnO_2 dissociation.

As for the consecutive transition $Mn(2) \rightarrow Mn(3)$ and $Mn(3) \rightarrow Mn(4)$ the carelation diagram for the complex active center involving $Mn(3)$ point to the energetic and symmetric difficulty of the transition $Mn(3) \rightarrow Mn(4)$ as in the Mn^{3+} state it is the undivided electrons on the molecular orbitals (Fig. 10). For the $Mn^{3+} \rightarrow Mn^{4+}$ transition and electron moving on the vacant lower level of the oxidation reaction center it is necessary to raise the electron level in the octahedral $Mn(3)$–complex.

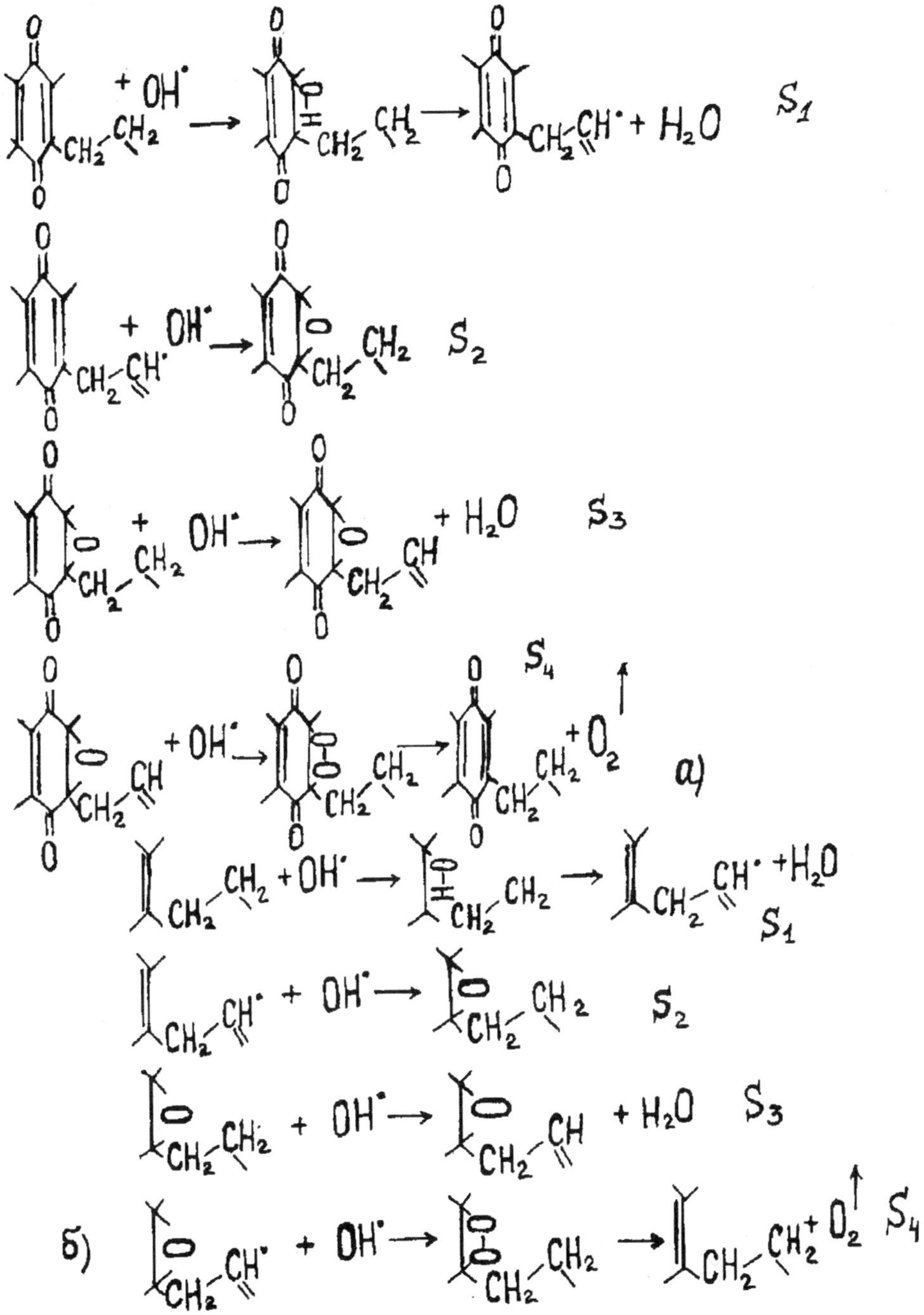

Figure 7. Scheme of the interaction OH°-radicals a) with quinon, b) with any organic, capable of formation the epoxyde groups.

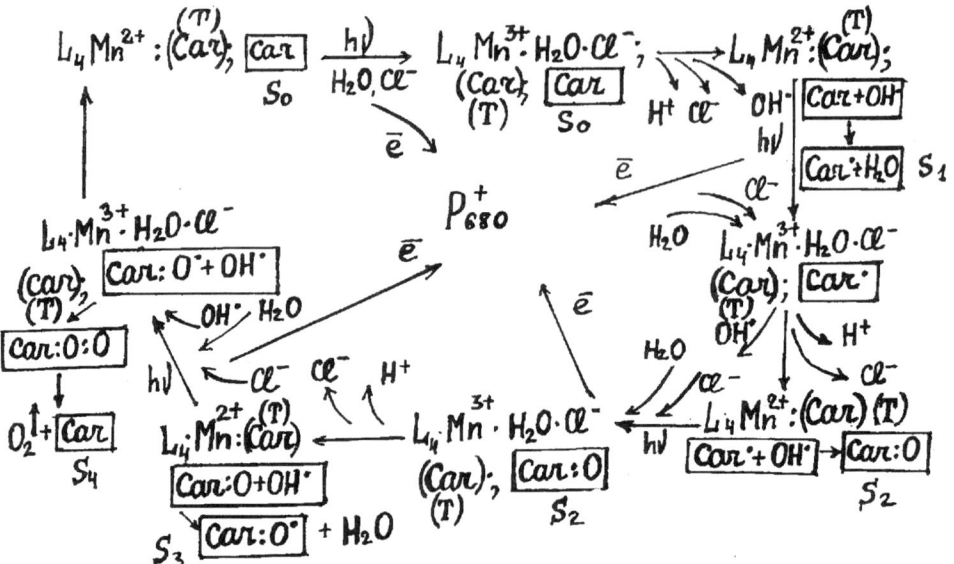

Figure 8. The total scheme of O_2 evolution. LMn^{n+} -manganese complex ($n = 2,3$), (Car)–carotinoid, (T)–tirosyn, Car, Car–xanthophyll or other organic (Car°-radical states), P^+_{680} – –oxidized state of the PS2 reaction center.

It may be realized by macromolecule conformational changes. The reverse conformation transition must be happen on the following stages. The reaction energy may be utilized for these transitions, as the whole process is exothermal. Considering examples demonstrate the possibility and efficiency of discussed method for biosystems. It is necessary to note that according to the last experimental dates [42,43] the active center is Mn^{2+}--Mn^{3+} dimer. It was supposed according to the discussed model that one of these ions Mn^{2+} carries out the functional role, and the second Mn^{3+} ion is locates in the coordination sphere of Mn^{2+} ion, stabilizing the stereochemistry of the active centre, simplifying the oxygen and redaction processes of the functional ion Mn^{2+}.

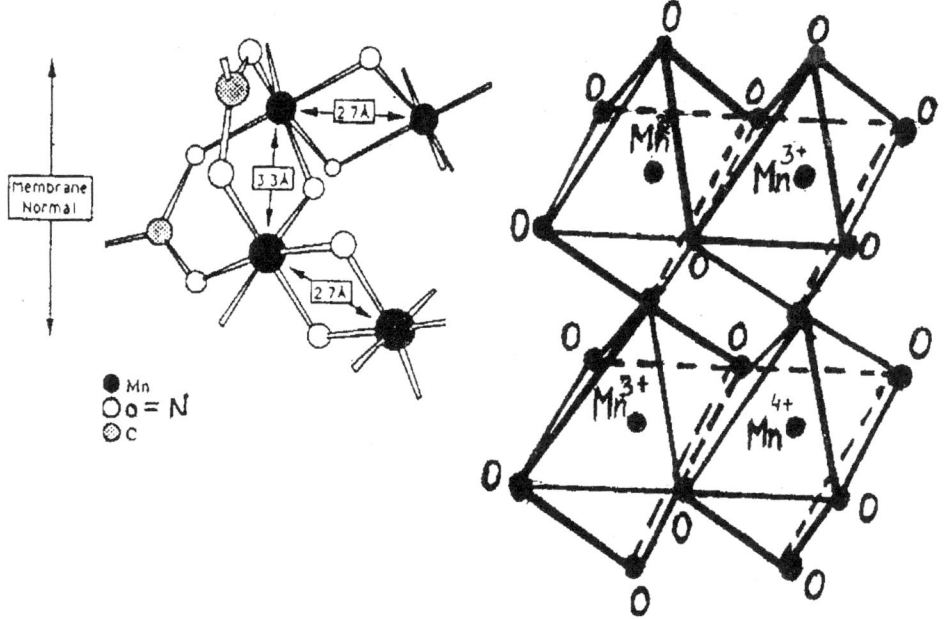

Figure 9. Scheme of four manganese active center: *a)* Klimov 40, Debus 41 *b)* the model of the cluster, constructing on the basis of the theory of the crystalline field.

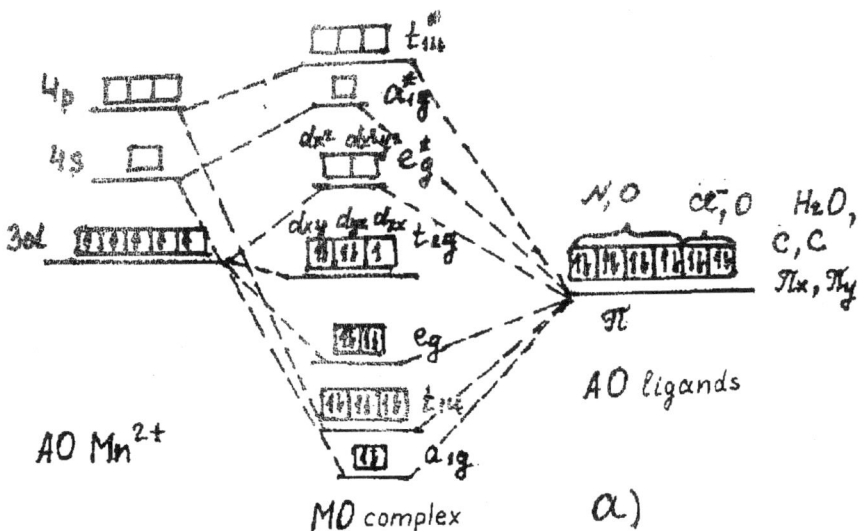

Figure 10. Correlation diagram for the active center of the Mn–complex: a)one Mn^{2+}-ion, b)cluster–dimer Mn^{2+}–Mn^{3+}, on the left –3d-, 4s- and 4p-orbitals of functioning Mn^{2+} ion, to the right-ligand orbitals, including Mn^{3+} orbitals.

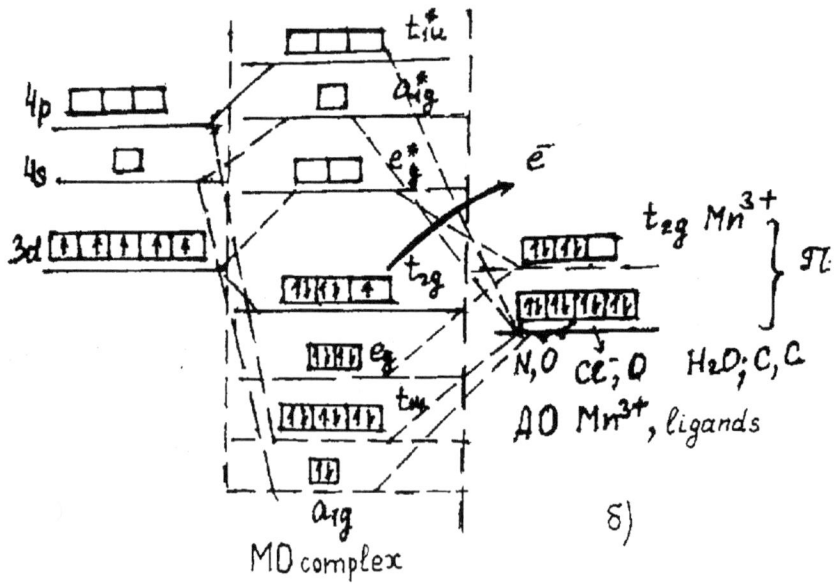

Figure 10. Continued.

REFERENCES.

1. I.B. Bersuker, The Electronic Structure and Properties of Coordination Compounds, Khimiya, Leningrad (1976)
2. Inorganic Biochemistry, G.B. Eichhorn, ed., Scientific Publishing company, Amsterdam (1978).
3. Techniques and topics in Bioinorganic Chemistry, C.A. McAuliffe, ed., Macmillan, London (1975).
4. B. Forbysh, B. Kok, M. McGloin, Cooperation of charges in photosynthetic O_2 evolution 1. Linear for step mechanism, *Photochem. and Photobiol.*, 11: 457 (1970).
5. M. Delrien, hight intensity saturation properties of O2 yields in a sequence of flashes in Chlorella, *Biochem. Biophys. Acta,*.592:478(1980).
6. Y. Lavorell, C. Limasson, Anomalies in Kinetics of flashes, *Biochem. Biopys. Acta, Abbr.*, 430: 510 (1976).
7. Y. Lavorell, An alternative to kok's model for the oxygen–evolving system in photosynthesis, *Febs. Lett.*, 66: 164 (1976).
8. V.M. Kutjiurin, About mechanism of water decomposition in photosynthetic process, *Izvestiya AN SSSR*, 4: 569 (1970) (in Russian).
9. V.M. Kutjurin, I.N. Anisimova, N.I. Zacharova, M.V. Ulubekova, Action of manganese deficit on O_2 evolution and composition of pigment–protein complexes, *Physiology of Plants*, 23: 932 (1976).
10. R. Randmer, G. Chenige, Photoactivation of the manganese catalyst of O_2 evolution, *Biochem. Biophys. Acta*, 253: 167 (1971).
11. G. Chenige, J. Martin, Site of manganese function in photosynthesis, *Biochem. Biophys. Acta*, 153: 819 (1968).
12. G. Chenige, J. Martin, Sits of function of manganese within photosystem 2. Roles in O_2 evolution and system 2, *Biochem. Biophys. Acta*, 197: 219 (1970).

13. Y. Siderer, S. Malkin, R. Poupko, Z. Lus, Electron Spin Resonance and photoreaction of MN(2) in lettice choroplasts, *Actives of Biochemistry and Biophysics*, 179: 174 (1977).
14. T. Kambara, G. Ovingee, Molecular mechanism of water oxidation in photosynthesis basen on the functioning of manganese in two different enveroment, *Proc. Nat. Acad. Sci. USA*,. 82: 6119 (1985).
15. G. Remger, The watersplitting system of photosynthesis. Postulated Model, *Z. Naturforch.*, 25: 966 (1970).
16. G. Renger, W. Weiss, Studies of the nature of the water–oxygizing enzyme, *Biochem. Biophys. Acta*, 252: 428 (1972).
17. J. Saleido, B. Ellis, Manganese–labeile pool and plant uptake, *Japan. Sci.*, 127: 227 (1979).
18. S. Izawa, A. Muallen, Inactivation of the O_2 evolving mechanism by exogenous Mn^{2+} in Cl– depleted Chroplasts, *Febs. Lett.*, 115: 49 (1980).
19. G. Lazar, Ligth induced valance change of a manganese–containing chlorophyllderivative, *Z. Naturforsch.*, 35e: 470 (1980).
20. T.S. Djabaev, A.V. Shiulov, Photosynthesis and its chemical models, *Journal of allunion chemical society by name D.I. Mendeleev*, 25: 503 (1980).
21. Y. Sauer, R.S. Blankenship, Manganese in photosynthetic Oxygen Evolution. 1 Electron Paramagnetic resonance study of the enviroment of manganese in tris–washed chloroplasts, *Biochem. Biophys. Acta*, 357: 254 (1974).
22. Y. Amesz, A role of manganese in photosynthetic oxygen evolution in photosynthesis, *Account. Chem Research*, 13: 250 (1080).
23. M. Goldfeld, L. Blumenfeld, Light dependent paramagnetic centres in the photosynthesis of higher plants, *Bulletin of Magnetic Resonance*, 1: 66, (1981).
24. T. Matsushita, Fujiwara, F. Shono, Reaction of dichloromanganese (4) shiff base, complexes with water as model for water oxigation in photosystem 2, *Chem. Lett.*, 3: 631 (1981).
25. G. Dismukes, B. Nair, Models for the photosynthetic water—oxidizing enzyme. A binuclear manganese (3)–B–cyclodextrincomplex, *Journal of Amer. Chem. Soc.* 105: 124.
26. A. Daniel, D. Abramovich, G. Dismakes, Manganese proteins isolated from spinach thylakoid membranes and their role in O_2 evolution, *Biochem. Biophys. Acta*, 765: 309 (1984).
27. I.Yu. Vasijeva, N.M. Chenavskaya, O.V. Napalkina, Manganese role in the work of PS2 in Properties of Substances and Structure of Malecules, Kalinin state university, Kalinin (1982).
28. G. Ovingee, T. Kambara, W. Coleman, The electron donor side of photosystem 2: the oxygen evolving complex, *Phtochem. and Photobiol.*, 42: 187 (1985).
29. N.M. Chernavskaya, D.S. Chernavskii, Electrons tunnel transport in photosynthesis, Moscow state University, Moscow (1977).
30. M.G. Goldfild, L. Blumenfeld, I.G. Dmitrovskii, V.D. Mikojan, Role of plastoquinone in PS2 reactions, *Molecular Biology*, 14: 804 (1980).
31. M. Kusunoki, K. Kitaura, K. Morokuma, C. Nagata, Molecular orbital study of photosynthetic water decomposition, *Febs. Lett.*, 117: 179 (1980).
32. V.M. Kudjirin, Wateris the sourse of oxygen in photosynthesis, *Biochemistry*, 1: 24 (1972).
33. D.O. Sapozhnikov, About aprticipation of xanthophylls, in: Biochemistry and Biophysies of Photosynthesis, Nauka, Moscow (1965).
34. A. Hager, Lichtbedingle *pH*–erniedrigug in ienem choloroplast–compartment als imvandlung, Beziehungen zur Photophosphoryllilung. *Platna* (Berl.), 89: 224 (1969).

35. Y. Yamomoto, S. Schimada, M. Nishimura, Purification and molecular properties of polypeptides reliesed from a highly active O_2 evolving photosystem 2 preparation by Tris–treatment, *Febs. Lett.*, 151: 49 (1983).
36. E. Pfundel, W. Bilder, Regulation and possible function of the violaxanthin cycle, *Photosynthesis Research*, 42: 89 (1994).
37. A.F. Harry, R.Y. Cogdell, Carotenoids in photosynthesis, *Photochemistry and Photobiology*, 63: 257 (1996).
38. Y.K. Yachndra, R.D. Guiles, A. Dermott, R.D. Britt, S.L. Dexheimer, K. Sauer, M.P. Klein, The structure in the photosynthetic apparatus. Structure of the manganese complex in PS2, *Biochem. Biophys. Acta*, 850: 324 (1986).
39. A. Nuijs, Y. Hans, Vangorkim, N.M. Duysens, Primary–charge separation and excitation of chlorophyll in Phtosydtem 2 particles from spinach as studied by picosecond absorbance–difference spectroscopy, *Biochem. Biophys. Acta*, 848: 167 (1986).
40. V.V. Klimov, Water oxygation and O_2 evolution in photosynthesis, *Soros Educational Journal*, 11: 9 (1996).
41. R.J. Debus, The manganese and calcium ions of photosynthetic oxygen evolution, *Biochim. Biophys. Acta*, 1002: 269 (1992).
42. N.Jshutilova, V.V. Klimov, T.M. Antropova, V.L. Shnirov, About thermoinactivation nuthanism of O_2 evolution complex of functional nucleus of PS2 chloroplasts, *Biochemistry*, 57: 1508 (1992).
43. B.K. Semin, I.I. Ivanov, A.B. Rubin, Ph. Parak, Specific binding of Fe_2-cations with *Mn*–binding site of O_2 evolution complex of PS2, International conference Bioenergetics of Photosynthesis, Pushchino (1996).

RELATION OF THE PROPERTIES OF SUBSTANCES TO MOLECULAR STRUCTURE: PHENOMENOLOGICAL STUDY OF SUBSTITUTED METHANES AND THEIR ANALOGS

Yurii G. Papulov, Marina G. Vinogradova

Tver State University
170002 Tver, Russia

1. INTRODUCTION

The establishment of the relation between the properties of substances and structures of their molecules constitutes a fundamental scientific problem of chemistry. In solution of this problem the development of quantitative methods of calculation and prediction is of great importance.

In the arsenal of theoretical chemistry there are the various calculation methods – methods of quantum chemistry, statistical thermodynamics, molecular mechanics etc. These methods do not except, but supplement each other.

Our aim to examine the theory and methods concerning the correlation between the properties of substances and their molecular structure on the basis of the physical phenomenological model: "molecule as an interacting atom unity"[1–6].

Here the substitutes of CH_4, SiH_4, GeH_4, ... are chosen as objects of research, being the key compounds for testing the methods of calculation.

2. STARTING POINT

From the phenomenological point of view, some molecular property (P) can be presented as sum of properties of atom-atomic interactions: one-body (p_α), two-body $(p_{\alpha\beta})$, three-body $(p_{\alpha\beta\gamma})$, etc.[7]

$$P = \sum_{\alpha} p_\alpha + \sum_{\alpha,\beta} p_{\alpha\beta} + \sum_{\alpha,\beta,\gamma} p_{\alpha\beta\gamma} + ... \tag{1}$$

(mathematical model). The summation in (1) is over all atoms, all pairs of atoms, all triads of atoms, etc.

Two partner interactions may be divided into bonded and nonbonded interactions separated by one, two, three and so on skeletal atoms of molecule. The interactions of triads of nonbonded atoms via one skeletal atom are taken into account also.

A correspondence is established between the phenomenological model and quantum mechanics [6,7].

3. OBJECTS OF RESEARCH

We shall consider the series $AH_{4-1}X_1$, $AH_{4-1-m}X_1Y_m$, ..., where $A = C, Si, Ge, Sn, ..., X, Y = D, T, F, Cl, Br, I, CH_3, NO_2, ...$.

The number of terms of the series is given by the number of combination[6]

$$\sigma(k) = C^4_{k+4} = (1/4!)(k+1)(k+2)(k+3)(k+4) \tag{2}$$

(k is the number of sorts of substituents).

For the series $AH_{4-1}X_1$ we have $\sigma(1) = 5$. This is $AH_4, AH_3X, AH_2X_2, AHX_3, AX_4$.

For the series $AH4$-l-$mXlYm$ we have $\sigma(2) = 15$. This is $AH_4, AH_3X, AH_3Y, AH_2X_2, AH_2XY, ... AY_4$, etc.

4. INTRAMOLECULAR INTERACTIONS

The interactions of atoms in substituted methane's, silicomethanes etc. may be divided into: bonded interactions $p_{AH}, p_{AX}, ...$; nonbonded binary interactions $p_{HH}, p_{HX}, ...$; nonbonded ternary interactions $p_{HHH}, p_{HHX}, ...$ and so on.

The numbers of interactions ($x_{AH}, x_{AX}, ... , x_{HH}, ...$) can be expressed through the numbers of substituents ($l, m , ...$). Thus, for the series $AH_{4-l}X_l$ we find

$$xAH = 4-l, \quad xAX = l, \tag{3}$$

$$xHH = (1/2)(4-l)(3-l), \quad xHX = (4-l)l, \quad xXX = (1/2)l(l-1) \tag{4}$$

and so on.

5. ADDITIVITY SCHEMES OF CALCULATION

Let us take into consideration in (1), moreover of atoms, only bonded interactions of atoms. Then

$$P_{AH_{4-l}X_l} = x_{AH}p^*_{AH} + x_{AX}p^*_{AX}, \tag{5}$$

here $p^*_{AH} = p_{AH} + (1/4)p_A + p_H$, $p^*_{AX} = p_{AX} + (1/4)p_A + p_X$.

Owing to (3), formula (5) may be written as

$$P_{AH_{4-l}X_l} = a_0 + a_1 l \quad (l = 0,1,2,3,4), \tag{6}$$

where $a_0 = 4p^*_{AH}$, $a_1 = -p^*_{AH} + p^*_{AX}$,

The property P of series $AH_{4-l}X_l$, as (6) shows, is a linear function of the number of substituents (linear approximation).

Let us take into account in (1), besides of atoms, the binary bonded and nonbonded interactions of atoms. Then

$$P_{AH_{4-l}X_l} = x_{AH} p^*_{AH} + x_{AX} p^*_{AX} + x_{HH} p_{HH} + x_{HX} p_{HX} + x_{XX} p_{XX} \tag{7}$$

Using (3) and (4), instead (7) we arrive at 1,5

$$P_{AH_{4-l}X_l} = a_0 + a_1 l + a_2 l^2 \quad (l = 0,1,2,3,4), \tag{8}$$

where $a_0 = 4 p^*_{AH}$,

$$a_1 = -p^*_{AH} + p^*_{AX} - (7/2) p_{HH} + 4 p_{HX} - (1/2) p_{XX}, \tag{9}$$

$$a_2 = (1/2) p_{HH} - p_{HX} + (1/2) p_{XX}.$$

The property P of the series $AH_{4-l}X_l$, as (8) shows, is a parabolic function of the number of substituents *(quadratic approximation)*.

If we postulate the arithmetic mean for nonbonded interaction, i.e.5, 6

$$p_{HX} = (1/2)(p_{HH} + p_{XX}), \tag{10}$$

Then, as will be seen from (9), $a_2 = 0$, and we return to (6).

If we take into account the ternary nonbonded interaction, then[8,9]

$$P_{AH_{4-l}X_l} = a_0 + a_1 l + a_2 l^2 + a_3 l^3 \quad (l = 0,1,2,3,4), \tag{11}$$

in which

$$a_3 = (1/6) p_{HHH} + (1/2) p_{HHX} - (1/2) p_{HXX} + (1/6) p_{XXX}. \tag{12}$$

The property $P_{AH_{4-l}X_l}$, as (11) shows, is a third degree algebraic function of the number of substituents (cubic approximation).

The parameter a^3 (12) vanishes, if[10, 11]

$$p_{HHX} = (1/3)(2 p_{HHH} + p_{XXX}), \quad p_{HXX} = (1/3)(p_{HHH} + 2 p_{XXX}). \tag{13}$$

With allowance for quaternary nonbonded interactions the property $P_{AH_{4-l}X_l}$ is a fourth-degree algebraic function of the number of substituents[8, 12]

$$P_{AH_{4-l}X_l} = a_0 + a_1 l + a_2 l^2 + a_3 l^3 + a_4 l^4 \quad (l = 0,1,2,3,4), \tag{14}$$

in which

$$a_4 = (1/24) p_{HHHH} - (1/6) p_{HHHX} - (1/2) p_{HHXX} - (1/6) p_{HXXX} + (1/24) p_{XXXX}. \tag{15}$$

The parameter a_4 (15) vanishes, if[12]

$$p_{HHHX} = (1/4)(3p_{HHHH} + p_{XXXX}), \quad p_{HXXX} = (1/4)(p_{HHHH} + 3p_{XXXX}),$$
$$p_{HHXX} = (1/4)(2p_{HHHH} + 2p_{XXXX}) \qquad (16)$$

The formulae (6), (8), (11), (14) are extended to other series. Thus, for the property P of series $AH_{4-l-m}X_lY_m$ in quatratic approximation we obtain[5, 6, 10, 11]

$$P_{AH_{4-l-m}X_lY_m} = a_0 + a_1l + a_2l^2 + b_1m + b_2m^2 + f(lm), \quad (l,m = 0,1,2,3,4; l \le m), \qquad (17)$$

in which a_0, a_1, a_2 are the same as in (9), and

$$b_1 = -p^*_{AH} + p^*_{AY} - (7/2)p_{HH} + 4p_{HY} - (1/2)p_{YY},$$
$$b_2 = (1/2)p_{HH} - p_{HY} + (1/2)p_{YY}, \qquad (18)$$
$$f = p_{HH} - p_{HX} - p_{HY} + p_{XY} \text{ etc.}$$

In an analogous way we may write the formulae for $P_{AH_{4-l-m}X_lY_m}$ in another approximations. The parameters of schemes expressed by bonded and nonbonded interactions similar to (9), (12), (15), (18). The assumptions, such as (10), (13), (16), can be made.

The formulae for the property X-, XY-, XYZ-, ... substitutes of AH_4 contain in linear approximation 2, 3, 4, ... k+1 parameters *(linear numbers)*, in quadratic approximation 3, 6, 10, ... (1/2)(k+1)(k+2) parameters *(trianqular number)*, in cubic approximation 4, 10, 20, ... (1/6)(k+1)(k+2)(k+3) parameters (tetrahedral numbers) and so on.

6. CONNECTIONS BETWEEN ADDITIVE SCHEMES

Linear Approximation. The equation (6) may be rewritten in the form

$$P_{AH_{4-l}X_l} = (4-l)p_{A-H} + lp_{A-X}, \qquad (19)$$

where $p_{A-H} = (1/4)a_0$, $p_{A-X} = (1/4)a_0 + a_1$.

We obtain the simple scheme (19) of Fajans[13], widespread in organic chemistry.
In the series of let us take P_{AH_4} и P_{AH_3X} as independent variables. Then, from (6)

$$P_{AH_{4-l}X_l} = (1-l)P_{AH_4} + lP_{AH_3X}. \qquad (20)$$

In a similar way we find

$$P_{AH_{4-l}X_l} = (1/2)[(2-l)P_{AH_4} + lP_{AH_2X_2}], \qquad (21)$$

$$P_{AH_{4-l}X_l} = (1/3)[(3-l)P_{AH_4} + lP_{AHX_3}], \qquad (22)$$

$$P_{AH_{4-l}X_l} = (1/4)[(4-l)P_{AH_4} + lP_{AX_4}], \text{ etc.} \qquad (23)$$

The formulae (20)–(23) represent the so-called additive-statistical method of Maslov[14].

Let us examine the series of $AH_{4-l}X_l$ and $_lAXY_{4-l}$ in conjunction. Then

$$P_{AX_lY_{4-l}} = A + BP_{AH_{4-l}X_l} \tag{24}$$

(A u B are constants). We have here (24) as an example of the application of comparative methods of Karapet'yants15.

Quadratic Approximation. Let us introduce

$$p_{A\text{-}H} = p^*_{AH} + (3/2)p_{HH},$$

$$p_{A\text{-}X} = p^*_{AX} + (3/2)p_{XX}, \tag{25}$$

$$\delta_{HX} = p_{HX} - (1/2)(p_{HH} + p_{XX}).$$

Then

$$P_{AH_{4-l}X_l} = (4-l)p_{A\text{-}H} + lp_{A\text{-}X} + (4-l)l\delta_{HX}. \tag{26}$$

This is the Zahn's scheme[16].
  Let

$$p_{AH_4} = 4p^*_{AH} + 6p_{HH},$$

$$\alpha = -p^*_{AH} + p^*_{AX} + 3p_{HH} + 3p_{HX}, \tag{27}$$

$$\Delta_{HX} = p_{HH} - 2p_{HX} + p_{XX}.$$

Then

$$P_{AH_{4-l}X_l} = p_{AH_4} + \alpha l + (1/2)l(l-1)\Delta_{HX}. \tag{28}$$

This is the Bernstein's scheme [1,9]
  The schemes (8), (26), (28) are equivalent, and it is not difficult to establish the links between the parameters (9), (25), (27)

$$p_{AH_4} = 4p_{A\text{-}H} = a_0,$$

$$\alpha = -p_{A\text{-}H} + p_{A\text{-}X} + 3\delta_{HX} = a_1 + a_2, \tag{29}$$

$$\Delta_{HX} = -2\delta_{HX} = 2a_0.$$

Cubic Approximation. Let us put into operation

$$p_{A\text{-}H} = p^*_{AH} + (3/2)p_{HH} + p_{HHH},$$

$$p_{A-X} = p^*_{AX} + (3/2)p_{XX} + p_{XXX};\qquad(30)$$

parameter δ_{HX}, as in (25), and

$$\delta_{HHX} = p_{HHX} - (1/3)(2p_{HHH} + p_{XXX}),$$

$$\delta_{HXX} = p_{HXX} - (1/3)(p_{HHH} + 2p_{XXX}).$$

Then

$$P_{AH_{4-l}X_l} = (4-l)p_{A-H} + lp_{A-X} + (4-l)l\delta_{HX} + (1/2)(4-l)(3-l)l\delta_{HHX} + (1/2)(4-l)(l-1)l\delta_{HXX}.\qquad(31)$$

This formula represents the scheme of Somayajulu-Kudchader-Zwolinski[17]. However scheme (31) contains "superfluous" parameter, and it can be transformed into[18]

$$P_{AH_{4-l}X_l} = (4-l)p_{A-H} + lp_{A-X} + (4-l)l\delta'_{HX} + (4-l)l^2\chi,\qquad(32)$$

in which

$$\delta'_{HX} = \delta_{HX} + (3/2)\delta_{HHX} - (1/2)\delta_{HXX},$$

$$\chi = -(1/2)\delta_{HHX} + (1/2)\delta_{HXX}.$$

Let add in (30) the two new parameters

$$\Gamma_{XX} = p_{HH} - 2p_{HX} + p_{XX} - 2p_{HHH} - 4p_{HHX} + 2p_{HXX},$$

$$\Delta_{XXX} = -p_{HHH} + 3p_{HHX} - 3p_{HXX} + p_{XXX}.$$

Then

$$P_{AH_{4-l}X_l} = (4-l)p_{A-H} + lp_{A-X} + (1/2)l(l-1)\Gamma_{XX} + (1/6)l(l-1)(l-2)\Delta_{XXX}.\qquad(33)$$

This is Allen's scheme[19].

The schemes (11), (32), (33) are equivalent; the relationships of these parameters are

$$p_{A-H} = (1/4)a_0,$$

$$p_{A-X} = (1/4)a_0 + a_1 + 4a_2 + 16a_3,$$

$$\Gamma_{XX} = -2\delta'_{HX} + 2\chi = 2a_2 + 6a_3,$$

$$\Delta_{XXX} = -2\chi = 6a_3.$$

7. PROCEDURE OF CONSTRUCTION OF CALCULATION SCHEMES

Let us write the initial scheme, presenting molecular property P as sum of properties of bonded and nonbonded interactions of atoms in the forms :

$$P = Ca, \tag{34}$$

where $a = \{p^*_{AH}, p^*_{AX}, p_{HH},\}$ is a set of intramolecular interactions and C is the matrix of the numbers of these interactions.
Let

$$P = Db \tag{35}$$

be the end-working scheme, where b is a set of parameters and D is the matrix of the numbers of these parameters. We have

$$b = \Lambda a, \tag{36}$$

where Λ is the matrix of coefficients equal to the numbers of intramolecular interactions through which the end-scheme parameters are expressed. Solving matrix equation

$$D\Lambda = C, \tag{37}$$

we find the elements of matrix Λ.

Let the initial scheme (34), for example, be scheme (7). The matrix of the numbers of intramolecular interactions is

$$C = \begin{bmatrix} 4 & 0 & 6 & 0 & 0 \\ 3 & 1 & 3 & 3 & 0 \\ 2 & 2 & 1 & 4 & 1 \\ 1 & 3 & 0 & 3 & 3 \\ 0 & 4 & 0 & 0 & 6 \end{bmatrix}. \tag{38}$$

The rank of C (38) is equal 3. We take into account as an example, the independent columns 3, 4, 5 (with p_{HH}, p_{HX}, p_{XX}). Then, equation (37) take the form

$$\begin{bmatrix} 6 & 0 & 0 \\ 3 & 3 & 0 \\ 1 & 4 & 1 \\ 0 & 3 & 3 \\ 0 & 0 & 6 \end{bmatrix} \begin{bmatrix} \lambda_{11} & \lambda_{12} & \lambda_{13} & \lambda_{14} & \lambda_{15} \\ \lambda_{21} & \lambda_{22} & \lambda_{23} & \lambda_{24} & \lambda_{25} \\ \lambda_{31} & \lambda_{32} & \lambda_{33} & \lambda_{34} & \lambda_{35} \end{bmatrix} = \begin{bmatrix} 4 & 0 & 6 & 0 & 0 \\ 3 & 1 & 3 & 3 & 0 \\ 2 & 2 & 1 & 4 & 1 \\ 1 & 3 & 0 & 3 & 3 \\ 0 & 0 & 0 & 0 & 6 \end{bmatrix}.$$

Therefore

$$\Lambda = \begin{bmatrix} 2/3 & 0 & 1 & 0 & 0 \\ 1/3 & 1/3 & 0 & 1 & 0 \\ 0 & 2/3 & 0 & 0 & 1 \end{bmatrix}.$$

Hence, the end-working scheme (35) is the following:

$$P_{AH_{4-l}X_l} = (1/2)(4-l)(3-l)\tilde{p}_{HH} + (4-l)l\tilde{p}_{HX} + (1/2)l(l-1)\tilde{p}_{XX}, \tag{39}$$

were

$$\tilde{p}_{HH} = (2/3)p^*_{AH} + p_{HH},$$

$$\tilde{p}_{HX} = (1/3)p^*_{AH} + (1/3)p^*_{AX} + p_{HX},$$

$$\tilde{p}_{XX} = (2/3)p^*_{AX} + p_{XX}.$$

The scheme (8), (26), (28), (39) are equivalent. There are the links similar to (29).

8. NUMBER OF PARAMETERS OF THEORY

Table 1 indicates the number of parameters required for the calculation of properties of substituted methane's, silicomethanes etc in different approximations [6,10–12,20,21].

Table 1. The parameters of additive schemes of calculation of substituted CH_4, SiH_4, etc

Substitutes of AH_4	Number of kinds	Number of parameters by the approximations		
		linear	quadratic	qubic
X	5	2	3	4
XY	15	3	6	10
XYZ	35	4	10	20
XYZU	70	5	15	35
...	...	...	...	...
XYZUV...	$\sigma(k)$ (2)	$k+1$	$(½)(k+1)(k+2)$	$(1/6)(k+1)(k+2)(k+3)$

The parameters of additive schemes may be calculated from some experimental values of P by using least-squares procedure.

9. NUMERAL CALCULATIONS

The schemes (6), (8), (11), (17) etc may be used for estimating the enthalpy of formation, entropy, Gibbs free energy, molal volume, polarizability, magnetic susceptibility, heat of vaporization, boiling point and so on[6, 20].

The formula (8) is tested here for the standard enthalpy of formation of gas phase and the standard absolute entropy of number of X-substituted methanes (see Table 2 and 3).

The least-squares procedure yields the enthalpy parameters (*kJ/mol*) and entropy parameters (*J/mol K*):

		a_0	a_1	a_2
$\Delta_f H^0_{298}(g)$	X = D	−74,829	−4,409	0,094
	X = T	−74,897	−5,851	0,086
	X = F	−73,377	−170,755	−11,282
	X = Cl	−72,565	−16,121	2,464
S^0_{298}	X = F	186,492	41,610	−5,730
	X = Cl	186,433	53,156	−5,571
	X = Br	186,519	64,213	−5,364
	X = I	186,421	72,410	−5,281

Table 2. The enthalpy of formation of the series $CH_{4-l}X_l$ for $X = D, T, F, Cl$

Substitutes of methane	$\Delta_f H^0_{298}$ (g), kJ/mol							
	$X = D$		$X = T$		$X = F$		$X = Cl$	
	Expt.	Calc.	Expt.	Calc.	Expt.	Calc.	Expt.	Calc.
CH_4	−74,68	−74,68	−74,68	−74,68	−74,68	−73,38	−74,68	−72,57
CH_3X	−79,46	−79,14	−81,15	−80,66	–	−255,41	−81,90	−86,22
CH_2X_2	−83,22	−83,27	−86,09	−86,26	452,20	−460,02	−95,23	−94,95
CHX_3	−86,96	−87,21	−91,41	−91,68	−697,60	−687,18	−102,70	−98,75
CX_4	−91,09	−90,96	−97,09	−96,93	−933,00	−936,91	−95,60	−97,62
$\lvert \overline{\varepsilon} \rvert$	0,18		0,26		5,86		2,54	
ε_{max}	0,32		0,49		10,42		−4,32	

Table 3. The entropy of the series $CH_{4-l}X_l$ for $X = F, Cl, Br, I$

Substitutes of methane	S^0_{298}, J/(mol·K)							
	$X = F$		$X = Cl$		$X = Br$		$X = I$	
	Expt.	Calc.	Expt.	Calc.	Expt.	Calc.	Expt.	Calc.
CH_4	186,31	186,49	186,31	186,43	186,31	186,52	186,31	186,42
CH_3X	222,80	222,37	234,39	234,02	245,85	245,37	253,84	253,55
CH_2X_2	246,60	246,79	270,08	270,46	293,30	293,49	309,91	310,12
CHX_3	259,58	259,75	295,89	295,76	330,66	330,89	356,10	356,12
CX_4	261,37	261,25	309,91	309,91	357,70	357,56	391,60	391,56
$\lvert \overline{\varepsilon} \rvert$	0,22		0,20		0,25		0,13	
ε_{max}	−0,43		0,38		−0,48		−0,29	

Tables 2 and 3 (bottom) show also the average calculation deviation ($\lvert \overline{\varepsilon} \rvert$) and the maximum deviation (ε_{max}). The calculations agree with the results of experiment.

10. CONCLUSION

In 1929, P. Dirac[22] stated: "The underlying physical laws necessary for the mathematical theory of large part of physics and the whole of chemistry are thus completely known and the difficulty is only that the exact application of these laws leads to equation much too complicated to be soluble".

The phenomenological approach may prove very useful in this case[23]. The phenomenological methods are not usually laborious and practice, convenient for realization of a lot of calculations.

The treatment outlined in this paper for calculating and predicting the properties may be extended to substituted ethanes, ethylenes, benzenes, cyclopropanes, etc.

REFERENCE

1. H.J. Bernstein, The physical properties of molecules in relations to their structure. I. Relations between additive molecular properties in several homologous series, *J. Chem. Phys.*, 20: 263, 1328 (1952).
2. K. Ito, On the heats of formation and potential barriers for internal rotation in hydrocarbon molecules, *J. Am. Chem. Soc.*, 75: 2430 (1953).
3. V.M. Tatevskii and Yu.G. Papulov, Relation of the energy of formation of molecule

from free atoms to her structure II, *Zh. Fiz. Khim.* 34: 489 (1960) (in Russian).
4. Yu.G. Papulov and V.M. Tatevskii, The energy of formation of molecules as sum of energies of pair interactions of atoms, *Vestn. Mosc. Univ. Ser.2 (Khimia)*, 5: 13 (1960) (in Russian).
5. Yu.G. Papulov, The pair interactions of atoms and properties of X–subsituted methanes and their radicals. *Dokl. Akad. Nayk SSSR*, 143: 1395 (1962) (in Russian).
6. Yu.G. Papulov, Relation of the physical properties and reactivities of substituted methanes to their structures, I–III. *Zh. Obshch. Khim.*, 37: 1183, 1191, 2591 (1967) (in Russian).
7. V.M. Tatevskii, *Theory of Physicochemical Properties of Molecules and Substances*, Moscow Univ. Press, Moskva (1987) (in Russian).
8. Yu.G. Papulov, About consideration of the multiple interactions of atoms in molecules, *Zh. Struct. Khim.*, 4: 462 (1963) (in Russian).
9. H.J. Bernstein, Bond and interaction contributions for calculating the heat of formation, diamagnetic suscepttibility, molar refraction and volume, and thermodynamic properties of some substituted methanes, *J. Phys. Chem.*, 69: 1550 (1965).
10. Yu.G. Papulov, Correlation between properties of substances and structure of molecules, *Properties of Substances and Structure of Molecules*, Kalinin Univ. Press. Kalinin, 3 (1974) (in Russian).
11. Yu.G. Papulov, Postulate the arithmetic mean and nonbonded interactions of atoms, *Calculation Methods in Physical Chemistry*, Kalinin Univ. Press. Kalinin, 3 (1983) (in Russian).
12. Yu.G. Papulov and M.G. Vinogradova, *Mathematics in Chemistry*, Tver Univ. Press. Tver (1997) (in Russian).
13. K. Fajans, Die energie der atombindungen in diamanten und in aliphatishen kohlenwasserstoffen, *Ber. Deut. Chem. Ges.*, 53B: 643 (1920).
14. P.G. Maslov, Vibrations of molecules and thermodynamics properties of organic substances, *Usp. Khim.*, 25: 1069 (1956) (in Russian).
15. M.Kh. Karapet'yants, *Methods for the Comparative Calculation of Physicochemical Properties*, Nayka. Moskva (1965) (in Russian).
16. C.T. Zahn, The significance of chemical bond energies, *J. Chem. Phys.*, 2: 671 (1934).
17. G.R. Somayajulu, A.P. Kudchadker, and B.J. Zwolinski, Thermodynamics, *Ann. Rev. Phys. Chem.*, 16: 213 (1965).
18. M.M. Kanovich, Yu.G. Papulov, V.M. Smoljakov, V.N. Poterin, and V.A. Kljuchnikov, Additive schemes of calculation of the enthalpies of formation of substituted silicomethanes, *Zh. Fiz. Khim.*, 56: 1766 (1982) (in Russian).
19. T.L. Allen, Bond energies and the interactions between next–nearest neighbours. I. Saturated hydrocarbons, diamond, sulfanes, S_8, and organic sulfur compounds, *J. Chem. Phys.*, 31: (1959).
20. Yu.G. Papulov, M.G. Vinogradova, N.Yu. Kuzina, and I.G. Davydova, Phenomenological methods of calculation of the properties of substituted methanes and their analogs for subgroup, *Mathematical Methods in Chemistry*, Tver Univ. Press. Tver, 3 (1994) (in Russian).
21. Yu.G. Papulov, M.G. Vinogradova, Investigation of relation of the properties of substances to molecular structure on the basis of the phenomenological model for molecule as system of interacting atoms, *Zh. Fiz. Khim.*, 70: 1059 (1996) (in Russian).
22. P.A.M. Dirac, Quantum mechanics of many–electron systems. *Proc. Roy. Soc.*, A123: 714 (1929).
23. W. Heisenberg, Role of the phenomenological theories in systems of theoretical physics, *Usp. Fiz. Nayk*, 91: 731 (1967) (in Russian).

INDEX

Absorption coefficient, 121
Abstract theory information, 27
Activity coefficient, 105
Additional-statistical Maslov method, 402
Aerosol particles, 325
Allen scheme, 404
Amplitude probability, 185
Arrhenius law, 318
Askoli-Arzela theorem, 54
Assembly drops, 355
Atomic operator formalism, 195
Attractors, 66, 69
Autoacceleration, 369

Backward Kolmogorov equation, 243
Balance Langevin equation, 106
Bernstein scheme, 403
Bessel function, 238
Bifurcation, 320, 321
Bimodal society structure, 103
Bistability, 320
Blister formation model, 242
"Block-conveyor" sweep algorithm, 225
Bridgeman anvils, 312
Brownian particles, 243
Boolean variables, 198
Boltzmann equation, 3
Boussinesq equation, 210, 212
Boundary values, 79
Bosonized Hamiltonian, 206

Characteristic equation, 8, 51
    curve, 79, 83

Chain length dependence, 370
Chapman-Enskog function, 6
Coagulation, 72

Complex representation, 269
Computer simulation schemes, 247
Concentration jump, 6
Conducting channels structure, 301
Convective (Stefan) vaporization, 339
Critical phenomena, 323
Cubic anharmonicity, 270

Debye radius, 176, 179
Deterministic chaos, 17
Deutch gate, 196–199, 203
Dielectric-metal switching effect, 301, 304
Diffusion-controlled reactions, 369
Diffusin Markov process, 243
Diffusive vaporization, 326
    growth, 326
Disperse system, 72
Dispersible medium, 129
Dispersion function, 9, 51
    matrix, 8, 9
Displacement vector, 96
Distribution function, 4
DNA dynamics, 93–95, 97
Dusty plasmas, 171, 172
    particles interaction, 171, 178
Dufour effect, 332, 339, 340

Echo-phenomenon, 229
"Effectively weak", 292, 293, 298
Eigenfunction, 196

Eigenvalue, 196
Eigenvectors, 8
Eikonal equation, 125
Einstein-Fokker equation, 245
Electric field, 122
    potential distribution, 251
Electroacoustic echo, 229
Electromagnetic radiation, 121
Elite expense, 106
Elliptic integral, 217
Emission microcathode, 222
Equilibrium distribution function, 4
Euler equations system, 157
Everyday necessary expenses, 106

Fermi operators, 207
Ferroelastic, 229
Field emission, 221
Finite elements method, 131
First price profit, 115
Fixed point, 58, 59, 65, 67
Flicker-noise, 17, 49
Fluid dynamics limit, 72
Fluctuation stage, 241
Fokker-Plank equation, 107
Fokker-Plank-Kolmogorov equation, 244
Fol'mer-Zel'dovich problem, 244
Frank-Condon factor, 189, 192, 193
Fragmentation, 312, 313
Free radical polymerization, 369
Frenkel definition, 19
    excitons, 196, 197
Frobenius inversion, 165

Galerkin method, 262
Gaussian chains, 373
Generalized coherent states, 208
Gibbs free energy, 242
    energy, 244
Glauber coherent states, 206
Gradient-type symmetric potential, 296
Gross income, 113

Heat, mass transfer, 121
Heisenberg magnet model, 205
Heisenberg-Frenkel magnet model, 206
Heterogeneity medium, 129

H-function, 51
Hill equation, 265
Hooge-phenomenological law, 23

Initial-boundary value problem, 79
"Intensive stress action such as compression & shear", 311
Intramolecular interaction, 400
Invariant manifolds, 64, 65
Inverse scattering method, 293
Inverse scattering transform method, 85

Jiny's index, 105

Karapet'yants method, 403
Kinetically finite difference schemes, 137
Klein-Gordon equation, 293, 299
Kok model, 386
Kolmogorov K-entropy, 22, 27
    theory, 23
Kolmogorov-Obukhov law, 23
Knudsen layer, 153–155, 325, 326
Kramers equation, 245
Kramers-Smoluchovsky equation, 245
Kutjurin model, 386

Landau-Lifshitz equation, 205, 210, 212
Large, moderately large particles, 339, 341, 42, 344
Larmor radius, 177, 178
Lessage model, 174
Levy approximation, 22
Linear transport equation, 51
Liquid crystals, 230, 235
Lone-pair semiconductor, 301
Low-molecular impurities, 313
Lorentz diagram, 105
    force, 176
Loss-free propagation, 85
Lyapunov instability, 295

Maekker method, 261
Magnet-acoustic resonance, 205
Magnetic field, 122
Magnetoelastic interaction, 207

Manganese complex, 387, 388
Mathieu equation, 265
Maxwell equations, 122, 123
Maxwell-Bloch system, 86
Metabolism process, 385
Metal/insulator/metal sandwich, 301
"Model caricatures", 22
Model collisions integral, 5
Model Peyrard, Bishop, 94–97
Modified Korteveg-de Vries equation, 297
Molecular weight distribution, 370
"Monochromatic echo", 238
Monte Carlo method, 38, 40, 41, 44, 45, 46, 49
Multiple disintegration, 311, 318, 322

Non-linear dissipative system, 17, 27–29
    self-compression pulse, 86
    spin waves, 205
    Schroedinger equation, 207, 214, 298, 299
Norm equivalency, 83

Order-disorder type ferroelectric, 233, 234

Pauli operator, 196, 198, 206
Phase transition, 241
Pikar method, 262
Photosyntetic oxygen evolution, 386
Poincare-Lighthill type transformation, 270
Poisson brackets, 209
    equation, 222, 223, 254, 255
Polymer chain, 369, 370
 composites, 311
Polymers, 311
"Potential" singularity, 19
Prandtl number, 14
Profit total income, 106, 114
Pseudospin operators, 229

Quasigasdynamic equation, 137
Quantum-chemical models, 385

Random Wiener process, 244
"Running away tail", 111
"Real" singularity, 19
Resonantly absorbing medium, 85

Riccati equation, 157, 159–161
Riemann-Hilbert problem, 10

Self-induced transparency, 85
Second price, 115
Sine-Gordon equation, 87, 304
Singular integral equation, 10
Small amplitude approximation, 213, 218
Smetic liquid crystals, 235
Smoluchowski problem, 3, 15
Solar activity, 33
Solitary waves, 35
Soliton pulse stability, 304
Somayajulu-Kudchader-Zwolinski scheme, 404
Spherical, spheroidal particles, 339, 341, 341, 344
Spin-photon dynamics equation, 215
Spontaneous solution formation, 86
Stefan-Boltzmann constant, 328
Stochastic simulation method, 241
Strouhal number, 141
"Structural function", 20

"Tail" of Paretto's type, 116
Termination reaction, 370
Thermal rupture, 301
Thin amorphous films, 301
Time-dependent transport problem, 83
Toffoli gate, 199
Total costs, 113, 114
Total family income, 106
Trace, 79
Transition coefficient, 183, 185
"Through-the-thickness" crack, 312, 318, 323
Tunnel transport, 181

Van der Pol equation, 281

Water-spliting reaction, 390
Weak evaporation, 5, 15
"White noise", 244
Wiener-Klinchin theorem, 20
Wigner-Seitz transformation, 207

Yajima-Oikawa potential, 214